Synchronization
in Digital
Communications
Volume 1

WILEY SERIES IN TELECOMMUNICATIONS

Donald L. Schilling, Editor
City College of New York

Synchronization in Digital Communications

Volume 1

Phase-, Frequency-Locked Loops, and Amplitude Control

Heinrich Meyr
Aachen University of Technology (RWTH)

Gerd Ascheid
CADIS GmbH, Aachen

WILEY

A Wiley-Interscience Publication
JOHN WILEY & SONS
New York • Chicester • Brisbane • Toronto • Singapore

Library of Congress Cataloging in Publication Data:

Meyr, Heinrich.
    Synchronization in digital communications / Heinrich Meyr, G.
Ascheid.
        p.   cm. -- (Wiley series in telecommunications)
    "A Wiley-Interscience publication."
    Includes bibliographical references.
    Contents: v.   1. Phase-, frequency-locked loops and amplitude
control.
    1. Digital communications.   2. Synchronization.   I. Ascheid, G.
II. Title.   III. Series.
TK5103.7.M49   1989   89-22445
621.382--dc20   CIP
    ISBN 0-471-50193-X (v. 1)

Printed in the United States of America

10 9 8 7 6

To Annemarie, Nicole, Claudia and
to Agnes, Friederike

CONTENTS

PREFACE

SCOPE

This book and its companion volume are concerned with synchronization in digital communications. Synchronization issues in point-to-point transmission and telecommunication networks are discussed.

Point-to-point Transmission

When a bit stream is digitally transmitted from one geographical location to another, the receiver must somehow reconstruct the time base of the transmitter in order to be able to convert the continuous-time received signal into a sequence of data symbols. In its most general definition, the reconstruction of this time base in the receiver is what is understood by synchronization. The various levels of synchronization in a point-to-point transmission are as follows.

Carrier synchronization. The function of aligning the phase and frequency of the receiver oscillator with that of the transmitter oscillator if the bit stream is modulated onto a carrier.

Clock synchronization. The function of adjusting the clock of the receiver, which controls the sampling of the input signal, to the clock of the transmitter running at symbol rate.

Word synchronization. The function of determining the beginning and end of a group (word) of symbols.

Digital Telecommunication Networks

In a digital telecommunication network where bit streams from different users are transported simultaneously, sharing facilities and media, several problems arise. Due to geographical separation, the clocks of the users run independently. Therefore, measures must be taken to ensure either the synchronization of the respective clocks or, alternatively, to handle asynchrony. The most common method for a single link to be shared by a number of users is to combine the data streams into a single high-rate data stream, such that the users transmit at different times. This is known as time-division multiplexing. The proper organization of access (multiplexing) of the users to the medium must be ensured.

When a bit stream is transmitted from one point to another, it passes a large number of regenerative repeaters which are necessary to compensate for signal attenuation and time dispersion. At each repeater, the clock is recovered from the preceding repeater, the repeater then retransmits the bit stream to the next repeater using this recovered clock. Thus, this accumulation of timing errors along the chain of repeaters must be limited by proper repeater design.

WHY WRITE A BOOK ON SYNCHRONIZATION?

In most technical books on digital communication, synchronization is addressed only superficially or not at all. This must give the reader the impression that the synchronization task is a trivial one and that the error performance is always close to the limiting case of (assumed) perfect synchronization. However, this would be a most unfortunate misconception for the following reasons:

1. *Error performance*
   The synchronization function is *critical* to error performance.
2. *Design efforts*
   A large amount of design time is spent on solving synchronization problems. Much more than on the topics treated extensively in most books.
3. *Hardware*
   A very large portion of the receiver hardware (and thus cost) is attributed to synchronizers.

We therefore, believe that the importance of the topic justifies these two volumes. Their purpose is to provide a theoretical framework of synchronization with which practical problems in digital communications can be solved. The theory should provide tools to

1. Analyze synchronizer structures and
2. Systematically derive synchronizer structures.

In order to have two manageable volumes it was necessary to be selective about the topics.

In point-to-point transmission over various channels of practical importance, we examine carrier and clock synchronization. We discuss terrestrial baseband transmission via optical fiber and wire, satellite transmission and mobile radio transmission. We show how the performance of synchronizers (analysis) can be determined as well as how the structure of synchronizers can be systematically derived (synthesis).

The advent of digital VLSI technology has and will continue to radically change the whole field of communications. This is also true for receivers. Digital receivers will have a fundamentally different structure than their analog counterparts. We will discuss digital algorithms for synchronization which take full advantage of the potential of VLSI in Volume 2. To our knowledge, this will be the first book to cover these topics.

We treat synchronization issues in telecommunication networks employing time-division multiple-access techniques. In particular, jitter accumulation due to chaining of asynchronous multiplexers and repeaters are discussed. We explain the basic concepts of frame synchronization in public domain and local area networks. A brief discussion of network synchronization principles are given.

FOR WHOM IS THIS BOOK INTENDED?

The book should appeal to three classes of readers. Primarily, it is intended for system and design engineers in the telecommunications industry. The book should also be useful for advanced courses in communications, teaching the fundamentals of synchronization. The second volume should also be of interest to researchers. The results, for example, on digital algorithms suitable for VLSI implementation are close to the current research frontier.

To understand most parts of the book, a working knowledge of stochastic processes, basic control, Laplace- and Fourier-transform, and linear systems is sufficient. A first glance at the book reveals that a vast amount of mathematics have been used. Since we stated that these volumes were primarily written for the engineer we must explain our reasons for this. We believe that the largest part of synchronization can only be expressed in mathematical terms. We have given the mathematical derivations in the main text if we believe they are important for understanding. Tedious algebraic manipulations are relegated to appendices. We have made an attempt to clearly outline problems before delving into a mathematical treatment. We have also summarized the key points of each section at the end of the section. We thus hope to provide the reader with enough information to quickly grasp the key points without first going through the detailed derivations.

We hope, that the reader will find this book useful in his work. Suggestions for improvements are welcome.

HEINRICH MEYR

Aachen University of Technology (RWTH)

GERD ASCHEID

CADIS GmbH, Aachen

ACKNOWLEDGMENT

We wish to express our gratitude to many individuals who have contributed towards the compilation of this book. In particular we wish to thank the following:

1. FOR CONTRIBUTIONS TO THE TEXT

Professor William C. Lindsey for many suggestions on the content and structure of the book. For contributing Section 3.5 on time and frequency stability of signal generators and providing unpublished notes on automatic gain controllers and limiters. Unfortunately important duties made it impossible for him to stay with us as full co-author.

Dr. Dieter Ryter for contributing the results of his mathematical research on eigenvalues and eigenfunctions of Fokker–Planck operators and for writing his results in a form accessible to the communications engineer (Chapter 12, Sections 1–3 and Appendix 12A).

Dr. Abbas Aghamohammadi for his detailed and careful review of the first five chapters, his many valuable suggestions and his diligent assistance in the preparation of the entire book.

Dr. Reinhold Häb for reviewing and complementing Chapter 8 on automatic frequency control.

2. FOR VALUABLE DISCUSSIONS AND COMMENTS

Dr. Floyd M. Gardner for sharing his expert knowledge on the subject with us and for freely spending so much of his time reviewing the manuscript. His comments, corrections, and suggestions for improvement were of great value to us.

Dr. Marc Moeneclay for reviewing the entire manuscript and for discovering many errors and omissions.

Gerhard Fettweis for reviewing Chapter 12 and for his contribution to the theory of the N-attractor model.

Stefan Fechtel for the numerical computation of the eigenvalues in Chapter 12 and a careful review of Chapter 12.

Heinz-Josef Schlebusch for many stimulating discussions and for his many suggestions to improve the presentation of the nonlinear theory.

To our former students Dr. Bernhard Zach and Dr. Martin Oerder for patiently wading through first drafts of the manuscript.

3. FOR LONG STANDING SUPPORT OF OUR RESEARCH WORK

The Deutsche Forschungsgemeinschaft.

4. FOR PREPARATION OF THE MANUSCRIPT

Wilma Vonhoegen and Anneliese Luchte patiently and skillfully typed the manuscript and all the mathematical formulae. Auguste Stüsser did an excellent job on the illustrations. Maureen Neumann's efforts in editing the manuscript so that it became an English language script are greatly appreciated.

HEINRICH MEYR
GERD ASCHEID

PART 1

INTRODUCTION

This book and its companion volume are concerned with *synchronization in digital communications*. Our goal is to develop a theoretical framework of synchronization which allows us to solve practical problems in many diverse areas of digital communications. In this chapter, four outlines of the material are presented. The first is a *topical outline* in which we develop a qualitative understanding of the synchronization function by examining some typical problems of interest. The second is a *methodical outline* in which we explore the various methods of solving a synchronization problem. In the third outline we discuss various *alternatives of realizing synchronizers*, and in the fourth outline we give an overview of the *content* of volume 1.

1.1. TOPICAL OUTLINE

An easy way to explain what is meant by synchronization is to examine typical examples. A simplified model of a baseband transmission system is shown in Figure 1.1-1.

Example: *Clock Synchronization*   The channel encoder puts out a symbol a_k every T seconds. In the simplest case the symbol a_k assumes one of two values, $a_k = \pm 1$. The information is then encoded into the amplitude values of a signal pulse shape $g_T(t)$, restricted to one symbol length T (Figure 1.1-2). Thus, the transmitted signal is written as the sum of nonoverlapping pulses

$$s(t) = \sum_k a_k g_T(t - kT - \varepsilon_0 T) \tag{1.1-1}$$

Notice that $s(t)$ contains a constant ε_0; this constant is best viewed as the (unknown) time shift of the transmitter time axis relative to the time reference of a hypothetical observer.

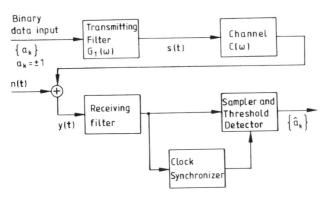

Figure 1.1-1. Simplified model of a point-to-point baseband digital transmission system.

The signal $s(t)$ is then transmitted over a channel. In the simplest case it arrives D seconds later essentially unchanged ($|C(\omega)| = 1$) at the receiver. We can always write the propagation delay D as the sum of multiples of T plus a fractional time delay $\varepsilon_c T$

$$D = MT + \varepsilon_c T$$

Since the receiver is only interested in the sequence of symbols, the delay MT is of no concern and may be omitted. The received signal is then written as

$$
\begin{aligned}
y(t) &= s(t - \varepsilon_c T) + n(t) \\
&= \sum_k a_k g_T(t - kT - \varepsilon_T T) + n(t)
\end{aligned}
\tag{1.1-2}
$$

where $\varepsilon_T = \varepsilon_0 + \varepsilon_c$ and $n(t)$ is the additive noise. In order to optimally retrieve the information $\{a_k\}$, the receiver should sample the arriving signal every T seconds at the instants when the amplitude of the pulse is maximum. But how does the receiver obtain this timing information? The only technically feasible solution to this problem is that the receiver must somehow extract this timing information from the received signal itself.

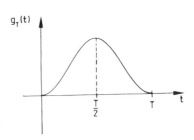

Figure 1.1-2. Example of a baseband symbol shaping pulse $g_T(t)$.

Let us assume the receiver has a perfect clock running at exactly the same clock frequency $1/T$ as the transmitter clock and a relative (unknown) time shift ε_R relative to the hypothetical observer. In order to optimally sample the signal, the alignment error between the received signal and the receiver clock

$$\varepsilon = \varepsilon_T - \varepsilon_R \qquad (1.1\text{-}3)$$

must be made known to the receiver (notice that only the difference $(\varepsilon_T - \varepsilon_R)$ is of concern, *not* the two time shifts relative to the hypothetical observer.) From this we may also conclude that we may use the receiver clock as the time reference, if we choose to do so. We are now in a position to define the task of *clock synchronization** (see Figure 1.1-3).

1. Estimating the misalignment

$$\varepsilon = \varepsilon_T - \varepsilon_R$$

   between the received signal and the receiver clock.
2. Using this estimate to generate a reference time axis in alignment with the received signal ($\varepsilon = 0$).

The various methods to accomplish this task are discussed in Section 1.2 and the various categories of implementation in Section 1.3.

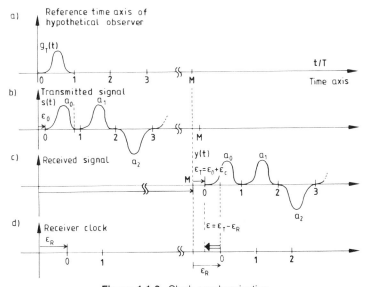

Figure 1.1-3. Clock synchronization.

---

*Also called timing/clock recovery or symbol/bit synchronization.

Example: *Carrier Synchronization in Carrier-Modulated Transmission*   If the transmitted signal is carrier modulated (Figure 1.1-4), the baseband signal is modulated onto a carrier $\cos(\omega_0 t)$

$$s(t) = \sqrt{2} A s_L(t) \cos(\omega_0 t - \theta_0) \tag{1.1-5}$$

where

$$s_L(t) = \sum_k a_k g_T(t - kT - \varepsilon_0 T)$$

The constant θ_0 describes the phase shift of the transmitter oscillator relative to the phase of the oscillator of a hypothetical observer.

Let us again assume that the signal arrives essentially undistorted D seconds later at the receiver. We can write for the received signal

$$y(t) = \sqrt{2} A s_L(t - \varepsilon_c T) \cos(\omega_0 t - \theta_T) \tag{1.1-6}$$

with

$$\theta_T = \theta_0 + \theta_c$$

where θ_c is the $\omega_0 D$ phase modulo-2π reduced. To retrieve the baseband signal $s_L(t - \varepsilon_c T)$, the signal $y(t)$ must be converted back to baseband. Let us assume that the receiver has a perfect oscillator running at ω_0 angular frequency with a phase θ_R relative to the hypothetical observer. Multiplying the received signal $y(t)$ by the locally generated oscillation $\sqrt{2} \cos(\omega_0 t - \theta_R)$ yields

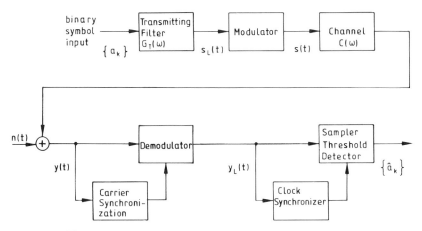

Figure 1.1-4. Example of carrier-modulated transmission link.

$$y(t)\sqrt{2}\cos(\omega_0 t - \theta_R) = y_L(t)\cos(\theta_T - \theta_R) + \text{high frequency terms}$$
$$(1.1\text{-}7)$$

Equation (1.1-7) clearly demonstrates that the baseband signal can be exactly retrieved *only* when the oscillations are perfectly aligned, i.e., $\theta_T = \theta_R$. Thus, the receiver must accomplish the task of *carrier synchronization* which involves

1. Estimating the phase difference

$$\phi = \theta_T - \theta_R$$

   between the received signal and a locally generated oscillation.
2. Producing a reference oscillation such that the phase error ϕ becomes zero.

Notice that again only the phase difference ϕ is involved (not θ_T and θ_R individually).

Once clock and carrier synchronizations have been accomplished one must be able to determine the beginning and the end of a code word or of a group of code words. These tasks of synchronization are called *word* and *frame synchronization*, respectively.

The above ideas deal with the various levels of synchronization in point-to-point transmission. An additional layer of the synchronization function is needed for *network synchronization*. In order to transmit and switch in an integrated digital format, the clock frequency in the various nodes must be synchronized, or, alternatively, proper measures to handle the asynchrony must be provided. A variety of techniques can be used for this purpose.

While the two examples serve well to introduce the synchronization function, for most practical applications the model is too idealized. In the idealized model, clock and carrier oscillators run at constant frequencies, identical in the transmitter and the receiver. However, even precision oscillators are affected by phase and frequency deviations of different origin. Depending on the specific application, these deviations must be properly taken into account. In addition to these oscillator-inherent perturbations, other deviations may result, e.g., from a relative motion between transmitter and receiver, known as Doppler shifts.

Mathematically, we model these phenomena as deviations from a nominal phase/frequency. For example, the output of an oscillator may be represented by

$$\sqrt{2}A\sin\Phi(t)$$

where $\Phi(t)$ is the total phase. The total phase $\Phi(t)$ is comprised of a term produced by a perfect oscillator with nominal angular frequency ω_0 and a time-varying phase deviation $\theta(t)$

$$\Phi(t) = \omega_0 t + \theta_0 + \theta(t) \tag{1.1-8}$$

The phase $\theta(t)$ contains both random and deterministic components. As a consequence, the receiver must accomplish the synchronization task with time-varying quantities, instead of estimating two constant parameters ε and ϕ.

Example: **Carrier Synchronization with a Practical Oscillator**   The frequency of a receiver oscillator is allowed to assume nominal values within a tolerance band of $\omega_0 \pm \Delta\omega$. If the random components of $\theta(t)$ are negligible the receiver must first estimate the frequency deviation $\Delta\omega$ before a phase coherent oscillation can be generated.

   In order to use bandwidth economically, complex modulation/coding schemes are employed for which synchronization is far more difficult to achieve than for the simple modulation used in the idealized model. In addition it has been realized for these schemes that synchronization can no longer be treated as a separate entity, but the receiver must be understood as a *joint estimator* of the symbols and the synchronization parameters.

1.2. POSSIBLE APPROACHES

We can divide the possible approaches into two categories, which we denote as ad-hoc structure and derived structure. An example illustrates what is meant by ad-hoc structure approach.

Ad-hoc Structure Approach

Example: **Clock Recovery for a Baseband Signal**   A practical and frequently-used method to recover the timing information from the data-carrying, signal is the square and filter approach which is explained below. Squaring the received signal given by (1.1-2) yields (Figure 1.2-1)

$$
\begin{aligned}
y^2(t) &= \left[\sum_k a_k g_T(t - kT - \varepsilon_T T) \right]^2 \\
&= \sum_k (a_k)^2 g_T^2(t - kT - \varepsilon_T T) \\
&\quad + \sum_k \sum_{m \neq 0} a_k a_{k+m} g_T(t - kT - \varepsilon_T T) g_T[t - (k+m)T - \varepsilon_T T]
\end{aligned}
\tag{1.2-1}
$$

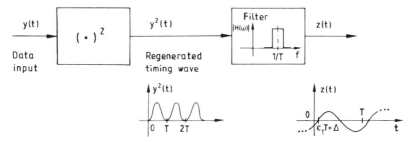

Figure 1.2-1. Square and filter clock synchronizer.

with the additive noise ignored here. Assuming that the data symbols are statistically independent and equiprobable the second term has a zero-mean value

$$E\left[\sum_{k}\sum_{m\neq0}(\cdot)\right]$$
$$=\sum_{k}\sum_{m\neq0}E(a_k)E(a_{k+m})g_T(t-kT-\varepsilon_TT)g_T[t-(k+m)T-\varepsilon_TT]=0$$
$$(1.2\text{-}2)$$

Since $a_k^2=1$, the first term of (1.2-1) becomes a periodic signal independent of the symbol. This signal carries the information to retrieve the clock. For example, if we filter the signal $y^2(t)$ with a narrow bandpass filter centered at $1/T$ the filter output will be an almost sinusoidal signal disturbed by a random fluctuation

$$z(t)=\sqrt{2}A_z\sin\left[\frac{2\pi}{T}(t-\varepsilon_TT-\Delta)\right]+\text{disturbance} \qquad (1.2\text{-}3)$$

(It is not difficult to show that the time shift Δ is completely determined by the pulse shape $g_T(t)$ and is independent of ε_T and thus may be set equal to zero for the present discussion.) Thus, the zero crossings of the signal $z(t)$ provide the time reference needed in the receiver to properly sample the incoming signal $y(t)$. The additive disturbance causes a fluctuation around the nominal zero crossings of $z(t)$.

The example illustrates three essential features of the ad-hoc structure approach to a synchronization problem.

Structure

The clock synchronizer was required to be built of the following two blocks:

1. A *memoryless nonlinearity* to regenerate a timing wave
2. A *linear time-invariant filter* to extract the clock rate component

Within this class of systems we look for the best solution. Systems that are not in this class are not studied.

Performance Criterion

Minimize the random fluctuations around the nominal zero crossings of the timing wave $z(t)$.

Information

To determine the random fluctuations of the zero crossings we have to know the pulse shape $g_T(t)$ and the statistical properties of the symbols (statistical independence and equiprobability).

Notice that the three features of the ad-hoc structure approach are closely related. For example, if we change the structure of the synchronizer, the information needed might be entirely different.

In the ad-hoc structure approach, a synchronizer consists of a few relatively simple, well-understood, basic building blocks, such as non-linearities, filters, voltage-controlled oscillators (VCO). Type and technology of these blocks greatly depend on the data rate. This approach has employed predominantly analog circuitry in the past and continues to be important, particularly for applications in the high data rate range where no alternative to analog technology exists.

The technique for solving ad-hoc structure problems is a straightforward concept. We allow the structure to vary within the class allowed and choose the solution which is optimum for a given performance criterion. There are two main advantages to this approach. On the one hand one uses well-understood building blocks to realize a synchronizer. Such a modular solution greatly simplifies design, testing, and manufacturing. On the other hand, since the fixed structure approach uses only partial information about the channel, it is usually quite robust to modeling errors.

The most serious disadvantage of having a fixed structure is that it is often impossible to tell whether the chosen structure is correct. We will encounter practical examples (fading channels, Volume 2) where choosing the wrong structure has led to the dismissal of a particular transmission technique superior to those employed in practice today.

Derived Structure Approach. The alternative to an ad-hoc structure approach is a derived structure approach in which no a priori guesses are made on the structure of the synchronizer. We establish a performance criterion and implement the algorithm that solves the problem on a suitable processor architecture. The derived structure approach has several important advantages:

1. We can systematically derive algorithms for a given modulation/coding scheme and channel. These algorithms are, with respect to the chosen criterion, the best of all possible algorithms.
2. Even if the algorithms found are not feasible (e.g., due to complexity reasons), they form the basis from which suboptimal, realizable solutions can be devised.
3. Lower bounds on performance can be obtained. This is of great importance since it is often very difficult to directly compute the performance of an algorithm. A lower bound is also of interest when comparing the performance of a suboptimal algorithm to the theoretically best solution.

The derived structure approach is particularly useful (if not indispensible) when advanced modulation/coding schemes are employed and/or when unknown channel parameters have to be extracted from the received signal. In these cases the various receiver functions such as detection and synchronization can no longer be treated separately. Rather, the receiver must be understood as an estimator for jointly estimated data, synchronization, and channel parameters.

1.3. IMPLEMENTATION OF SYNCHRONIZERS

We can divide the synchronizers into three categories denoted by:

1. Error tracking.
2. Maximum seeking + post filtering.
3. Nonlinearity + passive filtering.

The first two implementations can appear as the results of either ad-hoc structure or derived structure approach. The third type clearly belongs to the class of an ad-hoc structure approach.

An error tracking system (Figure 1.3-1) is characterized by the following principles of operation. An error signal as a function of the alignment error

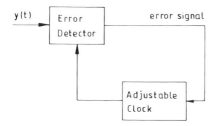

Figure 1.3-1. Basic building blocks of an error tracking system.

is computed in a functional block called an error detector. This error signal is then used in a feedback loop to adjust the clock. If properly designed, the feedback circuit forces the error signal to zero. The adjustable clock is then aligned with the received signal and may serve as reference time base in the receiver. Error tracking systems are the most widely used today. Their advantages and deficiencies are best illustrated by an example.

Example: *Carrier Synchronization by Error Tracking*   Let us assume that the received signal is a pure sinusoid

$$y(t) = \sqrt{2}A \sin(\omega_0 t - \theta_T) \tag{1.3-1}$$

Multiplying it by a locally generated signal

$$r(t) = \sqrt{2} \cos(\omega_0 t - \theta_R) \tag{1.3-2}$$

yields

$$y(t)r(t) = A \sin(\theta_T - \theta_R) + A \sin(2\omega_0 t - \theta_R - \theta_T) \tag{1.3-3}$$

The high-frequency contribution $2\omega_0 t$ can be filtered out by a low-pass filter. The remaining constant term is the alignment error signal. The error signal vanishes not only for zero error, $\phi = \theta_T - \theta_R = 0$, but also for $\phi = \pm\pi$. This points to a problem inherent in most tracking systems. A zero error signal is only a necessary condition but not a sufficient condition for synchronous operation. Additional circuitry is needed to distinguish between the correct tracking point and the false error detector output conditions.

The multiplier serves as a phase detector provided the frequency of the two oscillators is nearly identical. In this case a properly designed feedback loop will automatically correct the small initial frequency difference and subsequently achieve phase synchronization. However, if the initial frequency uncertainty is large (the usual case in practice), auxiliary circuits that generate a frequency sensitive signal become necessary to bring the phase tracking loop into its range of proper operation. In general, any tracking system has two distinct modes of operation. In the *tracking mode* all variables have values in the close vicinity of the steady-state values. The process of bringing the tracking system from its initial state to the tracking mode is called *acquisition*.

Since the error detector in a feedback system only needs a single value of an error function, the complexity of the electronic circuitry (measured in number of components) is low. If analog circuits are employed, error tracking synchronizers are the only ones feasible. They can be built for both low and high frequencies.

The functional shortcomings of the error tracking approach can be overcome by a *maximum seeking* approach which works basically as follows.

Example: *Symbol Synchronization by Maximum Seeking*   To achieve synchronous operations the transmitter initially sends a training sequence of known symbols $\{a_0, \ldots, a_m\}$ to the receiver. Let us assume that the pulse shape $g_T(t)$ is also known to the receiver. The useful part of the signal

$$y(t) = s(t, \varepsilon_T) + n(t) \tag{1.3-4}$$

where

$$s(t, \varepsilon_T) = \sum_k a_k g_T(t - kT - \varepsilon_T T)$$

is therefore known to the receiver with the exception of the fractional time shift ε_T which must be estimated during the training period. To estimate this unknown synchronization parameter, the receiver compares the received signal with a stored replica of the (known) signal $s(t, \varepsilon_i)$ for a set of values $\{\varepsilon_1, \varepsilon_2, \ldots, \varepsilon_N\}$ and chooses the parameter value $\hat{\varepsilon}$ for the signal that best matches the received signal. The optimum parameter value $\hat{\varepsilon}$ then defines the time base in the receiver (see Figure 1.3-2). The best match condition is evaluated by determining the maximum value of a cost function defined by the performance criterion. For this reason the device is called maximum seeking.

One important advantage of the maximum seeking approach is that it provides a necessary and sufficient condition for zero synchronization error. The penalty to be paid is an increase in signal processing complexity of orders of magnitude. Instead of computing a single value of an error function as in the error tracking system, we have to compute entire functions for the set of trial parameters $\{\varepsilon_1, \ldots, \varepsilon_N\}$.

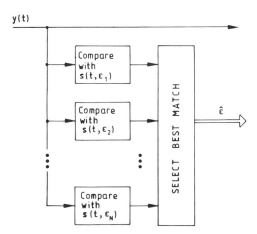

Figure 1.3-2. Maximum seeking synchronizer.

While maximum seeking synchronizers were of purely academic interest just a few years ago, with the advent of digital, very large scale integration (VLSI) technology they have become feasible today. It is safe to predict that they will replace the conventional (analog) tracking systems in many applications, thereby leading to fundamentally different receiver designs than are in use today. It is important to stress that digital signal processing is far more than a mere substitution of analog signal processing. Digital VLSI technology allows very complex algorithms to be implemented, providing functions totally inconceivable in conventional analog technology. However, these algorithms must be matched to the strength of VLSI technology. There exists a strong interaction between algorithm design and architectural issues, such as parallel processing, as will be demonstrated in Volume 2.

Example: *Clock Extraction by a Square-and-Filter Circuit*   This example belongs to the third class of implementations. It has already been discussed in Section 1.2.

1.4.   OUTLINE

Volume 1 deals with *error tracking systems* for phase and frequency control which are widely used in communications, ranging, radar, and instrumentation systems. Despite the many different forms of implementations these tracking systems can be modeled by a single, yet versatile mathematical model with a deceivingly simple structure (Figure 1.4-1). In Volume 1 we assume that this model is *given*. The task of deriving such models for the various forms of synchronization is described in Volume 2.

The model in Figure 1.4-1 is time-continuous. However, it also arises when we work with a *quasi-continuous* approximation to a sampled data, error tracking system. The reason for working with such a quasi-continuous model is that it enables us to make use of existing, powerful tools of analysis developed for time-continuous systems. A remark is in order here. The reader should be aware that one must not confuse the aspect of *modeling of a physical system* (time-continuous versus time-discrete) with the aspect of the *realization of a system* (analog versus digital technology). Thus, the results developed in Volume 1 are of much wider interest than for systems realized in analog technology.

In general, the behavior of the system shown in Figure 1.4-1 is complicated to analyze because it is nonlinear and involves random noise. Fortunately, for most engineering applications we need not worry about the general analysis but can (and should!) work with a simplified linearized model.

The modular bottom-up structure of this book reflects this fact. We start by explaining the basic principles of phase control by a phase-locked loop (PLL) in *Chapter 2*. We describe the realizations of the PLL which are of

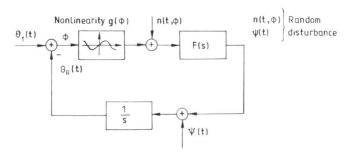

Figure 1.4-1. Baseband model of a phase tracking system.

practical importance, and show how to choose parameters in a design. We then gradually modify and extend the basic model to include the various disturbances.

In *Chapter 3*, the PLL performance in the presence of additive noise and oscillator imperfections is analyzed. We assume that the noise intensity is such that we are allowed to linearize the loop equation.

In Chapters 4 and 5 we are concerned with the acquisition process of a PLL. We first show (*Chapter 4*) that the PLL is a near optimum tracking device but displays a very poor performance to self-acquire tracking conditions. *Chapter 5* then describes auxiliary systems (acquisition aids) to bring the PLL into its tracking range.

When the signal-to-noise ratio becomes too small the PLL shows a distinct threshold behavior. Below the threshold, performance completely deteriorates and the PLL is no longer capable of producing a synchronous reference signal. The threshold behavior is due to a complicated interaction of noise and the nonlinearity. Based on experimental data, the fundamental principles of these interactions are explained in *Chapter 6*. Avoiding the intricacies of advanced mathematical techniques the experimental results are complemented by simple approximate analysis. A set of diagrams provides the reader with sufficient information for a large variety of design problem.

Chapter 7 describes the practical importance of controlling the level of the input signal by automatic gain control (AGC) and limiters.

An automatic frequency control (AFC) system estimates and tracks the frequency of a received signal. It is employed in various data links as well as acquisition aids for a PLL. Structure and performance of AFCs are discussed in *Chapter 8*.

The material covered in Chapters 1–8 is sufficient for most design problems involving PLLs, AGCs, and AFCs. The background required to understand the material is assumed to be provided by basic courses in random processes, control theory, and communications.

In *Chapters 9–12* we present a mathematical treatment of the *nonlinear theory of synchronizers*. The nonlinear theory is required for a quantitative characterization of the *threshold behavior* of an *error tracking system* (PLL, etc.) as discussed qualitatively in Chapter 6.

The theory provides us with the mathematical tools to analyze the model of Figure 1.4-1 where both nonlinearity and noise have to be taken into account. When discussing threshold phenomena of digital synchronization algorithms in Volume 2 we frequently work with a quasi-continuous approximation. The methods and results of Chapters 9–12 apply therefore to a wider class of systems than the classical analog PLL.

While we make use of the results of the nonlinear theory in various parts of Volume 2, we do not require knowledge of the mathematics involved to arrive at these results. Thus, the application oriented reader may omit Chapters 9–12 without loss of continuity. The content of Chapters 9–12 is more abstract than that of other chapters. The material is aimed at the research oriented reader familiar with advanced concepts of random processes and system theory, taught at engineering level. An effort has been made to make the material self-contained and to restrict the mathematical techniques to those accessible to the engineer. Where advanced techniques are inevitable, the emphasis is on explaining concepts and describing why these advanced mathematical tools are needed, rather than on mathematical rigor.

1.5. REFERENCES TO VOLUME 1

There must now be well over 1000 papers published in the area of phase-locked loops, automatic frequency and amplitude control. Due to the availability of computerized data bases it is no longer practical to include a comprehensive bibliography but sufficient to include a set of papers which can be used as a starting point in the search for material. A number of key papers have been reprinted in IEEE Press books edited by Lindsey and Simon [5] and Lindsey and Chie [6].

There are a number of excellent books with a different emphasis than this one. The pioneering work of Viterbi on the nonlinear theory of phase-locked loops has been collected in [1]. The fundamental research of Lindsey is documented in [2]. It contains a wealth of information on advanced topics of the theory of phase-locked loops that is found nowhere else. The book by Blanchard focuses on the application of phase-locked loops to coherent receiver design [3]. The book by Gardner is aimed at the practicing engineer [4]. It avoids long mathematical derivations and instead forcuses on explaining principles and concepts of phase-locked techniques, stressing practical results.

REFERENCES

1. A. J. Viterbi, *Principles of Coherent Communication*, McGraw-Hill, New York, 1966.

2. W. C. Lindsey, *Synchronization Systems in Communication and Control*, Prentice Hall, Englewood Cliffs, NJ, 1972.

3. A. Blanchard, *Phase-Locked Loops*, Wiley, New York, 1976.

4. F. M. Gardner, *Phaselock Techniques*, Wiley, New York, 1979.

5. W. C. Lindsey and M. K. Simon (eds.), *Phase-Locked Loops and their Application*, IEEE Press, New York, 1978.

6. W. C. Lindsey, C. M. Chie (eds.), *Phase-Locked Loops*, IEEE Press, New York, 1985.

PART 2

2

PHASE-LOCKED LOOP FUNDAMENTALS

2.1. AUTOMATIC PHASE CONTROL

A phase-locked loop is a control system used to automatically adjust the phase of a locally generated signal $\hat{s}(t)$ to the phase of an incoming signal $s(t)$. Let us assume that the two signals are sinusoids and given by

$$s(t) = \sin\left[\omega_0 t + \theta(t)\right], \quad \hat{s}(t) = \sin\left[\omega_0 t + \hat{\theta}(t)\right] \tag{2.1-1}$$

where the phase $\theta(t)$ is a slowly varying quantity with respect to ω_0, i.e.,

$$\left|\frac{d\theta(t)}{dt}\right| \ll \omega_0 \tag{2.1-2}$$

Without loss of generality we may assume the angular frequency ω_0 to be equal for both signals. Any difference in instantaneous frequency can be included in the time-varying function $\hat{\theta}(t)$. As an example, let us assume that θ is constant and that $\hat{s}(t)$ has an instantaneous angular frequency of ω_1. Then

$$\omega_1 = \omega_0 + \frac{d\hat{\theta}}{dt} = \omega_0 + (\omega_1 - \omega_0)$$

i.e., the difference in instantaneous frequencies $\omega_1 - \omega_0 = d\hat{\theta}(t)/dt$. We seek to adjust the phase of the signal $\hat{s}(t)$

$$\hat{\Phi}(t) = \omega_0 t + \hat{\theta}(t)$$

to that of the reference signal $s(t)$

$$\Phi(t) = \omega_0 t + \theta(t)$$

This adjustment procedure is conveniently illustrated by means of rotating phasors as shown in Figure 2.1-1 where the phasors $\underline{s} = \exp(j\Phi)$ and $\underline{\hat{s}} = \exp(j\hat{\Phi})$ are associated with the signals $s(t)$ and $\hat{s}(t)$, respectively*. The two phasors rotate with instantaneous angular velocity

$$\frac{d\Phi(t)}{dt} = \omega_0 + \frac{d\theta(t)}{dt} \; , \quad \frac{d\hat{\Phi}(t)}{dt} = \omega_0 + \frac{d\hat{\theta}(t)}{dt}$$

respectively. Ideally, at every time t the two phasors should coincide. The misalignment of the two phasors is described by the phase error $\phi(t)$

$$\phi(t) = \Phi(t) - \hat{\Phi}(t) = \theta(t) - \hat{\theta}(t) \qquad (2.1\text{-}3)$$

Note that the phase error $\phi(t)$ is not an absolute but a relative quantity defined only with respect to the phase of the reference signal.

The two phasors can be adjusted by an automatic control system if we are able to generate a control signal as a function of the phase error. One way to achieve this is to multiply the two signals. The result is

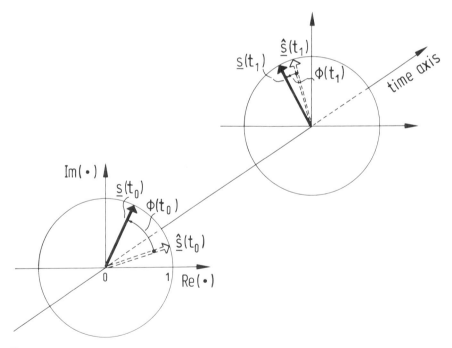

Figure 2.1-1. Phasor diagrams for describing the operation of an automatic phase control system.

*Complex numbers are underlined.

$$\sin\left[\omega_0 t + \theta(t)\right]\sin\left[\omega_0 t + \hat{\theta}(t)\right]$$

$$= \frac{1}{2}\cos\left[\theta(t) - \hat{\theta}(t)\right] - \frac{1}{2}\cos\left[2\omega_0 t + \theta(t) + \hat{\theta}(t)\right] \tag{2.1-4}$$

The first term in (2.1-4) provides a measure for the phase difference $\theta(t) - \hat{\theta}(t)$. Since the phase $\theta(t)$ is slowly varying compared to $2\omega_0$, the second term can be removed by a low-pass filter. However, since our objective is to adjust the phase $\hat{\theta}(t)$ to the phase of the incoming signal, this measure is not completely satisfactory. The cosine function is an even function of the phase error $\phi(t) = \theta(t) - \hat{\theta}(t)$ and we cannot tell from $\cos\phi(t)$ whether $\theta(t)$ is larger than $\hat{\theta}(t)$ or vice versa. Therefore, we must seek an error signal which is an *odd* function of the phase error $\phi(t)$. Such a signal can be obtained if we advance the locally generated signal by $\pi/2$ and then multiply this signal by the input signal. The resulting product is given by

$$\sin\left[\omega_0 + \theta(t)\right]\sin\left[\omega_0 t + (\pi/2) + \hat{\theta}(t)\right]$$

$$= \frac{1}{2}\sin\left[\theta(t) - \hat{\theta}(t)\right] + \frac{1}{2}\sin\left[2\omega_0 t + \theta(t) + \hat{\theta}(t)\right] \tag{2.1-5}$$

where as before, filtering effectively eliminates the sum-frequency term.

If the phase error $[\theta(t) - \hat{\theta}(t)]$ is nonzero, an error signal with the same sign as the phase error is produced. This error signal is filtered and subsequently applied to a device whose frequency can be varied according to the voltage applied to it. Such a device is called a *voltage controlled oscillator* (VCO). When the control voltage equals zero, the VCO runs at its quiescent frequency ω_0. A positive (negative) control voltage causes the VCO to increase (decrease) its instantaneous angular frequency $d\hat{\theta}/dt$ thereby forcing the phase error to decrease (increase). Note that zero phase error occurs when the locally generated VCO phasor $\underline{r}(t)$ and the signal phasor $\underline{s}(t)$ have a phase difference of $\pi/2$. But the phasor $\underline{\hat{s}}(t)$ which coincides with the signal phasor $\underline{s}(t)$ is obtained from the VCO phasor $\underline{r}(t)$ by a phase shift of $\pi/2$ which is easily realized.

2.2. THE PHASE-LOCKED LOOP

The block diagram of the automatic phase control system discussed in the previous section is shown in Figure 2.2-1. It is generally referred to as phase-locked loop (PLL). The major building blocks of a PLL are a multiplier, a filter and a voltage-controlled oscillator. These are shown in Figure 2.2-1 with the pertinent signals.

The peak amplitude of the incoming signals is $\sqrt{2}A$. The local oscillation has a peak voltage of $\sqrt{2}K_1$. The output of the multiplier is given by

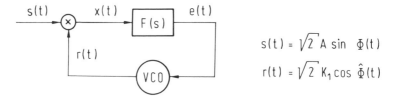

$$s(t) = \sqrt{2}\, A \sin \Phi(t)$$

$$r(t) = \sqrt{2}\, K_1 \cos \hat{\Phi}(t)$$

Figure 2.2-1. Block diagrams of the simplest form of phase-locked loop.

$$x(t) = A K_1 K_m \sin [\theta(t) - \hat{\theta}(t)] \qquad (2.2\text{-}1)$$

where K_m is the multiplier gain with dimension, V^{-1}. Since filters in practice do not respond to the sum-frequency term, this was discarded in (2.2-1).

The linear filter operates on its input $X(s)$ to produce the output

$$E(s) = F(s) X(s) \qquad (2.2\text{-}2)$$

The quantities in (2.2-2) are the Laplace transforms of the corresponding signals in the time domain, i.e.,

$$X(s) = \int_0^{\infty} x(t) e^{-st}\, dt$$

$$E(s) = \int_0^{\infty} e(t) e^{-st}\, dt \qquad (2.2\text{-}3)$$

$$F(s) = \int_0^{\infty} f(t) e^{-st}\, dt$$

The Laplace transform $F(s)$ is the transfer function of the filter. In addition, the initial condition of the filter is assumed to be zero.

Using $x(t)$ from (2.2-1) in (2.2-2) gives in the time domain

$$e(t) = f(t) * [K_m K_1 A \sin [\theta(t) - \hat{\theta}(t)]] \qquad (2.2\text{-}4)$$

where $*$ is the convolution operator. In arriving at (2.2-4) we have made use of the well-known fact that multiplication in the Laplace domain corresponds to convolution in the time domain

$$E(s) = F(s) X(s) \qquad \bullet\!\!-\!\!\circ \qquad e(t) = f(t) * x(t)$$

$$= \int_0^t f(t - u) x(u)\, du \qquad (2.2\text{-}5)$$

(notation: $\circ\!\!-\!\!\bullet$ denotes transform pairs).

The frequency of the VCO is a function of the error voltage $e(t)$. When $e(t)$ is removed, the VCO generates a signal with angular frequency ω_0*. This frequency is called the quiescent frequency of the VCO. When the control signal $e(t)$ is applied, the VCO frequency becomes $\omega_0 + K_0 e(t)$ where K_0 is the VCO gain factor with the dimension, $s^{-1}V^{-1}$. Since frequency is the derivative of phase we can write the following for the control law of the VCO

$$\frac{d\hat{\Phi}(t)}{dt} = \omega_0 + K_0 e(t) \qquad (2.2\text{-}6)$$

However, by definition

$$\hat{\Phi}(t) = \omega_0 t + \hat{\theta}(t) \qquad (2.2\text{-}7)$$

Differentiating (2.2-7) and comparing it with (2.2-6) we find that the frequency deviation of the VCO from the quiescent frequency is governed by

$$\frac{d\hat{\theta}(t)}{dt} = K_0 e(t) \qquad (2.2\text{-}8)$$

At this point, using (2.2-1), (2.2-5), and (2.2-8) yields

$$\frac{d\hat{\theta}(t)}{dt} = K_0 K_m K_1 A \int_0^t f(t-u)\sin\left[\theta(u) - \hat{\theta}(u)\right] du \qquad (2.2\text{-}9)$$

If we recall that the phase error is defined by $\phi(t) = \theta(t) - \hat{\theta}(t)$, (2.2-9) can be written as follows

$$\frac{d}{dt}\left[\theta(t) - \phi(t)\right] = K_0 K_m K_1 A \int_0^t f(t-u)\sin\left[\phi(u)\right] du$$

A slight rearrangement of this relation leads to the dynamic equation for the phase error

$$\frac{d\phi(t)}{dt} = \frac{d\theta(t)}{dt} - KA \int_0^t f(t-u)\sin\left[\phi(u)\right] du \qquad (2.2\text{-}10)$$

with

$$K = K_0 K_m K_1 \quad s^{-1}V^{-1}$$

*Throughout this book, phase $\theta(t)$ is measured in radians, which is taken as a dimensionless quantity. Therefore, angular frequency, defined as the derivative of $\theta(t)$ with respect to time, $d\theta(t)/dt$, has the physical dimension, s^{-1}. As usual, the number of cycles per second $f = (1/2\pi)(d\theta/dt)$ is measured in Hertz.

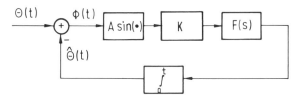

Figure 2.2-2. Baseband model of the phase-locked loop.

It can be easily verified that the control system depicted in Figure 2.2-2 obeys the dynamic equation (2.2-10). Note that the constants are moved around the loop such that the input and the output of the filter $F(s)$ have the unit s^{-1}.

This control system is therefore a mathematically equivalent model of the phase-locked loop. The multiplier is replaced by a subtractor and a sinusoidal memoryless nonlinearity, and the VCO by an integrator. It is quite interesting that the variables $\theta(t)$ and $\hat{\theta}(t)$ appear explicitly in the equivalent model while they appear only as arguments of time functions in the block diagram of the PLL. Several comments about the model of Figure 2.2-2 are in order. First of all, the model is independent of the quiescent frequency ω_0. The phasor diagram of Figure 2.1-1 shows that the model describes the *relative* motion between the two phasors, i.e., not with respect to a fixed coordinate system but with respect to one that is locked to the reference phasor $\exp(j\Phi)$.

Viewed in the frequency domain, the spectrum of the phase error signal $\phi(t)$ is a baseband spectrum and the model shown in Figure 2.2-2 is a *baseband model* of the PLL. The control system is nonlinear which makes an exact analysis mathematically difficult. This will be discussed in later chapters. Also note that the dynamics of the control system depend on the amplitude A of the incoming signal.

2.3. THE LINEAR APPROXIMATION

Let us assume now that the phase error $\phi(t)$ is small compared to 1 rad at all times. Then we may use the approximation

$$\sin \phi \approx \phi \tag{2.3-1}$$

and disregard the nonlinearity in Figure 2.2-2. The loop operation (2.2-10) is now described by a linear equation

$$\frac{d\phi}{dt} = \frac{d\theta(t)}{dt} - KA \int_0^\infty f(t-u)\phi(u)\, du \tag{2.3-2}$$

The constant KA is given by

$$KA = K_m K_1 K_0 A \qquad (2.3\text{-}3)$$

Quite frequently in the data books of the PLL manufacturers, the product in (2.3-3) is written as

$$KA = K_D K_0 \qquad (2.3\text{-}4)$$

where $K_D = K_m K_1 A$. K_D is then referred to as *phase detector gain* measured in volts and KA has units s^{-1}. This definition is reasonable as long as the amplitude A of the input signal is truly a constant. (There are operating conditions of a PLL where this is not the case.) It has the advantage that K_D can easily be measured in the laboratory.

2.3.1. Basic Transfer Functions

Analysis of linear systems is conveniently done using Laplace transform techniques. Assuming the existence of the Laplace transforms*

$$\theta(s) = \int_0^\infty \theta(t)\, e^{-st}\, dt$$

$$\phi(s) = \int_0^\infty \phi(t)\, e^{-st}\, dt$$

and vanishing initial conditions $\theta(t=0) = \phi(t=0) = 0$, (2.3-2) can be written as

$$s\phi(s) = s\theta(s) - KAF(s)\phi(s) \qquad (2.3\text{-}5)$$

The linearized model is shown in Figure 2.3-1.

Replacing $\phi(s)$ by $\theta(s) - \hat{\theta}(s)$ leads to the *closed-loop tranfer function* of the PLL

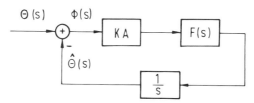

Figure 2.3-1. Linearized baseband model of the phase-locked loop.

*For the sake of simplicity in the notation we denote the Laplace transform of θ by the same Greek letter θ but with s as the argument. The same is true for other Greek symbols. Whether a time function or its Laplace transform is meant should always be clear from the context.

$$H(s) = \frac{\hat{\theta}(s)}{\theta(s)} = \frac{KAF(s)}{s + KAF(s)} \qquad (2.3\text{-}6)$$

To summarize, the basic transfer functions of the PLL are

$$H(s) = \frac{\hat{\theta}(s)}{\theta(s)} = \frac{KAF(s)}{s + KAF(s)}$$

$$\phi(s) = \theta(s) - \hat{\theta}(s) = [1 - H(s)]\theta(s) \qquad (2.3\text{-}7)$$

$$\frac{\phi(s)}{\theta(s)} = \frac{1}{1 + [KAF(s)/s]}$$

2.3.2. Steady-State Phase Error

In every control system, dynamic as well as steady-state, performance must be evaluated. We start with the steady-state phase error since it is much simpler to obtain than the transient phase error. As the name implies, the steady-state phase error is the value that $\phi(t)$ assumes after all transients have died away. It can be readily evaluated by means of the final value theorem in Laplace transformation which states that

$$\lim_{t \to \infty} \phi(t) = \lim_{s \to 0} s\phi(s) \qquad (2.3\text{-}8)$$

provided that a steady-state value exists. The final value theorem is very convenient since it allows us to compute the steady state directly from the Laplace transform $\phi(s)$ without going back into the time domain. The phase error in the Laplace domain given by (2.3-7) is

$$\phi(s) = \frac{1}{1 + [KAF(s)/s]} \, \theta(s) \qquad (2.3\text{-}9)$$

The transfer function

$$G_0(s) = KAF(s)/s \qquad (2.3\text{-}9a)$$

is called the *open-loop transfer function* of the system. The reason for this designation is readily understood from Figure 2.3-1. Suppose we disconnect the integrator in the path from the summer and compute the transfer function $\hat{\theta}(s)/\theta(s)$. We immediately find $KAF(s)/s$ as the result.

As can be seen from (2.3-9), the steady phase error depends on the open-loop transfer function $G_0(s)$ and on the type of input signal $\theta(s)$. It is customary to specify $\phi(t \to \infty)$ with respect to well-defined elementary signals such as

$$\theta(t) = \begin{cases} \Delta\theta \\ \Delta\omega t \\ \Delta\dot{\omega}t^2/2 \end{cases} \qquad \circ\!\!-\!\!\bullet \qquad \theta(s) = \begin{cases} \Delta\theta/s & \text{phase step} \\ \Delta\omega/s^2 & \text{frequency step} \\ \Delta\dot{\omega}/s^3 & \text{frequency ramp} \end{cases}$$

$$(2.3\text{-}10)$$

Using these elementary functions, many reasonable input phase signals $\theta(t)$ can be approximated with sufficient accuracy.

As a first example, let us consider the steady-state phase error resulting from a phase step of magnitude $\Delta\theta$ used in (2.3-9)

$$\phi(t\rightarrow\infty) = \lim_{s\rightarrow 0}\left\{ s\,\frac{\Delta\theta}{s}\,\frac{1}{1+[KAF(s)/s]} \right\} = 0, \quad F(0)\neq 0 \quad (2.3\text{-}11)$$

thus, the PLL will reduce any phase error to zero. This is true even if $F(s)=1$, the case in which the filter has degenerated to a constant. As another example, let us examine the steady-state error resulting from a change of frequency of magnitude $\Delta\omega$. The input phase $\theta(s)$ is then a ramp $\theta(s) = \Delta\omega/s^2$ for which we obtain in the limit $t\rightarrow\infty$

$$\phi(t\rightarrow\infty) = \lim_{s\rightarrow 0}\left\{ s\,\frac{\Delta\omega}{s^2}\,\frac{1}{1+[KAF(s)/s]} \right\} = \lim_{s\rightarrow 0}\left[\frac{\Delta\omega}{s+KAF(s)} \right]$$

$$(2.3\text{-}12)$$

In order to keep the state phase error small, the vlaue of the filter transfer function $F(s=0)$ should be large. If $F(s)$ has poles of order k at the origin $s=0$, $F(s)$ can be written as

$$F(s) = \frac{1}{s^k}\cdot F_1(s) \qquad\qquad (2.3\text{-}13)$$

with

$$0 < |F_1(0)| < \infty$$

In case $F(s)$ has at least one pole at $s=0$ $(k=1)$, the PLL is able to completely eliminate any phase errors resulting from the frequency step applied at $t=0$.

Let us go one step further and study what happens if a frequency ramp is to be tracked by a PLL. This happens, for example, in the case of the reception of a signal from a constant frequency oscillator aboard a vehicle traveling with a constant radial acceleration $\Delta\dot{\omega} = c/\omega_0$ (c is the velocity of light) relative to the receiver. Carrying out the limiting operations we find

$$\phi(t\rightarrow\infty) = \lim_{s\rightarrow 0}\left\{ s\,\frac{\Delta\dot{\omega}}{s^3}\,\frac{1}{1+[KAF(s)/s]} \right\}$$

$$= \lim_{s\rightarrow 0}\left[\frac{\Delta\dot{\omega}}{s^2+KAF(s)s} \right] \qquad\qquad (2.3\text{-}14)$$

Only if $F(s)$ has at least two poles at the origin can the PLL track the frequency ramp with zero steady-state error. If a filter with one pole at the origin were used to track this signal, a phase error equal to

$$\phi(t \rightarrow \infty) = \frac{\Delta \dot{\omega}}{KAF_1(0)} \tag{2.3-15}$$

would remain. If we try to use no filter at all then a steady state is never reached since $\phi(t)$ increases without limit. As a general rule we can state that, in order to track an input phase signal with Laplace transform of the form s^{-k}, a filter with $(k-1)$ poles at the origin is needed if zero steady-state phase error is required.

A PLL without a filter is known as a *first-order loop*, one with a filter that contains one pole as a *second-order loop*. In general, a loop with a filter having $(k-1)$ *poles* is called a *kth-order loop*. The order of a control system is often not sufficient to characterize a feedback system. In control theory terminology feedback systems are also distinguished by their *type*. The open loop transfer function $G_0(s)$ of a *type k* system has k-poles (integrators) *at the origin*, $s = 0$. Loop order can be higher than type since all poles contribute to order (but not to type). The transfer function of the VCO contains one pole at $s = 0$ which explains why the loop filter must have $(k-1)$ poles at the origin if the loop is of type k.

In summary, we see that for a given input signal $\theta(s)$, the steady-state phase error is completely determined by the value of the open-loop transfer function at the origin. The more detailed structure of the loop filter has an influence only on the transient behavior of the PLL. The number of poles needed at the origin, and consequently, the number of perfect integrators of the loop filter is determined by the dynamics of the input signal $\theta(t)$. For example, a first-order loop is sufficient to track a frequency step with finite phase error while for zero phase error in the steady state, the loop filter must have at least one perfect integrator (type 2 loop). If the input dynamics are a frequency ramp then the first-order loop is no longer able to track the input signal while the second-order loop can cope with the signal where a finite phase error is tolerable. In this case, to reduce the phase error to zero, type 3 loop is needed.

While the steady-state error of a loop is readily obtained, our present discussion is rather abstract in terms of poles at the origin. When studying the transient response of a loop we will give intuitive physical meaning to the abstract results derived here.

2.3.3. Design of Feedback Systems Using the Bode Diagram

In the following discussion we briefly recapitulate topics of classical control theory assuming that the reader is familiar with the basic concepts of this field as taught at undergraduate level. Any reader who does not have basic knowledge of concepts such as frequency response, Nyquist stability criterion, phase margin, should consult one of the numerous undergraduate level textbooks on control systems (see, e.g., [1]).

A very useful design tool for linear control systems is the *Bode plot*. The Bode plot is a representation of the phase (or argument) and the magnitude of the *frequency response* of a system. The frequency response is obtained if we replace s in the transfer function $G(s)$ by the pure imaginary number $(j\omega)$. Thus, $G(j\omega)$ is a complex number for any given frequency ω.

There is an important physical interpretation of the frequency response. If a linear time-invariant stable system is excited by a sinusoidal input signal, the steady-state response will be a sinusoid of the same frequency but with a different amplitude and phase. Formally, let the transfer function for $s = j\omega$ be $G(j\omega)$ or simply $G(\omega)$ and let the input signal be

$$x(t) = \text{Re}\,\{A_x \exp\,(j\omega t)\} = A_x \cos\,(\omega t) \tag{2.3-16}$$

The steady-state response is

$$y(t) = \text{Re}\,\{|G(\omega)| \exp\,[\,j\,\arg\,(G(\omega))] \cdot A_x \exp\,(j\omega t)\}$$
$$= A_y \cos\,(\omega t + \phi_y) \tag{2.3-17}$$

From (2.3-17) we find that the amplitude and phase of the output signal are related to the corresponding input signal parameters by

$$A_y = |G(\omega)|A_x\,, \quad \phi_y = \arg\,\{G(\omega)\}$$

Sinusoidal test signals are readily available. Thus the experimental determination of the frequency response of a system is easily accomplished and is a reliable and uncomplicated method for the experimental analysis of a system. This is particularly important in high-frequency applications where it is difficult to obtain exact mathematical models.

In the Bode plot, the logarithm of magnitude $|G(\omega)|$ and the argument of $G(\omega)$ are plotted against a logarithmic scale of frequency. The logarithm of the magnitude is normally expressed in terms of base 10 logarithms and we use

$$20 \log_{10} |G(\omega)|$$

where the units are decibels (dB). The argument $\arg\,\{G(\omega)\}$ is displayed on a linear vertical scale, usually in degrees. As an example, the Bode plot of

$$G(\omega) = \frac{1 + j\omega T_1}{(\,j\omega T)^2(1 + j\omega T_2)(1 + j\omega T_3)} \tag{2.3-18}$$

is shown in Figure 2.3-2.

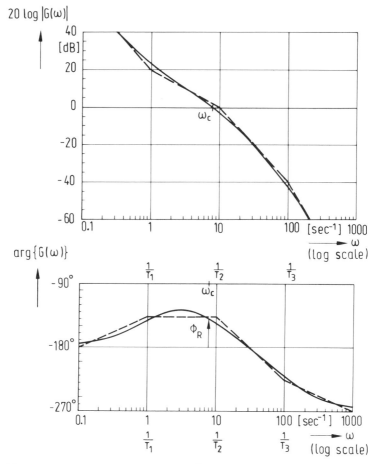

Figure 2.3-2. Bode diagram example ($T^2 = 0.1$, $T_1 = 1$, $T_2 = 0.1$, $T_3 = 0.01$).

The primary advantage of the Bode plot is that multiplicative factors are converted into additive terms by virtue of the logarithmic gain definition. Thus the contribution of the individual factors on the overall transfer function is clearly visible. This is a great advantage in designing a feedback system. A Bode plot is very simple to construct since for most practical purposes it is sufficient to work with the asymptotes, which are straight lines (dashed lines in Figure 2.3-2).

A basic requirement on any feedback system is stability. For our purposes, stability is defined in the sense that the system responds to any bounded input signal with a bounded output signal (BIBO stability). The basic transfer function of PLL was found to be

$$H(s) = \frac{G_0(s)}{1 + G_0(s)} \qquad (2.3\text{-}19)$$

where $G_0(s)$ is the open-loop transfer function (2.3-9a). Based on the Nyquist stability criterion, the Bode plot of the open-loop transfer function $G_0(s)$ provides for a simple determination of the stability of the system without the need for explicitly computing the poles of $H(s)$.

The simplified Nyquist criterion [1] states that a system is stable if the *phase margin* is positive. The phase margin ϕ_R is obtained by subtracting $-180°$ from the argument arg $\{G_0(\omega_c)\}$ (see Figure 2.3-2)

$$\phi_R := \arg \{G_0(\omega_c)\} - (-180°) \qquad (2.3\text{-}20)$$

where ω_c is the crossover frequency at which

$$|G_0(\omega_c)| = 1$$

Thus, using the Bode plot, stability is easily determined; simply read off the argument arg $\{G_0(\omega_c)\}$ at the point of intersection of the log magnitude function with the zero decibel line. Thus, the system with the open-loop Bode plot of Figure 2.3-2 has a phase margin of $+40°$ and hence is stable. It is remarkable that two fundamental performance measures of a closed-loop system, namely stability and steady-state phase error can be determined from knowledge of $G_0(\omega)$ alone.

In a control system not only these performance measures are of concern but equally important are measures for the transient response of the system. Typical measures include rise time T_r, settling time T_s, and peak overshoot M_p for a step input signal; they are illustrated in Figure 2.3-3. These measures can only be obtained by solving the dynamic equation given in Section 2.3.1. Fortunately, crossover frequency ω_c and phase margin ϕ_R can serve as good qualitative indicators for the settling time T_s and peak overshoot M_p as shown below.

Let us first determine an approximate relation between settling time T_s and crossover frequency. The open-loop plot of a typical control system is

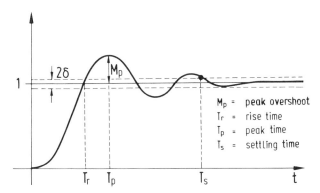

Figure 2.3-3. Step response performance measures (δ is a small number, e.g., 0.05).

displayed as a dashed line in Figure 2.3-4. Dividing the numerator and the denominator of the closed-loop transfer function (2.3-19) by G_0 yields

$$H(\omega) = \frac{1}{1 + [1/G_0(\omega)]} \tag{2.3-21}$$

From this we find the asymptotic relations

$$H(\omega) \begin{cases} \approx 1 & \omega \ll \omega_c \\ \approx G_0(\omega) & \omega \gg \omega_c \end{cases} \quad \text{for} \tag{2.3-22}$$

provided $|G_0(\omega)| \gg 1$ for frequencies much smaller than ω_c. Since a small steady phase error requires that $|G_0(0)| \gg 1$, this condition is always fulfilled in a practical control system. As will be seen shortly, to obtain a sufficiently damped transient response the phase margin must be at least 45°. But this implies that the slope of the magnitude curve at the crossover frequency ω_c must not be steeper than -20 dB/decade. Therefore, as a rather crude approximation we may write for the closed-loop transfer function

$$H(s) \approx \frac{1}{1 + (s/\omega_c)} \quad \bullet\!\!-\!\!-\!\!\circ \quad \omega_c \exp(-\omega_c t) \approx h(t), \quad t \geq 0 \tag{2.3-23}$$

Defining T_s arbitrarily as the time taken by the phase step-response error to settle within $\pm 5\%$ of its final value we obtain from (2.3-23)

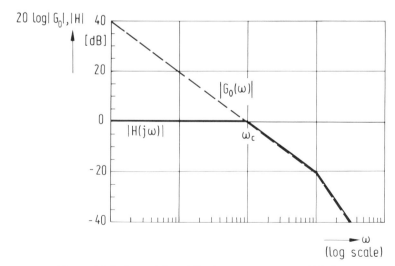

Figure 2.3-4. Asymptotes of the frequency response $|H(\omega)|$.

$$T_s \approx 3/\omega_c \qquad\qquad (2.3\text{-}24)$$

Zero phase margin means that the phase of $G_0(\omega_c)$ equals exactly $-180°$. By definition of the crossover frequency we have $|G_0(\omega_c)| = 1$. Combining the two findings we have $G_0(\omega_c) = -1$ for $\phi_R = 0$. In such a case the closed-loop system $H(s)$ has two purely imaginary poles at $s = \pm j\omega_c$. A pair of purely imaginary poles corresponds to an undamped sinusoidal oscillation, or equivalently, for $\phi_R = 0$ we have 100% peak overshoot. If we increase the phase margin to positive values the undamped oscillation turns into damped oscillation (i.e., with an exponentially decaying amplitude). Qualitatively, a small phase margin is coupled with a large peak overshoot M_p and vice versa

$$\phi_R \ \uparrow \quad \Leftrightarrow \quad M_p \ \downarrow$$

As a rule of thumb, for good relative stability one rarely allows less than 60° of phase margin. A more quantitative statement on the relation between phase margin and peak overshoot is given in Section 2.4 on second-order phase-locked loops.

In summary, static and dynamic performance measures can be obtained from the Bode plot of the open-loop transfer function $G_0(\omega)$. This makes it possible to arrive very quickly at a first draft design of a PLL for a given set of specifications.

Main Points of the Section

- For small phase errors the loop equations may be linearized using $\sin \phi \approx \phi$.
- Steady-state and transient performance measures can be obtained from the Bode plot of the open-loop frequency response $G_0(\omega)$

$$\phi(t \rightarrow \infty) = \lim_{s \rightarrow 0}\left[s\theta(s) \, \frac{1}{1 + G_0(s)} \right]$$

Settling time $T_s \approx 3/\omega_c$
Peak overshoot $M_p \quad \downarrow \quad \Leftrightarrow \quad \phi_R \quad \uparrow$, $\quad \phi_R = $ phase margin.

2.4. SECOND-ORDER PHASE-LOCKED LOOP

2.4.1. Transfer Functions

For a PLL, there exists almost always a difference $\Delta\omega$ between the incoming signal frequency and the free running (zero control voltage) frequency of the VCO. This difference may be due to an actual difference between the receiver and transmitter oscillators or it may be due to a Doppler shift. As (2.3-12) shows, for a second-order PLL, an integrator in the loop filter (which is equivalent to a pole at the origin) is capable of compensating this difference in such a way that no steady-state phase error remains. This is one of the reasons why second-order loops are by far the most popular in practice. As examples, two widely used filters are shown in Figure 2.4-1.

The passive filter realization approximates perfect integration by a low-pass filter with a pole at $s = -1/T_1$. This is quite simple and often satisfactory for many purposes. A PLL that incorporates this type of filter is called an imperfect second-order loop and yields a fixed phase error in response to an input frequency step.*

The (perfect) second-order loop requires a high-gain dc-amplifier. In contrast to the early days of electronics excellent integrated circuit dc-amplifiers with very little offset and drift are available today. Despite the extra amplifier the overall circuitry is usually simpler than that with a passive filter. The active filter should therefore be considered the norm today.

With the active filter, the overall closed-loop transfer function of (2.3-6) is (after accommodating the phase reversal of the operational amplifier)

$$H(s) = \frac{KA(1 + sT_2)}{s^2 T_1 + sKAT_2 + KA} \tag{2.4-1}$$

For the case of the passive filter, we obtain

$$H(s) = \frac{KA(1 + sT_2)}{s^2 T_1 + s(1 + KAT_2) + KA} \tag{2.4-2}$$

These closed-loop transfer functions may be rewritten as follows

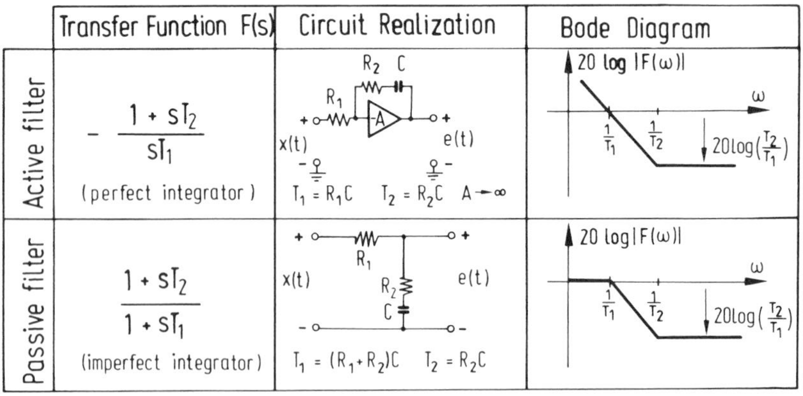

	Transfer Function F(s)	Circuit Realization	Bode Diagram
Active filter	$-\dfrac{1 + sT_2}{sT_1}$ (perfect integrator)	$x(t)$ — R_1 — [A] — $e(t)$, R_2 C $T_1 = R_1 C$ $T_2 = R_2 C$ $A \to \infty$	20 log $\|F(\omega)\|$ $\frac{1}{T_1}$ $\frac{1}{T_2}$ $20\log\left(\frac{T_2}{T_1}\right)$
Passive filter	$\dfrac{1 + sT_2}{1 + sT_1}$ (imperfect integrator)	$x(t)$ — R_1 — $e(t)$, R_2, C $T_1 = (R_1 + R_2)C$ $T_2 = R_2 C$	20 log $\|F(\omega)\|$ $\frac{1}{T_1}$ $\frac{1}{T_2}$ $20\log\left(\frac{T_2}{T_1}\right)$

Figure 2.4-1. Two widely used filters for second-order loops.

*A perfect second-order loop is a type 2 loop while an imperfect second-order loop is a type 1 loop.

$$H(s) = \frac{2\zeta\omega_n s + \omega_n^2}{s^2 + 2\zeta\omega_n s + \omega_n^2} \quad \text{(with active filter)} \tag{2.4-3}$$

and

$$H(s) = \frac{[2\zeta\omega_n - (\omega_n^2/KA)]s + \omega_n^2}{s^2 + 2\zeta\omega_n s + \omega_n^2} \quad \text{(with passive filter)} \tag{2.4-4}$$

where ω_n is called the natural frequency and ζ is the loop damping ratio. As will be discussed later, these two parameters are important descriptors of the transient response of the loop. They are tabulated as a function of T_1, T_2, and KA in Table 2.4-1. Note the difference in the definition of ζ for the two types of filters.

Similarly, using (2.3-7), the input error transfer functions can be found to be

$$\frac{\phi(s)}{\theta(s)} = \frac{s^2}{s^2 + 2\zeta\omega_n s + \omega_n^2} \quad \text{(with active filter)} \tag{2.4-5}$$

$$\frac{\phi(s)}{\theta(s)} = \frac{s^2 + (\omega_n^2/KA)s}{s^2 + 2\zeta\omega_n s + \omega_n^2} \quad \text{(with passive filter)} \tag{2.4-6}$$

The asymptotic approximation to the Bode diagram of the second-order PLL open-loop frequency response is depicted in Figure 2.4-2. The natural frequency is the frequency at which the extension of the line with $-40\,\text{dB}/$ decade slope crosses the 0 dB line. The crossover frequency ω_c is related to the natural frequency ω_n and the damping ratio ζ by

TABLE 2.4.1 Interrelations Between Loop and Filter Parameters of the Second-Order Phase-Locked Loop

Active Filter	Passive Filter
$\omega_n = \left(\dfrac{KA}{T_1}\right)^{1/2}$	$\omega_n = \left(\dfrac{KA}{T_1}\right)^{1/2}$
$\zeta = \dfrac{T_2}{2}\left(\dfrac{KA}{T_1}\right)^{1/2}$	$\zeta = \dfrac{T_2}{2}\left(\dfrac{KA}{T_1}\right)^{1/2}\left(1 + \dfrac{1}{KAT_2}\right)$
$\quad = \dfrac{T_2\omega_n}{2}$	$\quad = \dfrac{T_2\omega_n}{2} + \dfrac{\omega_n}{2KA}$
$T_1 = \dfrac{KA}{\omega_n^2}$	$T_1 = \dfrac{KA}{\omega_n^2}$
$T_2 = \dfrac{2\zeta}{\omega_n}$	$T_2 = \dfrac{2\zeta}{\omega_n}\cdot\left(1 - \dfrac{\omega_n}{KA}\cdot\dfrac{1}{2\zeta}\right)$
$KA = K_0 K_D$	$KA = K_0 K_D$

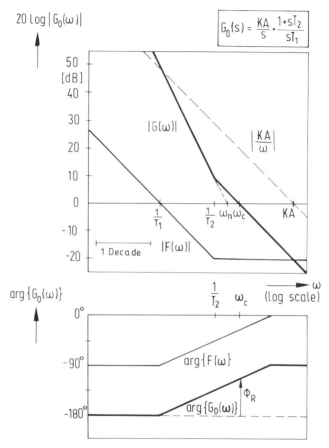

Figure 2.4-2. (a) Bode diagram of the second-order phase-locked loop open-loop frequency response with active filter.

$$\omega_c \approx 2\zeta\omega_n \qquad \text{(with active filter)}$$

$$\omega_c \approx 2\zeta\omega_n\left(1 - \frac{\omega_n}{KA}\frac{1}{2\zeta}\right) \quad \text{(with passive filter)} \tag{2.4-7}$$

Let us derive this result for the case with active filter. By definition, for $\omega = \omega_c$, the open-loop magnitude function equals 1, that is,

$$|G_0(\omega_c)| = \left|\frac{1 + j\omega_c T_2}{j\omega_c T_1}\right|\left|\frac{KA}{j\omega_c}\right| = 1 \tag{2.4-8}$$

Using the asymptotic approximation to (2.4-8) yields for $\omega_c > 1/T_2$

$$|G_0(\omega_c)| \approx \frac{T_2}{T_1}\frac{KA}{\omega_c} = 1 \tag{2.4-9}$$

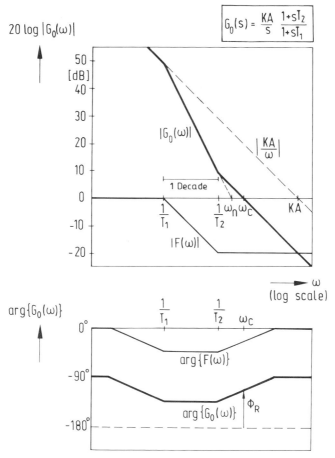

Figure 2.4-2. (*b*) Bode diagram of the second-order phase-locked loop open-loop frequency response with passive filter.

Thus solving (2.4-9) for ω_c and replacing T_1, T_2, and KA by the expressions given in Table 2.4-1 yields (2.4-7)*.

Let us now relate the phase margin of a second-order loop to the damping ratio ζ. Consider the open-loop frequency response

$$G_0(\omega) = \frac{1 + j\omega T_2}{j\omega T_1} \frac{KA}{j\omega}$$

$$= -\frac{1 + j(\omega/\omega_n)2\zeta}{(\omega/\omega_n)^2}$$

*Note that strictly speaking the crossover frequency ω_c is defined by (2.4-8). In our loose usage of the terminology we include both the exact (2.4-8) and the asymptotic (2.4-9) definition of ω_c.

The damping ratio ζ is plotted as a function of the phase margin ϕ_R in Figure 2.4-3. The actual curve of ζ versus ϕ_R may be approximated by the dashed line in Figure 2.4-3. This yields the simple formula

$$\zeta \approx 0.01\phi_R \tag{2.4-10}$$

We observe that the two transfer functions of the perfect (2.4-3) and imperfect second-order loop (2.4-4) are nearly the same if the relation

$$2\zeta\omega_n \gg \frac{\omega_n^2}{KA} \Rightarrow KA \gg \frac{\omega_n}{2\zeta} \approx \frac{1}{T_2} \tag{2.4-11}$$

holds for the imperfect loop. An imperfect second-order loop for which (2.4-11) holds is called a high-gain loop. Examining the Bode plots we find that the two transfer functions differ at very low frequencies only. In fact, the active filter is capable of tracking a frequency offset with zero steady-state phase error. (Remember that the steady state corresponds to $\omega = 0$ in the frequency domain). The passive filter leaves a phase error of

$$\phi(t \rightarrow \infty) = \frac{\Delta\omega}{KA}$$

just like a first-order loop. Why then use an imperfect second-order loop instead of a first-order one at all? The answer to the question can be found by examining the Bode plot of Figure 2.4-2b. A first-order loop with $G_0(s) = KA/s$ and the same KA as an imperfect second-order loop has the crossover frequency at $\omega_{c,1} = KA$, which is much larger than that of the second-order loop $\omega_{c,2}$. While both loops have identical steady-state perfor-

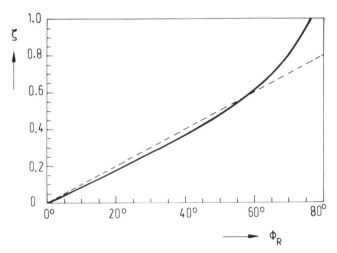

Figure 2.4-3. Damping ratio ζ versus phase margin ϕ_R.

mance they show entirely different transient responses and noise suppression performances. Since $\omega_{c,1} \gg \omega_{c,2}$ the first-order loop reacts much faster to a disturbance than its second-order counterpart. What may initially appear as an advantage is indeed a disadvantage. Until now we have completely ignored any noise disturbance. As will be demonstrated in a later chapter the noise suppression capability of the loop is inversely proportional to the crossover frequency. Therefore, the stronger the noise, the smaller ω_c should be chosen. Thus, the two requirements of a large dc-gain, $KAF(0)$ corresponding to good steady-state performance and a small ω_c

$$\omega_c \ll KA$$

implicating good noise suppression, are incompatible in a first-order but not in a second-order loop.

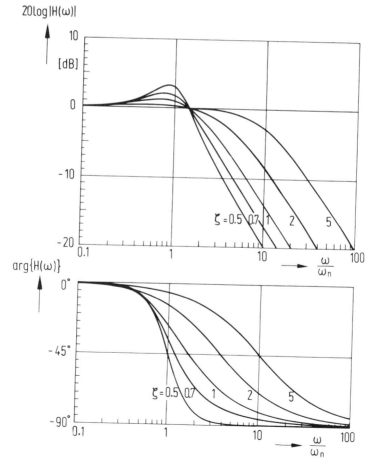

Figure 2.4-4. Closed-loop frequency response of a high-gain second-order loop.

The magnitude of the closed-loop frequency response (2.4-3) of a second-order loop (or an imperfect high-gain loop) is shown in Figure 2.4-4. It can be seen that the loop performs a low-pass filtering operation. The input error frequency response obtained from (2.4-5) and displayed in Figure 2.4-5 is also of interest. The high-pass characteristic obtained shows that the loop can follow slowly varying changes but is unable to track phase changes with high frequencies.

2.4.2. Stability Considerations

So far we have assumed that all closed-loop systems are stable. A necessary and sufficient condition for stability of linear systems is that all poles of the closed-loop system lie in the left-hand half-plane. The Nyquist criterion that

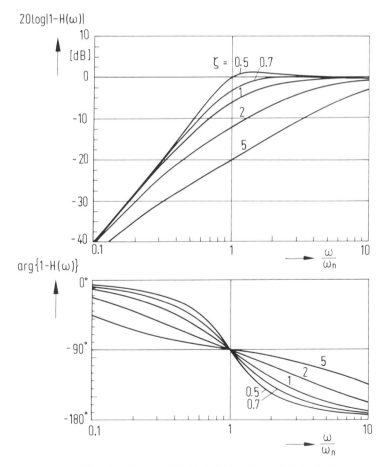

Figure 2.4-5. Error response of a high-gain second-order loop.

we have used to relate phase margin and stability is also based on this condition.

It is useful to observe the loci of the poles of $H(s)$ as a function of KA for a given open-loop transfer function $KAF(s)/s$. The root loci for the perfect and imperfect second-order loop are shown in Figure 2.4-6. The second-order loop is stable for all values of KA. This is of particular importance in the case of a time-varying amplitude $A(t)$ of the incoming waveform.

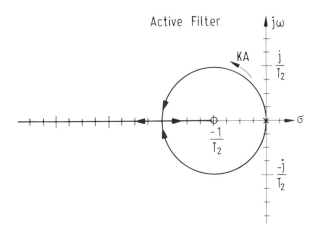

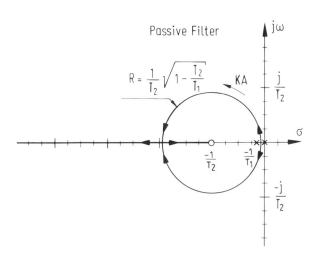

Figure 2.4-6. Root loci of the second-order loop.

2.4.3. State Variable Description

So far we have discussed an input-output (I/O) system description as expressed by a transfer function. Although for many cases an input-output description is sufficient, we will, however, encounter problems where we need a thorough understanding of the internal system variables, e.g., the voltage across the capacitor in the loop filter. In this case, a state variable approach is needed. Quite often state variables offer insight into the loop behavior which is not attainable from an I/O description.

We start by deriving a state variable representation of a second-order loop with perfect integrator. A detailed block diagram of the equivalent baseband model is given in Figure 2.4-7. Note that we have not linearized the $\sin(\cdot)$ nonlinearity. Labeling the output of the integrator as $y_1(t)$ we can write

$$\frac{d\hat{\theta}(t)}{dt} = KA \frac{T_2}{T_1} \sin \phi(t) + y_1(t)$$

Replacing $\hat{\theta}(t)$ by $\theta(t) - \phi(t)$ we obtain

$$\frac{d\phi(t)}{dt} = -KA \frac{T_2}{T_1} \sin \phi(t) - y_1(t) + \frac{d\theta(t)}{dt} \qquad (2.4\text{-}12)$$

By examining Figure 2.4-7 we readily obtain

$$\frac{dy_1(t)}{dt} = \frac{1}{T_1} KA \sin \phi(t) \qquad (2.4\text{-}13)$$

The two variables $\phi(t)$ and $y_1(t)$ are state variables of the system. This is not the only permissible choice. We could have equally well used $(\hat{\theta}(t), y_1(t))$. Then the state variables would be directly associated with the storage elements (integrators) of the system.

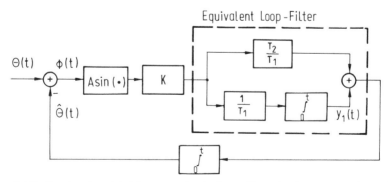

Figure 2.4-7. Baseband model of the second-order loop with perfect integrator and associated state variables $(\phi(t), y_1(t))$.

Clearly, knowing $(\phi(t_0), y_1(t_0))$ at $t = t_0$ the future state $(\phi(t), y_1(t))$ is completely specified given the input signal $\theta(t)$ for $t > t_0$. There is no need to know how the system has reached its initial state $(\phi(t_0), y_1(t_0))$. In other words, the future is independent of the past. This is true for any permissible set of state variables. A detailed discussion of the state variable concept can be found in any book on control theory, e.g. [1].

The physical meaning of $\phi(t)$ needs no explanation. But does $y_1(t)$ possess an intuitive interpretation? The answer is yes. The meaning of $y_1(t)$ can best be explained by examining the steady state of the system. Let us suppose the system eventually arrives at a steady state. It is then necessary that the state variables ϕ and y_1 be constant. This implies that their derivatives, with respect to time, equal zero. The state equations in the steady state read

$$\frac{d\phi_s}{dt} = 0 = -KA \frac{T_2}{T_1} \sin \phi_s - y_{1,s} + \frac{d\theta}{dt}$$

$$\frac{dy_{1,s}}{dt} = 0 = \frac{1}{T_1} KA \sin \phi_s$$

$$(2.4\text{-}14)$$

where the subscript s denotes steady state of the state variables.

Let us further assume that the input phase is a ramp

$$\theta(t) = \begin{cases} \Delta\omega t, & t \geq 0 \\ 0, & t < 0 \end{cases}$$

Then

$$\frac{d\theta}{dt} = \Delta\omega, \quad t \geq 0 \qquad (2.4\text{-}15)$$

From the second equation of (2.4-14) we find $\phi_s = 0$. Inserting this into the first state equation yields

$$y_{1,s} = \Delta\omega \qquad (2.4\text{-}16)$$

Correspondingly, in terms of the voltage across the capacitor in the loop filter we obtain

$$v_c(t \to \infty) = v_{c,s} = \frac{y_{1,s}}{K_0} \qquad (2.4\text{-}17)$$

where K_0 is the VCO gain. Hence $y_{1,s}/K_0$ equals the control voltage to retune the VCO by an amount of $\Delta\omega$.

The state equations as they stand in (2.4-12) and (2.4-13) are of little value. In reality the parameters can vary over orders of magnitude and it is

difficult to visualize their influence on loop performance. Introducing normalized and dimensionless variables yields a set of equations more easily
interpreted. We introduce the normalized time τ and replace y_1 by the
dimensionless quantity x_1 as follows

$$\tau = \omega_n t, \quad x_1 = \frac{y_1}{\omega_n} \tag{2.4-18}$$

By using the chain rule of calculus

$$\frac{d}{dt}(\cdot) = \frac{d}{d\tau}(\cdot)\frac{d\tau}{dt} \tag{2.4-19}$$

and the definition of ω_n and ζ from Table 2.4-1, in (2.4-12) and (2.4-13) we
arrive after some simple algebraic manipulations at

$$\frac{d\phi(\tau)}{d\tau} = -2\zeta \sin \phi(\tau) - x_1(\tau) + \frac{d\theta(\tau)}{d\tau}$$

$$\frac{dx_1(\tau)}{d\tau} = \sin \phi(\tau) \tag{2.4-20}$$

In (2.4-20) only the damping ratio ζ (besides the natural frequency absorbed
in τ) appears. The equivalent baseband model using normalized variables is
shown in Figure 2.4-8. The attentive reader will notice the use of ω_n and ζ in
a nonlinear state equation. This might be confusing since these quantities
have appeared in the context of a transfer function $H(s)$ which implies a
linear system. We should be well aware that ω_n and ζ are formally defined in
the same way for both linear and nonlinear equations but they have their
usual physical meaning in the linear case only.

We leave it as an exercise for the reader to show that for the second-
order loop with imperfect integrator (Figures 2.4-9 and 2.4-10) the state
equations assume the form

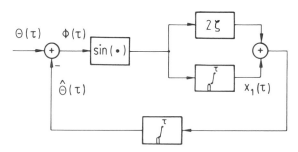

Figure 2.4-8. Baseband model of the second-order loop with perfect integrator and associated
normalized state variables ($\phi(\tau)$, $x_1(\tau)$).

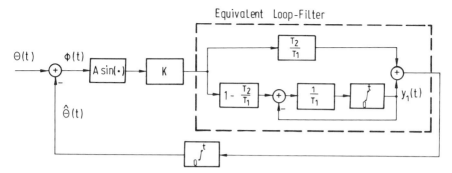

Figure 2.4-9. Baseband model of the second-order loop with imperfect integrator and associated state variables $(\phi(t), y_1(t))$.

$$\frac{d\phi(\tau)}{d\tau} = -2\zeta\left(\frac{1-\beta}{2\zeta}\right)\sin\phi(\tau) - x_1(\tau) + \frac{d\theta(\tau)}{d\tau}$$

$$\frac{dx_1(\tau)}{d\tau} = (1 - 2\zeta\beta + \beta^2)\sin\phi(\tau) - \beta x_1(\tau)$$

$(2.4\text{-}21)$

The variables and parameters are given by

$$x_1(t) = \frac{y_1(t)}{\omega_n} = \frac{K_0[1 - (T_2/T_1)]v_c(t)}{\omega_n}$$

$$= \frac{K_0(1 - 2\zeta\beta + \beta^2)v_c(t)}{\omega_n}$$

$(2.4\text{-}22)$

where v_c is the capacitor voltage and

$$\beta := \frac{\omega_n}{KA} = \frac{1}{\omega_n T_1}$$

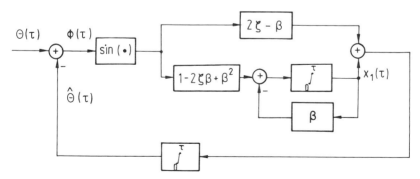

Figure 2.4-10. Baseband model of the second-order loop with imperfect integrator and associated *normalized* state variables $(\phi(\tau), x_1(\tau))$.

Note that for ω_n and ζ the relations on the right-hand side of Table 2.4-1 must be used.

The values of greatest interest for ζ lie in the range

$$0.7 < \zeta < 2.0 \qquad (2.4\text{-}23)$$

Values larger than 1 correspond to an overdamped loop used in certain applications.

We can also delimit values for β. Using the right-hand side of Table 2.4-1 it is readily shown that the ratio T_2/T_1 is given by

$$\frac{T_2}{T_1} = \beta(2\zeta - \beta) \geq 0 \qquad (2.4\text{-}24)$$

Since β and T_2/T_1 must always be nonnegative we find

$$0 \leq \beta \leq 2\zeta \qquad (2.4\text{-}25)$$

In the case of a commonly used high-gain loop we have (2.4-11)

$$\frac{\omega_n}{KA} \ll 2\zeta$$

Using the definition of β we can write the equivalent relation

$$\beta \ll 2\zeta \quad \text{(high-gain loop)} \qquad (2.4\text{-}26)$$

Typical values of β are within the range 0.001–0.1. If we actually set $\beta = 0$ we obtain the equations for the second-order loop with perfect integrator.

The advantage of using normalized variables now becomes clear. Originally, these parameters T_1, T_2, KA appear in the loop equations. The constants can vary within orders of magnitude depending on the specific application. The influence of the various parameters on the loop performance cannot be grasped easily. By scaling the time axis by ω_n only the parameter ζ (and β) remain. Since ζ assumes only values within a small interval this greatly facilitates the design. We also want to point out that the normalized state equations are in a form appropriate for computer simulation.

2.4.4. Transient Loop Response Under Linear Conditions

To study tracking we examine the transient response that results from a specified input $\theta(t)$. A small phase error $\phi(t)$ is usually desirable and is considered to be the criterion for good tracking performance. Here we assume that the phase error remains small enough to justify the linearization of the loop equations.

If we are interested in the phase error only, then an input-output description in the form the transfer function $H(s)$ is appropriate. To obtain a deeper understanding of the loop behavior we will solve the linearized state equation and examine the phase error as well as the other state variable x_1. For reasonably small β, the dynamic behavior of the imperfect second-order loop is practically indistinguishable from the perfect integrator loop. Thus, we only consider $\beta = 0$.

The linearized equations for a second-order loop with perfect integrator are easily found from (2.4-20) by replacing $\sin \phi$ by ϕ.

$$\frac{d\phi(\tau)}{d\tau} = -2\zeta\phi(\tau) - x_1(\tau) + \frac{d\theta(\tau)}{d\tau}$$

$$\frac{dx_1(\tau)}{d\tau} = \phi(\tau)$$

(2.4-27)

In the Laplace domain, (2.4-27) takes on the form

$$p\phi(p) - \phi(0) = -2\zeta\phi(p) - X_1(p) + p\theta(p)$$

$$pX_1(p) - x_1(0) = \phi(p)$$

(2.4-28)

Note that we have used $p = s/\omega_n$ as the "normalized Laplace variable" to take into account the time normalization $\tau = \omega_n t$. Solving for $\phi(p)$ yields (in matrix notation)

$$\begin{bmatrix} \phi(p) \\ X_1(p) \end{bmatrix} = \frac{1}{p^2 + 2\zeta p + 1} \begin{bmatrix} p & -1 \\ 1 & p + 2\zeta \end{bmatrix} \begin{bmatrix} \phi(0) \\ x_1(0) \end{bmatrix} +$$

$$\frac{1}{p^2 + 2\zeta p + 1} \begin{bmatrix} p & -1 \\ 1 & p + 2\zeta \end{bmatrix} \begin{bmatrix} p\theta(p) \\ 0 \end{bmatrix}$$

(2.4-29)

Using state variables it has now become clear that the response of the PLL consists of two distinct parts. The first part solely depends on the initial conditions $[\phi(0), x_1(0)]$ and is independent of the input signal $\theta(p)$, whereas the second part is a function of the input signal only and is independent of the initial conditions.

For vanishing initial conditions we obtain, for example,

$$\phi(p) = \frac{p^2}{p^2 + 2\zeta p + 1} \theta(p)$$

(2.4-30)

which is of course identical to the transfer function relating phase error and input phase given by (2.4-5) when the normalization $s = p\omega_n$ is used.

As examples, the time-varying state vectors for the signal phases (in normalized time) are

- a step of phase $\Delta\theta$
- a step of frequency $d\theta/d\tau = \Delta\omega/\omega_n$
- a frequency ramp $d^2\theta/d\tau^2 = \Delta\dot{\omega}/\omega_n^2$

and zero initial conditions are computed. The results are plotted in Figures 2.4-11–2.4-13.

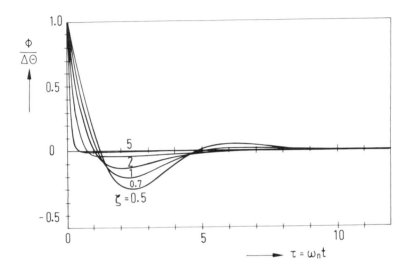

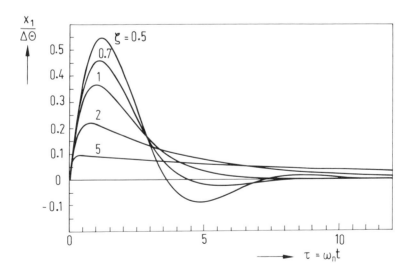

Figure 2.4-11. Transient state vector $[\phi, x_1]^T$ due to a phase step $\Delta\theta$ for a perfect second-order loop: (a) $\phi(\tau)$; (b) $x_1(\tau)$.

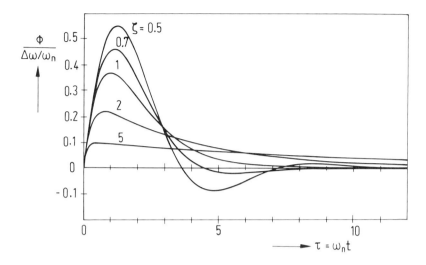

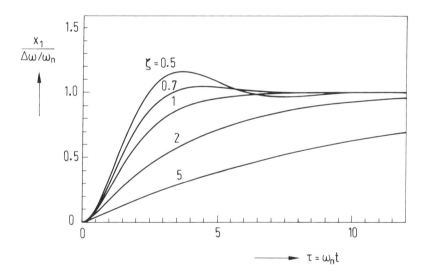

Figure 2.4-12. Transient state vector $[\phi, x_1]^T$ due to a step in frequency for a perfect second-order loop: (a) $\phi(\tau)$; (b) $x_1(\tau)$.

In Section 2.3 the step response performance measures, settling time T_s, rise time T_r, and peak overshoot M_p were introduced. These measures are calculated here more precisely and plotted in Figure 2.4-14 as a function of loop damping ratio ζ.

From Figure 2.4-14 we learn that small peak overshoot cannot be realized

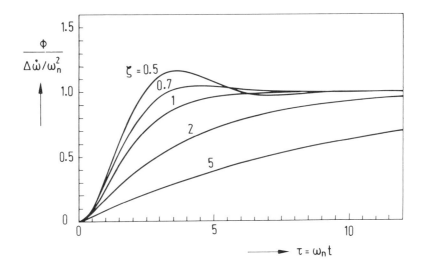

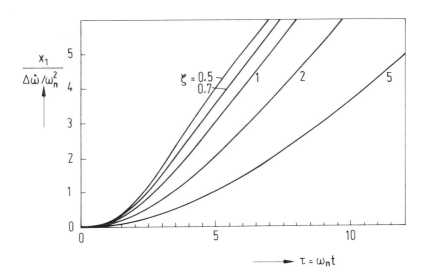

Figure 2.4-13. Transient state vector $[\phi, x_1]^T$ due to a frequency ramp for a perfect second-order loop: (a) $\phi(\tau)$; (b) $x_1(\tau)$.

simultaneously with a small settling time T_s. The curves also substantiate our claim of Section 2.3 concerning the usefulness of the crossover frequency ω_c and phase margin ϕ_R as indicators for the transient loop behavior. (Recall that loop damping ratio and phase margin are approximately related by $\zeta \approx 0.01\phi_R$ for $\zeta < 1$).

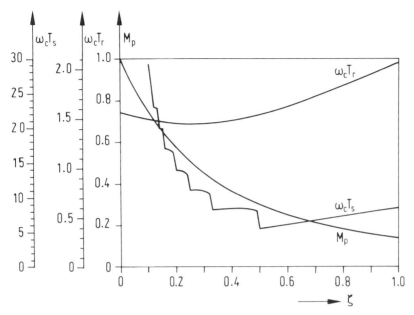

Figure 2.4-14. Transient performance measures for a perfect second-order phase-locked loop.

Main Points of the Section

- *Basic Transfer Functions of Second-order Loops*
  (*a*) Active filter: (type 2 loop)

$$G_0(s) = \frac{KAF(s)}{s} = \frac{KA(1 + sT_2)}{s^2 T_1} = \frac{2\zeta\omega_n s + \omega_n^2}{s^2}$$

$$H(s) = \frac{\hat{\theta}(s)}{\theta(s)} = \frac{2\zeta\omega_n s + \omega_n^2}{s^2 + 2\zeta\omega_n s + \omega_n^2} \tag{2.4-3}$$

$$1 - H(s) = \frac{\phi(s)}{\theta(s)} = \frac{s^2}{s^2 + 2\zeta\omega_n s + \omega_n^2} \tag{2.4-5}$$

  (*b*) Passive filter: (type 1 loop)

$$G_0(s) = \frac{KAF(s)}{s} = \frac{KA(1 + sT_2)}{s(1 + sT_1)} = \frac{[2\zeta\omega_n - (\omega_n^2/KA)]s + \omega_n^2}{s^2 + (\omega_n^2/KA)s}$$

$$H(s) = \frac{[2\zeta\omega_n - (\omega_n^2/KA)]s + \omega_n^2}{s^2 + 2\zeta\omega_n s + \omega_n^2} \tag{2.4-4}$$

$$1 - H(s) = \frac{s^2 + (\omega_n^2/KA)s}{s^2 + 2\zeta\omega_n s + \omega_n^2} \tag{2.4-6}$$

For both filters damping ratio ζ and phase margin ϕ_R are related by $\zeta \approx 0.01 \phi_R$

- *Stability*
  A second order PLL is *stable* for all values of KA.
- *State Variable Description*
  (a) Active filter: (type 2 loop)

$$\frac{d\phi}{d\tau} = -2\zeta \sin \phi(\tau) - x_1(\tau) + \frac{d\theta(\tau)}{d\tau}$$

$$\frac{dx_1(\tau)}{d\tau} = \sin \phi(\tau)$$

(2.4-20)

(b) Passive filter: (type 1 loop)

$$\frac{d\phi(\tau)}{d\tau} = -2\zeta \left(1 - \frac{\beta}{2\zeta}\right) \sin \phi(\tau) - x_1(\tau) + \frac{d\theta(\tau)}{d\tau}$$

$$\frac{dx_1(\tau)}{d\tau} = (1 - 2\zeta\beta + \beta^2) \sin \phi(\tau) - \beta x_1(\tau)$$

(2.4-21)

(c) Normalized variables:

$$\tau = \omega_n t, \quad x_1 = \frac{y_1}{\omega_n}, \quad \beta = \frac{\omega_n}{KA} = \frac{\omega_n}{K_0 K_D}$$

(Other normalizations such as $\tau = \beta_L t$ are also used on certain occasions.)

2.5. THIRD-ORDER, TYPE-3 PHASE-LOCKED LOOP

2.5.1. Transfer Functions

One reason to employ a type 3 or higher type loop is the resulting improved tracking capability of phase signals $\theta(t)$ with high dynamics. For example, to track a frequency ramp with zero steady-state phase error, a loop filter with two poles at the origin is required. The loop is then of type 3.

The open-loop transfer function of a type 3 loop has three poles at the origin causing a 270° phase shift. Two zeros are needed for a positive phase margin at the crossover frequency ω_c. Therefore, the open-loop transfer function assumes the general form

$$G_0(s) = \frac{KA(1 + sT_2)(1 + sT_3)}{s(sT_1)^2}$$

(2.5-1)

As an example, the Bode plot for coincident zeros is shown in Figure 2.5-1. The closed-loop transfer functions are

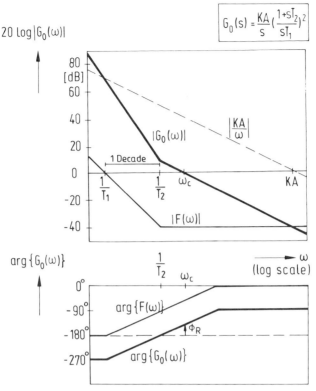

$$G_0(s) = \frac{KA}{s}\left(\frac{1+sT_2}{sT_1}\right)^2$$

Figure 2.5-1. Bode diagram of the open-loop transfer function of a third-order loop with two coincident zeros.

$$H(s) = \frac{\hat{\theta}(s)}{\theta(s)} = \frac{\omega_c(s + \alpha\omega_c)(s + \alpha_1\omega_c)}{s^3 + \omega_c(s + \alpha\omega_c)(s + \alpha_1\omega_c)}$$

(2.5-1a)

$$1 - H(s) = \frac{\phi(s)}{\theta(s)} = \frac{s^3}{s^3 + \omega_c(s + \alpha\omega_c)(s + \alpha_1\omega_c)}$$

with the parameters ω_c, α, and α_1 defined as

$$\omega_c = KA\left(\frac{T_2}{T_1}\right)\left(\frac{T_3}{T_1}\right), \quad \alpha = \frac{1}{\omega_c T_2}, \quad \alpha_1 = \frac{1}{\omega_c T_3}$$

As described in Section 2.5.3, ω_c is the (asymptotic) crossover frequency. The parameters α and α_1 relate the corner frequencies $1/T_2$ and $1/T_3$ of the loop filter to ω_c.

2.5.2. Stability Considerations

For a stable loop the two zeros of $G_0(s)$ must be located such that a positive phase margin results. It is common practice, though not necessary, to have the two zeros coincident

$$G_0(s) = \frac{KA}{s} \left(\frac{1 + sT_2}{sT_1} \right)^2 \qquad (2.5\text{-}2)$$

For $s = j/T_2$ the argument of $G_0(\omega)$ equals exactly $-180°$.

$$\arg \{G_0(j/T_2)\} = \arg \{-jT_2\} + 2 \arg \{1 + j\} + 2 \arg \{-jT_2/T_1\}$$
$$= -180° \qquad (2.5\text{-}3)$$

The magnitude of $G_0(\omega)$ for this value of s is

$$\left| G_0 \left(\frac{1}{T_2} \right) \right| = KAT_2 2 \left(\frac{T_2}{T_1} \right)^2$$

Now, as can be seen by inspection of the Bode diagram in Figure 2.5-1, a positive phase margin is possible if and only if

$$\left| G_0 \left(\frac{1}{T_2} \right) \right| > 1 \qquad (2.5\text{-}4)$$

The magnitude function $|G_0(\omega)|$ is a monotonically decreasing function of ω while $\arg \{G_0(\omega)\}$ is an increasing function of ω. Let us assume now that $|G_0(1/T_2)| > 1$. In this case, there must be a frequency $\omega_c > 1/T_2$ such that $|G_0(\omega_c)| = 1$. By definition, this frequency ω_c is the crossover frequency. Since $\arg \{G_0(\omega)\}$ increases we must have $\arg \{G_0(\omega_c)\} > -180°$ which means we have a positive phase margin ϕ_R corresponding to a stable system.

Conversely, using the same reasoning we find that for $|G_0(1/T_2)| < 1$ the crossover frequency ω_c is smaller than $1/T_2$. This, however, means that $\arg \{G_0(\omega_c)\} < -180°$. In this case, the system is unstable.

Combining our findings, we have a necessary and sufficient stability condition for the third-order loop, namely

$$\text{stability} \Leftrightarrow \left| G_0 \left(\frac{1}{T_2} \right) \right| = KAT_2 \, 2 \left(\frac{T_2}{T_1} \right)^2 > 1$$

$$\Leftrightarrow KA > \frac{1}{2T_2} \left(\frac{T_1}{T_2} \right)^2 \qquad (2.5\text{-}5)$$

This result is of great significance. While first- and second-order loops are unconditionally stable a third-order, type 3 loop requires that the gain KA be larger than the minimum value given in (2.5-5). In practical applications the amplitude A of the input signal must be kept above a minimum value by means of an automatic gain control (AGC) circuit or a limiter (see Chapter 7). The location of the three poles of $H(s)$ as a function of KA is shown in Figure 2.5-2.

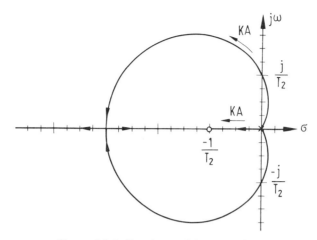

Figure 2.5-2. Root locus of the type 3 loop.

2.5.3. State Variable Description

A mathematically equivalent block diagram of the type 3 loop is shown in Figure 2.5-3.

We choose the outputs of the three integrators as state variables. Physically, the integrators correspond to capacitors in the loop filter and to the VCO. Among the many possible state variable representations, this particular one has the important advantage that the state variables have a describable physical meaning.

From Figure 2.5-3 we can directly write the following set of equations.

$$\frac{d\phi(t)}{dt} = -\frac{T_3 T_2}{T_1^2} KA \sin \phi(t) - y_1(t) - \frac{T_2}{T_1} y_2(t) + \frac{d\theta(t)}{dt}$$

$$\frac{dy_1(t)}{dt} = \frac{T_3 KA}{T_1^2} \sin \phi(t) + \frac{1}{T_1} y_2(t) \qquad (2.5\text{-}6)$$

$$\frac{dy_2(t)}{dt} = \frac{KA}{T_1} \sin \phi(t)$$

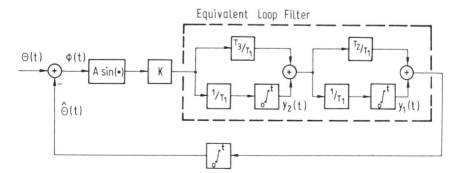

Figue 2.5-3. Equivalent baseband model of the type 3 loop.

As the equations stand they provide little insight into loop behavior. Much can be gained by using appropriate normalized variables as was demonstrated earlier for the second-order loop.

First of all, time t must be normalized. In Section 2.3 we learnt that the settling time T_s is closely related to the crossover frequency ω_c. We know that all system transients die away within a few multiples of $(1/\omega_c)$. It therefore makes a lot of sense to normalize time to $1/\omega_c$.

Recall that the second-order loop was normalized to the natural frequency ω_n. We could equally well have used $\omega_c = \omega_n 2\zeta$ (2.4-7) there. We define

$$\tau = \omega_c t \tag{2.5-7}$$

where (see footnote in Section 2.4.1)

$$\omega_c = KA\left(\frac{T_2}{T_1}\right)\left(\frac{T_3}{T_1}\right)$$

and can be found using a similar discussion which led to (2.4-9).

We relate the time constants T_2 and T_3 to ω_c by the dimensionless parameters

$$\alpha = \frac{1}{\omega_c T_2}, \quad \alpha_1 = \frac{1}{\omega_c T_3} \tag{2.5-8}$$

Physically, α and α_1 are the normalized break frequencies $1/T_2$ and $1/T_3$ of the loop filter.

We next have to replace y_1 and y_2 by dimensionless state variables. Proceeding analogously as for the second-order loop we choose the normalization constants such that $x_1(\tau)$ and $x_2(\tau)$ have an intuitive physical interpretation. We define

$$x_1(\tau) = \frac{y_1(\tau)}{\omega_c} \tag{2.5-9}$$

$$x_2(\tau) = \frac{y_2(\tau)}{\omega_c^2 T_1} \tag{2.5-10}$$

Using the chain rule of differentiation and the (2.5-6)–(2.5-10) we arrive after some simple manipulations, at

$$\frac{d\phi(\tau)}{d\tau} = -\sin\phi(\tau) - x_1(\tau) - \frac{1}{\alpha}x_2(\tau) + \frac{d\theta(\tau)}{d\tau}$$

$$\frac{dx_1(\tau)}{d\tau} = \alpha\sin\phi(\tau) + x_2(\tau) \tag{2.5-11}$$

$$\frac{dx_2(\tau)}{d\tau} = \alpha\alpha_1\sin\phi(\tau)$$

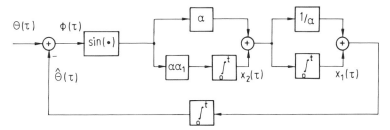

Figure 2.5-4. Block diagram of the type 3 loop with *normalized* variables.

The block diagram of the equivalent model with normalized variables is shown in Figure 2.5-4. Now, for a normalized frequency *step*

$$\frac{d\theta}{d\tau} = \frac{d\theta}{dt} \cdot \frac{dt}{d\tau} = \left(\frac{\Delta\omega}{\omega_c}\right) \tag{2.5-12}$$

the steady-state value of x_1 is readily found to be

$$x_{1,s} = \frac{\Delta\omega}{\omega_c}, \quad \phi_s = 0, \ y_{2,s} = 0$$

The state variable $x_1(\tau)$ has exactly the same physical meaning as in a second-order loop: in the steady state $x_{1,s}$ stores the (normalized) frequency $\Delta\omega/\omega_c$ of the input phase signal $\theta(\tau)$. Going one step further we examine the response of the loop to a normalized frequency *ramp*

$$\theta(\tau) = \left(\frac{\Delta\dot\omega}{\omega_c^2}\right) \cdot \frac{\tau^2}{2} \tag{2.5-13}$$

The steady-state phase error for the phase signal (2.5-13) equals zero. Solving the state equations (2.5-6) we find the following for the steady-state value of x_2

$$x_{2,s} = \left(\frac{\Delta\dot\omega}{\omega_c^2}\right) \tag{2.5-14}$$

Thus, in the steady state, the state variable x_2 stores the (normalized) rate of change of frequency of the input phase signal $\theta(\tau)$. (This explains the normalization used in (2.5-10).)

2.5.4. Transient Loop Response Under Linear Conditions

The linearized state equation for $\alpha = \alpha_1$ (i.e., coincident zeros) are obtained from (2.5-11) by replacing $\sin(\phi)$ by its argument ϕ.

$$
\frac{d}{d\tau}
\begin{bmatrix} \phi \\ x_1 \\ x_2 \end{bmatrix}
=
\begin{bmatrix}
-1 & -1 & -\dfrac{1}{\alpha} \\
\alpha & 0 & 1 \\
\alpha^2 & 0 & 0
\end{bmatrix}
\begin{bmatrix} \phi \\ x_1 \\ x_2 \end{bmatrix}
+
\begin{bmatrix} \dfrac{d\theta}{d\tau} \\ 0 \\ 0 \end{bmatrix}
\tag{2.5-15}
$$

In the Laplace domain the solution of (2.5-15) for zero initial conditions equals

$$
\begin{bmatrix} \phi(p) \\ x_1(p) \\ x_2(p) \end{bmatrix}
=
\frac{1}{p^3 + (p+\alpha)^2}
\begin{bmatrix}
p^2 & -p & -\dfrac{p+\alpha}{\alpha} \\
\alpha p + \alpha^2 & p^2 + p + \alpha & p \\
\alpha^2 p & -\alpha^2 & p^2 + p + \alpha
\end{bmatrix}
\begin{bmatrix} p\theta(p) \\ 0 \\ 0 \end{bmatrix}
\tag{2.5-16}
$$

For the signal phases

a step of phase $\Delta\theta$
a step of frequency (phase ramp) $d\theta/d\tau = \Delta\omega/\omega_c$
a step of acceleration $d^2\theta/d\tau^2 = \Delta\dot{\omega}/\omega_c^2$

and zero initial conditions, the components of the state vector as a function of τ are computed. The results are plotted in Figures 2.5-5 to 2.5-7.

2.5.5. Comparison of the Transient Response of a Second-Order and a Third-Order Loop

As we know the crossover frequency ω_c and the phase margin ϕ_R are useful descriptions of the transient reponse of a control system. It would now be interesting to compare the transient responses of a second- and a third-order phase-locked loop with equal crossover frequency ω_c and phase margin ϕ_R.
    The open-loop transfer function of the third-order loop is

$$
G_{0,3}(p) = \frac{(p+\alpha)^2}{p^3}, \qquad p = \frac{s}{\omega_c}
\tag{2.5-17}
$$

The second-loop transfer function when normalized to ω_c equals

$$
G_{0,2}(p) = \frac{p + (1/2\zeta)^2}{p^2}
\tag{2.5-18}
$$

with

$$
p = \frac{s}{\omega_c}, \qquad \omega_c = 2\zeta\omega_n
$$

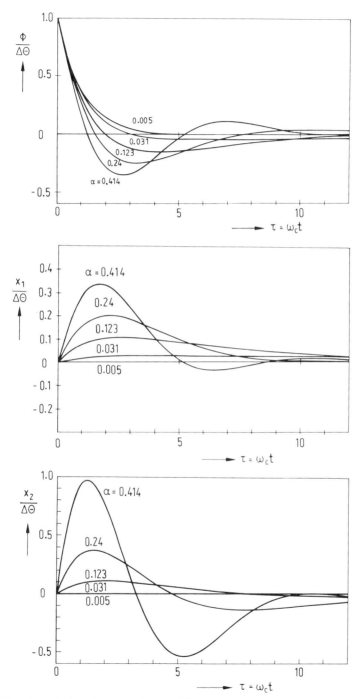

Figure 2.5-5. Transient state vector $[\phi, x_1, x_2]^T$ due to a phase step: (a) $\phi(\tau)$; (b) $x_1(\tau)$; (c) $x_2(\tau)$.

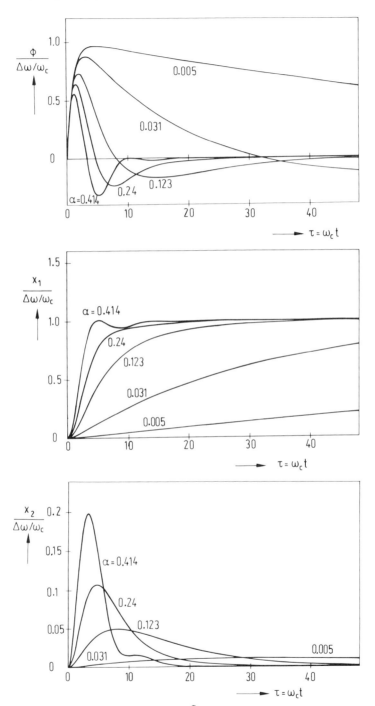

Figure 2.5-6. Transient state vector $[\phi, x_1, x_2]^T$ due to a frequency step: (a) $\phi(\tau)$; (b) $x_1(\tau)$; (c) $x_2(\tau)$.

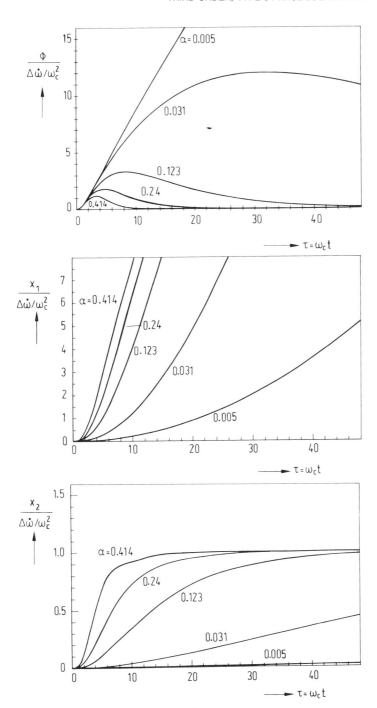

Figure 2.5-7. Transient state vector $[\phi, x_1, x_2]^T$ due to a frequency ramp: (a) $\phi(\tau)$; (b) $x_1(\tau)$; (c) $x_2(\tau)$.

The two loops have (approximately) the same phase margin if the arguments of (2.5-17) and (2.5-18) are the same for $p = j$ (i.e., for $s = j\omega_c$)

$$\arg\{G_{0,2}(j)\} = \arg\{G_{0,3}(j)\} \Rightarrow \arg\left\{\frac{j + (1/2\zeta)^2}{j^2}\right\} = \arg\left\{\frac{(j + \alpha)^2}{j^3}\right\}$$

(2.5-19)

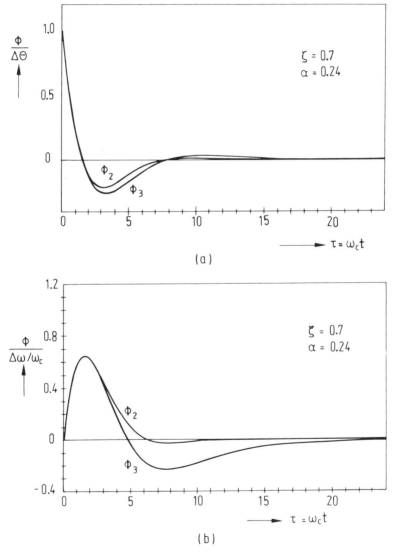

Figure 2.5-8. Phase error ϕ of a type 2 and 3 loop with equal phase margin ϕ_R and crossover frequency ω_c: (a) response to a phase step; (b) response to a frequency step.

From (2.5-19) we find, after an easy calculation, that ζ and α must be related by

$$\zeta = \frac{1}{2} \left(\frac{1 - \alpha^2}{2\alpha} \right)^{1/2} \tag{2.5-20}$$

in order to have an (approximately) equal phase margin in the two loops.

The phase errors as functions of time are plotted for a phase and a frequency step input in Figure 2.5-8. The curves are remarkably similar for the two loops and we conclude that the transient response of a third-order (type 3) loop is well approximated by that of a second-order (type 2) loop having the same crossover frequency ω_c and the same phase margin ϕ_R. This is not only so for a third-order system but also holds for systems of arbitrary order if one can assume that the response is primarily due to a dominant pair of complex poles [1: p. 165]. In general, a complex pair of poles is dominant if at least one of the following conditions is fulfilled:

The absolute value of the real parts of the additional poles of the closed-loop transfer function $H(s)$ are at least ten times larger than the absolute value of the real part of the dominant pole pair.

f a pole is located near the dominant pair, then there must be a zero if the closed-loop transfer function close to this pole (zero-pole doublet).

The situation is depicted in Figure 2.5-9.

The approximation of a higher order system by a second-order (type 2) system is a simple, yet fairly accurate technique of setting the specifications of a PLL. For example, given settling time T_s and the peak overshoot M_p, the designer can read off the loop damping factor ζ from Figure 2.4-14. The open-loop transfer function $G_{0,3}(p)$ of the type 3 loop given by (2.5-17) is specified by ω_c and α. Thus, using the value of ζ obtained and solving (2.5-20) for α completes the design of the desired type 3 system.

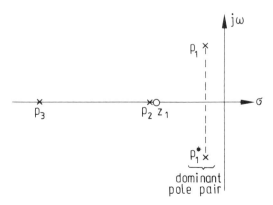

Figure 2.5-9. Dominant pole pair with respect to a "zero-pole doublet" and a "far away" pole.

Main Points of the Section

- *Basic Transfer Functions*

$$G_0(s) = \frac{KA(1 + sT_2)(1 + sT_3)}{s(sT_1)^2} \tag{2.5-1}$$

$$H(s) = \frac{\hat{\theta}(s)}{\theta(s)} = \frac{\omega_c(s + \alpha\omega_c)(s + \alpha_1\omega_c)}{s^3 + \omega_c(s + \alpha\omega_c)(s + \alpha_1\omega_c)}$$

$$\tag{2.5-1a}$$

$$1 - H(s) = \frac{\phi(s)}{\theta(s)} = \frac{s^3}{s^3 + \omega_c(s + \alpha\omega_c)(s + \alpha_1\omega_c)}$$

$$\omega_c = KA\left(\frac{T_2}{T_1}\right)\left(\frac{T_3}{T_1}\right), \quad \alpha = \frac{1}{\omega_c T_2}, \quad \alpha_1 = \frac{1}{\omega_c T_3}$$

- *Stability condition for* $\alpha = \alpha_1$

$$KA > \frac{1}{2T_2}\left(\frac{T_1}{T_2}\right)^2$$

The amplitude A of the input signal must be kept above a minimum level by means of an AGC.

- *State equations*

$$\frac{d\phi(\tau)}{d\tau} = -\sin\phi(\tau) - x_1(\tau) - \frac{1}{\alpha}x_2(\tau) + \frac{d\theta(\tau)}{d\tau}$$

$$\frac{dx_1(\tau)}{d\tau} = \alpha\sin\phi(\tau) + x_2(\tau) \tag{2.5-11}$$

$$\frac{dx_2(\tau)}{d\tau} = \alpha\alpha_1\sin\phi(\tau)$$

- For a first draft design a third-order (type 3) PLL may be approximated by a second-order (type 2) PLL. Damping ratio ζ and α are related by

$$\zeta = \frac{1}{2}\left(\frac{1 - \alpha^2}{2\alpha}\right)^{1/2} \tag{2.5-20}$$

2.6. PHASE DETECTORS

The purpose of a phase detector is to generate an output to function as a measure of phase error. Two broad categories of phase detectors, multiplier type circuits and sequential logic circuits, can be distinguished.

2.6.1. Multiplier Type Phase Detectors

In our previous investigations the phase detector (PD) was modeled as an ideal multiplier. The phase detector characteristic (dc-error voltage versus phase error) for sinusoidal input and VCO signal was found to be

$$x(t) = K_m K_1 A \sin \phi \qquad (2.6\text{-}1)$$

where K_m is the multiplier gain, $\sqrt{2}\, A$, the input signal peak amplitude, and $\sqrt{2}\, K_1$, the VCO signal peak amplitude. In general, the phase detector characteristic depends on the phase detector circuit as well as on the signal waveform. For example, if both signals applied to an ideal multiplier are square waves, a triangular detector characteristics results (see Figure 2.6-1).

   An ideal multiplier is a useful analytical model for a phase detector, but is rarely found in actual equipment. The reason is that costs for its implementation are high and usable frequencies tend to be low. Instead, modifications to the ideal multiplier are used. A widely used circuit is the switching phase detector. Functionally, a switching phase detector is identical to an ideal multiplier when the VCO drive to the multiplier is replaced by a unit amplitude square wave. The square wave can be expanded into a Fourier series.

$$r(t) = \frac{4}{\pi} \left[\cos\left(\omega_0 t + \hat{\theta}\right) - \frac{1}{3} \cos\left(3\omega_0 t + 3\hat{\theta}\right) + \cdots \right] \qquad (2.6\text{-}2)$$

Output of the ideal multiplier for a sinusoidal input signal is then

$$x(t) = K_m s(t) r(t) = \frac{4}{\pi\sqrt{2}}\, K_m A [\sin \phi + \text{terms with } (n\omega_0 t)] \qquad (2.6\text{-}3)$$

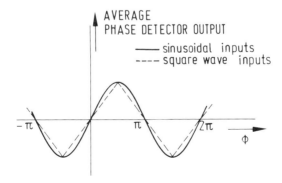

Figure 2.6-1. Average phase detector output (PD characteristic) of an ideal multiplier for sinusoidal and square wave inputs.

Ignoring terms with $(n\omega_0 t)$ we obtain the same error signal as for two sinusoidal inputs to the multiplier. The terms with $n\omega_0 t$, $n \geq 1$, only contribute to the high frequency ripple. However, multiplication by a unit amplitude square wave is equivalent to periodically switching the polarity of the input signal; the expensive multiplier can be replaced without degradation in performance by a switch. Therefore, the most common multiplier type PD is actually a switching circuit. Circuit details are given by Gardner [2] or Blanchard [3], together with pertinent literature references.

Multiplier type phase detectors can be realized up to very high frequencies. In addition, they are the only ones capable of operation on an input signal deeply buried in noise.

Harmonic locking. A multiplier type phase detector is capable of generating a usable control signal (dc-output) if both input signals have a component at the same frequency. It is *not* necessary that both signals have the same fundamental frequency. To see this let us express the two input signals as Fourier series

$$s(t) = \sqrt{2} \sum_n c_n \exp\left(jn\omega_s t \right)$$

and

$$r(t) = \sqrt{2} \sum_m d_m \exp\left(jm\omega_{VCO} t \right)$$

where ω_s and ω_{VCO} are the *fundamental frequencies* of the respective input signals. Multiplying the two signals yields

$$s(t)r(t) = 2 \sum_n \sum_m c_n d_m \exp\left[j(n\omega_s t + m\omega_{VCO} t) \right]$$

The product contains a dc-component only if there exists combinations of n and m such that

$$N\omega_s - M\omega_{VCO} = 0, \quad N, M > 0 \tag{2.6-4}$$

If $M = 1$ only *integer harmonic* locks are possible

$$N\omega_s = \omega_{VCO}$$

It is however entirely possible to have $M \neq 1$ and then we obtain *fractional locks*.

Usually, fractional harmonic lock is an *unwanted* effect which can be very troublesome. For example, in digital data transmissions a PLL is supposed to track the fundamental frequency ω_s ($N = 1$) of the incoming signal in order to recover the clock $T = 2\pi/\omega_s$ of the signal transmitted. If, instead the phase-locked loop locks to a higher harmonic the VCO frequency

$$\omega_{\text{VCO}} = \omega_s \left(\frac{N}{M} \right)$$

(for large N, M) will be slightly different from ω_s. Data reception is then impossible because the sender and the receiver do not operate at the same frequency. Notice that fractional harmonic lock is avoided if one of the input signals to the phase detector is sinusoidal.

For the sake of completeness we only mention more complex phase detectors of the multiplier type such as Tanlock, PD-product and phase feedback [2: p. 117]. These phase detectors have nonsinusoidal phase detector characteristics. They are very rarely, if ever, used in practice, because the additional complexity results in an insignificant improvement in performance when compared to the ordinary multiplier type phase detector. This is particularly true when the signal is buried deeply in noise. As will be explored in a later chapter any such multiplier characteristics tend toward a sinusoidal form obtainable by an ordinary multiplier type phase detector.

2.6.2. Sequential Logic Phase Detectors*

Sequential logic phase detectors operate on the zero crossings of the signal waveform; any other characteristic of the waveform is ignored. For reliable operation the input signal is usually hard limited before being applied to the phase detector. The simplest sequential logic phase detector is an RS (reset-set) flip-flop. Its operation is illustrated in Figure 2.6-2.

The negative going edge of the input signal sets the flip-flop to a true state ($Q = $ high) while the negative transition of the VCO resets the flip-flop ($Q = $ low). The on-time of the Q terminal is proportional to the phase difference. Therefore, the duty ratio of Q is a linear function of the phase error within $[0, 2\pi]$ and periodic with a period of 2π.

The linear range of this characteristic is centered around a phase difference of π (see Figure 2.6-3). Therefore, the stable tracking point is at π and the dc-offset must be cancelled by appropriate circuits. Exactly as we have done with the multiplier type phase detector we define the phase error $\phi = 0$ to coincide with the stable tracking point. This means that the phase difference between VCO and signal in the equilibrium point equals π, rather than $\pi/2$ as for the multiplier type phase detector. The reader should be aware that this type of phase detector also exhibits (unwanted) fractional harmonic lock.

The block diagram of another widely used detector is shown in Figure 2.6-4. It is available in several versions as an integrated circuit. As will be shown, this circuit is not only capable of delivering a phase sensitive signal but also a *frequency sensitive* signal when the loop is out-of-lock. By

---

*Because these PDs are built up from digital circuits they are often called digital phase detectors. However, this terminology is incorrect.

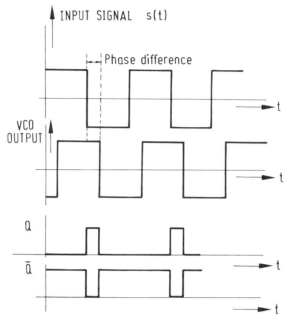

Figure 2.6-2. Signals of a sequential logic phase detector.

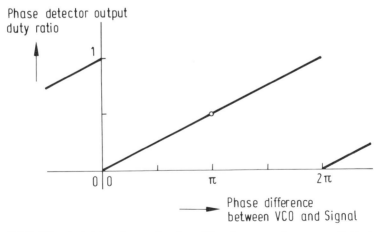

Figure 2.6-3. PD output duty ratio as a function of the phase error for a sequential logic phase detector.

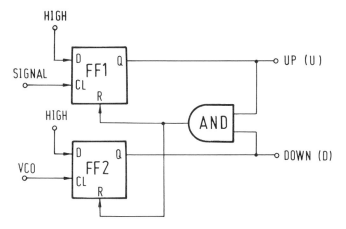

Figure 2.6-4. Block diagram of phase/frequency detector.

out-of-lock, is meant the situation where the VCO and signal frequency are significantly different.

The working principle of this phase detector is explained using a phasor diagram. The signal is taken as the real part of the phasor

$$\underline{s}(t) = \exp\left(j\Phi(t)\right) \tag{2.6-5}$$

with

$$\Phi(t) = \omega_0 t + \theta(t)$$

and similarly for the VCO signal

$$\underline{r}(t) = \exp\left(j\hat{\Phi}(t)\right) \tag{2.6-6}$$

with

$$\hat{\Phi}(t) = \omega_0 t + \hat{\theta}(t)$$

The negative going transitions occur when the phasors cross the positive imaginary axis.

Let us assume for the moment that the instantaneous angular frequency of both phasors is ω_0. Then θ and $\hat{\theta}$ are constant and the phase error $\phi = \theta - \hat{\theta}$. If the signal phasor $\underline{s}(t)$ leads $\underline{r}(t)$ as in Figure 2.6-5a, a negative going transition of $\underline{s}(t)$ sets FF1 (flip-flop 1), i.e., output UP (U) is set high. The next transition of $\underline{r}(t)$ sets the output DOWN (D) high. But the two signals U and D can never be high simultaneously since the condition U and D resets the two flip-flops. Conversely, if the VCO phasor $\underline{r}(t)$ leads $\underline{s}(t)$ (see

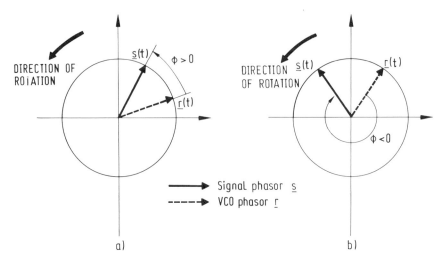

Figure 2.6-5. Relative position of phasors for positive (*a*) and negative (*b*) phase error.

Figure 2.6-5*b*) then $\underline{r}(t)$ sets D true while $\underline{s}(t)$ resets *D*. A convenient way to illustrate the working principle of this phase detector is by means of a *state diagram* in the manner of Figure 2.6-6. The PD has three different states labeled HL, LL, and LH. The first letter describes the state of FF2 and the second letter the state of FF1. Since the two flip-flops can never be high simultaneously, there is no state HH. Transition from one state to another is conditioned on the negative transitions of signal or VCO waveform appropriately. For example, transition from LL to LH occurs for S ↓ .

The on-time of either of the two flip-flops is proportional to the absolute value of the phase error ϕ. The fractional time a terminal is high during a full phasor rotation equals the duty ratio

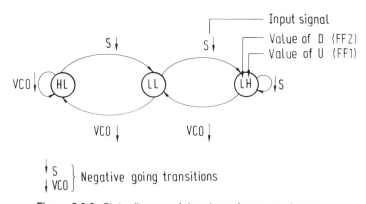

Figure 2.6-6. State diagram of the phase/frequency detector.

$$d := \frac{|\phi|}{2\pi} \tag{2.6-7}$$

and is plotted in Figure 2.6-7 as a function of the phase error ϕ. If the output of each flip-flop is viewed as a function of time, a periodic sequence of unipolar, rectangular pulses is observed.

The dc component of this waveform is the useful control signal. It is directly proportional to the duty ratio. The ac components of the periodic pulse sequence cause high-frequency disturbance (ripple) which must be filtered out. Notice, that for this phase detector the VCO phase and the signal phase coincide for zero phase error; compare this with the previous characteristics where in the stable tracking point, a fixed phase difference of π (RS flip-flop) and of $\pi/2$ (multiplier) existed.

The phase detector characteristic of Figure 2.6-8 has several unique features. It is linear over the entire range of $(-2\pi, 2\pi)$. Both outputs are quiescent at the equilibrium tracking point. In the near vicinity of equilibrium either one of the two flip-flops is set high with a very small duty ratio, thus minimizing the high-frequency ripple.

For every phase error ϕ in the phase detector characteristic of the Figure 2.6-8 there exist two different output signals. To understand why let us start initially with zero phase error and increase the phase error to a value slightly larger than 2π. Since in this case the signal phasor always leads the VCO phasor (Figure 2.6-5a), only the UP-terminal is active. If we now decrease the phase error below a value of 2π, the VCO phasor will lead the signal phasor, exactly the opposite from before. But this means, that the DOWN-terminal is active and we observe a negative phase detector output (shown as a dashed line in Figure 2.6-8) instead of a positive signal for the same phase error when approached from $\phi = 0$. As a result, we find the hysteresis-type nonlinearity of Figure 2.6-8.

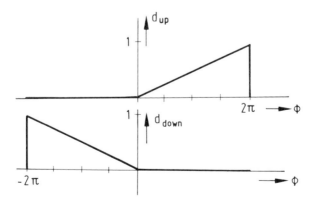

Figure 2.6-7. Duty ratio as a function of the phase error.

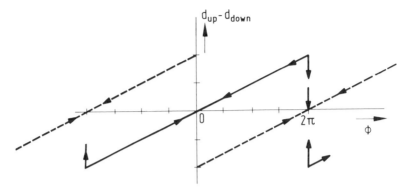

Figure 2.6-8. Phase detector characteristic of phase/frequency detector.

Operation of the phase detector is more difficult to understand if there exists a frequency difference between signal and VCO. A specific phasor diagram arrangement for the case where

$$\omega_{VCO} = \omega_0 + \Delta\omega > \omega_0$$

is depicted in Figure 2.6-9a. Because of the frequency difference, the

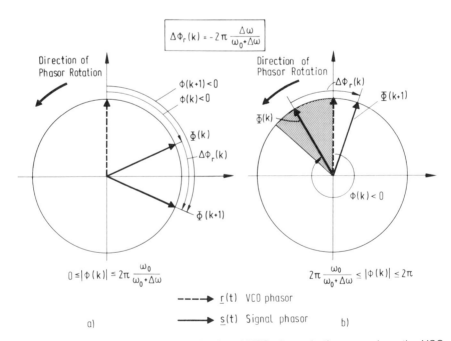

Figure 2.6-9. Relative position of the signal and VCO phasor in the case where the VCO frequency is larger than the signal frequency.

relative position of the two phasors is time-varying. If we take the faster phasor $\underline{r}(t)$ as reference phasor, then for every full rotation of $\underline{r}(t)$ we observe an increment in the absolute value of ϕ. This change in ϕ, denoted by $\Delta\phi_r$, is

$$\Delta\phi_r = -2\pi \frac{\Delta\omega}{\Delta\omega + \omega_0} < 0, \quad \Delta\omega > 0 \tag{2.6-8}$$

The minus sign expresses the fact that the increment is clockwise, i.e., mathematically negative. For convenience, we set the time origin to coincide with an instant of a negative going transition of $\underline{r}(t)$, i.e., when the phasor $\underline{r}(t)$ coincides with the top portion of the imaginary axis. By $\Phi(k)$ we denote the phase $\Phi(t)$ taken at

$$t = k \frac{2\pi}{\omega_0 + \Delta\omega} \tag{2.6-9}$$

where $2\pi/(\omega_0 + \Delta\omega)$ is the period of reference phasor $\underline{r}(t)$, and similarly for the other variables. During the kth cycle

$$k \frac{2\pi}{\omega_0 + \Delta\omega} \leq t < (k+1) \frac{2\pi}{\omega_0 + \Delta\omega} \tag{2.6-10}$$

the on-time of the DOWN-terminal equals $\phi(k)/\omega_0$. The duty ratio is obtained by dividing this time by the period of the reference phasor $\underline{r}(t)$, namely $2\pi/(\omega_0 + \Delta\omega)$. Thus

$$d_{\text{DOWN}} = \frac{|\phi(k)|}{\omega_0} \frac{\omega_0 + \Delta\omega}{2\pi}, \quad d_{\text{UP}} = 0 \tag{2.6-11}$$

The result (2.6-11) is conditional on the fact that $\phi(k)$ lies within the interval

$$-(2\pi - |\Delta\phi_r|) \leq \phi(k) \leq 0$$

or using $\Delta\phi_r$ from (2.6-8)

$$-2\pi \frac{\omega_0}{\omega_0 + \Delta\omega} \leq \phi(k) \leq 0 \tag{2.6-12}$$

In the other specific arrangement for the case $\omega_{\text{VCO}} = \omega_0 + \Delta\omega > \omega_0$, where $\Phi(k)$ is within the shaded sector of Figure 2.6-9b, the duty ratio equals 1 because the $\underline{s}$-phasor does not cross the imaginary axis during the kth cycle. From the state diagram of Figure 2.6-6 we note that the state HL is maintained if a VCO transition is directly followed by another VCO transition. Combining the results for the two arrangements of Figure 2.6-9 we obtain the duty ratio as a function of the phase error (see Figure 2.6-10).

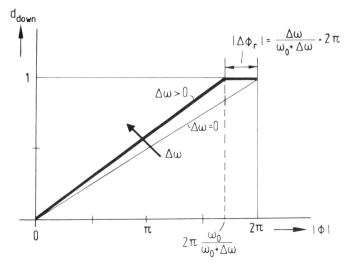

Figure 2.6-10. Duty ratio d_{DOWN} as a function of $|\phi|$ with $\Delta\omega > 0$ as parameter (Case 1: VCO frequency is larger than signal frequency).

$$d_{DOWN} = \begin{cases} \dfrac{|\phi|}{\omega_0} \dfrac{\omega_0 + \Delta\omega}{2\pi}, & \phi \in \left[-\left(2\pi\dfrac{\omega_0}{\omega_0 + \Delta\omega}\right), 0\right] \\[4mm] 1, & \phi \in \left[-2\pi, -\left(2\pi\dfrac{\omega_0}{\omega_0 + \Delta\omega}\right)\right] \end{cases} \tag{2.6-13}$$

It is important to understand that for

$$\omega_{VCO} > \omega_0$$

only the DOWN-terminal is active. (Initially, for a limited number of cycles the UP-terminal can become active. If, however, the $\underline{r}(t)$ phasor *sets* the down terminal for the first time and the phasor $\underline{s}(t)$ resets it, the UP-terminal will never become active again.)

If the VCO frequency is smaller than the signal frequency, can we simply take $\Delta\omega < 0$ and use the previous results? The answer is no! But what we can do is to draw a similar phasor diagram with $\underline{s}(t)$ as reference phasor. Using the same reasoning as before we have phase error increment per cycle rotation of $\underline{s}(t)$

$$\Delta\phi_s = 2\pi\frac{\Delta\omega}{\omega_0}, \quad \Delta\omega < 0$$

with

$$\omega_{VCO} = \omega_0 + \Delta\omega, \quad \Delta\omega < 0$$

$$\omega_s = \omega_0$$

From this, the duty ratio of the (only) active UP-terminal is found to be (see Figure 2.6-11)

$$
d_{UP} = \begin{cases} \dfrac{|\phi|}{\omega_0 + \Delta\omega} \dfrac{\omega_0}{2\pi}, & \phi \in [0, 2\pi - |\Delta\phi_s|] \\[4mm] 1, & \phi \in [2\pi - |\Delta\phi_s|, 2\pi] \end{cases} \tag{2.6-14}
$$

$$
d_{DOWN} = 0
$$

Now, similar to our representation in Figure 2.6-8 of the phase detector characteristic by the function of duty ratio versus ϕ, here we try to define a frequency detector characteristic to be the function of duty ratio versus frequency difference $\Delta\omega$.

Since the duty ratio for $\Delta\omega \neq 0$ depends on ϕ and $\Delta\omega$ the best we can do is to take an average of all possible values of the phase error. We define

$$
\bar{d} = \bar{d}_{UP} - \bar{d}_{DOWN} = \frac{1}{2\pi} \int_0^{2\pi} d_{UP}(\phi)\, d\phi - \frac{1}{2\pi} \int_0^{2\pi} d_{DOWN}(\phi)\, d\phi \tag{2.6-15}
$$

Using (2.6-14) we obtain

$$
\begin{aligned}
\bar{d}_{UP} &= \frac{1}{2\pi} \int_0^{2\pi - |\Delta\phi_s|} \frac{\phi}{\omega_0 + \Delta\omega} \frac{\omega_0}{2\pi}\, d\phi + \frac{1}{2\pi} \int_{2\pi - |\Delta\phi_s|}^{2\pi} d\phi \\[2mm]
&= \frac{1}{2}\left[1 - \frac{\Delta\omega}{\omega_0}\right], \quad \Delta\omega < 0
\end{aligned} \tag{2.6-16}
$$

and

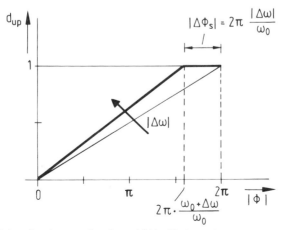

Figure 2.6-11. Duty ratio d_{up} as a function of $|\phi|$ with $\Delta\omega < 0$ as parameter (Case 2: VCO frequency is smaller than signal frequency).

$$\bar{d}_{\text{UP}} = 0, \quad \Delta\omega > 0$$

Similarly, from (2.6-13)

$$\bar{d}_{\text{DOWN}} = \frac{1}{2\pi} \int_0^{2\pi - |\Delta\phi_r|} \frac{\phi}{\omega_0} \frac{\omega_0 + \Delta\omega}{2\pi} \, d\phi + \frac{1}{2\pi} \int_{2\pi - |\Delta\phi_r|}^{2\pi} d\phi$$

$$= \frac{1}{2} \left[1 - \frac{\Delta\omega/\omega_0}{1 + (\Delta\omega/\omega_0)} \right], \quad \Delta\omega > 0 \qquad (2.6\text{-}17)$$

and

$$\bar{d}_{\text{DOWN}} = 0, \quad \Delta\omega < 0$$

The (averaged) duty ratio as a function $\Delta\omega$ is plotted in Figure 2.6-12. It is discontinuous at $\Delta\omega = 0$ and not symmetric for positive and negative values of $\Delta\omega$. The discontinuity at $\Delta\omega = 0$ is easily explained. For zero frequency difference, the duty ratio is proportional to ϕ; averaged over ϕ $\bar{d}_{\text{UP}} - \bar{d}_{\text{DOWN}} = \frac{1}{2} - \frac{1}{2} = 0$. Maximum average duty of 1 is reached for $\Delta\omega/\omega_0 = -1$ because the VCO frequency can never become negative, i.e., we must have

$$\omega_{\text{VCO}} = \omega_0 + \Delta\omega \geq 0$$

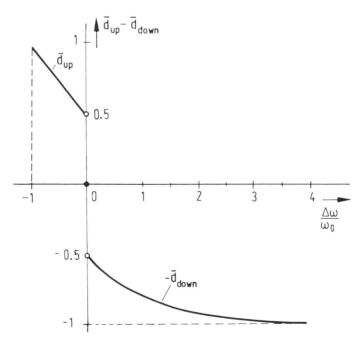

Figure 2.6-12. Average duty ratio as a function of the frequency difference.

For this limit value the UP-terminal remains active all the time once it has been initiated.

For $\Delta\omega > 0$, the case when the VCO frequency is larger than the signal frequency ω_0, the average duty ratio asymptotically reaches its maximum value as $\Delta\omega/\omega_0 \to \infty$. This means that the phasor $\underline{r}(t)$ can have an infinite angular frequency. With respect to such reference, the signal phasor $\underline{s}(t)$ would stand still.

Due to gate delays in the flip-flops and other logic circuits, application of the sequential logic phase detector is restricted to much lower frequencies than switching phase detectors used to implement multiplier type PDs. Furthermore, since they extract their useful control signal from the edges of the waveform only, sequential logic phase detectors are very sensitive to missing or extra transitions. In consequence, they must not be used with very noisy signals.

Main Points of the Section

- Two broad categories of phase detections can be distinguished, multiplier type circuits and sequential logic circuits.
- Multiplier type phase detector fully utilizes the waveform. It is the only one capable of operating on signals deeply buried in noise.
- Sequential logic circuits operate on level crossings of the signal only. They have a linear phase detector characteristic. Due to delays in the gates, sequential logic circuits are restricted to lower frequencies. A certain class is capable of providing a frequency sensitive signal in addition to a phase sensitive one.

2.7. CHARGE PUMP PHASE-LOCKED LOOPS

Phase-locked loops incorporating sequential logic phase/frequency detectors are widely used. The reasons for their popularity are availability as low-cost integrated circuits (IC), an extended tracking range which is linear from -2π to 2π, and a frequency sensitive error signal. From the frequency detector characteristics of Figure 2.6-12 we know that a phase/frequency detector (PFD) is capable of compensating any frequency difference $\Delta\omega$ that might initially exist between VCO and signal frequency. This is a unique property of the sequential logic PFD not shared by other PDs. Multiplier type PDs can handle only a small initial frequency difference; should the difference be large, the loop will never reach its tracking point. This is discussed more fully in the chapters on acquisition.

2.7.1. Principles of Charge Pump Phase-Locked Loops

A sequential logic PFD is usually accompanied by a *charge pump*. The purpose of the charge pump is to convert the logic states of the PFD into clearly specified analog signal levels. This is necessary because the signal amplitude in digital circuits can vary within a large tolerance band. For example, the high state of a transistor-transistor logic (TTL) circuit is specified by manufacturers to be anywhere between 2.4 and 5.5 V. A charge pump is nothing but a three position electronic switch controlled by the three states of the PFD (see Figure 2.7-1). When the UP-terminal of the PFD is high, a positive current of magnitude $+I_p$ is delivered to the impedance $Z_F(s)$, while a negative current $-I_p$ is delivered if the DOWN-terminal is high. If both terminals are low, corresponding to the LL state in the state diagram of Figure 2.6-6, the impedance $Z_F(s)$ is isolated from the charge pump. We call this the neutral state N. This open condition is not encountered in a conventional PLL and creates important novel characteristics. We will see that charge pumps' PLLs are capable of extremly accurate phase tracking of the input signal.

Let us assume that the PLL is in-lock (both phasors have the same angular frequency) and that the VCO phase leads the signal phase. For convenience we have the time origin coincide with the instant when the VCO phasor crosses the reference axis (see Figure 2.7-2a). Then at time $t = 0$ the negative current source is connected to the loop filter network and starts to deliver a current $-I_p$. This is done during an interval of time t_p exactly proportional to the phase error taken at $t = 0$, namely

$$t_p = \frac{|\phi(0)|}{\omega_0} , \quad \phi(0) < 0 \qquad (2.7\text{-}1)$$

Note, that the active time t_p depends only on $\phi(0)$. Corrections of the VCO phase made during the active period of the charge pump have no influence on the present cycle.

When the signal phase θ leads the VCO phase, the positive current source is switched on at the instant when the signal phasor crosses the reference axis and tries to advance the VCO phasor $\underline{r}(t)$ relative to the signal

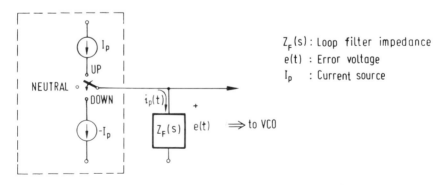

Figure 2.7-1. Charge pumps and loop filter impedance $Z_F(s)$.

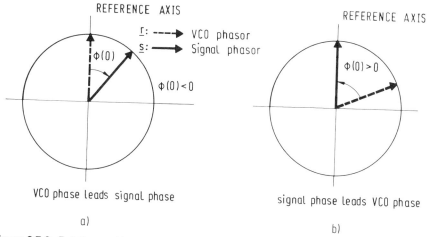

Figure 2.7-2. Relative position of signal and VCO phasor for positive/negative phase error at the instant of switching on off the charge pumps.

phasor. The active time-period of the charge pump is equal to the time it takes the VCO phasor to advance by an amount of $\phi(0)$ rad and cross the reference axis. Due to the acceleration of the VCO phasor during the active time of the positive current source, this time-period turns out to be smaller than $\phi(0)/\omega_0$

$$t_p < \frac{\phi(0)}{\omega_0}, \quad \phi(0) > 0 \tag{2.7-2}$$

Therefore, response of the charge pump PLL is different for positive and negative phase errors. This is a new phenomenon not encountered in a conventional PLL. The puzzled reader might well ask the question why this asymmetry did not show up in the PD characteristic of Figure 2.6-8. The answer lies in the definition of this characteristic; both phases θ and $\hat{\theta}$ are assumed constant with respect to time, corresponding to an open loop. As soon as the loop is closed, $\hat{\theta}(t)$ is time-dependent and the assumption underlying the PD characteristic is no longer valid.

In many applications the phase error of the PLL changes by only a very small amount during each cycle of the input. This means that the settling time of the PLL is much longer than the period $2\pi/\omega_0$ of the input signal. Since the settling time T_s is inversely proportional to ω_c, the crossover frequency ω_c is much smaller than ω_0, i.e.,

$$\frac{2\pi}{\omega_0} \ll T_s \approx \frac{3}{\omega_c} \Rightarrow \omega_c \ll \omega_0$$

and the loop is called a *narrowband loop*. For a narrowband loop we may not be interested in the fine details of the dynamic response within a cycle and work with a time-average behavior over many cycles. By applying an averaged analysis, the time-varying operation can be bypassed and the powerful tool of transfer function remains for our usage.

There is, however, an important feature which cannot be ignored, even for a narrow-bandwidth loop. Any actuation of the charge pump gives rise to voltage transients in the loop filter impedance. Since the output voltage of the loop filter impedance is the input to the VCO, these transients cause frequency excursions (ripples which can be very troublesome). Amplitudes of these transients can be much larger than the averaged change over one cycle; in a second-order loop they can easily become so large as to overload the VCO. As a remedy, more filter elements are added to reduce the frequency ripple. The loop is then at least of order three.

An exact analysis of the charge pump PLL must take into account the switching times of the current sources. The analysis is much more complex than the familiar quasi-continuous analysis. This is so because the switching times are not equidistant but are functions of the time-varying phase of the VCO and the signal phase. The result of the analysis is a sequence of state variables $x(k)$. For every k, $x(k)$ represents the state of the PLL taken at the beginning of a pump pulse. Stability of the charge pump PLL is of great importance. Again, the nonequidistant nature of the switching makes an exact stability analysis very complicated. Stated briefly, stability problems arise [5] which are not present in continuous-time analog PLLs.

2.7.2. Quasi-Continuous Analysis of the Charge Pump Phase-Locked Loop

If the phase error of the PLL changes only by a very small amount during each cycle of the input signal we may approximate the phase detector output by its average over one cycle. For example, for the case of Figure 2.7-2a and (2.7-1) we have

$$i_d(t) = \frac{1}{2\pi/\omega_0} \int_0^{t_p} -I_p \, dt = -\frac{\omega_0}{2\pi} \cdot I_p \frac{-\phi(0)}{\omega_0} = \frac{I_p}{2\pi}\phi(0)$$

Then, using the above assumption about $\phi(t)$, we can write in general

$$i_d(t) = \frac{I_p}{2\pi}\phi(t) \tag{2.7-3}$$

We remark that (2.7-3) can be interpreted as the output of a multiplier-type phase detector with linear characteristic. The VCO input control voltage is given by

$$E(s) = Z_F(s)I_d(s) \tag{2.7-4}$$

where $Z_F(s)$ is the impedance of Figure 2.7-1 that converts the current $i_d(t)$ into a control voltage. The control law of the VCO is

$$\hat{\theta}(s) = \frac{K_0}{s} E(s) \qquad (2.7\text{-}5)$$

Expressions (2.7-4) and (2.7-5) lead to the transfer functions

$$G_0(s) = \frac{\hat{\theta}(s)}{\theta(s)} = \frac{I_p}{2\pi} K_0 \frac{1}{s} Z_F(s) \quad \text{(open loop)}$$

$$\qquad (2.7\text{-}6)$$

$$H(s) = \frac{G_0(0)}{1 + G_0(s)} = \frac{K_0 I_p Z_F(s)}{2\pi s + K_0 I_p Z_F(s)} \quad \text{(closed loop)}$$

These transfer functions are completely analogous to those derived in Section 2.3.1 for a conventional PLL, with one exception; instead of the loop filter transfer function $F(s)$, which is dimensionless, the impedance $Z_F(s)$ with dimension volt/amp is used.

Second-Order Loop. We consider an impedance

$$Z_F(s) = R + \frac{1}{sC_2} \qquad (2.7\text{-}7)$$

corresponding to a series connection of a resistor and a capacitor. The transfer functions (2.7-6) can be written for this impedance in the form

$$G_0(s) = \left[\frac{\omega_n}{s}\right]^2 \left[1 + \frac{s}{\omega_n} 2\zeta\right]$$

$$\qquad (2.7\text{-}8)$$

$$H(s) = \frac{2\zeta\omega_n s + \omega_n^2}{s^2 + 2\zeta\omega_n s + \omega_n^2}$$

Here, the natural frequency ω_n and the damping ratio ζ are related to the circuit parameters as follows

$$\omega_n = \left(\frac{K_0 I_p}{2\pi C_2}\right)^{1/2}, \quad \zeta = \frac{RC_2}{2} \omega_n = \frac{T_2}{2} \omega_n, \quad T_2 = RC_2 \qquad (2.7\text{-}9)$$

The crossover frequency ω_c is given by

$$\omega_c = \frac{K_0 I_p R}{2\pi} = 2\zeta\omega_n \qquad (2.7\text{-}10)$$

Interestingly, using a passive filter we obtain the same transfer function realizable in a conventional PLL only with an active filter with infinite dc-gain (compare (2.4-3) and (2.7-8)). This effect arises because of the open-circuit input during the neutral state of the charge pump.

Let us briefly pause in our quasi-continuous analysis and turn our attention to the development of the error voltage during one cycle. Upon each cycle of the phase detector, the pump current is driven into the filter impedance which responds with an instantaneous jump of $\Delta e = RI_p$. At the end of each cycle the current switches off causing a jump of equal magnitude but opposite direction (see Figure 2.7-3). The frequency of the VCO follows the voltage step so there will be frequency excursion (ripple) of $K_0 \Delta e$

$$|\Delta \omega| = K_0 |\Delta e| = K_0 RI_p = 2 \pi \omega_c \qquad (2.7\text{-}11)$$

for each pulse. We compare this frequency ripple $\Delta \omega$ with the average angular frequency increment given by the net voltage increment of one cycle of the input signal

$$
\underbrace{K_0 \frac{1}{C_2} I_p t_p}_{\substack{\text{average} \\ \text{frequency} \\ \text{increment}}}
\quad = \quad
\underbrace{2 \pi \omega_c \left[\frac{\omega_c}{\omega_0} \left(\frac{1}{2\zeta} \right)^2 \phi \right]}_{\substack{|\Delta \omega| \\ \text{frequency} \\ \text{ripple}}}
\qquad (2.7\text{-}12)
$$

with a "net voltage increment" bracket over the left-hand term.

Since by assumption $\omega_c / \omega_0 \ll 1$ and since $(2\zeta)^2$ is usually of the order of magnitude 1, we conclude that the average frequency increment per cycle of the signal input is much smaller than the transient-induced frequency ripple of (2.7-11).

Some applications (e.g., bit synchronizer) may be able to tolerate this

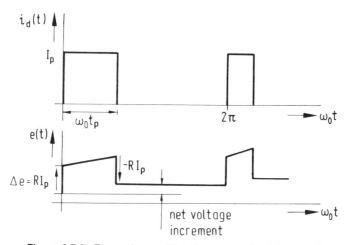

Figure 2.7-3. Error voltage $e(t)$ during one cycle of the signal.

frequency ripple, but others may require much better spectral purity. A more serious consequence of the jumps is the possible overload of the VCO. Any real VCO has a limited frequency range over which it can be tuned. We must require that the frequency jumps remain within the allowable tuning range of the VCO under all conditions. As an extreme instance, the frequency jump must not exceed the input frequency. A larger jump would imply a negative VCO frequency, a meaningless status. For this extreme condition the overload bound is

$$|\Delta\omega| < \omega_0 \tag{2.7-13}$$

Most oscillators will tolerate only a much smaller frequency deviation than given by (2.7-13).

Third-Order Phase-Locked Loop. For many applications the frequency ripple of a second-order loop is unacceptable. Additional filtering is needed to mitigate the frequency ripple. The simplest filter component to be added is a capacitor C_3 in parallel with the impedance of the second-order loop (see Figure 2.7-4). The impedance $Z_F(s)$ of this filter is

$$Z_F(s) = \left[\frac{b-1}{b}\right] \frac{1+sT_2}{sC_2[1+(sT_2/b)]} \tag{2.7-14}$$

where $b = 1 + (C_2/C_3)$ and $T_2 = RC_2$. Using (2.7-6) and (2.7-14) the open-loop transfer function is found to be

$$G_0(s) = \frac{\omega_c(1+sT_2)}{s^2 T_2[1+(sT_2/b)]} \tag{2.7-15}$$

with asymptotic crossover frequency $\omega_c{}^*$ given by

$$\omega_c = \frac{K_0 I_p R}{2\pi}\left(\frac{b-1}{b}\right) \tag{2.7-16}$$

Compared to the open-loop transfer function of a second-order loop (2.7-8)

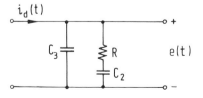

Figure 2.7-4. Loop filter impedance of a third-order loop.

*See footnote in Section 2.4.1.

there is an additional pole at $s = -b/T_2$. The location of this pole must be chosen such that good frequency ripple suppression is achieved. This calls for a pole at the lowest possible frequency. On the other hand, the additional pole is introduced solely for the purpose of ripple reduction. Therefore, the steady-state and dynamic response should not be significantly different from that of a second-order loop with the same crossover frequency ω_c and time constant T_2. (A second-order loop is completely specified by ω_c and T_2). This requires the break frequency of the additional pole to be far to the right of ω_c. Clearly this conflicts with the requirement of good frequency ripple reduction. To obtain a starting point for a design we can argue as follows: By inspection of the Bode diagram in Figure 2.7-5 we conclude that if the pole $s = -b/T_2$ is placed one decade to the right of ω_c, then the decrease in phase margin is only about 6° while the magnitude remains essentially unchanged. Thus, we will have a third-order system with a pair of closed-loop complex poles virtually identical to those of the corresponding second-order system and in addition, a real pole far to the left of the dominant complex pole pair.

The phase margin ϕ_R as a function of b with $\omega_c T_2$ as parameter is shown in Figure 2.7-6. For small values of b the phase margin becomes small indicating a poorly damped system with a conjugate complex pole pair close to the imaginary axis.

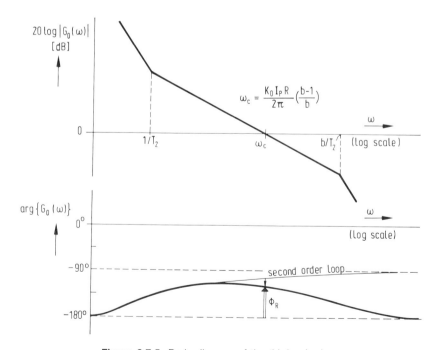

Figure 2.7-5. Bode diagram of the third-order loop.

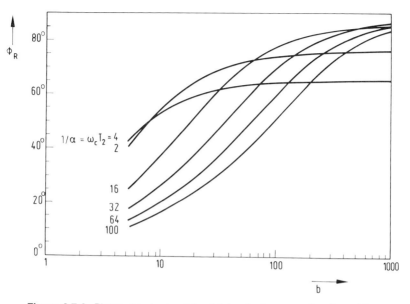

Figure 2.7-6. Phase margin ϕ_R of the third-order loop as a function of b.

The closed-loop transfer function of the third-order PLL can be found from (2.7-15) to be

$$H(s) = \frac{\hat{\theta}(s)}{\theta(s)} = \frac{1 + sT_2}{s^3(T_2^2/\omega_c b) + s^2(T_2/\omega_c) + sT_2 + 1} \qquad (2.7\text{-}17)$$

Using the normalized variable $p = s/\omega_c$ and introducing the parameter

$$\alpha = \frac{1}{\omega_c T_2} \qquad (2.7\text{-}18)$$

we can write (2.7-17) in the form

$$H(p) = \frac{1 + (p/\alpha)}{p^3(1/\alpha^2 b) + (p^2/\alpha) + (p/\alpha) + 1} \qquad (2.7\text{-}19)$$

From the Bode diagram it is easy to delimit values for the two parameters α and b. Let us assume that a second-order loop with damping ratio ζ fulfills the design specifications. Then, from the second-order loop relation between ω_c and T_2 (obtainable from (2.7-9) and (2.7-10))

$$\omega_c T_2 = (2\zeta)^2 \qquad (2.7\text{-}20)$$

we obtain

$$1/\alpha \approx (2\zeta)^2 \qquad (2.7\text{-}21)$$

In order not to decrease the phase margin ϕ_R, the additional pole at $s = -b/T_2$ is placed one decade above ω_c

$$b/T_2 \geq 10\omega_c \qquad (2.7\text{-}22)$$

that is

$$b \geq 10(\omega_c T_2) = 10/\alpha = 10 (2\zeta)^2$$

Typical values used in practice are $\alpha < 1/3$ and $b = 4$ to 9. For the sake of completeness the exact root locus of $H(p)$ for $\alpha = 1/2$ ($\zeta = 0.707$) as a function of b is shown in Figure 2.7-7. As predicted from the Bode diagram for large b the dominant pair of poles is practically indistinguishable from the pair of poles of the second-order loop ($b \to \infty$). The third pole has a real part much larger in magnitude than that of the complex pole pair.

Let us compute the frequency ripple reduction of a third-order loop compared to a second-order loop. The extra capacitor C_3 smoothes the discontinuous rectangular voltage jumps of Figure 2.7-3 into a ramp-like, exponential function.

Referring back to Figure 2.7-4, for an empty capacitor C_3, the response of the impedance to a current pulse of duration t_p is found to be

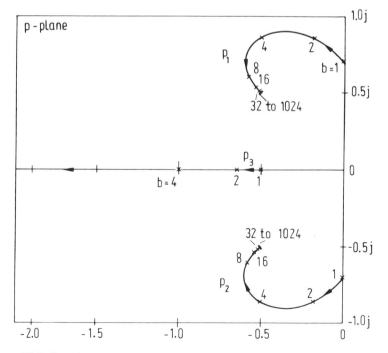

Figure 2.7-7. Root locus for the third-order loop as a function of b for $\alpha = 1/(\omega_c T_c) = 0.5$.

$$e(t_p) = I_p R_2 \left(\frac{b-1}{b}\right) \left\{ \frac{b-1}{b} \left[1 - \exp\left(-\frac{|\phi| b}{\omega_0 T_2}\right) \right] + \frac{|\phi|}{\omega_0 T_2} \right\}$$

$$\approx I_p R_2 \frac{b |\phi|}{\omega_0 T_2}, \quad b \gg 1 \tag{2.7-23}$$

The resulting frequency deviation equals

$$|\Delta\omega|_3 = K_0 e(t_p) \tag{2.7-24}$$

Define as the ripple suppression factor γ

$$\gamma = \frac{|\Delta\omega|_3}{|\Delta\omega|_2} \tag{2.7-25}$$

where the indices in $|\Delta\omega|_k$ correspond to the order of the loop. Combining (2.7-11) and (2.7-23) yields for $b \gg 1$

$$\gamma = \frac{b}{\omega_0 T_2} |\phi| \tag{2.7-26}$$

In all cases of practical interest we have $b/T_2 \ll \omega_0$ and when the loop is tracking near equilibrium $|\phi|$ is small. Therefore the frequency ripple suppression is substantial.

2.7.3. Phase Accuracy of a Practical Second-Order Charge Pump Phase-Locked Loop

We now discuss the steady-state accuracy of a charge pump PLL taking into account the nonideal behavior of the components. A second-order charge pump is capable of tracking a frequency step $\Delta\omega$ with zero static phase error. This is achieved with a passive filter provided that in the neutral state of the charge pump, the current $i_p(t)$ is zero and the input impedance of the VCO is infinite. In a real circuit there is always some resistive loading (finite input impedance) of the VCO and also, leakage current into the impedance $Z_F(s)$ during the neutral state. Thus, in the neutral mode the model shown in Figure 2.7-8 applies.

We first want to investigate the effect of resistive loading. During the interval when the charge pump is in the neutral state the capacitor voltage decreases by an amount of

$$\Delta V_c = v_c(t_p)[1 - \exp(-\Delta t/T)] \tag{2.7-27}$$

where $\Delta t = (2\pi/\omega_0) - t_p$ and $T = (R + R_s)C_2$. For good components ($R_s \gg R$) the capacitor voltage $v_c(t_p)$ equals the input voltage to the VCO and we can write

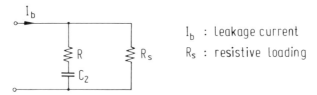

Figure 2.7-8. Model of the impedance faced by the charge pump in the neutral mode taking resistive loading of the VCO and leakage current into account.

$$\Delta \omega \approx K_0 v_c(t_p), \quad R_s \gg R \tag{2.7-28}$$

Now, since

$$t_p \ll 2\pi/\omega_0, \quad R_s \gg R$$

that is

$$\Delta t \approx \frac{2\pi}{\omega_0}, \quad T \approx R_s C_2 \quad \text{and} \quad \Delta t/T \ll 1$$

we can use the approximation $\exp(-\Delta t/T) \approx 1 - (\Delta t/T)$ in (2.7-27) to obtain

$$\Delta V_c \approx v_c(t_p)\left[\frac{\Delta t}{T}\right] = v_c(t_p)\frac{2\pi}{\omega_0}\cdot\frac{1}{R_s C_2}$$

Then inserting (2.7-28) into the above result gives

$$\Delta V_c \approx \frac{2\pi \Delta\omega}{K_0 R_s C_2 \omega_0} \tag{2.7-29}$$

The voltage decrease ΔV_c would be compensated in the following cycle by actuating the charge pump for t_p seconds such that

$$\Delta V_c = \frac{1}{C_2}I_p t_p = \frac{1}{C_2}I_p\frac{\phi_s}{\omega_0} \tag{2.7-30}$$

Replacing the left-hand side of (2.7-30) by (2.7-29) and solving for ϕ_s yields the static phase error due to resistive loading

$$\phi_s = \frac{2\pi \Delta\omega}{K_0 I_p R_s} \tag{2.7-31}$$

It is very instructive to insert typical numbers into (2.7-31). As (2.7-28) shows, the ratio $\Delta\omega/K_0$ is the control voltage used to drive the VCO for which we can assign 10 V in our calculation. The current I_p is typically

10 mA and the resistor R_s in the order of $10^9 \, \Omega$ [6]. Using these values we find

$$\phi_s = 2\pi \times 10^{-6}$$

i.e., one micro cycle.

Using similar reasoning we find that the static phase error due to a leakage current is

$$\phi_s = \frac{2\pi I_b}{I_p} \qquad (2.7\text{-}32)$$

Leakage currents as low as 1 nA are readily achievable. Thus, for $I_b = 1\,\text{nA}$ and $I_p = 10\,\text{mA}$, (2.7-32) predicts

$$\phi_s = 2\pi \times 10^{-7}$$

It is interesting to compare these numbers with the phase error of a conventional PLL. DC offset will always be present in the phase detector and in the operational amplifier of the active filter. With a total bias voltage V_b at the phase detector output the static phase error due to DC offset is given by

$$\phi_s = \sin^{-1} \frac{V_b}{K_D} \approx \frac{V_b}{K_D}$$

For $K_D = 1\,\text{V}$ (a typical value) a static phase error of $\phi_s < 2\pi \times 10^{-6}$ requires $V_b < 0.63 \, \mu\text{V}$ which is far below typical DC offsets.

Note that unequal current sources $\pm I_p$ do not introduce a steady-state phase error when a phase/frequency detector is used. Unequal current sources will merely make the loop respond more rapidly in one direction than the other.

In conclusion the analog circuit portion of the charge pump PLL can be designed to introduce an extremly small phase error. The dominant sources of error are likely to be the unequal switching times in the logic circuits. For example, take a delay difference of 1 ns as a good value for a PFD realized in TTL technology. Then for $f_0 = 100\,\text{kHz}$ the phase error introduced is 100 times larger than that introduced by the analog components.

2.7.4. Exact Analysis of a Second-Order Charge Pump Phase-Locked Loop

The second-order charge pump PLL of Figure 2.7-9 obeys the state equations

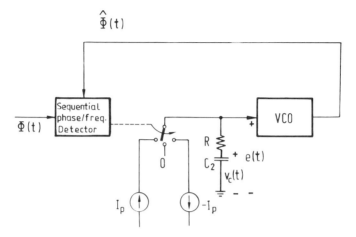

Figure 2.7-9. Equivalent circuit of charge pump phase-locked loop.

$$\frac{d\hat{\Phi}(t)}{dt} = \omega_0 + K_0 v_c(t) + K_0 R i_p(t)$$

$$\frac{dv_c(t)}{dt} = \frac{1}{C_2} i_p(t)$$

(2.7-33)

where

$$i_p(t) = \begin{cases} \pm I_p , & \text{if the current sources are active} \\ 0 , & \text{if the current sources are inactive} \end{cases}$$

(2.7-34)

Note that the state variable $\hat{\Phi}(t)$ of the first state equation (2.7-33) equals the *total phase* of the VCO oscillator and not the phase difference $\phi(t)$ as usual.

For the phase of the input signal we assume

$$\Phi(t) = (\omega_0 + \Delta\omega)t + \Phi(0)$$

(2.7-35)

An exact analysis must take into account the switching of the current between $(-I_p, 0, +I_p)$. To complicate the matter, the switching times are not equidistant, but a function of the time-varying phases of the VCO and the input signal. However, during any one switching condition the state equations (2.7-33) are linear with constant coefficients and constant input. Thus, given the initial condition at the start of a switching interval they are straightforward to solve.

If we define the state vector $x^T = [\hat{\Phi}, v_c]$ at the start of a current pulse as a discrete time state vector $x(k)$ then it is possible to write a difference state equation that describes the sequence of the states. We briefly outline the procedure and the reader is referred to Gardner [5] for more details.

Let us denote by $t_{on}(k)$ the beginning of the kth cycle initiated from signal or VCO, as the case may be. By $t_{off}(k)$ we mean the instant when the charge pump is shut off. The active time of the charge pumps during the kth cycle is then

$$t_p(k) = t_{off}(k) - t_{on}(k) \qquad (2.7\text{-}36)$$

As illustrated in Figure 2.7-10, to obtain $x(k+1)$ recursively from $x(k)$ we must first compute the active time $t_p(k)$ of the current pump together with the correspondent state vector $x(t)$ at $t = t_{on} + t_p(k)$. As discussed earlier, $t_p(k)$ depends on whether the VCO leads the signal phase (negative phase error) or vice versa. In the first case $t_p(k)$ is directly proportional to the phase error

$$t_p(k) = \left. \frac{|\phi(t)|}{\omega_0} \right|_{t=t_{on}(k)} \qquad \text{(VCO leads signal phase)}$$

while in the other case

$$t_p(k) < \left. \frac{|\phi(t)|}{\omega_0} \right|_{t=t_{on}(k)}$$

is the solution of a quadratic equation which is now derived. The state equations (2.7-33) during the on-time of the negative charge pump $(i_p(t) = -I_p)$ are easily integrated. Integrating the second state equation first

$$v_c(t) = v_c[t_{on}(k)] - \frac{1}{c_2} I_p[t - t_{on}(k)]] \qquad (2.7\text{-}37)$$

and the inserting into the first state equation gives

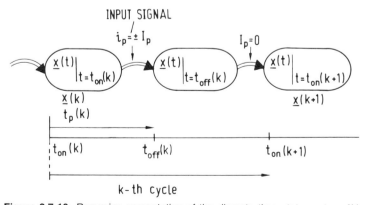

INPUT SIGNAL

Figure 2.7-10. Recursive computation of the discrete time state vector $x(k)$.

$$\hat{\Phi}(t) - \hat{\Phi}[t_{on}(k)] = \omega_0[t - t_{on}(k)] + k_0\left\{v_c[t_{on}(k)] \cdot [t - t_{on}(k)]\right.$$
$$\left. - \frac{1}{c_2} I_P \frac{[t - t_{on}(k)]^2}{2}\right\} - K_0 RI_p[t - t_{on}(k)]$$

$$(2.7\text{-}38)$$

Now, the negative charge pump is shut-off when $\hat{\Phi}(t) - \Phi[t_{on}(k)]$ has increased by $\phi(k) = \phi[t_{on}(k)]$, the value of the initial phase error of the kth cycle. Replacing the left-hand side of (2.7-38) by $\phi(k)$ leads to a quadratic equation for $t_p(k) = t_{off}(k) - t_{on}(k)$

$$\phi(k) = \omega_0 t_p(k) + k_0\left\{v_c[t_{on}(k)] \cdot t_p(k) - \frac{1}{c_2} I_P \frac{t_p^2(k)}{2}\right\} - k_0 RI_p t_p(k)$$

$$(2.7\text{-}39)$$

Thus, knowing the initial state vector $[\hat{\Phi}(t), v_c(t)]_{t=t_{on}(k)}^T$ and the i_p choice $(-I_p, 0, +I_p)$ for the kth cycle, and also the corresponding active time-period $t_p(k)$, the state equation (2.7-33) gives the state vector at $t = t_{off}(k) = t_{on}(k) + t_p(k)$. This state vector can then be used as the initial state vector for the $(k + 1)$th cycle whose beginning time $t_{on}(k + 1)$ can also be computed from this state vector by an analogous procedure.

Numerical results by Gardner [5] show that for values of $(\omega_0/\omega_C) > 20$ the response of the charge pump PLL is very close to that of the continuous loop. Finally, the response of third-order loops can be obtained in much the same way and is left as an exercise for the reader.

Stability. An analog second-order loop is unconditionally stable for any loop gain while its sampled equivalent will become unstable if the gain is made too large. Gardner [5] has performed a linearized time-discrete stability analysis and shown that the inequality

$$(2\zeta)^2 < \frac{1}{[\pi/(2\zeta)^2](\omega_c/\omega_0)\{1 + [\pi/(2\zeta)^2](\omega_c/\omega_0)\}}$$

$$(2.7\text{-}40)$$

must be fulfilled for a stable system. Fortunately, in all practical applications the crossover frequency ω_c is much smaller than the signal frequency ω_0 and hence the inequality is fulfilled.

Main Points of the Section

- Charge pumps are used to convert the logical levels of sequential logic phase detectors into precisely defined analog signals.
- PLLs employing charge pumps are capable of extremly accurate phase tracking.

- With a passive filter with impedance $Z_F(s) = R + 1/sC$ the transfer function of a type 2 loop is obtained (a conventional PLL requires an active filter).
- To reduce high-frequency frequency ripple an additional pole is often introduced into the loop filter. Placement of this pole is a compromise between good ripple suppression, which requires a pole placement at the slowest possible frequency, and loop stability which demands placement well above the crossover frequency.
- If ω_c/ω_0, i.e., ratio of the crossover frequency and the free-running frequency of the VCO is smaller than 0.05, the response of the charge pump PLL is practically indistinguishable from the quasi-continuous approximation in which the phase detector output is approximated by its average over one cycle.

2.8. RANDOM PHASE AND FREQUENCY MODULATION

So far we have considered the response of the PLL to deterministic phase signals. We now see how the PLL reacts when the signal is phase or frequency modulated. Often, the modulating signal $\theta(t)$, where

$$\theta(t) = \begin{cases} K_p m(t) & \text{phase modulation (PM)} \\ K_f \int_0^t m(v)\, dv, & \text{frequency modulation (FM)} \end{cases} \tag{2.8-1}$$

is a random process. Usually, $\theta(t)$ is modeled as a wide-sense stationary process characterized by its mean, $E[\theta(t)]$, and its autocorrelation function

$$R_\theta(\tau) = E[\theta(t + \tau)\theta(t)] \tag{2.8-2}$$

where $E[\cdot]$ is the statistical expectation. Equivalently, the spectral density $S_\theta(\omega)$, may be specified by the following Fourier transform pair

$$S_\theta(\omega) = \int_{-\infty}^{+\infty} R_\theta(\tau) \exp(-j\omega\tau)\, d\tau$$

$$\tag{2.8-3}$$

$$R_\theta(\tau) = \frac{1}{2\pi} \int_{-\infty}^{\infty} S_\theta(\omega) \exp(j\omega\tau)\, d\omega$$

In the case of a random modulation, the most we can hope for are the corresponding statistics of the output process $\hat{\theta}(t)$. The output $\hat{\theta}$ is related to the spectral density of the input function by

$$S_{\hat{\theta}}(\omega) = |H(\omega)|^2 \cdot S_{\theta}(\omega)$$

$$s_{\phi}(\omega) = |1 - H(\omega)|^2 \cdot S_{\theta}(\omega)$$

(2.8-4)

where the spectrum $S_{\theta}(\omega)$ can be found to be

$$S_{\theta}(\omega) = \begin{cases} K_p^2 S_m(\omega), & \text{PM} \\ K_f^2 \dfrac{S_m(\omega)}{\omega^2}, & \text{FM} \end{cases}$$

The variance of the phase error σ_{ϕ}^2 given by

$$\sigma_{\phi}^2 = R_{\phi}(\tau = 0) = \frac{1}{2\pi} \int_{-\infty}^{+\infty} |1 - H(\omega)|^2 \cdot S_{\theta}(\omega)\, d\omega$$

is particularly important since it represents the mean square random phase error, which must be maintained small almost all the time if the approximation sin $\phi \approx \phi$ is to be reasonably valid.

Note that if the mean and variance of the phase error ϕ are small, the PLL acts as a demodulator, as the phase of the VCO is then a good replica of the phase of the received signal. Since the VCO error signal is proportional to $d\hat{\theta}/dt$, i.e.,

$$\frac{d\hat{\theta}}{dt} = K_0 e(t)$$

$e(t)$ is a good replica of $d\theta/dt$. Thus, if the received signal is frequency modulated, the voltage $e(t)$ may be taken as the demodulator output. If $\theta(t)$ is to represent phase modulation, then the error signal must be integrated to obtain $m(t)$ (see Figure 2.8-1).

PLL demodulators for analog PM and FM and for coherent amplitude modulation (AM) are widely used. PLL-FM demodulators have an improved performance (lower threshold) when compared to conventional demodulators. For a detailed treatment of PLL-demodulators for analog mod-

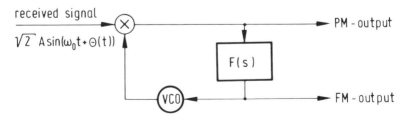

Figure 2.8-1. Phase-locked loop as FM and PM demodulator when $\theta(t)$ is given by (2.8-1).

ulated signals, the reader is referred to Gardner's book [2, Chapter 9]. Circuit details can be found in the application notes of the PLL manufacturers.

REFERENCES

1. R. C. Dorf, *Modern Control Systems*, 2nd edn., Addison-Wesley, Amsterdam, London, Manila, 1974.

2. F. M. Gardner, *Phaselock Techniques*, Wiley, New York, 1979.

3. A. Blanchard, *Phase-Locked Loops*, Wiley, New York, 1978.

4. R. C. Halgren, J. T. Harvey, and I. R. Peterson, Improved Acquisition in Phase-Locked Loops with Sawtooth Phase Detectors, *IEEE Transactions on Communications*, Vol. 30, No. 10, pp. 2364–2375, October 1982.

5. F. M. Gardner, Charge-Pump Phase-Lock Loops, *IEEE Transactions on Communications*, Vol. 28, No. 11, pp. 1849–1858, November 1980.

6. F. M. Gardner, Phase Accuracy of Charge pump PLLs, *IEEE Transactions on Communications*, Vol. 30, No. 10, pp. 2362–2363, October 1982.

3

PHASE-LOCKED LOOP TRACKING PERFORMANCE IN THE PRESENCE OF NOISE

In this chapter we study how the PLL responds if the signal applied to the phase detector is disturbed by additive noise. We assume that the noise process has a spectral density symmetrical about the frequency ω_0 and is negligibly small everywhere except in the frequency range $|\omega - \omega_0| < \pi B_{\mathrm{IF}}$. Such a spectrum is obtained if wideband noise is passed through a bandpass filter which always precedes the PLL in a practical application.

In Section 3.1 we first give a mathematical description of the noise process and then we study its effects on the phase detector operation in Section 3.2. In Section 3.3 the dynamical equations of the PLL in the presence of noise are derived and the ability of the PLL to cope with a large amount of noise is demonstrated. In Section 3.4 the state equations in the presence of noise are derived.

So far we have taken the VCO to be a perfect oscillator where the frequency ω_0 is constant in the absence of a control signal. This idealization is valid for a large majority of applications, but it is no longer valid if the bandwidth of the loop is very small. In Sections 3.5 and 3.6 we introduce a model of the VCO where the oscillations of the VCO is disturbed by an angular perturbation generated by an internal disturbance source and study the effect of this phase noise disturbance on the loop performance.

Finally, the optimization of the loop parameters in the presence of additive noise and oscillator phase noise is discussed in Section 3.7.

3.1. NARROWBAND GAUSSIAN NOISE PROCESS

A narrowband process is described in the frequency domain by the property that its spectrum is negligibly small except in the frequency range

$|\omega - \omega_0| < \pi B_{IF} \ll \omega_0$. This is always the case, as bandpass filters are always present in any practical application.

Suppose we are given two zero-mean stationary gaussian random processes with sample functions $n_c(t)$ and $n_s(t)$ from which we form the zero-mean gaussian noise process with sample function

$$n(t) = \sqrt{2}n_c(t) \cos(\omega_0 t) - \sqrt{2}n_s(t) \sin(\omega_0 t) \qquad (3.1\text{-}1)$$

To obtain the autocorrelation function of $n(t)$ we multiply (3.1-1) by the similar expression corresponding to time $t + \tau$ and statistically average the result:

$$
\begin{aligned}
E[n(t)n(t+\tau)] = 2\{ &E[n_c(t)n_c(t+\tau)] \cos(\omega_0 t) \cos[\omega_0(t+\tau)] \\
&- E[n_c(t)n_s(t+\tau)] \cos(\omega_0 t) \sin[\omega_0(t+\tau)] \\
&- E[n_s(t)n_c(t+\tau)] \sin(\omega_0 t) \cos[\omega_0(t+\tau)] \\
&+ E[n_s(t)n_s(t+\tau)] \sin(\omega_0 t) \sin[\omega_0(t+\tau)]\}
\end{aligned}
$$

This can be rewritten as

$$
\begin{aligned}
E[n(t)n(t+\tau)] = &E[n_c(t)n_c(t+\tau)] \cos(\omega_0 \tau) \\
&- E[n_c(t)n_s(t+\tau)] \sin(\omega_0 \tau) \\
&+ E[n_s(t)n_c(t+\tau)] \sin(\omega_0 \tau) \\
&+ E[n_s(t)n_s(t+\tau)] \cos(\omega_0 \tau) \\
&+ E[n_c(t)n_c(t+\tau)] \cos[\omega_0(2t+\tau)] \\
&- E[n_c(t)n_s(t+\tau)] \sin[\omega_0(2t+\tau)] \\
&- E[n_s(t)n_c(t+\tau)] \sin[\omega_0(2t+\tau)] \\
&- E[n_s(t)n_s(t+\tau)] \cos[\omega_0(2t+\tau)] \qquad (3.1\text{-}2)
\end{aligned}
$$

From (3.1-2) it is clear that $n(t)$ is nonstationalry, as $E[n(t)n(t+\tau)]$ depends on t. In fact, $n(t)$ is wide-sense stationary (and since gaussian it is strict-sense stationary) if and only if

$$E[n_c(t)n_c(t+\tau)] = E[n_s(t)n_s(t+\tau)] \qquad (3.1\text{-}3)$$

and

$$E[n_c(t)n_s(t+\tau)] = -E[n_s(t)n_c(t+\tau)] \qquad (3.1\text{-}3a)$$

Assuming that these relations hold, then the autocorrelation function of the noise becomes

$$R_n(\tau) = 2[R_{n_c}(\tau) \cos(\omega_0 \tau) + R_{n_c n_s}(\tau) \sin(\omega_0 \tau)] \qquad (3.1\text{-}4)$$

where we have introduced the notation

$$R_{n_c}(\tau) = E[n_c(t+\tau)n_c(t)] = R_{n_s}(\tau)$$

$$R_{n_c n_s}(\tau) = E[n_c(t+\tau)n_s(t)] = -R_{n_s n_c}(\tau) \qquad (3.1\text{-}5)$$

To show that the cross-correlation function $R_{n_c n_s}(\tau)$ is an odd function of τ we write first

$$E[n_s(t+\tau)n_c(t)] = E[n_s(t')n_c(t'-\tau)], \quad t' = t + \tau$$

$$\Rightarrow R_{n_s n_c}(\tau) = R_{n_c n_s}(-\tau)$$

Then, in the above equation we use (3.1-3a) to replace $R_{n_c n_s}(\tau)$ by $(-R_{n_s n_c}(\tau))$. Hence

$$R_{n_s n_c}(\tau) = -R_{n_s n_c}(-\tau) \qquad (3.1\text{-}6)$$

The autocorrelation function of a real stationary process is, however, easily shown to be an even function of τ. Therefore,

$$R_{n_c}(\tau) = R_{n_c}(-\tau) \qquad (3.1\text{-}6a)$$

The top part in (3.1-5) implies that the two zero-mean processes $n_c(t)$ and $n_s(t)$ have equal variances

$$\sigma_{n_c}^2 = \sigma_{n_s}^2 = E[n_c^2(t)] = R_{n_c}(\tau)|_{\tau=0}$$

Then the average power in our stationary random process $n(t)$ equals

$$P_n = E[n^2(t)] = \sigma_{n_c}^2 + \sigma_{n_s}^2 = 2\sigma_{n_c}^2 \qquad (3.1\text{-}7)$$

which follows from (3.1-4) by taking $\tau = 0^*$.

Now (3.1-4) can be written in complex notation as

$$R_n(\tau) = 2 \, \mathrm{Re} \, \{[R_{n_c}(\tau) - jR_{n_c n_s}(\tau)] \exp(j\omega_0 \tau)\}$$

$$= [R_{n_c}(\tau) - jR_{n_c n_s}(\tau)] \exp(j\omega_0 \tau)$$

$$+ [R_{n_c}(\tau) + jR_{n_c n_s}(\tau)] \exp(-j\omega_0 \tau)$$

Taking Fourier transforms to obtain the spectral density, we have

---

*The variance $E[n_c^2(t)]$ has dimension V^2. The average noise power (in watts) also equals $E[n_c^2(t)]$ if we assume a $1\,\Omega$ resistor as load.

$$S_n(\omega) = [S_{n_c}(\omega - \omega_0) + S_{n_c}(\omega + \omega_0)]$$

$$- j[S_{n_c n_s}(\omega - \omega_0) - S_{n_c n_s}(\omega + \omega_0))] \quad V^2 s \qquad (3.1\text{-}8)$$

where

$$S_{n_c}(\omega) = \int_{-\infty}^{\infty} R_{n_c}(\tau) \exp(-j\omega\tau) \, d\tau$$

$$S_{n_c n_s}(\omega) = \int_{-\infty}^{\infty} R_{n_c n_s}(\tau) \exp(-j\omega\tau) \, d\tau$$

Using (3.1-6) and (3.1-6a) it is easy to show that $S_{n_c}(\omega)$ is real and even, whereas $S_{n_c n_s}(\omega)$ is purely imaginary and odd (an example is shown in Figure 3.1-1).

Only if $S_{n_c n_s}(\omega)$ equals zero everywhere, are both portions of the spectrum $S_n(\omega)$ symmetrical about their center. This requires that $n_c(t)$ and $n_s(t)$ be uncorrelated, i.e.,

$$E[n_c(t + \tau)n_s(t)] = 0 \quad \text{for all } \tau \qquad (3.1\text{-}9)$$

Then since they are gaussian they are also independent.

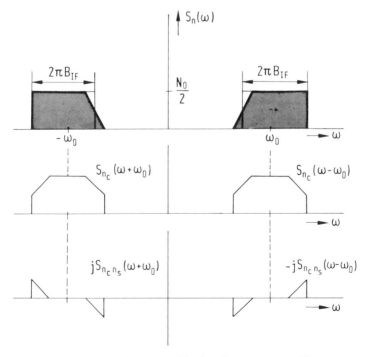

Figure 3.1-1. Spectrum of the bandpass process $n(t)$.

Using the inverse Fourier transformation, the average power of $n(t)$ can be related to the spectral density

$$P_n = R_n(0) = \frac{1}{2\pi} \int_{-\infty}^{\infty} S_n(\omega)\, d\omega \qquad (3.1\text{-}10)$$

It proves to be convenient to introduce the term *equivalent, noise bandwidth* B_{IF}. We define B_{IF} (in Hertz) in such a way that if we approximate $S_n(\omega)$ by two rectangles for the positive and negative frequency axes of width $(2\pi B_{IF})$ and amplitude $S_{n_c}(0)$ the area will be the same as the area under $S_n(\omega)$. Thus,

$$\int_{-\infty}^{\infty} S_n(\omega)\, d\omega = \underset{\substack{\text{negative} \\ \text{frequencies}}}{S_{n_c}(0)2\pi B_{IF}} + \underset{\substack{\text{positive} \\ \text{frequencies}}}{S_{n_c}(0)2\pi B_{IF}} \qquad (3.1\text{-}11)$$

Hence

$$B_{IF} = \frac{1}{2} \cdot \frac{1}{2\pi} \int_{-\infty}^{\infty} \left(\frac{S_n(\omega)}{S_{n_c}(0)} \right) d\omega$$

$$= \frac{1}{2\pi} \int_{0}^{\infty} \left(\frac{S_n(\omega)}{S_{n_c}(0)} \right) d\omega \quad \text{Hz} \qquad (3.1\text{-}11a)$$

Note that B_{IF} is measured in cycles per second (Hz) and *not* in radians per second. This is often confusing but should be clear from the $1/2\pi$ factor in (3.1-11a). The value $S_{n_c}(0)$ has been given a special symbol, namely

$$\frac{N_0}{2} := S_{n_c}(0) \qquad (3.1\text{-}12)$$

Using (3.1-10), (3.1-11a), and (3.1-12) we can write for the average noise power

$$P_n = N_0 B_{IF} \qquad (3.1\text{-}13)$$

The *signal-to-noise ratio* at the input of the PLL is then defined by

$$\text{SNR}_i = \frac{P_s}{P_n}$$

where P_s, the average signal power, is given by

$$P_s = \frac{1}{2\pi} \int_{0}^{2\pi} s^2(t)\, d(\omega_0 t) = A^2$$

$$s(t) = \sqrt{2} A \sin(\omega_0 t + \theta) \qquad (3.1\text{-}14)$$

Hence

$$\text{SNR}_i = \frac{A^2}{N_0 B_{\text{IF}}} \tag{3.1-15}$$

The slowly varying processes $n_c(t)$ and $n_s(t)$ are related to the original process through (3.1-1). However, instead of $n_c(t)$ and $n_s(t)$ it is sometimes of interest to consider the envelope process $N(t)$ and the phase process $\arg\{n_L(t)\}$ of $n(t)$. The envelope and phase processes are defined by

$$N(t) = [n_c^2(t) + n_s^2(t)]^{1/2}, \quad \arg\{n_L(t)\} = \tan^{-1}\left(\frac{n_s(t)}{n_c(t)}\right) \tag{3.1-16}$$

Then the sample function of $n(t)$ can be written as

$$n(t) = \sqrt{2}N(t)\cos\{\omega_0 t + \arg\{n_L(t)\}\} \tag{3.1-17}$$

If we assume that $S_{n_c n_s}(\omega) = 0$, that is, both portions of $S_n(\omega)$ are symmetrical about their center, then the cross-correlation $R_{n_c n_s}(\tau)$ equals zero for all τ. Hence, the random variables $n_c(t)$ and $n_s(t + \tau)$ are statistically independent for all t and τ and their joint probability density function (PDF) is given by

$$p(n_c, n_s) = \frac{1}{2\pi\sigma_{n_c}^2}\exp\left(-\frac{1}{2}\frac{n_c^2 + n_s^2}{\sigma_{n_c}^2}\right) \tag{3.1-18}$$

Introducing the polar coordinates $n_c = N\cos\{\arg\{n_L(t)\}\}$ and $n_s = N\sin\{\arg\{n_L(t)\}\}$ and integrating over all possible values of $\arg\{n_L(t)\} \in [0, 2\pi)$ we obtain the Rayleigh PDF

$$p(N) = \begin{cases} \dfrac{N}{\sigma_{n_c}^2}\exp\left(-\dfrac{N^2}{2\sigma_{n_c}^2}\right), & N > 0 \\[3mm] 0, & \text{elsewhere} \end{cases} \tag{3.1-19}$$

This PDF is of interest in the characterization of the amplitude of the additive noise voltage present in a communication receiver front end.

Main Points of the Section

- The spectral density of a narrowband process $n(t)$ is negligibly small outside a frequency band $|\omega - \omega_0| < 2\pi B_{\text{IF}} \ll \omega_0$
- *Mathematical Representation*
  (a) Time domain:
     Sample function $n(t)$

$$n(t) = \sqrt{2}n_c(t)\cos(\omega_0 t) - \sqrt{2}n_s(t)\sin(\omega_0 t) \tag{3.1-1}$$

Conditions for stationarity of $n(t)$

$$R_{n_c}(\tau) = R_{n_s}(\tau)$$

$$R_{n_c n_s}(\tau) = -R_{n_s n_c}(\tau) \tag{3.1-5}$$

Correlation function for a stationary process

$$R_n(\tau) = 2[R_{n_c}(\tau)\cos(\omega_0\tau) + R_{n_c n_s}(\tau)\sin(\omega_0\tau)] \tag{3.1-4}$$

(b) Frequency domain:

Power spectrum $S_n(\omega)$ for a stationary process $n(t)$ (Figure 3.1-1)

$$S_n(\omega) = S_{n_c}(\omega - \omega_0) + S_{n_c}(\omega + \omega_0) \\ - j[S_{n_c n_s}(\omega - \omega_0) - S_{n_c n_s}(\omega + \omega_0)]$$

Conditions for stationarity of $n(t)$

$$S_{n_c}(\omega) = S_{n_s}(\omega)$$

$$S_{n_c n_s}(\omega) = -S_{n_s n_c}(\omega)$$

Symmetry of $S_n(\omega)$ about ω_0 and $-\omega_0$ when

$$S_{n_c n_s}(\omega) = 0 \quad \text{for all } \omega \Leftrightarrow n_c(t) \text{ and } n_s(t) \text{ uncorrelated}$$

- *Noise Parameters*

  (a) Equivalent noise bandwidth B_{IF}:

$$B_{IF} = \frac{1}{2\pi} \int_0^\infty \frac{S_{n_c}(\omega)}{S_{n_c}(0)} \, d\omega \quad \text{Hz} \tag{3.1-11a}$$

$$S_{n_c}(0) =: \frac{N_0}{2}$$

  (b) Input signal-to-noise ratio SNR_i:

$$\text{SNR}_i = \frac{P_s}{P_n} = \frac{A^2}{N_0 B_{IF}} \tag{3.1-15}$$

  (c) Average power:

$$P_n = R_n(0) = 2R_{n_c}(0) = \frac{1}{2\pi} \int_{-\infty}^{\infty} S_n(\omega) \, d\omega = N_0 B_{IF}$$

- *Complex Representation of n(t)*

$$n(t) = \sqrt{2} \, \mathrm{Re} \, [\underbrace{(n_c(t) + jn_s(t)) \exp(j\omega_0 t)}]$$

$$\underline{n}_L(t) \quad \text{complex noise envelope}$$

$$R_n(\tau) = 2 \, \mathrm{Re} \, \{[R_{n_c}(\tau) - jR_{n_c n_s}(\tau)] \exp(j\omega_0\tau)\}$$

$$= 2 \, \mathrm{Re} \, \{[R_{n_c}(\tau) + jR_{n_s n_c}(\tau)] \exp(j\omega_0\tau)\}$$

3.2. PHASE DETECTOR OPERATION IN THE PRESENCE OF ADDITIVE NOISE

3.2.1. Sinusoidal Phase Detector Characteristics

We model the phase detector as an ideal multiplier, partly because of its analytical convenience and partly because many practical phase detectors are good approximations to multipliers. Another equally important reason is that the optimum phase estimator in the maximum likelihood sense (to be discussed later) has the form of a cross-correlator, i.e., consists of a multiplier plus integrator.

The first step in obtaining the dynamical equations of the PLL in the presence of additive noise is to calculate the phase detector output. We assume that the VCO frequency is constant and exactly equals the frequency of the incoming signal. Physically this means that the feedback loop is open; for example, the phase detector is disconnected from the loop filter. Later the loop will be closed.

The received signal corrupted by the additive gaussian noise introduced in Section 3.1 is

$$y(t) = \sqrt{2}A \sin(\omega_0 t + \theta) + \sqrt{2}n_c(t) \cos(\omega_0 t) - \sqrt{2}n_s(t) \sin(\omega_0 t) \quad (3.2.1)$$

The output of the multiplier equals

$$u(t) = K_m[\sqrt{2}A \sin(\omega_0 t + \theta) + \sqrt{2}n_c(t) \cos(\omega_0 t)$$
$$- \sqrt{2}n_s(t) \sin(\omega_0 t)]\sqrt{2}K_1 \cos(\omega_0 t + \hat{\theta}) \quad (3.2-2)$$

where K_m is the multiplier gain and $\sqrt{2}K_1$ is the amplitude of VCO signal.

Carrying out the multiplication in (3.2-2) and discarding the terms with the double frequency $2\omega_0 t$ we obtain

$$x(t) = K_D\left[\sin\phi + \underbrace{\frac{n_c(t)}{A}\cos\hat{\theta} + \frac{n_s(t)}{A}\sin\theta}\right] \quad (3.2-3)$$

$$n'(t, \hat{\theta})$$

where $\hat{\theta}$ is a constant, $\phi = \theta - \hat{\theta}$ and $K_D = K_m K_1 A$ is the phase detector gain. In practice, the double frequency terms can be very troublesome and substantial effort is needed to suppress them.

The first term on the right-hand side of (3.2-3) represents the sinusoidal characteristic of the multiplier type PD and is a useful control signal which would be obtained experimentally as the dc-term of the phase detector output. The other part, $n'(t, \hat{\theta})$ is a zero-mean, rapidly fluctuating disturbance which must be filtered out by the PLL. It is one of the major attractions of the PLL that it can cope with a large amount of noise. An exact equivalent model of the phase detector is shown in Figure 3.2-1b.

Note that $n'(t, \hat{\theta})$ is a dimensionless quantity which can be given a graphic interpretation. For a small phase error we may approximate $\sin \phi$ by its argument, ϕ. The phase detector output can now be written

$$x(t) = K_D[\phi(t) + n'(t, \hat{\theta})] \qquad (3.2\text{-}4)$$

Thus, $n'(t, \hat{\theta})$ can be viewed as angular phase disturbance which replaces the additive bandpass noise $n(t)$ when we come to the baseband model. The statistical properties of $n'(t, \hat{\theta})$ will be of key importance in our analysis. The autocorrelation function of $n'(t, \hat{\theta})$, given $\hat{\theta}$, is

$$R_{n'}(\tau|\hat{\theta}) = E[n'(t + \tau, \hat{\theta})n'(t, \hat{\theta})] =$$

$$E\left[\left(\frac{n_c(t + \tau)}{A} \cos \hat{\theta} + \frac{n_s(t + \tau)}{A} \sin \hat{\theta}\right)\left(\frac{n_c(t)}{A} \cos \hat{\theta} + \frac{n_s(t)}{A} \sin \hat{\theta}\right)\right]$$

$$(3.2\text{-}5)$$

Assuming $n_c(t)$ and $n_s(t)$ satisfy (3.1-5), the stationarity condition for $n(t)$, the expression above simplifies to

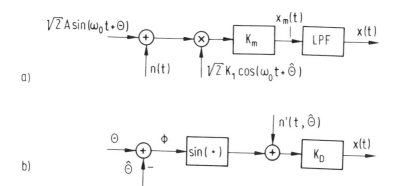

Figure 3.2-1. (a) Multiplier type phase detector in the presence of noise; (b) equivalent baseband model.

$$R_{n'}(\tau|\hat{\theta}) = \frac{R_{n_c}(\tau)}{A^2} \cos^2 \hat{\theta} + \frac{R_{n_s}(\tau)}{A^2} \sin^2 \hat{\theta}$$

$$= \frac{R_{n_c}(\tau)}{A^2} \tag{3.2-6}$$

From (3.2-6) we observe the important property that the conditional autocorrelation function $R_{n'}(\tau|\hat{\theta})$ is independent of $\hat{\theta}$. This is in agreement with the general fact that a random phase rotation applied to a stationary, complex zero-mean gaussian process does not change the statistics (here, specifically, mean and autocorrelation) of that process if it is a white process or if the applied phase rotation is constant.

Equation (3.2-6) implies that the noise variance $\sigma_{n'}^2$ is the same for any fixed phase $\hat{\theta}$ of the VCO

$$\sigma_{n'}^2 = \frac{R_{n_c}(0)}{A^2}$$

Then using the average noise power P_n and the signal power given by (3.1-7) and (3.1-14)

$$P_n = 2R_{n_c}(0), \quad P_s = A^2$$

we obtain

$$\sigma_{n'}^2 = \frac{P_n}{2P_s} = \frac{1}{2(\text{SNR}_i)} \tag{3.2-7}$$

In view of our interpretation of $n'(t)$ as an angular disturbance*, the inverse of twice the signal-to-noise ratio gives the variance of the phase difference between a clean signal and a signal corrupted by additive noise. The power spectrum $S_{n'}(\omega)$ has the same low-pass characteristics of $S_{n_c}(\omega)$, as (3.2-6) gives

$$S_{n'}(\omega) = \frac{S_{n_c}(\omega)}{A^2} \tag{3.2-8}$$

It might be surprising that only half the noise power appears in $\sigma_{n'}^2$ as given by (3.2-7). Again, the concept of rotating phasors yields a graphic interpretation of this result. A simple trigonometric identity shows that the sum of signal and noise can be written in the form

---

*For the sake of a compact notation we will omit $\hat{\theta}$ in $n'(t, \hat{\theta})$ and in similar expressions that are independent of this parameter.

$$y(t) = \mathrm{Re}\left\{\sqrt{2}A\underbrace{\left[\frac{N_c(t)}{A} - j\left(1 - \frac{N_s(t)}{A}\right)\right]}_{\underline{y}_L}\exp\left[j(\omega_0 t + \theta)\right]\right\} \quad (3.2\text{-}9)$$

with

$$N_c(t) = n_c(t)\cos(\theta) + n_s(t)\sin(\theta)$$

$$N_s(t) = n_s(t)\cos(\theta) - n_c(t)\sin(\theta)$$

For a constant input phase θ, $N_c(t)$ and $N_s(t)$ have the same statistical properties as $n_c(t)$ and $n_s(t)$, respectively. From (3.2-9) we learn that the component $-N_s(t)$ of the noise is in phase with the useful signal $s(t)$ while $N_c(t)$ is perpendicular to $s(t)$. $N_s(t)$ is therefore called the *in-phase* component while $N_c(t)$ is called the *quadrature* component *with respect to* $s(t)$.

The noise causes a random perturbation of the amplitude and the phase of the signal phasor. The random phase fluctuations are given by (see Figure 3.2-2)

$$\phi_i(t) = \tan^{-1}\frac{N_c(t)/A}{1 - [N_s(t)/A]} \quad (3.2\text{-}10)$$

If the input signal-to-noise ratio SNR$_i$ is large we may neglect the random amplitude modulation of the in-phase component in the denominator to obtain $(\tan^{-1} x \approx x)$

$$\phi_i(t) \approx \frac{N_c(t)}{A} \quad (3.2\text{-}11)$$

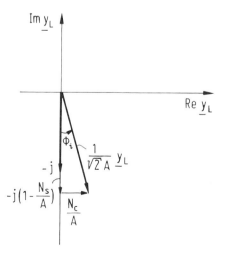

Figure 3.2-2. Phasor diagram for the received signal $\underline{y}_L(t)$ (complex baseband equivalent of $\underline{y}(t)$) corrupted by additive noise.

We thus have the important result that for such SNR_i, the phase perturbation is practically only due to the quadrature component of the noise. This explains the factor $1/2$ in (3.2-7).

The statistical properties of $n'(t)$ were computed under the condition of a constant VCO phase. If we close the loop, the VCO phase is no longer constant but undergoes random fluctuations as the noise frequency modulates the VCO. As a consequence, the results derived here are strictly speaking not applicable. We are forced to make further simplifying assumptions in order to be able to proceed.

The purpose of a PLL is to generate a replica of the incoming signal. It will, therefore, always be designed in such a way that it will smooth the phase detector fluctuations in order to extract the (slowly varying) control signal sin $\phi(t)$. Therefore, the phase $\hat{\theta}(t)$ can be considered nearly constant over the short time interval, $1/B_{\mathrm{IF}}$, where a significant correlation between two noise samples $n(t)$ and $n[t + (1/B_{\mathrm{IF}})]$ exists. The statistics of $n'(t)$ will, therefore, be approximately the same as in the case of an open loop. This is the key assumption we have to make in our analysis. The condition just given is usually stated in terms of a bandwidth ratio; if the noise bandwidth B_{IF} is much larger than the loop bandwidth B_L (defined in (3.3-5)), then the closed-loop statistics of $n'(t)$ can be approximated by those obtained under open-loop conditions. (For details the reader is referred to Appendix 3.2A.)

3.2.2. Nonsinusoidal Phase Detector Characteristics

Phase-locked loops with hard limiters (cf. Figure 3.2-3) preceding the phase detectors are of great practical interest. A hard limiter preceding the phase-locked loop is often used in order to keep the input signal range down to a reasonable level and also to protect the phase detector from damage under high signal or noise level. In addition, a multiplier type phase detector is realized as a switching phase detector which is functionally equivalent to an ideal multiplier when the VCO signal is hard-limited before being applied to the multiplier (cf. Section 2.6.1). When both inputs to a multiplier type phase detector are hard-limited, the phase detector characteristic (i.e., the plot of the average of the PD output vs. phase error ϕ) becomes triangular if no noise is present. When two hard-limited signals are applied to an RS type flip-flop we obtain a sawtooth phase detector characteristic.

While a sinusoidal characteristic of a multiplier type phase detector is unaffected by noise (cf. Figure 3.2-1b where $E[x] = K_D \sin \phi$), this is not true for nonsinusoidal phase detectors in which the characteristics change their form as a function of the input signal-to noise ratio. This subsection is devoted to the analysis of nonsinusoidal phase detector characteristics in the presence of noise. As an example, consider the phase detector of Figure 3.2-3c. The input to the limiter on the top is modeled by

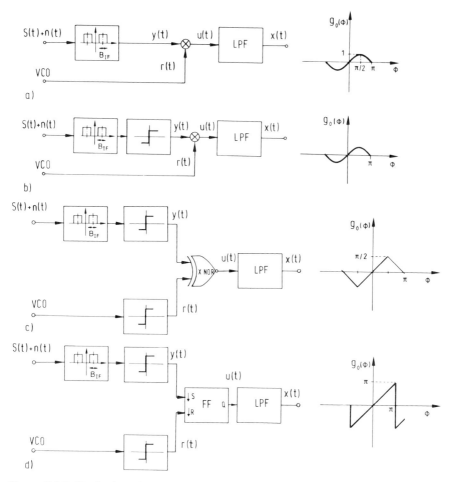

Figure 3.2-3. Realization of sinusoidal phase detector characteristic (a) without limiters and (b) with signal path limiter; (c) and (d), limiter phase detectors. ($g_0(\cdot)$) is the normalized noise-free PD characteristic with $g_0'(0) = 1$). (a) Multiplier without limiters; (b) multiplier with limiter in the signal path; (c) multiplierr with both inputs hard-limited; (d) Sequential logic PD arrangement with one RS flip-flop.

$$s(t) + n(t) = \sqrt{2}A \sin (\omega_0 t + \theta) + \sqrt{2}n_c(t) \cos (\omega_0 t)$$
$$- \sqrt{2}n_s(t) \sin (\omega_0 t) \qquad (3.2\text{-}12)$$

Using complex notation we can write for (3.2-12)

$$s(t) + n(t) = \sqrt{2} \, \text{Re} \, \{-j[(A - N_s(t)) + jN_c(t)] \exp [j(\omega_0 t + \theta)]\}$$
$$= \sqrt{2} \, \text{Re} \, \{-j\sqrt{[A - N_s(t)]^2 + N_c^2(t)} \exp [j(\omega_0 t + \theta + \theta_n(t))]\}$$
$$= \sqrt{2[A - N_s(t)]^2 + 2N_c^2(t)} \sin [\omega_0 t + \theta + \theta_n(t)] \qquad (3.2\text{-}13)$$

with
$$\theta_n(t) = \arg \{[A - N_s(t)] + jN_c(t)\}$$

$$= \tan^{-1} \left\{ \frac{N_c(t)}{A - N_s(t)} \right\} \tag{3.2-14}$$

where

$$N_c(t) = n_c(t) \cos \theta(t) + n_s(t) \sin \theta(t)$$

$$N_s(t) = n_s(t) \cos \theta(t) - n_c(t) \sin \theta(t)$$

The limiter, modeled as a signum function, removes the amplitude variations so that

$$y(t) = \text{sgn} \{\sin [\omega_0 t + \theta + \theta_n(t)]\} \tag{3.2-15}$$

The signal $y(t)$ is a square wave with unit amplitude and randomly varying zero crossings due to $\theta_n(t)$. The square wave $r(t)$ which is obtained from the VCO output has the form

$$r(t) = \text{sgn} [\cos (\omega_0 t + \hat{\theta})] \tag{3.2-16}$$

The multiplier output $u(t)$ equals the product of the two square waves

$$u(t) = K_m \, \text{sgn} \{\sin [\omega_0 t + \theta + \theta_n(t)]\} \cdot \text{sgn} [\cos (\omega_0 t + \hat{\theta})] \tag{3.2-17}$$

We now seek a decomposition of $u(t)$ into a useful control signal and a rapidly varying noise term. Due to the limiters, this goal is not as readily achieved as before. An attempt to define the useful control signal as the expected value of $u(t)$ fails since this value depends on time t. This time-dependence was encountered previously in the form of the double-frequency term $\sin (2\omega_0 t + \theta + \hat{\theta})$ at the multiplier output. There, we argued that the multiplier was followed by a low-pass filter which rejected high-frequency portions of the multiplier output and thus the double-frequency term could be neglected.

The problem we must now solve is to develop a mathematical approach to extract the baseband component of $u(t)$ in the general case. But before we can do so we need to introduce the mathematical concepts of a cyclostationary process and of time randomization.

We claim that the random process $u(t)$ is *cyclostationary*. A process is called cyclostationary if its statistics are invariant to a shift of the time origin by integral multiples of a constant T called the period. This is indeed true for $u(t)$ with $T = 2\pi/\omega_0$. Evaluating (3.2-17) at $t + mT$ yields

$$u(t + mT) = K_m \, \text{sgn} \, (\sin [\omega_0(t + mT) + \theta$$
$$+ \theta_n(t + mT)]) \, \text{sgn} \, [\cos (\omega_0(t + mT) + \hat{\theta})]$$
$$= K_m \, \text{sgn} \, (\sin [\omega_0 t + \theta + \theta_n(t + mT)$$
$$+ 2\pi m]) \, \text{sgn} \, [\cos (\omega_0 t + \hat{\theta} + 2\pi m)]$$
$$= K_m \, \text{sgn} \, (\sin [\omega_0 t + \theta + \theta_n(t + mT)])$$
$$\cdot \text{sgn} \, [\cos (\omega_0 t + \hat{\theta})] \qquad (3.2\text{-}18)$$

From (3.2-14), it follows that $\theta_n(t)$ is a stationary process. Hence, $u(t + mT)$ has the same statistics as $u(t)$, which proves that $u(t)$ is cyclostationary.

We now show how one can extract the (wide-sense stationary) baseband component of any cyclostationary process $u(t)$. We introduce a new process $x(t) := u(t - t_0)$ which is obtained by a random t_0 shift of the origin of $u(t)$. The random variable t_0 is assumed to be uniformly distributed over the period T. We claim that $x(t) := u(t - t_0)$ is a wide-sense stationary process (i.e., the first two moments are independent of time) which contains only the baseband component of $u(t)$ and, hence, equals the phase detector output after the low-pass filter.

To prove the above claim, we use the well-known theorem on conditional expectations where the expected value with respect to the random variables t_0 and u, $E[u(t - t_0)]$, is obtained by first computing the expected value with respect to u, of $u(t - t_0)$ conditioned on a fixed $t_0{}^*$ and by subsequently averaging the result over all possible values of t_0

$$E[u(t - t_0)] = E[E[u(t - t_0) \mid t_0 = t_0] \qquad (3.2\text{-}19)$$

where t_0 is the fixed value of the random variable t_0. Since the random shift t_0 is statistically independent of the prcess $u(t)$ the (inner) conditional expected value equals the (time-varying) expected value of the original process $u(t)$ taken at $(t - t_0)$

$$E[u(t - t_0) \mid t_0 = t_0] = E[u(t - t_0) \mid t_0 = t_0] = E[u(t - t_0)]$$

The outer expectation in (3.2-19) averages $E[u(t - t_0)]$ over all values of the uniformly distributed random variable t_0.

$$p(t_0) = \begin{cases} \dfrac{1}{T}, & \text{for } t_0 \in [0, T] \\ 0, & \text{otherwise} \end{cases}$$

Therefore

---

*For clarity in the following derivation we must notationally distinguish between random quantities and their realization.

$$E[u(t - t_0)] = \frac{1}{T} \int_0^T E[u(t - t_0)] \, dt_0 \qquad (3.2\text{-}20)$$

Since $u(t)$ is cyclostationary, the above integrand is a periodic function of t_0. The integral can therefore be taken over an arbitrary interval of length $T = 2\pi/\omega_0$

$$E[x(t)] = E[u(t - t_0)] = \frac{1}{T} \int_0^T E[u(t)] \, dt \qquad (3.2\text{-}21)$$

Reasoning similarly, it is left as an exercise to the reader to verify that the correlation function of $x(t) = u(t - t_0)$ equals the time average of the correlation function $R_u(t, t + \tau)$ of the cyclostationary process $u(t)$, i.e.,

$$R_x(\tau) = \frac{1}{T} \int_0^T R_u(t, t + \tau) \, dt \qquad (3.2\text{-}22)$$

with

$$R_u(t, t + \tau) = E\{u(t + \tau - t_0)u(t - t_0) \mid t_0 = t_0\}$$

As (3.2-21) and (3.2-22) show, it is interesting that the mean $E[x(t)]$ and the correlation function $R_x(\tau)$ are, respectively, the *time averages* of the mean and correlation function of the original cyclostationary process $u(t)$. Clearly, $E[x(t)]$ and $R_x(\tau)$ are independent of time t and $x(t)$ is a wide-sense stationary process as claimed, and hence can contain only the baseband component of $u(t)$.

Finally, we must relate the mathematical operations of taking expected values to the operations of the physical circuits. The reader must be aware of the fact that the actual circuit comprised of a memoryless phase detector followed by a low-pass filter does not perform a time-averaging; rather, the time-averaging operation is approximated by the low-pass filter (Time-averaging is the sole function of the filter, not noise suppression!). We examplify these rather subtle arguments by the example of the multiplier type phase detector discussed earlier. Of course the results are identical to those derived previosly in a much simper way.

Example: The output of the multiplier phase detector without limiters can be found to be ((3.2-2) and (3.2-3))

$$u(t) = K_D \left[\sin \phi + \sin (2\omega_0 t + \theta + \hat{\theta}) + n'(t, \hat{\theta}) + \frac{n_c(t)}{A} \cos (2\omega_0 t + \hat{\theta}) \right.$$

$$\left. - \frac{n_s(t)}{A} \sin (2\omega_0 t + \hat{\theta}) \right] \qquad (3.2\text{-}23)$$

The conditional expected value of $u(t - t_0)$ for a fixed t_0 equals

$$E[u(t - t_0) \,|\, t_0 = t_0] = E[u(t - t_0)] = K_D \{\sin \phi + \sin [2\omega_0(t - t_0) + \theta + \hat{\theta}]\}$$
$$(3.2\text{-}24)$$

which is periodic in t. The expected value of $x(t) := u(t - t_0)$ (3.2-11) is

$$E[x(t)] = \frac{1}{T} \int_0^T K_D[\sin \phi + \sin (2\omega_0 t + \theta + \hat{\theta})] \, dt = K_D \sin \phi$$

which is the useful control signal.

We now have all the necessary tools at our disposal to solve the problem of how to decompose the phase detector output into a useful control signal plus disturbance. As we have demonstrated, by time randomization, the cyclostationary process $u(t)$ of (3.2-17) becomes a wide-sense stationary process $x(t)$ whose expected value equals the useful control signal. Continuing from (3.2-17) we determine the statistical properties of the process $x(t)$ defined by

$$x(t) := K_m \, \mathrm{sgn} \{\sin [\omega_0(t - t_0) + \theta + \theta_n(t)]\} \, \mathrm{sgn} \{\cos [\omega_0(t - t_0) + \hat{\theta}]\}$$
$$(3.2\text{-}25)$$

Since the amplitude variations have been suppressed by the limiter, the only random quantities left in (3.2-25) are t_0 and $\theta_n(t - t_0)$. Since $\theta_n(t - t_0)$ is a stationary narrowband process we may neglect its variations within a period T. (Also, since a time shift t_0 is immaterial in a stationary process it is omitted from now on).

However[*], for a slowly varying $\theta_n(t)$, $x(t)$ is determined by the *noise-free phase detector characteristic* $g_0(\cdot)$ with the phase difference $\theta + \theta_n(t) - \hat{\theta} = \phi + \theta_n(t)$

$$x(t) \approx K_D g_0[\phi + \theta_n(t)] \tag{3.2-26}$$

where K_D is the phase detector gain and $g_0(\cdot)$, the normalized noise-free phase detector characteristic with unit slope at the origin ($g_0'(0) = 1$). Notice that the above relation is not only true for the triangular characteristic but also for any memoryless phase detector characteristic when the input signal is hard-limited.

Having laid the mathematical basis, we are now ready to derive a *general phase detector model*. We mathematically divide the phase detector output into a useful control signal and a random disturbance

$$x(t) = K_D[g(\phi) + n'(t, \phi)] \tag{3.2-27}$$

---

[*]Random quantities and their realizations are notationally no longer distinguished.

where $g(\phi)$ denotes the expected value of $x(t)/K_D$ and $n'(t, \phi)$ is defined as the zero mean, fluctuating part of $x(t)$.

$$K_D n'(t, \phi) := x(t) - E[x(t)]$$

Thus, we obtain the model shown in Figure 3.2-4 as a generalization of the model previously considered in Figure 3.2-1b. From (3.2-26) we learn that the phase noise adds to the phase error ϕ. Therefore, unless $\theta_n(t) = 0$ by chance, the control signal to the VCO is either too large or two small, depending on the actual value $\theta_n(t)$ assumes. The *effective phase detector characteristic* $g(\phi)$ is, therefore, the average over all possible values of the random phase variable $\theta_n(t)$

$$K_D g(\phi) = K_D E\{ g_0[\phi + \theta_n(t)]\} = K_D \int_{-\pi}^{\pi} g_0(\phi + \theta_n) p(\theta_n)\, d\theta_n \tag{3.2-28}$$

with $p(\theta_n)$ the PDF of θ_n in $[-\pi, \pi]$.

If $p(\theta_n) = p(-\theta_n)$, as is true for gaussian input additive noise and if $g_0(-x) = -g_0(x)$, then the above integrand evaluated at $\phi = 0$ is an odd function and the null $g(\phi) = 0$ at $\phi = 0$ is fixed and will not shift with the varying input SNR_i. If at least one of these conditions is violated then the null can shift, which is a highly unsatisfactory occurrence.

Example: To numerically evaluate $g(\phi)$ the PDF $p(\theta_n)$ must be known. As an example, we compute $g(\phi)$ for the triangular nonlinearity $g_0(\cdot)$ of Figure 3.2-3c and gaussian symmetric additive noise. We expand this $g_0(\cdot)$ into a Fourier series

$$g_0(x) = \frac{4 K_m}{\pi} \sum_{k=0}^{\infty} (-1)^{2k+1} \frac{\sin\left[(2k+1)x\right]}{(2k+1)^2} \tag{3.2-29}$$

Using obvious trigonometric identities we find

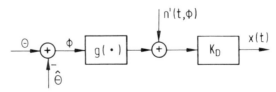

Figure 3.2-4. Equivalent baseband model of a memoryless phase detector in the presence of noise.

$$g_0(\phi + \theta_n) = \frac{4K_m}{\pi} \sum_{k=0}^{\infty} (-1)^{2k+1} \frac{1}{(2k+1)^2} \{\sin[(2k+1)\phi]\cos[(2k+1)\theta_n]$$
$$+ \cos[(2k+1)\phi]\sin[(2k+1)\theta_n]\} \tag{3.2-30}$$

Taking expected values yields*

$$g(\phi) = E[g_0(\phi + \theta_n)] = \frac{4K_m}{\pi} \sum_{k=0}^{\infty} (-1)^{2k+1} \frac{1}{(2k+1)^2}$$
$$\times \{\sin[(2k+1)\phi]\langle\cos[(2k+1)\theta_n]\rangle$$
$$+ \cos[(2k+1)\phi]\langle\sin[(2k+1)\theta_n]\rangle\} \tag{3.2-31}$$

The expected values $\langle\cos[2k+1)\theta_n]\rangle$ and $\langle\sin[(2k+1)\theta_n]\rangle$ are evaluated in Appendix 7.1B and the resulting PD characteristics are given in Table 3.2-1. The effective phase detector characteristic is plotted in Figure 3.2-5b with the input SNR_i as parameter. For comparison, in the same figure the effective characteristics for other phase detectors of Figure 3.2-3 are included.

The slope g' at the origin also varies as a function of the input SNR_i. The effective phase detector gain, therefore, equals

$$\text{Effective PD gain} = K_D g'(0), \quad g'(0) := \frac{dg}{d\phi}(0) \tag{3.2-32}$$

For the phase detectors of Figure 3.2-3, $g'(0)$ is given in Table 3.2-1 and plotted in Figure 3.2-6 as a function of SNR_i.

The phase noise $n'(t, \phi)$ in (3.2-27) generally is a function of ϕ. Comparison of (3.2-26) and (3.2-27) yields

$$n'(t, \phi) := g_0[\phi + \theta_n(t)] - g(\phi) = g_0[\phi + \theta_n(t)] - E\{g_0[\phi + \theta_n(t)]\} \tag{3.2-33}$$

The variance $\sigma_{n'}^2(\phi)$ of this noise term has been evaluated for a specific $g(\cdot)$ as a function of ϕ by Simon and Springett [20] with the result that $\sigma_{n'}^2(\phi)$ does not vary much as a function of ϕ even if ϕ is as large as $\pi/4$. Moreover, the smaller the input SNR_i (the case when larger ϕ are expected at all), the smaller the deviation of $\sigma_{n'}^2(\phi)$ from its maximum at $\phi = 0$. Extrapolating this result to other phase detector characteristics we simplify further analysis by neglecting the dependence of $n'(t, \phi)$ on ϕ

*Notation: we interchangeably use $\langle\cdot\rangle$ or $E[\cdot]$ to denote statistical expectation.

TABLE 3.2-1 Characteristic Quantities of Various Memoryless Phase Detectors in the Presence of Symmetric, Additive Gaussian Noise

	Sinusoidal PD	Sinusoidal Limiter PD	Triangular Limiter PD	Sawtooth Limiter PD
$g(\phi)$	$\sin \phi$	$\mu_1 \sin \phi$	$\dfrac{4}{\pi} \sum\limits_{k=1,3,5}^{\infty} (-1)^{\frac{1}{2}(k-1)} \dfrac{1}{k^2} \mu_k \sin k\phi$	$2\sum\limits_{\nu=1,2,3}^{\infty} (-1)^{\nu+1} \dfrac{1}{\nu} \mu_\nu \sin \nu\phi$
$g'(0)$	1	μ_1	$\dfrac{4}{\pi} \sum\limits_{k=1,3,5}^{\infty} (-1)^{\frac{1}{2}(k-1)} \dfrac{1}{k} \mu_k$	$2\sum\limits_{\nu=1,2,3}^{\infty} (-1)^{\nu+1} \mu_\nu$
$\sigma_{n'}^2$	$\dfrac{1}{2SNR_i}$	$\dfrac{1}{2SNR_i}[1 - \exp(-SNR_i)]$	$\dfrac{\pi^2}{12} + 4 \sum\limits_{l=2,4,6}^{\infty} (-1)^{l/2} \dfrac{1}{l^2} \mu_l$	$\dfrac{\pi^2}{3} + 4\sum\limits_{\nu=1,2,3}^{\infty} (-1)^{\nu} \dfrac{1}{\nu^2} \mu_\nu$
$\dfrac{B_1}{B_{IF}}$	1	$1 + 0.098 \exp[-SNR_i(1 - \pi/4)]$	$1 + 0.098 \exp[-SNR_i(1 - \pi/4)]$	$1 + 0.35 \exp[-SNR_i(1 - \pi/4)]$
μ_ν		$\mu_\nu = \dfrac{1}{2}\sqrt{\pi SNR_i}\, \exp\left(\dfrac{-SNR_i}{2}\right)\left[I_{(\nu-1)/2}\left(\dfrac{SNR_i}{2}\right) + I_{(\nu+1)/2}\left(\dfrac{SNR_i}{2}\right)\right]$		

$I_m(\cdot) = $ Modified Bessel Function of order m

Source: Werner Rosenkranz, Phase-Locked Loops with Limiter Phase Detectors in the Presence of Noise, *IEEE Transactions on Communications*, Vol. 30, No. 10 © 1982 IEEE.

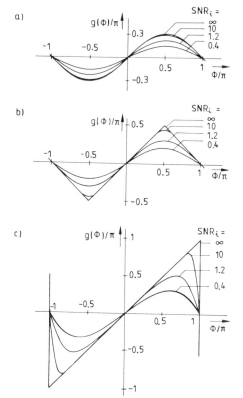

Figure 3.2-5. Effective phase detector characteristic $g(\phi)$ with input SNR_i as parameter. Input noise is symmetric, additive gaussian noise (From Werner Rosenkranz, Phase-Locked Loops with Limiter Phase Detectors in the Presence of Noise, *IEEE Transactions on Communications*, Vol. 30, No. 10 © 1982 IEEE.) (*a*) Multiplier PD with limiter in signal path; (*b*) multiplier with both input signals hard-limited; (*c*) RS flip-flop phase detector.

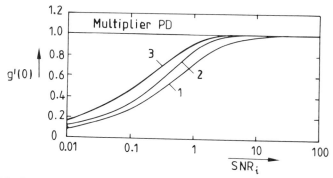

Figure 3.2-6. Slope of the effective phase detector characteristic at the origin. Same data as in Figure 3.2-5 (*a*) Multiplier PD with limiter in signal path; (*b*) multiplier with both input signals hard-limited; (*c*) RS flip-flop phase detector.

$$n'(t, \phi) \approx g_0(\theta_n) \tag{3.2-34}$$

If $p(\theta_n)$ and $g_0(\theta_n)$ have the symmetries previously noted, the variance $\sigma_{n'}^2$ is given by

$$\sigma_{n'}^2 = E[g_0^2(\theta_n)] = \int_{-\pi}^{\pi} g_0^2(\theta_n) p(\theta_n)\, d\theta_n \tag{3.2-35}$$

$\sigma_{n'}^2$ has been evaluated by Rosenkranz [19]. Analytical expressions are listed in Table 3.2-1 and numerical evaluations of these formulas are plotted in Figure 3.2-7. Various interesting properties are revealed by this figure. For large SNR$_i$ the variance $\sigma_{n'}^2$ is inversely proportional to SNR$_i$ for all types of phase detectors.

$$\sigma_{n'}^2 = \frac{1}{2 \mathrm{SNR}_i}, \quad \mathrm{SNR}_i \gg 1 \tag{3.2-36}$$

Equation (3.2-36) has been previously obtained for the multiplier phase detector without limiter. Since all $g_0(\phi)$ by definition have unity slope (see (3.2-26)) and, for high SNR$_i$, linearization of $g_0(\theta_n) \approx \theta_n$ is permissible in (3.2-35) the above finding is no surprise.

For any phase detector with limiter, $\sigma_{n'}^2$ converges to a finite bounded value (specific to the particular phase detector as given in Figure 3.2-7) as SNR$_i \to 0$. This does not mean that a limiter phase detector yields a better performance since on the other hand the effective phase detector gain (3.2-32) and, thus, the useful control signal decreases significantly for low SNR$_i$. Further discussion of this important point is deferred to Section 7.1 on limiters.

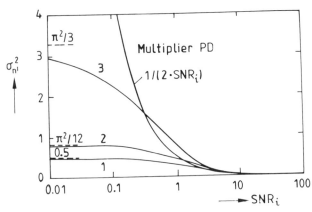

Figure 3.2-7. Variance $\sigma_{n'}^2$ of the effective phase noise $\theta_n(t)$ for different types of phase detectors. Same data as in Figure 3.2-5 (a) Multiplier PD with limiter in signal path; (b) multiplier with both input signals hard-limited; (c) RS flip-flop phase detector.

An analytical evaluation of the spectral density of $n'(t)$ does not seem possible. Based on experimental results one concludes that $S_{n'}(\omega)$ may be approximated by a rectangular spectrum with *two-sided* bandwidth B_1.

$$S_{n'}(\omega) = \begin{cases} \dfrac{\sigma_{n'}^2}{B_1}, & \text{for } \left|\dfrac{\omega}{2\pi}\right| < \dfrac{B_1}{2} \\ 0, & \text{otherwise} \end{cases} \qquad (3.2\text{-}37)$$

Figure 3.2-8 shows the ratio of B_1/B_{IF} as a function of SNR_i where B_{IF} is the equivalent noise bandwidth of the additive noise $n(t)$ before the limiter. A curve-fit to the experimental data of Figure 3.2-8 yields the empirical law

$$\frac{B_1}{B_{IF}} = 1 + A_i \exp\left[-SNR_i(1 - \pi/4)\right] \qquad (3.2\text{-}38)$$

where A_i is a constant depending on phase detector type (Table 3.2-1).

As a final topic we reconsider the frequency sensitive sequential logic phase detector of Figure 2.6-4. At first glance, we might be tempted to apply the theory just developed to this case. Such a procedure is erroneous since the phase detector has a memory which extends over more than one cycle, a property we explicitly ruled out. We recall that the PFD of Figure 2.6-4 operates on the zero crossings of the signal only. If the signal-to-noise ratio is large the zero crossings of the received signal will exhibit small random fluctuations around the nominal zero crossings. If the noise intensity increases, false zero crossings will occur. A missing signal transition or an extra signal transition has a particularly destructive effect on the PFD operation. Since the circuit has its own memory (to generate the frequency sensitive signal) the effect of such an event will propagate over many cycles.

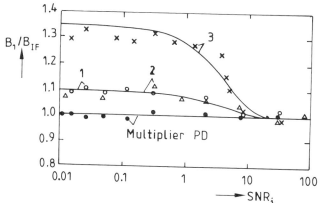

Figure 3.2-8. Bandwidth ratio of the phase noise θ_n to the bandwidth of the additive noise $n(t)$. Same data as in Figure 3.2-5 (a) Multiplier PD with limiter in signal path; (b) multiplier with both input signals hard-limited; (c) RS flip-flop phase detector.

If the loop is tracking with a small phase error, a missing signal transition is interpreted as a large phase error and it takes many cycles before the loop reaches equilibrium again.

Main Points of the Section

- The filtered output of a multiplier type phase detector in the presence of additive noise is

$$x(t) = K_D[\sin \phi(t) + n'(t, \hat{\theta})] \tag{3.2-3}$$

(note the sinusoidal characteristic of this PD) with

$$n'(t, \hat{\theta}) = \frac{n_c(t)}{A} \cos \hat{\theta} + \frac{n_s(t)}{A} \sin \hat{\theta} = \frac{1}{A} \operatorname{Re}[\underline{n}_L e^{-j\hat{\theta}}]$$

$$\underline{n}_L(t) = n_c(t) + jn_s(t)$$

At low signal-to-noise ratios multiplier type phase detectors are employed almost exclusively.

- Additive noise ($|\phi| \ll 2\pi$ for most of the time) generates a random phase disturbance $n'(t, \hat{\theta})$

$$x(t) = K_D[\phi(t) + n'(t, \hat{\theta})] \tag{3.2-4}$$

- Statistical properties of $n'(t, \hat{\theta})$: for a slowly varying VCO phase $\hat{\theta}$ we find (see Appendix 3.2A for conditions)

$$R_{n'}(\tau) = \frac{R_{n_c}(\tau)}{A^2} \tag{3.2-6}$$

and

$$S_{n'}(\omega) = \frac{S_{n_c}(\omega)}{A^2}$$

The statistics of $n'(t, \hat{\theta})$ are independent of $\hat{\theta}$, when the auto- and cross-correlation function of $n_c(t)$ and $n_s(t)$ satisfy (3.1-5).

- The variance $\sigma_{n'}^2$ equals

$$\sigma_{n'}^2 = \frac{1}{2(\text{SNR}_i)}, \quad \text{SNR}_i = \frac{P_s}{P_n} \tag{3.2-7}$$

Only *half* of the noise power appears in $\sigma_{n'}^2$. This is due to the fact that in practice only the quadrature component $N_c(t)$ causes the random phase fluctuation.

- Under the condition that the limiter phase detector has no memory extending over more than one cycle, the output of the nonsinusoidal phase detector equals

$$x(t) = K_D[g(\phi) + n'(t, \phi)] \qquad (3.2\text{-}27)$$

Remark: The phase/frequency detector does *not* fulfill the above condition of being memoryless.

- The effective *phase detector characteristic* $g(\phi)$ is a function of SNR$_i$ which equals the average of the noiseless phase detector characteristic over all possible values of the phase noise $\theta_n(t)$ (hard-limited input signal is assumed).

$$g(\phi) = \int_{-\pi}^{\pi} g_0(\phi + \theta_n) p(\theta_n)\, d\theta_n \qquad (3.2\text{-}28)$$

where $g_0(\cdot)$ is the noise-free phase detector characteristic (normalized to $g_0'(0) = 1$) and $p(\theta_n)$ is the PDF of θ_n in $[-\pi, \pi]$.

- In the presence of strong noise, $g(\phi)$ tends toward a sinusoidal form (Figure 3.2-5).
- Spectral density $S_{n'}(\omega)$ is approximately flat with bandwidth B_1. The bandwidth B_1 is only slightly larger than the bandwidth B_{IF}.

3.3. ADDITIVE NOISE IN THE LINEAR MODEL

If the phase error remains small most of the time we may linearize the nonlinearity in the phase detector model. The linearized model of the PLL in the presence of noise is shown in Figure 3.3-1. It is evident that $n'(t)$ is an additive disturbance to the phase $\theta(t)$, the quantity to be tracked by the PLL.

As long as we are dealing with a linear system the supersposition principle holds and we may determine the effect of noise and of the useful signal independently. Thus, assuming $\theta = 0$ and using n' instead of θ in (2.8-4) we obtain for the spectrum of the VCO phase

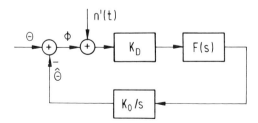

Figure 3.3-1. Linearized model of the phase-locked loop in the presence of additive noise.

$$S_{\hat{\theta}}(\omega) = |H(\omega)|^2 S_{n'}(\omega) \tag{3.3-1}$$

where $H(\omega)$ is the closed-loop transfer function. Since the phase error ϕ is equal to $-\hat{\theta}$, when $\theta = 0$ we obtain from (3.3-1) for the phase error variance σ_ϕ^2

$$\sigma_\phi^2 = \frac{1}{2\pi} \int_{-\infty}^{\infty} |H(\omega)|^2 S_{n'}(\omega) \, d\omega \tag{3.3-2}$$

In many cases of practical interest $S_{n'}(\omega)$ is nearly constant for all frequencies where $|H(\omega)|^2$ is significantly different from zero and we are allowed to approximate $S_{n'}(\omega)$ by a constant for all frequencies of interest

$$S_{n'}(\omega) \approx \frac{S_{n_c}(0)}{A^2} = \frac{N_0}{2A^2} \tag{3.3-3}$$

N_0 is called the *one-sided spectral density*, $N_0/2$ is the *two-sided spectral density* (see (3.1-12)). Using (3.3-3) we obtain

$$S_\phi(\omega) = \frac{N_0}{2A^2} |H(\omega)|^2$$

$$\sigma_\phi^2 = \frac{N_0}{2A^2} \frac{1}{2\pi} \int_{-\infty}^{\infty} |H(\omega)|^2 \, d\omega \tag{3.3-4}$$

Proceeding exactly as in the case for the noise bandwidth B_{IF} (see (3.1-11)), we define the *loop noise bandwidth* B_L measured in Hertz

$$B_L = \frac{1}{2\pi} \frac{\int_0^{\infty} |H(\omega)|^2 \, d\omega}{|H(0)|^2} \quad Hz \tag{3.3-5}$$

so that the phase error variance can be written as

$$\sigma_\phi^2 = \frac{N_0 B_L}{A^2} \tag{3.3-6}$$

Note that throughout this section, the symbol ϕ represents the *additional* phase error generated by insertion of the noise source $n'(t)$ in the model of Figure 3.3-1. The loop noise bandwidth is tabulated for various configurations in Table 3.3-1.

The inverse of the phase error variance σ_ϕ^2 is often called *signal-to-noise ratio in the loop* and given the symbol ρ

$$\rho = \frac{1}{\sigma_\phi^2} = \frac{A^2}{N_0 B_L} \tag{3.3-7}$$

TABLE 3.3-1 Loop noise bandwidth B_L*

Loop Filter $F(s)$	Closed Loop Transfer Function $H(s)$	Loop Noise Bandwidth B_L Hz
1	$\dfrac{K_0 K_D}{s + K_0 K_D}$	$\dfrac{K_0 K_D}{4}$
$\dfrac{1 + sT_2}{sT_1}$	$\dfrac{1 + 2\zeta(s/\omega_n)}{(s/\omega_n)^2 + 2\zeta(s/\omega_n) + 1}$	$\dfrac{\omega_n \zeta}{2}\left(1 + \dfrac{1}{4\zeta^2}\right)$
$\dfrac{1 + sT_2}{1 + sT_1}$	$\dfrac{1 + (2\zeta - \beta)(s/\omega_n)}{(s/\omega_n)^2 + 2\zeta(s/\omega_n) + 1}$	$\dfrac{\omega_c \zeta}{2}\left(1 - \dfrac{\beta}{\zeta} + \dfrac{1 + \beta^2}{4\zeta^2}\right)$
$\dfrac{(1 + sT_2)^2}{(sT_1)^2}$	$\dfrac{\alpha^2 + 2\alpha(s/\omega_c) + (s/\omega_c)^2}{\alpha^2 + 2\alpha(s/\omega_c) + (s/\omega_c)^2 + (s/\omega_c)^3}$	$\dfrac{\omega_c}{4} \dfrac{2\alpha + \alpha^2}{2\alpha - \alpha^2}$

*Recall that for second-order loops, $\beta = 1/\omega_n T_1$ (imperfect), ω_n and ζ as given in Table 2.4-1. For third-order loops with $T_3 = T_2$, $\omega_c = KA(T_2/T_1)^2$ and $\alpha = 1/\omega_c T_2$.

The definition is arbitrary, since ρ has no direct physical meaning (there is no signal with amplitude $\sqrt{2}A$ inside the loop). Nevertheless, it is quite a useful parameter. Because of its arbitrary character, the definition is not unique in the PLL literature. It is used by Viterbi [4] and Blanchard [2]. In the form (3.3-7) Gardner [3] has a factor of $1/2$ in his definition. There also exist various definitions of the noise bandwidth and of N_0. In this book, spectra are always defined from $-\infty$ to $+\infty$, as in mathematics. For convenience, the definitions are summarized in Figure 3.3-2.

Example: Second-Order PLL with Imperfect Integrator
Assume the following PLL parameters
Loop Filter

$$F(s) = \frac{1 + sT_2}{1 + sT_1}$$

$$T_1 = 25.7 \text{ ms}, \quad K_0 = 66\,881 \; s^{-1}\,V^{-1}$$

$$T_2 = 1.09 \text{ ms}, \quad K_D = 0.587 \text{ V}$$

Natural frequency ω_n, damping ratio ζ, β and loop bandwidth B_L can then be found

$$\omega_n = 1236 \; s^{-1}$$

$$\zeta = 0.69$$

$$B_L = 631.4 \text{ Hz}$$

$$\beta = 0.031$$

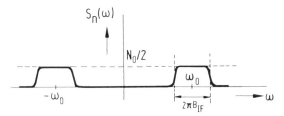

Figure 3.3-2a. Spectrum of the noise process $n(t)$ and related parameters for the multiplier phase detector. Equivalent bandwidth, $B_{IF} = [1/(2\pi)] \int_0^\infty S_{n_c}(\omega)/S_{n_c}(0) \, d\omega$; noise power, $P_n = [1/(2\pi)] \int_{-\infty}^\infty S_n(\omega) \, d\omega = N_0 B_{IF}$; Signal power, $P_S = A^2$; Signal-to-noise ratio, $SNR_i = P_S/P_n = A^2/(N_0 B_{IF})$.

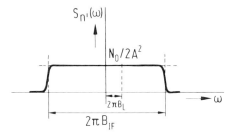

Figure 3.3-2b. Spectrum of $n'(t)$ and related parameters. Noise variance, $\sigma_{n'}^2 = N_0 B_{IF}/(2A^2) = 1/(2SNR_i)$; phase error variance, $\sigma_\phi^2 = N_0 B_L/A^2 = 1/\rho = (1/SNR_i)(B_L/B_{IF})$; loop noise bandwidth, $B_L = [1/(2\pi)] \int_0^\infty |H(\omega)|^2 \, d\omega/|H(0)|^2$ Hz; Noise suppression of the loop, $\rho/SNR_i = B_{IF}/B_L$.

IF-filter bandwidth and input signal-to-noise ratio are

$$B_{IF} = 15\,000 \text{ Hz}$$

$$SNR_i = 0 \text{ dB}$$

The ratio of the loop bandwidth B_L to the noise bandwidth B_{IF} is

$$\frac{B_{IF}}{B_L} = 23.7$$

From (3.1-15)

$$SNR_i = \frac{A^2}{N_0 B_{IF}}$$

we obtain for the two-sided noise spectral density $N_0/2$

$$\frac{N_0}{2} = \frac{A^2}{2SNR_i B_{IF}}$$

Inserting $N_0/2$ into the definition of the signal-to-noise ratio in the loop (3.3-7) yields

$$\rho = \frac{A^2}{N_0 B_L} = \mathrm{SNR}_i\left(\frac{B_{\mathrm{IF}}}{B_L}\right) = 13.75 \text{ dB}$$

Therefore an SNR of 0 dB at the PLL input corresponds to an SNR in the loop of 13.75 dB, showing the ability of the PLL to cope with large amounts of noise.

As a final topic we derive the linear model for the general phase detector characteristic $g(\phi)$. Linearizing the effective phase detector characteristic (3.2-27) about $\phi = 0$ yields*

$$x(t) = K_D g'(0)\left[\phi + \frac{n'(t)}{g'(0)}\right] \tag{3.3-8}$$

All the previous relations apply if the following replacement takes place

1. $K_D \to K_D g'(0)$

2. $S_{n'}(0) = \dfrac{N_0}{2A^2} \to \dfrac{\sigma_{n'}^2}{B_1}\dfrac{1}{[g'(0)]^2}$ (Recall (3.2-37)) $\tag{3.3-9}$

The values of $g'(0)$, B_1 and $\sigma_{n'}^2$ are found by numerically evaluating the respective formulas in Table 3.2-1 or by using Figures 3.2-5–3.2-8.

Example: Phase Error Variance of a First-Order PLL
Using (3.3-3)–(3.3-6) we have

$$\sigma_\phi^2 - S_{n'}(0)2B_L$$

Inserting the second expression on the right-hand side of (3.3-9) for $S_{n'}(0)$ and for $B_L = [K_0 K_D g'(0)]/4$ yields

$$\sigma_\phi^2 = \frac{\sigma_{n'}^2}{g'^2(0)B_1}\frac{K_0 K_D g'(0)}{4} = \frac{\sigma_{n'}^2 K_0 K_D}{2g'(0)B_1} \tag{3.3-10}$$

It is interesting to relate σ_ϕ^2 of (3.3-10) to that of the multiplier phase detector without limiters

$$(\sigma_\phi^2)_{\mathrm{Mult}} = \frac{1}{2\mathrm{SNR}_i}\left(\frac{K_0 K_D}{2B_{\mathrm{IF}}}\right) =: \frac{1}{\rho_0} \tag{3.3-11}$$

---

*It should be clear from the context whether a primed letter represents a derivative or not.

where ρ_0 is the signal-to-noise ratio in the loop for the multiplier phase detector. Combining (3.3-10) and (3.3-11) yields

$$\sigma_\phi^2 = \frac{\sigma_{n'}^2}{1/(2\text{SNR}_i)} \left(\frac{B_{\text{IF}}}{B_1} \right) \frac{1}{g'(0)} \frac{1}{\rho_0} = \frac{1}{\rho} \qquad (3.3\text{-}12)$$

The phase error variance σ_ϕ^2 is plotted in Figure 3.3-3 for the phase detectors of Figure 3.2-3 as a function of $1/\rho_0$. The variance is computed for the linear model and for the exact nonlinear model employing Fokker–Planck techniques (to be discussed in Chapters 9–12). The plot reveals

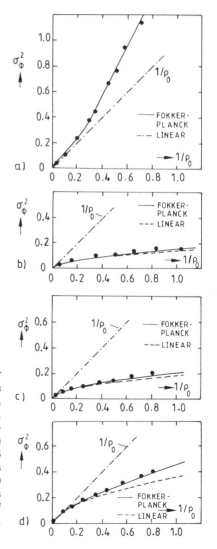

Figure 3.3-3. Phase error variance of a first-order PLL as a function of ρ_0 (SNR in the loop for a multiplier type PD without limiters). Dashed line, linear model; solid line, exact nonlinear model; •, experimental values. (a) Multiplier PD without limiter; (b) multiplier PD with limiter in signal path; (c) multiplier with both inputs hard-limited; (d) RS type flip-flop phase detector. Noise is modeled as symmetric, additive gaussian $K_0 K_D / 4B_{\text{IF}} = 1/50$ (From Werner Rosenkranz, Phase-Locked Loops with Limiter Phase Detectors in the Presence of Noise, *IEEE Transactions on Communications*, Vol. 30, No. 10 © 1982 IEEE).

several interesting features about the impact of limiters on loop tracking performance. For low ρ_0 the variance σ_ϕ^2 increases much more slowly than with $1/\rho_0$. The reason is that the effective phase detector gain decreases (signal suppression) and, thus, the effective bandwidth B_1 of the loop becomes narrower. The reduction of the phase error variance may also be achieved with a multiplier phase detector by an appropriate gain control. For limiter phase detectors the reduction is partly compensated by nonlinear effects which increase the noise variance $\sigma_{n'}^2$ (see the differences between the curves for linear and nonlinear models in Figure 3.3-3). These important facts are discussed further in the section on amplitude control.

Main Points of the Section

- If the phase error remains small most of time we may linearize the phase detector characteristic.
- The phase error variance σ_ϕ^2 equals

$$\sigma_\phi^2 = \frac{1}{2\pi} \int_{-\infty}^{\infty} |H(\omega)|^2 S_{n'}(\omega)\, d\omega \tag{3.3-2}$$

For many applications $S_{n'}(\omega)$ is constant within the range where $|H(\omega)|^2$ is significantly different from zero. Thus

$$S_{n'}(\omega) = S_{n'}(0)$$

for all frequencies of interest. The spectral density at the origin equals

$$S_{n'}(0) \begin{cases} \dfrac{N_0}{2A^2}, & \text{multiplier PD} \\[2ex] \dfrac{\sigma_{n'}^2}{B_1} \dfrac{1}{[g'(0)]^2}, & \text{for a general PD, see (3.2-37) for } B_1 \end{cases}$$

and the phase error variance becomes

$$\sigma_\phi^2 = \frac{A^2}{N_0 B_L} = \left(\frac{B_L}{B_{\mathrm{IF}}}\right) \frac{1}{\mathrm{SNR}_i}, \quad \text{multiplier PD}$$

For the general PD the phase detector gain equals $K_D g'(\phi)$ and σ_ϕ^2 becomes

$$\sigma_\phi^2 = S_{n'}(0) 2B_L = \frac{\sigma_{n'}^2}{B_1 [g'(0)]^2}\, 2 K_0 K_D g'(0)$$

3.4. STATE VARIABLE EQUATIONS IN THE PRESENCE OF ADDITIVE NOISE

To obtain the nonlinear baseband model of the PLL in the presence of additive noise we only have to replace the model of the phase detector in the noiseless case by the model shown in Figure 3.2-1b. The resulting model for the second-order PLL with perfect integrator and phase detector characteristic $g(\phi)$ is shown in Figure 3.4-1.

Proceeding exactly as in the noiseless case, the dynamic equations for the state vector $x^T = [\hat{\theta}, y_1]$ are found from Figure 3.4-1 to be

$$\frac{d\hat{\theta}(t)}{dt} = \frac{T_2}{T_1} K_0 K_D \{ g[\phi(t)] + n'(t) \} + y_1(t)$$

$$\frac{dy_1(t)}{dt} = \frac{K_0 K_D}{T_1} \{ g[\phi(t)] + n'(t) \}$$

$$(3.4\text{-}1)$$

Introducing the usual normalized and dimensionless variables (see Section 2.4.3)

$$\tau = \omega_n t, \quad x_1 = \frac{y_1}{\omega_n}$$

with

$$\omega_n = \left(\frac{K_0 K_D}{T_1} \right)^{1/2}, \quad \zeta = \frac{T_2}{2} \omega_n \qquad (3.4\text{-}2)$$

we arrive after some simple algebraic manipulations at

$$\frac{d\phi(t)}{d\tau} = -2\zeta g[\phi(\tau)] - x_1(\tau) - 2\zeta n'(\tau) + \frac{d\theta}{d\tau}$$

$$\frac{dx_1(t)}{d\tau} = g[\phi(\tau)] + n'(\tau)$$

$$(3.4\text{-}3)$$

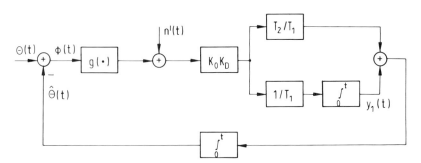

Figure 3.4-1. Nonlinear baseband model of the perfect second-order phase-locked loop in the presence of additive noise.

We repeat that ω_n and ζ are formally defined in the same way as for a linear system, but no longer have their physical meaning.

Let us now see how the time normalization affects the statistical properties of the noise. In order to simplify the notation we have written $n'(\tau)$ instead of the correct notation $n'(\tau/\omega_n)$. This abuse of notation will also hold throughout the following chapters. But to derive the relation between the statistical parameters of the noise process in real time and in normalized time we must distinguish them notationally and therefore for the moment introduce the new symbol $m(\tau)$ for the noise process in normalized time

$$m(\tau) = n'\left(\frac{\tau}{\omega_n}\right) \tag{3.4-4}$$

The correlation function of $m(\tau)$ is defined as

$$R_m(\tau_1) = E[m(\tau + \tau_1)m(\tau)] \tag{3.4-5}$$

In (3.4-5) we had to use τ_1 for the time shift variable since the usual variable τ has already been used for the normalized time. Replacing $m(\tau)$ by the right-hand side of (3.4-4) yields

$$R_m(\tau_1) = E\left[n'\left(\frac{\tau + \tau_1}{\omega_n}\right)n'\left(\frac{\tau}{\omega_n}\right)\right] = R_{n'}\left(\frac{\tau_1}{\omega_n}\right) \tag{3.4-6}$$

The correlation function of the noise $m(\tau)$ in normalized time equals the correlation function in real time, $R_{n'}(t_1)$, taken at $t_1 = \tau_1/\omega_n$.

The spectral density $S_{n'}(\omega)$ of the original process is

$$S_{n'}(\omega) = \int_{-\infty}^{\infty} R_{n'}(t_1) \exp(-j\omega t)\, dt_1 \tag{3.4-7}$$

Replacing t_1 by τ_1/ω_n we find, using (3.4-6)

$$\begin{aligned} S_{n'}(\omega) &= \int_{-\infty}^{\infty} R_{n'}\left(\frac{\tau_1}{\omega_n}\right) \exp\left[-j\left(\frac{\omega}{\omega_n}\right)\tau_1\right] d\left(\frac{\tau_1}{\omega_n}\right) \\ &= \frac{1}{\omega_n} \int_{-\infty}^{\infty} R_m(\tau_1) \exp\left[-j\left(\frac{\omega}{\omega_n}\right)\tau_1\right] d\tau_1 \end{aligned}$$

Introducing the normalized frequency

$$\Omega = \frac{\omega}{\omega_n} \tag{3.4-8}$$

yields

$$S_{n'}(\Omega\omega_n) = \frac{1}{\omega_n} \int_{-\infty}^{\infty} R_m(\tau_1) \exp(-j\Omega\tau_1)\, d\tau_1 \tag{3.4-9}$$

But the integral on the right-hand side of (3.4-9) is, by definition, the spectral density corresponding to $R_m(\tau_1)$. Thus, the spectral density in the normalized frequency Ω is related to $S_{n'}(\omega)$ by

$$S_{n'}(\Omega \omega_n) = \frac{1}{\omega_n} S_m(\Omega) \qquad (3.4\text{-}10)$$

The correspondence between the noise process descriptions in real and normalized time is summarized in Table 3.4-1 where we have used a general time normalization $\tau = Bt$.

Example: Let $n'(t)$ be a white noise process with

$$S_{n'}(\omega) = \frac{N_0}{2A^2}$$

The corresponding spectrum $S_m(\Omega)$ is, with $B = \omega_n$

$$S_m(\Omega) = \omega_n \frac{N_0}{2A^2} \qquad (3.4\text{-}12)$$

Using

$$B_L = \frac{\omega_n \zeta}{2} \left(1 + \frac{1}{4\zeta^2}\right)$$

from Table 3.3-1 to solve for ω_n and introducing the result into (3.4-12) yields

TABLE 3.4-1 **Correspondence between noise process descriptions in real and normalized time***

	Real Time	Normalized Time
Variables	t	$\tau = Bt$
	ω	$\Omega = \dfrac{\omega}{B}$
Noise process	$n'(t)$	$m(\tau) = n'\left(\dfrac{\tau}{B}\right)$
Correlation function	$R_{n'}(t_1)$	$R_m(\tau_1) = R_{n'}\left(\dfrac{\tau_1}{B}\right)$
	(t_1, ω)	(τ_1, Ω)
Spectral Density	$S_{n'}(\omega)$	$S_m(\Omega) = BS_{n'}(\Omega B)$

*The general normalization factor B is best given by: $B = K_0 K_D = 4B_L$ (first-order loop); $B = \omega_n$ (second-order loop); $B = \omega_c$ (third-order loop).

$$S_m(\Omega) = \frac{1}{\rho\zeta} \frac{1}{1 + (1/4\zeta^2)} \qquad (3.4\text{-}13)$$

where ρ is the signal-to-noise ratio in the loop given by (3.3-7). We notice that ρ is formally defined as for a linear system without having a physical meaning in nonlinear theory.

For $d\theta/d\tau = \text{constant}$, (3.4-3) is a nonlinear differential equation with a random driving term. For the moment it suffices to say that as long as $n'(\tau)$ is a physically realizable process, we can obtain a solution of (3.4-3) for every possible sample function of $n'(\tau)$. The driving term is treated in the same manner as a deterministic function. The exact mathematical theory can be found in the literature under mean square calculus and is treated in most elementary books on stochastic processes (see for example, Papoulis [1: Chapter 9.6]). The mathematical sophistication is greatly increased if we try to replace $n'(\tau)$ by a white noise process. The pertinent theory is treated in Chapters 9–12 of this book.

It should be emphasized that for a great part of this book the reader does not require knowledge of these advanced topics. We will, however, often quote results attainable using these mathematical tools only.

Main Points of the Section

- The operation of a PLL is described (3.4-3) by a nonlinear system of first-order differential equations with a random driving term $n'(\cdot)$.
- The spectral density and the correlation function of the noise process in real time and normalized time are related by

$$R_m(\tau_1) = R_{n'}\left(\frac{\tau_1}{B}\right)$$

$$S_m(\Omega) = B S_{n'}(\Omega B)$$

Time normalization: $\tau = Bt$

3.5. TIME AND FREQUENCY STABILITY OF SIGNAL GENERATORS

So far we have taken the VCO as a perfect oscillator. We assumed ω_0 and $\hat{\theta}$ to be constant in the absence of a control signal $e(t)$. However, even precision frequency sources such as quartz oscillators, masers, and passive atomic frequency standards are affected by phase and frequency instabilities including both random and determinsitic components. It is of importance to have a characterization of these instabilities in order to be able to assess the potential use of each source. This section is devoted to introducing a mathematical model for the oscillator quasi-sinusoidal output signal since frequency instability immediately implies that the signal is no longer a pure sine wave. Before establishing this model, it is useful to make some comments on determinisitic and, in contrast, random variations of the oscillator output frequency.

Systematic variations, also known as drifts or trends, may be due to aging of the resonator material (e.g., in quartz oscillators). These sometimes extremely slow changes are often referred to as "long-term" instabilities and are expressed in terms of parts of 10^x (e.g., $x = 9$) of frequency change per hour, day, month or year. No statistical treatment is needed for the evaluation.

Random fluctuations due to noise sources such as thermal, shot and flicker noise are encountered in electronic components. The related frequency fluctuations are often referred to as "short-term" instabilities since they become more and more significant when shorter time intervals are considered. Due to their random nature, statistical treatment is needed for their characterization.

In what follows, characterization of time interval and time interval stability are given for the models of a perfect and an imperfect oscillator. These basic concepts will then lead us to the definition of useful performance measures for frequency sources in both the time and frequency domain.

3.5.1. Characterization of Time Properties

Time Process Generated by Oscillators. The output of an oscillator may be represented by

$$s(t) = \sqrt{2}A \sin \Phi(t) \qquad (3.5\text{-}1)$$

where $\Phi(t)$ is the total phase. The instantaneous frequency equals

$$f(t) = \frac{1}{2\pi} \frac{d\Phi(t)}{dt} \qquad (3.5\text{-}2)$$

which is usually modeled in the form

$$\frac{1}{2\pi} \frac{d\Phi(t)}{dt} = f_0 + \Delta f + \Delta \dot{f} t + \frac{1}{2\pi} \dot{\Psi}(t) \qquad (3.5\text{-}3)$$

A perfect oscillator has a constant frequency f_0. For an imperfect oscillator the desired frequency is distorted by additional deterministic terms representing the oscillator *frequency settability* Δf (in Hz) and *drift rate* $\Delta \dot{f}$ (Hz/s) and the oscillastor random phase noise process $\Psi(t)$ (rad). We assume that the radom process $\Psi(t)$ has zero mean.

These distortions are all intrinsic instabilities of an oscillator. The deviations caused by environmental variations (e.g., temperature, accelerations, magnetic field) are not represented in this model.

We will require a few definitions in what follows. The *instantaneous normalized frequency* is defined by

$$u(t) := \frac{1}{2\pi f_0} \frac{d\Phi}{dt} \tag{3.5-4}$$

The *normalized fractional frequency deviation* $y(t)$ describes the random fluctuations of the instantaneous frequency caused by the phase noise

$$y(t) := \frac{1}{2\pi f_0} \frac{d\Psi}{dt} \tag{3.5-5}$$

Finally we introduce the *time function* $T(t)$

$$T(t) := \frac{\Phi(t)}{2\pi f_0} \tag{3.5-6}$$

The zero crossings of an oscillator may be used to mark a set of points along $T(t)$ which defines units of time and thus generates an *observable time axis*. A perfect oscillator generates the time

$$T_0(t) = t - t_0 \tag{3.5-7}$$

The observable time axis $T_0(t)$ is the same as that of the reference source (i.e., the zero crossings are $1/f_0$ apart) apart from a constant which takes into account the difference between the time origins. On the other hand, an imperfect oscillator described by (3.5-3) generates

$$T(t) = t - t_0 + \frac{\Delta f}{f_0} t + \frac{\Delta \dot{f}}{f_0} \frac{t^2}{2} + \frac{\Psi(t) - \Psi(t_0)}{2\pi f_0} \tag{3.5-8}$$

The instantaneous normalized frequency and the normalized fractional frequency deviation for this imperfect oscillator is easily obtained by inserting (3.5-3) into the respective definition

$$u(t) = 1 + \frac{\Delta f}{f_0} + \frac{\Delta \dot{f}}{f_0} t + \frac{\dot{\Psi}(t)}{2\pi f_0} \tag{3.5-9}$$

and

$$y(t) = \frac{1}{2\pi f_0} \dot{\Psi}(t) \tag{3.5-10}$$

Time Interval (TI). Consider time intervals of length τ marked along the time axis (Figure 3.5-1). The mapping of these segments on the observed time axis generated by $T(t)$ leads to the definition of the generated time interval (TI)

$$\Delta^1 T(t; \tau) := T(t + \tau) - T(t) \tag{3.5-11}$$

The time interval TI is also called the first increment $\Delta^1 T(t; \tau)$ of the time function $T(t)$. Notice that

$$\lim_{\tau \to 0} \frac{\Delta^1 T(t; \tau)}{\tau} = \lim_{\tau \to 0} \left[\frac{T(t + \tau) - T(t)}{\tau} \right] = \frac{dT(t)}{dt} = u(t) \tag{3.5-12}$$

relates TI to the instantaneous normalized frequency $u(t)$ of (3.5-4).

For the perfect oscillator

$$\Delta^1 T_0(t; \tau) = (t + \tau - t_0) - (t - t_0) = \tau \tag{3.5-13}$$

which says that the length of the TIs generated are independent of t, i.e., the frequency is stable ($u(t) = 1$ as shown before). The zero crossings are equidistant and

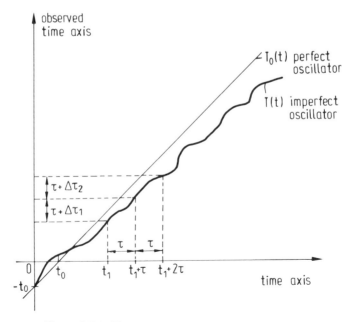

Figure 3.5-1. Time process generated by oscillators.

$$\Delta^1 T_0(t; 1/f_0) = 1/f_0$$

apart and thus define a unit of time on the time axis generated by the oscillator.

For the imperfect oscillator of (3.5-8)

$$\Delta^1 T(t; \tau) = \tau + \frac{\Delta f}{f_0}\,\tau + \frac{\Delta \dot{f}}{2 f_0}\,(2\tau t + \tau^2) + \frac{\Delta^1 \Psi(t; \tau)}{2\pi f_0} \qquad (3.5\text{-}14)$$

where $\Delta^1 \Psi(t; \tau)$ represents the first increment of the phase noise process. The mean of (3.5-14) characterizes the average length of the time interval generated by the oscillator during the time interval $[t, t + \tau]$ and is given by

$$E[\Delta^1 T(t; \tau)] = \tau + \frac{\Delta f}{f_0}\,\tau + \frac{\Delta \dot{f}}{2 f_0}\,(2\tau t + \tau^2) \qquad (3.5\text{-}15)$$

It is time-varying and its first term relates to the idealization depicted in (3.5-13). The other terms represent an average expansion or contraction of the TI generated as time passes. Note that in (3.5-15) the term related to $\Psi(t)$ does not appear since we assumed that $\Psi(t)$ is a zero-mean process. Now the mean-squared value of (3.5-14)

$$D_T^{(1)}(t; \tau) := E\{[\Delta^1 T(t; \tau)]^2\} \qquad (3.5\text{-}16)$$

is defined as *first structure function* of the generated time process and is found to be

$$D_T^{(1)}(t; \tau) = \frac{D_\Psi^{(1)}(\tau)}{(2\pi f_0)^2} + \{E[\Delta^1 T(t; \tau)]\}^2 \qquad (3.5\text{-}17)$$

The variance of the TI is therefore

$$\text{Var}\,[\Delta^1 T(t; \tau)] = \frac{D_\Psi^{(1)}(\tau)}{(2\pi f_0)^2} \qquad (3.5\text{-}18)$$

Where $D_\Psi^{(1)}(\tau)$ is the first structure function of the phase noise process $\Psi(t)$.

If $\Delta^1 T(t; \tau)$ of (3.5-14) were stationary (e.g., $\Delta^1 T(\tau) = \tau + (\Delta f/f_0)\tau$) then $D_T^{(1)}(t; \tau) = D_T^{(1)}(\tau)$ would be time-independent. This is equivalent to saying that the generated frequency of the source would be constant. Recall also from (3.5-12) that the normalized first increment $[\Delta^1 T(t; \tau)]/\tau$ is related to the instantaneous normalized frequency $u(t)$ of the oscillator.

Time Interval Error (TIE). In order to define time interval error one must assume a reference source (or standard), say $T_r(t)$, from which the error may be derived. Then the reference time interval is $\Delta^1 T_r(t; \tau)$ and the time interval error (TIE) generated during the time interval $[t, t + \tau]$ is defined by

$$\Delta^1 T_e(t; \tau) := \Delta^1 T(t; \tau) - \Delta^1 T_r(t; \tau) \qquad (3.5\text{-}19)$$

$$\underbrace{\hspace{4cm}}_{\substack{\text{TI gene-}\\\text{rated by}\\\text{source}}} \qquad \overbrace{\hspace{3cm}}^{\text{Reference TI}}$$

If we take our perfect oscillator as the reference source, then using (3.5-13) in (3.5-19) gives

$$\Delta^1 T_e(t; \tau) = \Delta^1 T(t; \tau) - \tau \qquad (3.5\text{-}20)$$

For the perfect oscillator $\Delta^1 T_e(t; \tau) = \tau - \tau = 0$. This says that an ideal oscillator generates zero TIE with the passage of time.

For the imperfect oscillator, substitution of (3.5-14) into (3.5-20) gives the TIE as

$$\Delta^1 T_e(t; \tau) = \frac{\Delta f}{f_0}\, \tau + \frac{\Delta \dot{f}}{2 f_0}\, (2\tau t + \tau^2) + \frac{\Delta^1 \Psi(t; \tau)}{2\pi f_0} \qquad (3.5\text{-}21)$$

The mean TIE

$$E[\Delta^1 T_e(t; \tau)] = \frac{\Delta f}{f_0}\, \tau + \frac{\Delta \dot{f}}{2 f_0}\, (2\tau t + \tau^2) \qquad (3.5\text{-}22)$$

is time-varying as is its mean-squared value

$$D_e^{(1)}(t; \tau) = \frac{D_\Psi^{(1)}(\tau)}{(2\pi f_0)^2} + \{E[\Delta^1 T_e(t; \tau)]\}^2 \qquad (3.5\text{-}23)$$

The variance of the TIE is then

$$\mathrm{Var}\,[\Delta^1 T_e(t; \tau)] = \frac{D_\Psi^{(1)}(\tau)}{(2\pi f_0)^2} \qquad (3.5\text{-}24)$$

It is interesting to evaluate the average amount of time it takes for the TIE to accumulate τ seconds. Setting the right-hand side of (3.5-22) equal to τ the amount of time required for the mean TIE to reach τ seconds is

$$t = \frac{1 - (\Delta f / f_0) - (\Delta \dot{f} \tau / 2 f_0)}{\Delta \dot{f} / f_0} \qquad (3.5\text{-}25)$$

Assuming that $\tau = 1 / f_0$ (one period), for a good oscillator the normalized settability, $\Delta f/f_0$, and drift rate, $\Delta \dot{f}/f_0^2$, are both much smaller than one and the numerator of (3.5-25) can be approximated to 1 so that

$$t \approx \left(\frac{\Delta \dot{f}}{f_0} \right)^{-1}$$

(3.5-26)

which is just the inverse of the oscillator normalized drift rate.

From (3.5-22) we observe that the build-up in the normalized mean TIE (assuming $t \gg \tau$)

$$E\left[\frac{\Delta^1 T_e(t; \tau)}{\tau} \right] \approx \frac{\Delta f}{f_0} + \frac{\Delta \dot{f}}{f_0} t$$

(3.5-27)

grows up linearly with time and is determined by the settability and the drift rate of the oscillator.

Time Interval Stability (TIS). Consider now the mapping of consecutive intervals $[t, t + \tau]$, $[t + \tau, t + 2\tau]$ of length τ along the time axis onto the corresponding time intervals $\Delta^1 T(t; \tau)$ and $\Delta^1 T(t + \tau; \tau)$, respectively, generated along the $T(t)$ axis. The difference in the length of these intervals

$$\Delta^2 T(t; \tau) = \Delta^1 T(t + \tau; \tau) - \Delta^1 T(t; \tau)$$

$$= T(t + 2\tau) - 2T(t + \tau) + T(t)$$

(3.5-28)

represents second-order increments of the function $T(t)$ versus time, characterizing *time interval stability* (TIS). Notice from (3.5-28) that the operation

$$\lim_{\tau \to 0} \frac{\Delta^2 T(t; \tau)}{\tau^2} = \lim_{\tau \to 0} \left[\frac{T(t + 2\tau) - 2T(t + \tau) + T(t)}{\tau^2} \right]$$

$$= \frac{d^2 T(t)}{dt^2}$$

(3.5-29)

when the limit exists, relates the TIS to the instantaneous normalized frequency drift

$$\frac{1}{2\pi f_0} \frac{d^2 \Phi(t)}{dt^2} = \frac{d^2 T(t)}{dt^2} = \ddot{u}(t)$$

(3.5-30)

the above equation following from (3.5-4) and (3.5-6).

For the perfect oscillator we trivially have

$$\Delta^2 T(t; \tau) = 0 - 0 = 0$$

(3.5-31)

indicating that the frequency of the perfect oscillator does not drift with the passage of time and the generated time intervals are stable.

For the imperfect oscillator substitution of (3.5-14) into (3.5-28) gives the TIS as

$$\Delta^2 T(t; \tau) = \frac{\Delta \dot{f} \tau^2}{f_0} + \frac{\Delta^2 \Psi(t; \tau)}{2 \pi f_0} \tag{3.5-32}$$

where $\Delta^2 \Psi(t; \tau)$ is the second increment of the phase noise process $\Psi(t)$. The mean TIS is

$$E[\Delta^2 T(t; \tau)] = \frac{\Delta \dot{f} \tau^2}{f_0} \tag{3.5-33}$$

while its mean-squared value (called the second structure function of $T(t)$) and variance are

$$D_T^{(2)}(t; \tau) = \frac{D_\Psi^{(2)}(\tau)}{(2 \pi f_0)^2} + \{ E[\Delta^2 T(t; \tau)] \}^2$$

$$\mathrm{Var}\,[\Delta^2 T(t; \tau)] = \frac{D_\Psi^{(2)}(\tau)}{(2 \pi f_0)^2} \tag{3.5-34}$$

Notice that for our imperfect oscillator the mean and the variance of the TIS do not depend on the time t. This indicates that the time process has a stationary second increment. One can also easily conclude that the third structure function of $T(t)$ will not contain any systematic terms, i.e., $E[\Delta^3 T(t; \tau)] = 0$. Thus we have found a method of annihilating the effect of deterministic oscillator imperfections on the characterization of the time process by using higher order structure functions. The interested reader is referred to [10] for details.

Before leaving Section 3.5.1, we recall that the comments given about (3.5-12) and (3.5-29) indicate that statistical characterizations of TI and TIS characterize the frequency and frequency stability of the oscillator.

3.5.2. Standard Parameters Characterizing Random Fluctuations of Oscillators

The random disturbance $\Psi(t)$ in the time process $T(t)$ of (3.5-8) is the only contribution to the variances of the generated TI, TIE, and TIS given by (3.5.18), (3.5-24), and (3.5-34). In this section we concentrate on characterizing oscillator fluctuations caused by the phase noise process $\Psi(t)$.

RMS Fractional Frequency Deviation and Allan Variance. The root mean square (RMS) *fractional frequency deviation* is an IEEE accepted standard which accounts for the fluctuations in the TI and TIE due to the oscillator phase noise process $\Psi(t)$. In essence it specifies the standard deviation of the length of the generated time intervals relative to the length of the time interval generated by an ideal oscillator. The quotient $\{E[(\Psi(t+\tau) - \Psi(t))^2]\}^{1/2}/\tau$ is, for small τ, a measure for the angular frequency deviation. Dividing by 2π yields the frequency and finally normalizing on the "nominal" frequency f_0 gives the following definition for the normalized RMS fractional frequency deviation

$$\frac{\delta f(\tau)}{f_0} = \left\{ \frac{E[(\Psi(t+\tau) - \Psi(t))^2]}{(2\pi f_0)^2} \right\}^{1/2} \frac{1}{\tau} = \left[\frac{D_\Psi^{(1)}(\tau)}{(2\pi f_0 \tau)^2} \right]^{1/2} \quad (3.5\text{-}35)$$

The *Allan variance* is used to account for fluctuations in the time interval stability due to the oscillator phase noise $\Psi(t)$. It specifies the variance of the length of the difference in two adjacent time intervals relative to the square of the length 2τ which would be generated by an ideal oscillator. With the same normalization on f_0 as in (3.5-35) we find for the Allan variance

$$\sigma^2(\tau) = \frac{E\{[\Psi(t+2\tau) - \Psi(t+\tau)] - [\Psi(t+\tau) - \Psi(t)]\}^2}{(2\tau)^2 (2\pi f_0)^2}$$

$$= \frac{D_\Psi^{(2)}(\tau)}{4(2\pi f_0 \tau)^2} \quad (3.5\text{-}36)$$

We can also write

$$\frac{1}{\tau^2} \frac{D_\Psi^{(2)}(\tau)}{\tau^2} = E\left\{ \frac{[(\Psi(t+2\tau) - \Psi(t+\tau)/\tau] - [(\Psi(t+\tau) - \Psi(t))/\tau]}{\tau} \right\}^2 \quad (3.5\text{-}37)$$

Equation (3.5-37) indicates that the Allan variance is a measure for the normalized instantaneous frequency drift associated with $\Psi(t)$.

The Allan variance is measured in the time domain by frequency counter type measurements. Table 3.5-1 provides a comparison of current precision oscillator technologies giving their advantages and disadvantages. Figure 3.5-2 is a performance comparison in terms of the Allan variance.

3.5.3. Time Domain to Frequency Domain Interconnections

The RMS fractional frequency deviation and the Allan variance are time domain measures of imperfections of the oscillator. However, communication engineers are often concerned with frequency domain characterization of systems.

Since the RMS fractional frequency deviation and the Allan variance are defined in terms of the structure function of $\Psi(t)$ it is not surprising that the introduction of the spectral density of the phase noise process, $S_\Psi(\omega)$, will relate the time domain measures to frequency domain measures.

TABLE 3.5-1 Comparison of Precision oscillators

	Precision Quartz Oscillator	Efratom Rubidium Oscillator	Commercial Cesium-Beam	Passive Hydrogen Maser
Primary standard	No	Secondary 6.834, 682, 613 GHz	Yes 9.192, 631, 770 GHz	Yes 1.420, 450, 751 GHz
Fundamental wearout mechanism	Basically none	Basically none	CS contamination of electron multiplier	Ion pumps H_2 source depletion
Maintenance	Basically None	Basically none	Replace CS beam tube every 2–5 years	Ion pumps (5 years?) H_2 depletion (10 years + ?)
Allan variance $\sigma^2(\tau)$				
$\tau = 1$ s	$\approx 1 \times 10^{-22}/10^{-24}$	$\approx 1 \times 10^{-22}$	$\approx 2 \times 10^{-21}$	2 to 4×10^{-24}
$\tau = 1$ day	$\approx 1 \times 10^{-20}$	$\approx 1 \times 10^{-24}/10^{-26}$	4×10^{-26}	Parts in 10^{30}
Drift $\Delta f/f_0$				
Per day	$\approx 1 \times 10^{-10}$	$\approx 1 \times 10^{-12}/10^{-13}$	None	$<1 \times 10^{-15}$
Per year	$\approx 1 \times 10^{-7}$	Parts in 10^{10}	None	$<2 \times 10^{-13}$
Warm-up time to parts in 10^{10}	Hours	<2 min	<1 h	Hours

Source: From [22], with permission.

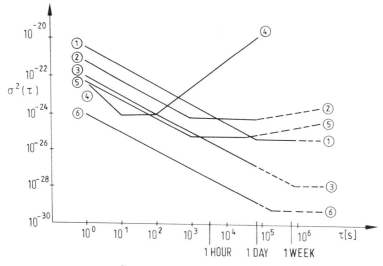

Figure 3.5-2. Allan variance $\sigma^2(\tau)$ versus averaging time τ (From [22], with permission). (1) Standard cesium; (2) standard efratom rubidium; (3) high-performance cesium; (4) standard quartz oscillator; (5) high-performance rubidium; (6) National Bureau of Standards (NBS)/ efratom passive hydrogen maser.

Consider the mean-squared value of the first increment of $\Psi(t)$

$$D_\Psi^{(1)}(\tau) = E[\Psi(t+\tau) - \Psi(t)]^2 = 2R_\Psi(0) - 2R_\Psi(\tau) \qquad (3.5\text{-}38)$$

where $R_\Psi(\tau)$ is the correlation function of the phase process $\Psi(t)$. This can be written in terms of the power spectral density of $\Psi(t)$ via

$$D_\Psi^{(1)}(\tau) = 2\,\frac{1}{2\pi} \int_{-\infty}^{+\infty} S_\Psi(\omega)\, d\omega - 2\,\frac{1}{2\pi} \int_{-\infty}^{+\infty} S_\Psi(\omega) \exp\,(j\omega\tau)\, d\omega$$

$$= 4\,\frac{1}{2\pi j} \int_{-\infty}^{+\infty} \exp\left(\frac{j\omega\tau}{2}\right) \sin\frac{\omega\tau}{2}\, S_\Psi(\omega)\, d\omega \qquad (3.5\text{-}39)$$

and since the spectral density is an even function

$$D_\Psi^{(1)}(\tau) = 4\,\frac{1}{2\pi} \int_{-\infty}^{+\infty} \sin^2\left(\frac{\omega\tau}{2}\right) S_\Psi(\omega)\, d\omega \qquad (3.5\text{-}40)$$

Introducing the spectral density of the first derivative of the phase process

$$S_{\dot\Psi}(\omega) = \omega^2 S_\Psi(\omega) \qquad (3.5\text{-}41)$$

which corresponds to the frequency process, we can rewrite equation (3.5-40)

$$D_{\Psi}^{(1)}(\tau) = 4 \frac{1}{2\pi} \int_{-\infty}^{+\infty} \sin^2 \left(\frac{\omega\tau}{2} \right) \frac{S_{\dot{\Psi}}(\omega)}{\omega^2} \, d\omega \tag{3.5-42}$$

In a similar manner the second structure function can be related to $S_{\dot{\Psi}}(\omega)$

$$D_{\Psi}^{(2)}(\tau) = 16 \frac{1}{2\pi} \int_{-\infty}^{+\infty} \sin^4 \left(\frac{\omega\tau}{2} \right) S_{\Psi}(\omega) \, d\omega$$

$$= 16 \frac{1}{2\pi} \int_{-\infty}^{+\infty} \sin^4 \left(\frac{\omega\tau}{2} \right) \frac{S_{\dot{\Psi}}(\omega)}{\omega^2} \, d\omega \tag{3.5-43}$$

Thus through the use of (3.5-40) and (3.5-43) in (3.5-35) and (3.5-36) we can relate the time domain parameters $(\delta f(\tau)/f_0)^2$ and $\sigma^2(\tau)$ to the frequency domain representation $S_{\Psi}(\omega)$

$$\left(\frac{\delta f(\tau)}{f_0} \right)^2 = \frac{D_{\Psi}^{(1)}(\tau)}{(2\pi f_0 \tau)^2} = \frac{1}{(2\pi f_0 \tau)^2} \int_{-\infty}^{+\infty} \frac{4}{2\pi} \sin^2 \left(\frac{\omega\tau}{2} \right) S_{\Psi}(\omega) \, d\omega \tag{3.5-44}$$

$$\sigma^2(\tau) = \frac{D_{\Psi}^{(2)}(\tau)}{4(2\pi f_0 \tau)^2} = \frac{1}{4(2\pi f_0 \tau)^2} \int_{-\infty}^{+\infty} \frac{16}{2\pi} \sin^4 \left(\frac{\omega\tau}{2} \right) S_{\Psi}(\omega) \, d\omega \tag{3.5-45}$$

3.5.4. Frequency Domain Model of Oscillator Phase Noise

In this section we will introduce a model of $S_{\Psi}(\omega)$. In practice, there are several noise components which spectrally combine to produce the total power spectral density of the phase noise process. These components possess spectral densities of the form $|\omega|^{-\alpha}$ where α is some real number. Such components possess infinite power when considered to exist for all frequencies and this has led to model pathologies. These pathologies arise because experimental results are extrapolated beyond their range of applicability and direct observation. The first such pathology arises from extrapolating measured behavior into the high-frequency region while the other pathology arises due to extrapolating measured data into the low-frequency region. Especially, measurement data taken over a time interval of $T_E = 1/f_l$ seconds with a frequency analyzer with high-frequency cut-off of f_h Hz leads one to the conclusion that the Fourier transform of the data is only meaningful in the frequency interval $[f_l, f_h]$. In short, the frequencies below f_l (no low-frequency cut-off in the model) are unobservable because of finite measurement times while frequencies larger than f_h (no high-frequency cut-off in the model) are not real because of finite bandwidth of any electronic measurement. In what follows it is therefore convenient to postulate that the form of the spectral density of the phase noise process and its derivative $\dot{\Psi}(t)$ be valid in the frequency interval $[f_l, f_h]$. This guarantees the wide-sense stationarity of $\Psi(t)$ and $\dot{\Psi}(t)$.

Based on a wealth of published data [12, 16], the (*two-sided!*) spectral density of the fractional frequency derivation $y(t)$ generally follows the form (see (3.5-5))

$$S_y(\omega) = \frac{1}{(2\pi f_0)^2} \, S_\psi(\omega) = h_{-2}|\omega|^{-2} + h_{-1}|\omega|^{-1} + h_0 + h_1|\omega| + h_2|\omega|^2$$

$$(3.5\text{-}46)$$

Thus we find for the spectral density of the phase noise process

$$S_\psi(\omega) \frac{1}{(2\pi f_0)^2} = \frac{1}{(2\pi f_0)^2} \, \frac{S_\psi(\omega)}{\omega^2}$$

$$= \frac{1}{\omega^2} \, (h_{-2}|\omega|^{-2} + h_{-1}|\omega|^{-1} + h_0 + h_1|\omega| + h_2|\omega|^2)$$

$$(3.5\text{-}47)$$

for $0 < 2\pi f_l \le |\omega| \le 2\pi f_h < \infty$.

The h_α-coefficients are oscillator-type dependent and are generally optimized in the design of the oscillator so as to minimize the output phase noise power. Figure 3.5-3 illustrates the components of the spectral density of the phase noise process versus ω. (Frequently in the literature, the one-sided spectrum as a function of the frequency f (Hz) is used.)

Through the use of (3.5-47) in (3.5-44) and (3.5-45), one can relate the frequency domain description of oscillator instabilities to the time domain description. These results are given, for example, by Walls and Allan [17]. We will not go into details here but only state the general result as

$$\sigma^2(\tau) \approx \begin{cases} \sim \tau^{-\alpha-1} \\ \sim \tau^{-2} \end{cases} \quad \text{for} \quad \begin{array}{l} \alpha \le 1 \\ \alpha \ge 1 \end{array} \qquad (3.5\text{-}48)$$

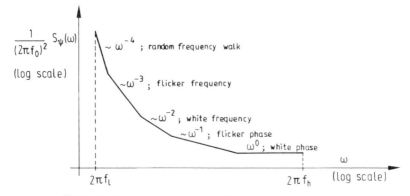

Figure 3.5-3. Power spectral density of phase noise process.

Hence the kind of noise (value of α) can be deduced from the τ-dependence of the Allan variance if $\alpha \leq +1$. Fortunately, this range of α is the usual range for most oscillators if $\tau \leq 1$ s.

Main Points of the Section

- The output of a physically realizable oscillator is distorted by deterministic and random disturbances

  Deterministic disturbances

  Frequency setting Δf Hz

  Frequency drift $\Delta \dot{f}$ Hz/s

  Random disturbances

  Phase noise $\Psi(t)$ rad

- The deterministic disturbances are often called "long term instabilities." The random disturbances $\Psi(t)$ are referred to as "short term instabilities" since they become more significant the shorter the observation intervals become.

- Short term instabilities can be expressed and measured in the time and frequency domain

  (a) Time domain

  RMS fractional frequency deviation

  $$\frac{\delta f(\tau)}{f_0} = \left\{ E\left[\frac{\Psi(t+\tau) - \Psi(t)}{2\pi f_0} \right]^2 \right\}^{1/2} \frac{1}{\tau}$$

  Allan variance

  $$\sigma^2(\tau) = \frac{E\{[\Psi(t+2\tau) - \Psi(t-\tau)] - [\Psi(t+\tau) - \Psi(t)]\}^2}{(2\tau)^2 (2\pi f_0)^2}$$

  The phase noise increments are measured by frequency counter type measurements [17].

  (b) Frequency domain

  The spectral density $S_\Psi(\omega)$ is experimentally determined with the aid of a frequency analyzer [17]. Based on a wealth of data, $S_\Psi(\omega)$ follows the form

  $$\frac{S_\Psi(\omega)}{(2\pi f_0)^2} = \frac{1}{(2\pi f_0)^2} \frac{S_\Psi}{\omega^2}$$

  $$= \frac{1}{\omega^2} [h_{-2}|\omega|^{-2} + h_{-1}|\omega|^{-1} + h_0 + h_1|\omega| + h_2|\omega|^2]$$

  for

$$2\pi f_l \leq |\omega| \leq 2\pi f_h$$

Time and frequency domain representations are linked via the Fourier transform of the structure functions and the spectral density.

3.6. EFFECT OF OSCILLATOR PHASE NOISE ON THE PHASE-LOCKED LOOP TRACKING PERFORMANCE

In the last section we introduced both a time domain and a frequency domain description of the imperfections of an oscillator. We now want to study the performance impairments of a phase-locked loop due to a nonideal VCO. We assume that the VCO output is given by (see Figure 3.6-1)

$$\sqrt{2}K_1 \cos\left[\omega_0 t + \hat{\theta}_p + \Psi(t)\right] \tag{3.6-1}$$

where $\hat{\theta}_p$ is the phase of the perfect oscillator and $\Psi(t)$ is a random angular perturbation generated by an internal disturbance source. In the following we will use the frequency domain description of the oscillator instabilities since this representation is very useful when determining the phase error statistics in a phase-locked loop. We now need to compute the phase error variance of a second-order PLL with a perfect integrator for frequency flicker

$$\frac{S_\Psi(\omega)}{\omega_0^2} = \frac{h_{-1}}{|\omega|^3}, \quad \omega_l \leq |\omega| \leq \omega_h \tag{3.6-2}$$

The phase $\hat{\theta}_p$ at the output of the perfect VCO is

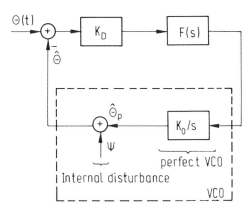

Figure 3.6-1. Linearized model of the PLL when the VCO has an internal disturbance source $\Psi(t)$.

$$\hat{\theta}_p(s) = K_D K_0 \frac{F(s)}{s} \phi(s) \qquad (3.6\text{-}3)$$

The phase error $\phi(s)$ is, as always, defined as the difference between $\theta(s)$ and its estimate, $\hat{\theta}(s)$

$$\phi(s) = \theta(s) - \hat{\theta}(s) = \theta(s) - [\hat{\theta}_p(s) + \Psi(s)] \qquad (3.6\text{-}4)$$

Inserting $\hat{\theta}_p(s)$ from (3.6-3) into (3.6-4) yields

$$\phi(s) = \theta(s) - \left[\frac{K_D K_0 F(s)}{s} \phi(s) + \Psi(s) \right]$$

solving for $\phi(s)$ gives

$$\phi(s) = \frac{1}{1 + [K_0 K_D F(s)/s]} [\theta(s) - \Psi(s)] \qquad (3.6\text{-}5)$$

From (3.6-5), the phase error variance due to oscillator instability is given by

$$\sigma_\phi^2 = \frac{1}{\pi} \int_0^\infty |1 - H(\omega)|^2 S_\Psi(\omega)\, d\omega \qquad (3.6\text{-}6)$$

with

$$1 - H(s) = \frac{1}{1 + [K_0 K_D F(s)/s]}, \quad s = j\omega$$

For a second-order loop with perfect integrator and for the noise power spectrum given by (3.6-2), the variance (3.6-6) can be obtained in closed form. (Due to the factor $|1 - H|^2$ the limits $\omega_l \to 0$, $\omega_h \to \infty$ in (3.6-2) can be used). The PLL phase error variance due to VCO frequency flicker is

$$\sigma_\phi^2 = \frac{\omega_0^2 h_{-1}}{4\pi(2B_L)^2}\, r(\zeta) = \frac{\omega_0^2 h_{-1}}{4\pi\omega_n^2}\, f(\zeta) \qquad (3.6\text{-}7)$$

with

$$f(\zeta) = \begin{cases} \dfrac{1}{4\zeta(\zeta^2 - 1)^{1/2}} \ln \dfrac{2\zeta^2 - 1 + 2\zeta(\zeta^2 - 1)^{1/2}}{2\zeta^2 - 1 - 2\zeta(\zeta^2 - 1)^{1/2}}, & \zeta > 1 \\[2ex] \dfrac{1}{2\zeta(1 - \zeta^2)^{1/2}} \left[\dfrac{\pi}{2} - \tan^{-1}\left(\dfrac{2\zeta^2 - 1}{2\zeta(1 - \zeta^2)^{1/2}} \right) \right], & \zeta < 1 \\[2ex] 1, & \zeta = 1 \end{cases}$$

$$(3.6\text{-}8)$$

The functions $r(\zeta)$ and $f(\zeta)$ are plotted in Figure 3.6-2. With a specified loop noise bandwidth, the minimum variance of the phase error due to oscillator frequency flicker is found for a loop damping ratio of $\zeta = 1.14$.

Equation (3.6-7) shows that the phase error variance is inversely proportional to the square of the bandwidth. Thus, to minimize the phase error variance due to oscillator instabilities, the loop bandwidth should be made as wide as possible. On the other hand, as (3.3-6) shows, the phase error variance due to the additive noise of the PLL received signal is proportional to B_L, and hence B_L should be made as narrow as possible. These two goals are directly opposed to one another and a compromise as discussed in the next section on optimum loop parameters must be made.

Numerical Example: Consider the performance of a standard quartz oscillator in terms of the Allan variance as illustrated in Figure 3.5-2. Now assume that for averaging time τ between 10 and 100, where the Allan variance $\sigma^2(\tau)$ is constant, the oscillator instabilities are only due to flicker frequency. From Figure 3.5-2 we find

$$\sigma^2(\tau) \approx 10^{-24}, \quad 10 < \tau < 100$$

Using this in the formula for the oscillator h-coefficient [22] we obtain

$$h_{-1} = \frac{(2\pi)^3}{2 \ln 2}\, \sigma^2(\tau) = 1.8 \times 10^{-22}$$

For a PLL with

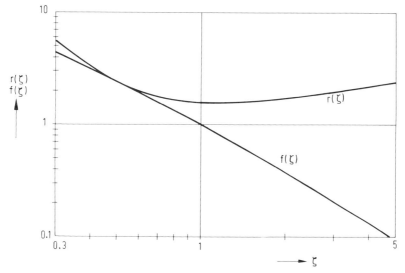

Figure 3.6-2. Flicker variance factors $r(\zeta)$ and $f(\zeta)$.

$$B_L = 1 \, \text{Hz} \,, \quad \zeta = 1.0 \,, \quad \text{and} \quad f_0 = 100 \, \text{MHz}$$

we find for the phase error variance due to VCO frequency flicker (3.6-7)

$$\sigma_\phi^2 = 2.1 \times 10^{-6}$$

3.7. OPTIMIZATION OF THE TRACKING PERFORMANCE IN THE PRESENCE OF NOISE

The optimization of a system for a fixed structure means finding a set of parameters such that a given performance criterion is maximized. The choice of this criterion depends on the application. A meaningful performance criterion in the presence of random disturbances is the minimization of mean square phase error.

As an example we consider the case where a PLL is disturbed by received signal additive noise and frequency flicker noise (see Figure 3.7-1). Since the model is linear and since the noise disturbances $n'(t)$ and $\psi(t)$ are statistically independent we may evaluate the contributions of the noise sources independently. The total phase error variance σ_ϕ^2 equals the sum of the individual contributions

$$\sigma_\phi^2(\omega_n, \zeta) = \underbrace{\frac{N_0}{2A^2}\left(\frac{1+4\zeta^2}{4\zeta}\right)\omega_n}_{\substack{\text{additive noise} \\ \text{contribution}}} + \underbrace{\frac{\omega_0^2 h_{-1}}{4\pi\omega_n^2}f(\zeta)}_{\substack{\text{frequency flicker} \\ \text{contribution} \\ \text{(3.6-7)}}} \qquad (3.7\text{-}1)$$

where the additive noise contribution is obtained from (3.3-6) and Table 3.3-1 (for the case of a perfect second-order loop).

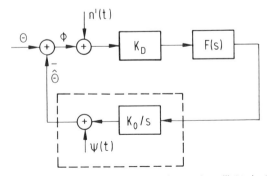

Figure 3.7-1. PLL disturbed by additive noise and oscillator instabilities.

There are two parameters free for optimization in (3.7-1), namely ω_n and ζ. A necessary condition for a minimum phase error variance is that the partial derivatives with respect to ω_n and ζ equal zero

$$\frac{\partial \sigma_\phi^2}{\partial \omega_n}(\omega_n, \zeta) = 0, \quad \frac{\partial \sigma_\phi^2}{\partial \zeta}(\omega_n, \zeta) = 0 \qquad (3.7\text{-}2)$$

The optimum parameter values (ω_n, ζ) are found by solving the (nonlinear) system of equations (3.7-2) for ω_n and ζ.

In our example it is easy to verify that from the first equation of (3.7-2) it follows that

$$\frac{N_0}{2A^2}\left(\frac{1 + 4\zeta^2}{4\zeta}\right)\omega_n = 2\,\frac{\omega_0^2 h_{-1}}{4\pi\omega_n^2}\,f(\zeta) \qquad (3.7\text{-}3)$$

Solving (3.7-3) for ω_n yields

$$\omega_n = \left(\frac{\omega_0^2 h_{-1} A^2}{\pi N_0}\,\frac{4\zeta}{1 + 4\zeta^2}\,f(\zeta)\right)^{1/3} \qquad (3.7\text{-}4)$$

Introducing this optimum value of ω_n into (3.7-1) and using the fact that the two terms in (3.7-1) are related by (3.7-3) yields

$$\sigma_\phi^2(\omega_{n,\mathrm{opt}}, \zeta) = \frac{3}{4}\left(\frac{N_0}{A^2}\,\frac{1 + 4\zeta^2}{4\zeta}\right)^{2/3}\left(\frac{\omega_0^2 h_{-1}}{\pi}\,f(\zeta)\right)^{1/3} \qquad (3.7\text{-}5)$$

from which the optimum damping factor can be found. Using the sketch of Figure 3.7-2 we read off $\zeta_{\mathrm{opt}} = 1.2$.

As another example, we consider Wiener-optimized loops. This optimization method which has received much attention in the literature is based on the optimum filter theory of N. Wiener. The performance criterion to be minimized is the mean square phase error subject to the constraint of finite ε_T^2. The corresponding lagrangian is

$$\sigma_\phi^2 + \lambda^2 \varepsilon_T^2 \qquad (3.7\text{-}6)$$

where σ_ϕ^2 is the variance due to noise, $\varepsilon_T^2 = \int_0^\infty \phi_T^2(t)\,dt$ is the integral of the square of deterministic error due to transients and λ is the Lagrange multiplier. In the context of the Wiener-filter theory the PLL is viewed as a linear system with transfer function $H(s)$. The known quantities in the Wiener-filter theory are the spectrum of the noise and the input signal. The Lagrange multiplier λ can be considered to be a factor that determines the relative weight of the noise and the transient phase error.

The Wiener-filter theory specifies the structure, $H(s)$, as well as the optimum parameter values for a given input. This is the major difference

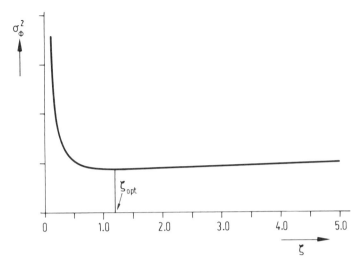

Figure 3.7-2. Optimum damping ratio for the perfect second-order PLL of Figure 3.7-1.

from the first example where the optimum values of the parameters (ω_n, ζ) had to be evaluated for a fixed structure. Unfortunately, application of the Wiener theory is difficult. We state without proof a few important results taken from Jaffe and Rechtin [13]. The noise spectrum is white and the input signal is a phase step, a frequency step and a frequency ramp (see Table 3.7-1).

For the three input types, the Wiener filter theory specifies a first-, a second- and a third-order loop, respectively. The type of the loop (this is also true for other inputs) is the lowest type for which the steady-state phase error vanishes, which is a necessary condition for a finite value of ε_T^2. The optimum bandwidth is a function of the signal-to-noise ratio. If the noise is

TABLE 3.7-1   Wiener Optimized Loops

Input $\theta(t)$	Loop Filter $F(s)$	Optimum $H(s)$	
$\Delta\theta$	$\dfrac{\omega_1}{K_0 K_D}$	$\dfrac{\omega_1}{s + \omega_1}$	$\omega_1 = \Delta\theta \; \lambda \left(\dfrac{2A^2}{N_0}\right)^{1/2}$
$\Delta\omega \, t$	$\dfrac{\omega_n^2 + \sqrt{2}\,\omega_n s}{K_0 K_D s}$	$\dfrac{\omega_n^2 + \sqrt{2}\,\omega_n s}{\omega_n^2 + \sqrt{2}\,\omega_n s + s^2}$	$\omega_n^2 = \Delta\omega \; \lambda \left(\dfrac{2A^2}{N_0}\right)^{1/2}$
			$\zeta = \dfrac{\sqrt{2}}{2}$
$\dfrac{\Delta\dot\omega \, t^2}{2}$	$\dfrac{\omega_3^3 + 2\omega_3^2 s + 2\omega_3 s^2}{K_0 K_D s^2}$	$\dfrac{\omega_3^3 + 2\omega_3^2 s + 2\omega_3 s^2}{\omega_3^3 + 2\omega_3^2 s + 2\omega_3 s^2 + s^3}$	$\omega_3^3 = \Delta\dot\omega \; \lambda \left(\dfrac{2A^2}{N_0}\right)^{1/2}$

weak and σ_ϕ^2 is small, the bandwidth is widened to make the transient phase error contribution small. On the other hand, if the signal is deeply immersed in noise, a smaller bandwidth to make σ_ϕ^2 small is chosen, thereby accepting a large transient phase error contribution.

For the type 2 loop (the most important one in practice), the Wiener-filter theory specifies an optimum loop damping factor of $\zeta = \sqrt{2}/2$, a value which is often considered to be *the* optimum value. This is, of course, not always the case. We will encounter practical examples where only highly overdamped loops can be used.

Since the bandwidth depends on the signal-to-noise ratio, a PLL should ideally have additional circuitry in order to estimate the signal-to-noise ratio and to adaptively adjust the loop parameters. This is not done in practice because Jaffe and Rechtin discovered that a bandpass limiter preceding the phase detector causes the loop bandwidth to vary as a function of SNR. The bandwidth adaptation is not strictly optimal, but sufficiently close to optimality for practical purposes.

Main Points of the Section

We have given two examples of how to find a PLL that minimizes the mean square phase error. In both cases, the phase error variance was assumed small and linearization was permissible. If the phase error variance is plotted as a function of the parameters to be optimized, it is found that the minimum is rather flat. Therefore, a moderate departure from the optimum values does not significantly degrade loop performance. This property is also found if other performance criteria are used.

REFERENCES

1. A. Papoulis, *Probability, Random Variables, and Stochastic Processes*, Kogakusha, Tokyo, and McGraw-Hill, New York, 1965.
2. A. Blanchard, *Phase-Locked Loops: Application to Coherent Receiver Design*, Wiley, New York, 1976.
3. F. M. Gardner, *Phaselock Techniques*, Wiley, New York, 1979.
4. A. J. Viterbi, *Principles of Coherent Communications*, McGraw-Hill, New York, 1966.
5. J. A. Barnes, *Models for the Interpretation of Frequency Stability Measurements*, NBS Technical Note 683, U.S. Department of Commerce, August, 1976.
6. D. W. Allan, J. H. Shoaf and D. Halford, Statistics of Time and Frequency Analysis. Time and Frequency: Theory and Fundamentals, *NBS Monograph* 140, U.S. Dept. of Commerce, Chap. 8, May 1974.
7. D. Halford, A. Wainwright and J. Barnes, Flicker Noise of Phase in R. F. Amplifiers and Frequency Multiplies: Characterization, Cause, and Cure, In *Proceedings of the 22th Symposium on Frequency Control*, pp. 340–341, 1968.

8. W. R. Attkinson, L. Fey and J. Newmann, Spectrum Analysis of Extremely Low Frequency Variations of Quartz Oscillators, *Proceedings of the Institution of Radio Engineers* Vol. 51, p. 379, February 1963.

9. R. M. Gray and R. C. Tausworthe, Frequency-Counted Measurements and Phase Locking to Noise Oscillators, IEEE *Transactions on Communications*, Vol. 19, pp. 21–30, February 1971.

10. W. C. Lindsey and C. M. Chie, Theory of Oscillator Instabilities Based Upon Structure Functions, *Proceedings of the IEEE* Vol. 64, pp. 1652–1666, December 1976.

11. W. C. Lindsey and C. M. Chie, Theory of Oscillator Phase Noise in Correlative Tracking Systems, *IEEE Transactions on Information Theory*, Vol. 26, No. 4, pp. 432–440, July 1980.

12. V. F. Kroupa, Noise Properties of PLL-Systems, *IEEE Transactions on Communications*, Vol. 30, No. 10, pp. 2244–2252, October 1982.

13. R. Jaffe and E. Rechtin, Design and Performance of Phase-Lock Loops Capable of Near-optimum Performance over a Wide Range of Input Signal and Noise Levels, IRE *Transactions on Information Theory*, Vol. 1, pp. 66–76, March 1955.

14. J. A. Barnes et al., Characterization of Frequency Stability, *IEEE Transactions on Instrumentation and Measurement*, Vol. 20, pp. 105–120, May 1972.

15. W. C. Lindsey and C. M. Chie, *Phase-Locked Loops*, IEEE Press, New York, 1986.

16. W. C. Lindsey and C. M. Chie, Identification of Power-Law Type Oscillator Phase Noise Spectra from Measurements, *IEEE Transactions on Instrumentation and Measurement*, Vol. 27, No. 1, pp. 46–53, March, 1978.

17. F. L. Walls and D. W. Allan, Measurements of Frequency Stability, *Proceedings of the IEEE*, Vol. 74, No. 1, pp. 162–168, January 1986.

18. J. A. Barnes and D. W. Allan, A Statistical Model of Flicker Noise, *Proceedings of the IEEE*, Vol. 54, pp. 176–178, February 1966.

19. W. Rosenkranz, Phase-Locked Loops with Limiter Phase Detectors in the Presence of Noise, *IEEE Transactions on Communications*, Vol. 30, No. 10, October 1982.

20. J. C. Springett and M. K. Simon, An Analysis of the Phase Coherent-Incoherent Output of the Bandpass Limiter, *IEEE Transactions on Communication Technology*, Vol. 19, No. 1, pp. 42–49, February 1971.

21. L. E. Franks, *Signal Theory*, Prentice Hall, Englewood Cliffs, NJ, 1969.

22. W. C. Lindsey, unpublished internal report, LinCom Corporation, P.O. Box 2793D, Pasadena, CA 91105, 1985.

23. H. J. Kushner, Diffusion Approximations to Output Processes of Nonlinear Systems with Wide-Band Inputs and Applications, *IEEE Transactions on Information Theory*, Vol. 26, No. 6, November 1980.

APPENDIX   3.2A
THE CLOSED-LOOP PHASE-LOCKED LOOP WITH MULTIPLIER-TYPE PHASE DETECTOR AND THE EXACT NOISE MODEL $n'(t, \hat{\theta}(t))$

The received signal and the output of the VCO are of the form

$$s(t) = \sqrt{2}A \sin [\omega_0 t + \theta(t)]$$
$$r(t) = \sqrt{2}K_1 \cos [\omega_0 t + \hat{\theta}(t)]$$

$$(3.2A-1)$$

where $\theta(t)$ and $\hat{\theta}(t)$ are random phase modulation to be subsequently characterized.

As shown in Section 3.1 a narrowband gaussian random process $n(t)$ can be written in the form

$$n(t) = \sqrt{2}n_c(t) \cos (\omega_0 t) - \sqrt{2}n_s(t) \sin \omega_0 t = \sqrt{2} \text{ Re } [\underline{n}_L(t) \exp (j\omega_0 t)]$$
$$(3.2A-2)$$
$$\underline{n}_L(t) := n_c(t) + jn_s(t)$$

Then filtering the result of the product $r(t)[s(t) + n(t)]$ gives

$$x(t) = K_D \{\sin \phi + n'[t, \hat{\theta}(t)]\} \qquad (3.2A-3)$$

with

$$n'[t, \hat{\theta}(t)] = \frac{1}{A} \text{ Re } \{\underline{n}_L(t) \exp [-j\hat{\theta}(t)]\} = \frac{n_c(t)}{A} \cos \hat{\theta}(t) + \frac{n_s(t)}{A} \sin \hat{\theta}(t)$$

We now compute the autocorrelation of the angular phase disturbance $n'[t, \hat{\theta}(t)]$.

$$R_{n'}(t, \tau, \hat{\theta}(t)) = \frac{1}{A^2} E\{[n_c(t) \cos \hat{\theta}(t) + n_s(t) \sin \hat{\theta}(t)]$$
$$\times [n_c(t + \tau) \cos \hat{\theta}(t + \tau) + n_s(t + \tau) \sin \hat{\theta}(t + \tau)]\}$$
$$= \frac{1}{A^2} E[n_c(t)n_c(t + \tau) \cos \hat{\theta}(t) \cos \hat{\theta}(t + \tau)$$
$$+ n_s(t)n_s(t + \tau) \sin \hat{\theta}(t) \sin \hat{\theta}(t + \tau)$$
$$+ n_c(t)n_s(t + \tau) \cos \hat{\theta}(t) \sin \hat{\theta}(t + \tau)$$
$$+ n_s(t)n_c(t + \tau) \sin \hat{\theta}(t) \cos \hat{\theta}(t + \tau)] \qquad (3.2A-4)$$

Since the correlation time (defined at the end of this Appendix) of $n_c(t)$.

$n_s(t)$ is much shorter* than that of $\hat{\theta}(t)$ we may assume here that $\hat{\theta}(t)$ is (approximately) independent of $n_c(t)$ and $n_s(t)$ for time periods of the order of the VCO-phase correlation time $\tau_{\mathrm{corr},\hat{\theta}}$. Then

$$R_{n'}(\tau, \hat{\theta}(t)) = \frac{1}{A^2} \left[R_{n_c}(\tau) \langle \cos \Delta\hat{\theta}(\tau) \rangle + R_{n_c n_s}(\tau) \langle \sin \Delta\hat{\theta}(\tau) \rangle \right],$$

$$|\tau| < \tau_{\mathrm{corr},\hat{\theta}} \qquad (3.2A\text{-}5)$$

where $\Delta\hat{\theta}(\tau) := \hat{\theta}(t) - \hat{\theta}(t + \tau)$. Now assuming $|\Delta\hat{\theta}(\tau)| \ll 1$ gives

$$R_{n'}(\tau, \hat{\theta}) = \frac{R_{n_c}(\tau)}{A^2}, \qquad |\tau| < \tau_{\mathrm{corr},\hat{\theta}} \qquad (3.2A\text{-}5)$$

which is the same as (3.2-6) obtained for the open-loop case.

Another interpretation of $n'[t, \hat{\theta}(t)]$ which brings out the phase error $\phi(t) = \theta(t) - \hat{\theta}(t)$ explicitly is also sometimes used and is explained here in passing. Introducing the new phasor

$$\underline{N}_L(t) := N_c(t) + jN_s(t) = \underline{n}_L \exp\left[-j\theta(t)\right] \qquad (3.2A\text{-}6)$$

the angular phase disturbance can be written as

$$n'[t, \hat{\theta}(t)] = \frac{1}{A} \operatorname{Re} \{\underline{n}_L \exp\left[-j\hat{\theta}(t)\right]\}$$

$$= \frac{1}{A} \operatorname{Re} \{\underline{n}_L \exp\left[-j\theta(t)\right] \cdot \exp\left[j\phi(t)\right]\}$$

$$= \frac{1}{A} \operatorname{Re} \left[\underline{N}_L(t) \exp\left[j\phi(t)\right]\right] \qquad (3.2A\text{-}7)$$

with

$$N_c(t) = n_c(t) \cos\theta(t) + n_s(t) \sin\theta(t) = \operatorname{Re} \{\underline{n}_L(t) \cdot \exp\left[-j\theta(t)\right]\}$$

$$\qquad (3.2A\text{-}8)$$

$$N_s(t) = n_s(t) \cos\theta(t) - n_c(t) \sin\theta(t) = \operatorname{Im} \{\underline{n}_L(t) \cdot \exp\left[-j\theta(t)\right]\}$$

The correlation functions of $N_c(t)$ and $N_s(t)$ can be expressed (after some simple algebraic manipulations) as

$$R_{N_c}(\tau) = R_{N_s}(\tau) = R_{n_c}(\tau) \langle \cos \Delta\theta(\tau) \rangle$$

$$\qquad (3.2A\text{-}9)$$

$$R_{N_c N_s}(\tau) = -R_{N_c N_s}(-\tau) = R_{n_c}(\tau) \langle \sin \Delta\theta(\tau) \rangle - R_{n_s n_c}(\tau) \langle \cos \Delta\theta(\tau) \rangle$$

---

*This is only true if the loop bandwidth B_L (defined by (3.3-5)) is small compared to B_{IF}. Situations exist (e.g. FM demodulators) where this inequality is not true. The result applies therefore for narrow-band loop only.

where $R_{n_c}(\tau) = R_{n_s}(\tau)$ is the correlation function of $n_c(t)$ and $\langle \cos \Delta\theta(\tau) \rangle := E\{\cos[\theta(t) - \theta(t + \tau)]\}$. This is important as it shows how the correlation function of the angular phase disturbance is modified in the presence of a random phase modulation $\theta(t)$ of the transmitter oscillator. Usually the *correlation time* of $\theta(t)$, defined at the end of this Appendix, is much larger than that of the process $n_c(t)$ so that the processes $N_c(t)$ and $N_s(t)$ have approximately the same statistical properties as $n_c(t)$ and $n_s(t)$. This approximation can be seen by setting $\langle \cos \Delta\theta(\tau) \rangle = 1$ in (3.2A-9). When $n_L(t)$ and $N_L(t)$ have approximately the same statistical characteristics then $n'[t, \hat{\theta}(t)]$ is statistically equivalent to $n'[t, \phi(t)]$. A rigorous mathematical treatment of the topics involved in this modeling can be found in the paper by Kushner [23].

Finally we define the correlation time τ_x of a stochastic process $x(t)$ and its relation to its equivalent noise bandwidth B_x. *The correlation time* τ_x of a stochastic process $x(t)$ gives some measure of the time interval over which statistical dependency exists between different samples of the process $x(t)$.

$$\tau_x := \frac{1}{R_x(0)} \int_0^\infty |R_x(\tau)| \, d\tau \qquad (3.2A\text{-}10)$$

When $R_x(\tau) > 0$ for all τ, then using the formulas

$$S_x(0) = \int_{-\infty}^\infty R_x(\tau) \, d\tau \, , \quad R_x(0) = \frac{1}{2\pi} \int_{-\infty}^\infty S_x(\omega) \, d\omega \qquad (3.2A\text{-}11)$$

we easily arrive at the expression which relates the equivalent noise bandwidth

$$B_x = \frac{1}{2\pi} \int_0^\infty \frac{S_x(\omega)}{S_x(0)} \, d\omega \qquad (3.2A\text{-}12)$$

to the correlation time τ_x

$$B_x \tau_x = \frac{1}{4} \qquad (3/2A\text{-}13)$$

APPENDIX 3.2B
COMPLEX ENVELOPE REPRESENTATION OF SIGNALS

A straightforward extension of the familiar two-dimensional *phasor* representation has proven to be of great convenience for the analysis of communication systems. The basic properties are described here.

An arbitrary signal $y(t)$ can be represented by a *complex envelope* $\underline{y}_L(t)$ relative to a center frequency ω_0

$$y(t) = \sqrt{2} \, \text{Re} \, [\underline{y}_L(t) \exp(j\omega_0 t)] \qquad (3.2B\text{-}1)$$

The complex envelope $\underline{y}_L(t)$ can be represented either in polar or in rectangular form.

$$\underline{y}_L(t) = |\underline{y}_L(t)| \exp[j \arg\{\underline{y}_L(t)\}] = \text{Re}[\underline{y}_L(t)] + j\,\text{Im}[\underline{y}_L(t)]$$

The narrowband noise has the complex envelope representation

$$n(t) = \text{Re}\,(\sqrt{2}\underbrace{[n_c(t) + jn_s(t)]}_{\underline{n}_L(t)} \exp(j\omega_0 t)\} \qquad (3.2\text{B-}2)$$

where $\underline{n}_L(t)$ is the complex noise envelope and the combination of signal and noise is written as

$$
\begin{aligned}
y(t) &= \sqrt{2}A\sin(\omega_0 t + \theta) + \sqrt{2}n_c(t)\cos\omega_0 t - \sqrt{2}n_s(t)\sin\omega_0 t \\
&= \sqrt{2}\,\text{Re}\,\{-jA\exp[j(\omega_0 t + \theta)]\} + \sqrt{2}\,\text{Re}\,\{\underline{n}_L(t)\exp(j\omega_0 t)\} \\
&= \sqrt{2}\,\text{Re}\,\{\underbrace{[-jAe^{j\theta} + \underline{n}_L(t)]}_{\underline{y}_L(t)}\exp(j\omega_0 t)\} \qquad (3.2\text{B-}3)
\end{aligned}
$$

Sometimes it is convenient to decompose the complex noise envelope into a component in phase with the envelope of the signal $s(t)$ and a component perpendicular to it. Taking the factor $e^{j\theta}$ out of the squared bracket in the previous equation yields

$$
\begin{aligned}
y(t) &= \sqrt{2}\,\text{Re}\,\{[-jA + \underline{n}_L(t)\,e^{-j\theta}]\exp[j(\omega_0 t + \theta)]\} \\
&= \sqrt{2}\,\text{Re}\,\{[N_c(t) - j[A - N_s(t)]]\exp[j(\omega_0 t + \theta)]\} \quad (3.2\text{B-}4)
\end{aligned}
$$

with

$$\underline{N}_L = N_c(t) + jN_s(t) = \underline{n}_L\,e^{-j\theta} \qquad (3.2\text{B-}5)$$

where $N_s(t)$ is the in-phase component and $N_c(t)$ is the quadrature component. From (3.2B-4) we learn that $-N_s(t)$ is in phase with the envelope of the useful signal while $N_c(t)$ is perpendicular to it. $N_s(t)$ is therefore called the *in-phase component* while $N_c(t)$ is called the *quadrature component*. Notice that $N(t)$ is obtained by rotating the envelope of $n(t)$ by an angle $(-\theta)$ in the complex plane. The correspondence between $n(t)$ and $N(t)$ is one-to-one (3.2B-5).

Multiplier Output. Next we consider the output $x_m(t)$ of a multiplier when the two inputs are expressed in the envelope representation.

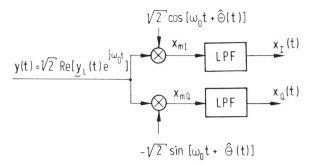

Figure 3.2B-1. Low-pass components $x_I(t)$ and $x_Q(t)$.

$$x_m(t) = \text{Re}\,[\sqrt{2}\underline{y}_{L1}(t)\exp\,(j\omega_0 t)]\,\text{Re}\,[\sqrt{2}\underline{y}_{L2}(t)\exp\,(j\omega_0 t)]$$

$$= \text{Re}\,[\underline{y}_{L1}(t)\underline{y}_{L2}^*(t)] + \text{Re}\,[\underline{y}_{L1}(t)\underline{y}_{L2}(t)\exp\,(j2\omega_0 t)] \quad (3.2B\text{-}6)$$

The multiplier output consists of two terms, representing low-frequency components and components around $2\omega_0$.

We apply the above results to compute $x_I(t)$ and $x_Q(t)$ as defined in Figure 3.2B-1.

$$x_{mI}(t) = \sqrt{2}\,\text{Re}\,[\underline{y}_L(t)\exp\,(j\omega_0 t)]\sqrt{2}\,\text{Re}\,\{\exp\,[j\hat{\theta}(t)]\exp\,(j\omega_0 t)\}$$

$$\tag{3.2B-7}$$

$$x_I(t) = \text{Re}\,[\underline{y}_L(t)\exp\,[-j\hat{\theta}(t)]]$$

$$x_{mQ}(t) = \sqrt{2}\,\text{Re}\,[\underline{y}_L(t)\exp\,(j\omega_0 t)]\sqrt{2}\,\text{Re}\,\{j\exp\,[j\hat{\theta}(t)]\exp\,(j\omega_0 t)$$

$$\tag{3.2B-8}$$

$$x_Q(t) = \text{Re}\,[-j\underline{y}_L(t)\exp\,[-j\hat{\theta}(t)]] = \text{Im}\,\{\underline{y}_L(t)\exp\,[-j\hat{\theta}(t)]\}$$

It follows from the last two equations that $x_I(t)$ and $x_Q(t)$ are the real and imaginary parts of the complex envelope $\underline{x}_L(t)$

$$\underline{x}_L(t) = \underline{y}_L(t)\exp\,[-j\hat{\theta}(t)]$$

$$= x_I(t) + jx_Q(t) \tag{3.2B-9}$$

The relation (3.2B-9) is a key result as it demonstrates that the complex envelope of a received signal can be recovered by multiplying it by $\sin\,(\omega_0 t)$ and $\cos\,(\omega_0 t)$. The complex equivalent representation for obtaining $\underline{x}_L(t)$ is shown in Figure 3.2B-2.

Example: For $\underline{y}_L(t)$ defined as in (3.2B-3) $x_I(t)$ equals the output of a multiplier phase detector of which one input is a signal disturbed by additive noise. Then using (3.2B-7) yields

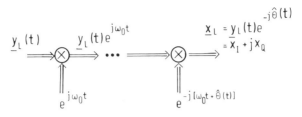

Figure 3.2B-2. Complex equivalent representation for obtaining $\underline{x}_L(t)$.

$$x_I(t) = \mathrm{Re}\,\{[-jAe^{j\theta} + \underline{n}_L(t)]\,e^{-j\hat{\theta}}\}$$

$$\sim K_D\left[\sin\phi + \underbrace{\frac{n_c(t)}{A}\cos\hat{\theta} + \frac{n_s(t)}{A}\sin\hat{\theta}}_{n'(t,\,\hat{\theta})}\right]$$

where $\phi = \theta - \hat{\theta}$ and K_D is the phase detector gain. This has been discussed in Section 3.2 and shown in Figure 3.2-1. The output $x_Q(t)$ equals the output of a quadrature phase detector

$$x_Q(t) = \mathrm{Re}\,\{[-jA\,e^{j\theta} + \underline{n}_L(t)](-j)\,e^{-j\hat{\theta}}\}$$

$$\sim K_D\left[\cos\phi + \underbrace{\frac{n_c(t)}{A}\sin\hat{\theta} - \frac{n_s(t)}{A}\cos\hat{\theta}}_{\hat{n}'_Q(t,\,\hat{\theta})}\right]$$

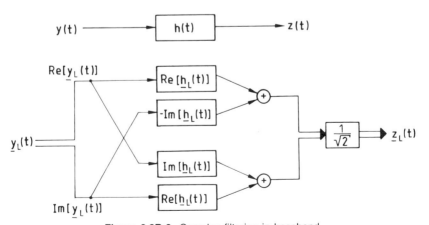

Figure 3.2B-3. Complex filtering in baseband.

Complex Filtering. An arbitrary real filter can be represented by the impulse response

$$h(t) = \sqrt{2} \, \text{Re} \, [\underline{h}_L(t) \exp(j\omega_0 t)] \qquad (3.2B\text{-}10)$$

where $\underline{h}_L(t)$ is the complex low-pass equivalent impulse response. The result of passing the real signal $y(t)$ of (3.2B-1) through the filter $h(t)$ is

$$z(t) = h(t) * y(t)$$
$$=: \sqrt{2} \, \text{Re} \, [\underline{z}_L(t) \exp(j\omega_0 t)] \qquad (3.2B\text{-}11)$$

The complex low-pass output $\underline{z}_L(t)$ can be shown to have the following quadrature components (Figure 3.2B-3)

$$\text{Re} \, [\underline{z}_L(t)] = \frac{1}{\sqrt{2}} \, \{\text{Re} \, [\underline{h}_L(t)] * \text{Re} \, [\underline{y}_L(t)] - \text{Im} \, [\underline{h}_L(t)] * \text{Im} \, [\underline{y}_L(t)]\}$$

$$(3.2B\text{-}12)$$

$$\text{Im} \, [\underline{z}_L(t)] = \frac{1}{\sqrt{2}} \, \{\text{Re} \, [\underline{h}_L(t)] * \text{Im} \, [\underline{y}_L(t)] + \text{Im} \, [\underline{h}_L(t)] * \text{Re} \, [\underline{y}_L(t)]\}$$

4

UNAIDED ACQUISITION

4.1. INTRODUCTION

The phase-locked loop has two distinct modes of operation. In the *tracking mode* all state variables of the loop have values in the close vicinity of the steady-state values. Initially, however, the difference between the steady-state values and the initial values of the state variables can be large. The process of bringing the loop from its initial state to the tracking mode is called *acquisition* and is the subject of this and the next chapter.

There are several ways to accomplish acquisition. If the loop acquires lock itself, we call the process unaided acquisition; if auxiliary circuits are employed to perform the task the process is called aided acquisition.

Unaided acquisition is usually a slow and unreliable process. While the PLL is a near optimum tracking device it performs poorly during acquisition. Therefore, the vast majority of applications employ some sort of aided acquisition.

An nth order loop is described by n state variables. During acquisition all these state variables must be brought from their initial values to their steady-state tracking values. A first-order loop has only one state variable (phase $\hat{\theta}$). Therefore, only phase has to be acquired. A second-order loop has two state variables and accordingly, we speak of phase and frequency acquisition. Similarly, in a third-order loop we have to distinguish between phase, frequency, and frequency rate acquisition.

Acquisition is inherently a nonlinear phenomenon and we must use nonlinear analysis throughout.

4.2. FIRST-ORDER LOOP

4.2.1. Phase Acquisition in the Absence of Noise

For a frequency offset $\Delta\omega$, the equation of the PLL in the absence of noise is

$$\frac{d\phi(t)}{dt} = -(K_0 K_D) \sin \phi(t) + \Delta\omega , \quad t \geq 0 \qquad (4.2\text{-}1)$$

Introducing the normalized time τ

$$\tau = (K_0 K_D)t = 4B_L t$$

where B_L is the loop noise bandwidth, (4.2-1) reads

$$\frac{d\phi(\tau)}{d\tau} = -\sin \phi(\tau) + \left(\frac{\Delta\omega}{4B_L}\right), \quad \tau \geq 0 \qquad (4.2\text{-}2)$$

Let us assume the loop has reached an equilibrium. Then, by definition $d\phi/d\tau$ must be zero and the steady-state phase error, ϕ_s, is found by solving (4.2-2) for $d\phi/d\tau = 0$

$$\phi_s = \sin^{-1}\left(\frac{\Delta\omega}{K_0 K_D}\right) \qquad (4.2\text{-}3)$$

Since the sine function cannot exceed unity, the maximum frequency difference $\Delta\omega$ that the loop can track is found when

$$\left|\frac{\Delta\omega}{K_0 K_D}\right| = 1 \qquad (4.2\text{-}4)$$

The frequency difference

$$\Delta\omega_H = K_0 K_D = 4B_L \qquad (4.2\text{-}5)$$

is called the *hold-in range* of a first-order loop.

It is useful to plot $d\phi/d\tau$ from (4.2-1) versus the phase error ϕ, as in Figure 4.2-1. From this figure it may be seen that there are two points in each interval of 2π for which $d\phi/d\tau$ is zero, i.e., we have two points of equilibrium. Point A represents stable equilibrium while point B corresponds to an unstable equilibrium.

The fundamentally different nature of the two equilibrium points is most conveniently understood by linearizing equation (4.2-2) around these two points. For the stable point A, the slope of the sine function is positive (this slope in Figure 4.2-1 is negative because we plotted $-\sin \phi$).

$$\sin (\phi_s + \phi) \approx \sin \phi_s - K_\phi \phi \qquad (4.2\text{-}6)$$

with

$$K_\phi = \frac{d}{d\phi} \sin \phi \bigg|_{\phi = \phi_s} > 0 , \quad \text{Point } A$$

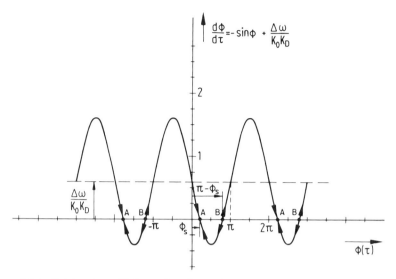

Figure 4.2-1. Points of equilibrium of a first-order loop with sinusoidal phase detector characteristic.

Inserting (4.2-6) into (4.2-2) and using equation (4.2-3) yields for a small displacement ϕ from ϕ_s

$$\frac{d\phi}{d\tau} = -\sin \phi_s - K_\phi \phi + \frac{\Delta\omega}{K_0 K_D} = -K_\phi \phi \qquad (4.2\text{-}7)$$

Let us assume that a slight disturbance at $\tau = 0$ produces a phase error $\phi(0) \neq 0$. For $\tau > 0$ the phase error decreases exponentially towards zero

$$\phi(\tau) = \phi(0) \exp\left(-K_\phi \tau\right), \quad K_\phi > 0 \qquad (4.2\text{-}8)$$

and the loop returns to its stable equilibrium point.

For point B the slope of the sine function is negative (opposite sign of point A)

$$\sin\left[(\pi - \phi_s) + \phi\right] \approx \sin(\pi - \phi_s) - K_\phi \phi \qquad (4.2\text{-}9)$$

with

$$K_\phi = \left.\frac{d}{d\phi} \sin \phi\right|_{\phi = \pi - \phi_s} < 0$$

The dynamic equation now reads

$$\frac{d\phi}{d\tau} = -K_\phi \phi, \quad K_\phi < 0 \qquad (4.2\text{-}10)$$

Suppose we again have a phase error $\phi(0) \neq 0$. Instead of approaching zero, the phase error grows exponentially.

$$\phi(\tau) = \phi(0) \exp{(-K_\phi \tau)}, \quad K_\phi < 0 \tag{4.2-10a}$$

Since $\phi(0)$ can be arbitrarily small, the slightest disturbance will drive the state of the loop away from equilibrium; point B is an unstable point.

The different nature of the two equilibrium points is illustrated by arrows in Figure 4.2-1 which show the direction of the phase change. We see, that for a phase error smaller than $(\pi - \phi_s)$ the loop will settle at the stable equilibrium point A while for $\phi > (\pi - \phi_s)$ the loop will settle at the stable point in the next 2π interval.

Having qualitatively discussed the phase acquisition process we next want to calculate the exact phase transients $\phi(\tau)$ for an arbitrary initial phase error $\phi(0)$. Clearly, the solutions given in (4.2-8) or (4.2-10a) are valid only for small phase errors about the equilibrium points. For a large phase error we must use nonlinear analysis and cannot ease the way with linear approximations.

The differential equation (4.2-2) can be solved analytically by the standard technique of separation of variables.

$$\int_0^\tau d\tau' = \int_{\phi(0)}^{\phi(\tau)} \frac{d\phi}{-\sin\phi + (\Delta\omega/4B_L)} \tag{4.2-11}$$

Consider first the case $|\Delta\omega/4B_L| < 1$, i.e., when $|\Delta\omega| < \Delta\omega_H$. Using integration tables yields

$$\tau = \frac{1}{[1-(\Delta\omega/4B_L)^2]^{1/2}} \ln \frac{(\Delta\omega/4B_L)\tan{[\phi(\tau)/2]} - 1 - [1-(\Delta\omega/4B_L)^2]^{1/2}}{(\Delta\omega/4B_L)\tan{[\phi(\tau)/2]} - 1 + [1-(\Delta\omega/4B_L)^2]^{1/2}}$$

$$- \frac{1}{[1-(\Delta\omega/4B_L)^2]^{1/2}} \ln \frac{(\Delta\omega/4B_L)\tan{[\phi(0)/2]} - 1 - [1-(\Delta\omega/4B_L)^2]^{1/2}}{(\Delta\omega/4B_L)\tan{[\phi(0)/2]} - 1 + [1-(\Delta\omega/4B_L)^2]^{1/2}},$$

$$|\Delta\omega| < \Delta\omega_H \tag{4.2-12}$$

In the case of zero loop detuning $\Delta\omega = 0$, (4.2-11) has the solution

$$\tau = \ln \frac{\tan{[\phi(0)/2]}}{\tan{[\phi(\tau)/2]}} \tag{4.2-12a}$$

or

$$\phi(\tau) = 2\tan^{-1}\left[e^{-\tau} \tan\frac{\phi(0)}{2}\right], \quad \Delta\omega = 0$$

Typical phase error transients are shown in Figure 4.2-2 for $\Delta\omega = 0$ and several values of $\phi(0)$. If $\phi(0)$ is small, the loop operation is almost linear and the phase error decreases exponentially with a normalized time constant of 1 (corresponding to $1/K_0 K_D$ in real time). If the initial phase error is large the phase transients are no longer simple exponential functions and the settling times are much longer than those obtained in the linear case with the same initial condition.

Since it takes an infinite amount of time to reach the steady state $d\phi/d\tau = 0$, we define the *phase-acquisition time* in the absence of noise (deterministic case) T_D as the time it takes the loop to decrease the initial phase error $\phi(0)$ to a small amount ϕ_ε. From (4.2-12a) this time is found to be

$$T_D = \frac{1}{4B_L} \ln \left\{ \frac{\tan [\phi(0)/2]}{\tan [\phi_\varepsilon/2]} \right\}, \quad \Delta\omega = 0 \qquad (4.2\text{-}13)$$

If the initial phase error is close to the unstable equilibrium point, the phase can dwell near that point for a long time. This phenomenon has been called hang-up by Gardner [1]. It can be very troublesome in applications where rapid phase acquisition is essential.

If the magnitude of frequency difference $|\Delta\omega|$ is smaller than the hold-in range $\Delta\omega_H$, the loop will always lock to the next stable tracking point without slipping a cycle. In a first-order loop the *pull-in range*, which represents the largest frequency offset $\Delta\omega$ for which the loop will lock,

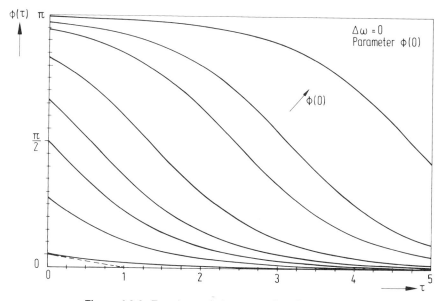

Figure 4.2-2. Transients of phase error in a first-order PLL.

equals the hold-in range. As we will see, this is not true for higher order loops.

Outside the hold-in range, i.e., for $|\Delta\omega| > \Delta\omega_H$, the phase error as a function of the normalized time τ is found (using integration tables) from (4.2-11)

$$\tau = \frac{2}{[(\Delta\omega/4B_L)^2 - 1]^{1/2}} \tan^{-1}\left\{\frac{(\Delta\omega/4B_L)\tan(\phi/2) - 1}{[(\Delta\omega/4B_L)^2 - 1]^{1/2}}\right\}\Bigg|_{\phi(0)}^{\phi(\tau)}$$

$$|\Delta\omega| > \Delta\omega_H \tag{4.2-14}$$

Equation (4.2-14) represents a periodic oscillation characterized by its waveform during one period and its fundamental frequency. Figure 4.2-3 illustrates an example of this periodic oscillation (*beat note*). We have plotted $\sin\phi(\tau)$ and $\phi(\tau)$ versus τ. The period T_{bn} of the beat note is found by noting that, if time t increases by T_{bn}, then the phase ϕ increases by 2π; therefore, from (4.2-14) we have

$$\tau_{bn} = (4B_L T_{bn}) = \frac{2\pi}{[(\Delta\omega/4B_L)^2 - 1]^{1/2}} \tag{4.2-15}$$

Notice that the periodic waveform $\sin\phi(\tau)$ is *not* symmetric but has a dc-component. For example, the phase error in Figure 4.2-3b traverses the

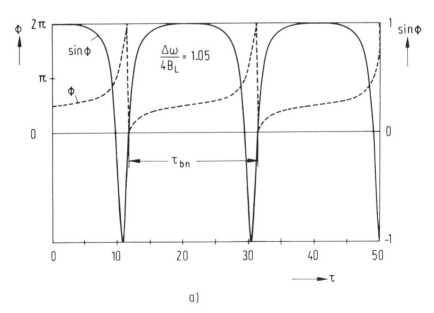

Figure 4.2-3. Examples of a beat note in a first-order PLL. (a) $\Delta\omega/4B_L = 1.05$; (b) $\Delta\omega/4B_L = -1.4$; (c) $\Delta\omega/4B_L = 3$.

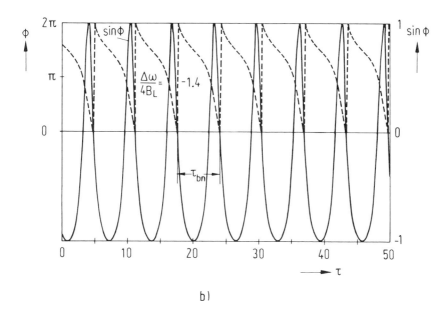

b)

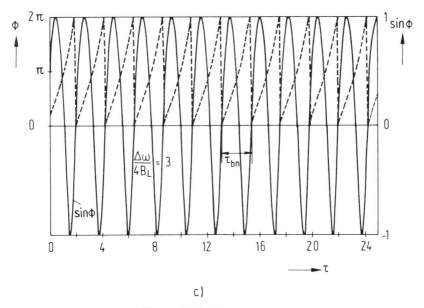

c)

Figure 4.2-3. (Continued)

region $\phi \in (-\pi, -2\pi)$ faster than the interval from $(0, -\pi)$ because the restoring force $\sin \phi$ for $\phi \in (0, -\pi)$ has the correct polarity (counteracting the phase error) while in the other interval $\phi \in (-\pi, -2\pi)$, the polarity is such that the loop increases the phase error (instead of decreasing it). The phase error, therefore, traverses the unstable region much faster than the region of stable loop operation.

The dc-component $\overline{\sin \phi}$ can be obtained by integrating the differential equation (4.2-2) over a full cycle. For a positive frequency offset $\Delta\omega > \Delta\omega_H$ the phase increment over a cycle equals

$$\phi(\tau_{bn}) - \phi(0) = +2\pi$$

Thus

$$\int_{\phi(0)}^{\phi(\tau_{bn})} d\phi = \phi(\tau_{bn}) - \phi(0) = 2\pi = \int_0^{\tau_{bn}} \left[-\sin \phi(\tau) + \frac{\Delta\omega}{4B_L} \right] d\tau$$

$$= -\int_0^{\tau_{bn}} \sin \phi(\tau) \, d\tau + \frac{\Delta\omega}{4B_L} \tau_{bn}, \quad \Delta\omega > \Delta\omega_H \qquad (4.2\text{-}16)$$

The dc-component of the beat note is

$$\overline{\sin \phi} = \frac{1}{\tau_{bn}} \int_0^{\tau_{bn}} \sin \phi(\tau) \, d\tau = \frac{\Delta\omega}{4B_L} - \frac{2\pi}{\tau_{bn}} \qquad (4.2\text{-}17)$$

Inserting (4.2-15) into (4.2-17) yields

$$\overline{\sin \phi} = \frac{\Delta\omega}{4B_L} - \left[\left(\frac{\Delta\omega}{4B_L} \right)^2 - 1 \right]^{1/2}, \quad \frac{\Delta\omega}{4B_L} > 1 \qquad (4.2\text{-}18)$$

If the frequency detuning is negative, i.e., $\Delta\omega < 0$, the phase increment in (4.2-16) is also negative (-2π) and we obtain (verify!)

$$\overline{\sin \phi}\big|_{\Delta\omega < 0} = -\overline{\sin \phi}\big|_{\Delta\omega > 0}, \quad |\Delta\omega| > \Delta\omega_H \qquad (4.2\text{-}18a)$$

The dc-term $(4B_L)\overline{\sin \phi}$ is the average generated frequency shift of the VCO which is maximum for $\Delta\omega/4B_L = 1$. For frequency offsets $\Delta\omega/4B_L > 1$ the average frequency shift is smaller than $\Delta\omega$.

$$4B_L \overline{\sin \phi} = 4B_L \left\{ \frac{\Delta\omega}{4B_L} - \left[\left(\frac{\Delta\omega}{4B_L} \right)^2 - 1 \right]^{1/2} \right\} < \Delta\omega, \quad \Delta\omega > \omega_H$$

This means that there is not enough voltage generated to shift the VCO frequency by $\Delta\omega$ and the loop oscillates forever. For $|\Delta\omega/4B_L| < 1$, however, the dc-voltage is sufficient to shift the VCO by $\Delta\omega$. The steady-state phase error is then given by

$$\frac{d\phi}{d\tau} = 0 = -\sin \phi_s + \frac{\Delta\omega}{4B_L} \tag{4.2-19}$$

and the loop settles at the next stable tracking point. Figure 4.2-4 shows the dc-term $\overline{\sin \phi}$ as a function of the normalized frequency detuning.

4.2.2. Phase Acquisition in the Presence of Additive Noise

The dynamic equation of a first-order PLL in the presence of noise and zero loop detuning* is

$$\frac{d\phi(\tau)}{d\tau} = -\sin \phi(\tau) + m(\tau) \tag{4.2-20}$$

where $\tau = K_0 K_D t$, $m(\tau)$ is the normalized white gaussian noise process with normalized spectral density $S_m(0) = [N_0/(2A^2)]K_0 K_D = 2/\rho$ (see (3.3-3) and Table 3.4-1). Since $m(\tau)$ is a random disturbance the phase error $\phi(\tau)$ is a random process. Typical trajectories are ploted in Figure 4.2-5.

The trajectory starts with an initial phase error $\phi(0)$. In the noise free case the phase acquisition time, T_D, is defined as the time it takes the phase

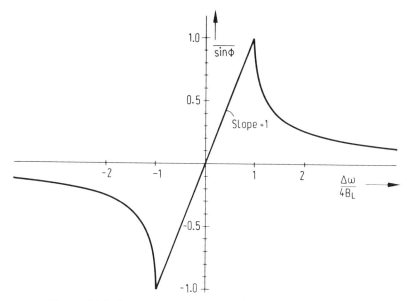

Figure 4.2-4. Average dc-component at the phase detector output.

*The restriction to zero loop detuning is not essential but merely used to simplify the discussion.

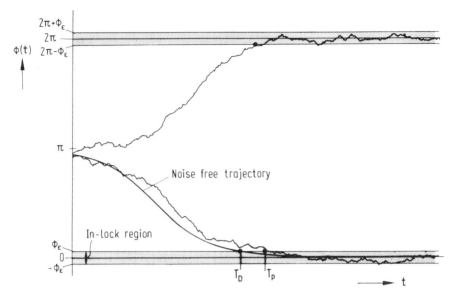

Figure 4.2-5. Typical phase trajectories in the presence of noise for the case of steady-state phase error $\phi_s = 0, 2\pi, \dots$ (i.e., for $\Delta\omega = 0$).

error to decay below a certain level, say ϕ_ε. In the presence of noise things are more complicated.

In contrast to the undisturbed PLL, the fact that the phase error enters the in-lock interval does not guarantee that the phase error subsequently remains within this interval. Due to noise there always exists the possibility that the phase error will become larger than $|\phi_\varepsilon|$ at some later point in time. Also, the loop will not always settle at the *nearest* equilibrium point but may settle at the equilibrium point of the adjacent 2π interval as shown in Figure 4.2-5. Therefore, we first have to find a suitable definition of the acquisition time.

We define the phase acquisition process to be completed when the trajectory enters one of the two intervals $\phi \in [-\phi_\varepsilon, \phi_\varepsilon]$ or $\phi \in [2\pi - \phi_\varepsilon, 2\pi + \phi_\varepsilon]$ for the first time. Then the phase acquisition time, T_p, is a random variable. The matter of most interest is the probability distribution of this random variable. We denote by

$$P[T_p \le t \mid \phi(0)]$$

the conditional probability that the phase error process reaches either of the two boundaries located at $\phi = \phi_\varepsilon$ or $\phi = 2\pi - \phi_\varepsilon$ within the time interval $[0, t]$ under the condition that the initial phase error is equal to $\phi(0)$. There are two different approaches to obtain the cumulative probability distribution $P[T_p \le t \mid \phi(0)]$. We could simulate the stochastic differential equation

(4.2-20) on a digital computer to obtain statistics on T_p. This approach is feasible if we are interested in probabilities of the order of magnitude 0.9–0.99. Very often, the probabilities of interest are of the order of magnitude $1 - 10^{-8}$; i.e., the probability that the loop has not acquired lock within a given interval of time

$$1 - P[T_p \le t \mid \phi(0)] \le 10^{-8}$$

is a very small number. Such small probabilities cannot be obtained by computer simulation since it would take far too much computer time to obtain reliable data of such extremely rare events.

Using the Fokker–Planck technique it is possible to compute the acquisition probability for the range of interest, say $1 - 10^{-8}$. We do not want to go into the mathematical details but merely discuss the results. The reader familiar with the Fokker–Planck technique is referred to Appendix 4.2.

If the initial phase error $\phi(0)$ is very close to the unstable null, the phase can dwell near the null for an extended time. Clearly, this hang-up of the loop is worst for $\phi(0) = \pi$ when $\Delta\omega = 0$. Figure 4.2-6 shows the probability of synchronization failure during the first t seconds, $1 - P[T_p \le t \mid \phi(0) = \pi]$,

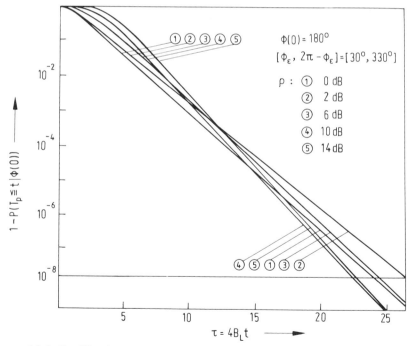

Figure 4.2-6. Conditional probability of not acquiring lock during the first t seconds $(1 - P[T_p \le t \mid \phi(0) = \pi])$ for $\Delta\omega = 0$ (From Heinrich Meyr and Luitjens Popken, Phase Acquisition Statistics for Phase-Locked Loops, *IEEE Transactions on Communications*, Vol. 28, No. 8 © IEEE 1980).

with signal-to-noise ratio in the loop ρ as parameter. A typical example for an initial phase error $\phi(0) + \pi$ is shown in Figure 4.2-7. The higher ρ, the closer the probability function $P[T_p \leq t \mid \phi(0) = 120°]$ approximates the unit step function $u(\tau - \tau_D)$ where τ_D stands for the normalized acquisition time in the deterministic case from 4.2-13

$$\tau_D = \ln \left(\frac{\tan [\phi(0)/2]}{\tan [\phi_\varepsilon/2]} \right), \quad \Delta\omega = 0 \quad (4.2\text{-}21)$$

In the case of greatest practical interest the initial phase error is uniformly distributed over $[\phi_\varepsilon, 2\pi - \phi_\varepsilon]$. The synchronization failure probability distribution

$$1 - P\{T_p \leq t \mid \phi(0) \in [\phi_\varepsilon, 2\pi - \phi_\varepsilon]\}$$

$$= 1 - \frac{1}{2(\pi - \phi_\varepsilon)} \int_{\phi_\varepsilon}^{2\pi - \phi_\varepsilon} P[T_p \leq t \mid \phi(0)] \, d\phi(0) \quad (4.2\text{-}22)$$

is plotted in Figure 4.2-8. The curve for $\rho \to \infty$ can be obtained in closed form by integrating $u\{\tau - \tau_D[\phi(0)]\}$ over $\phi(0)$.

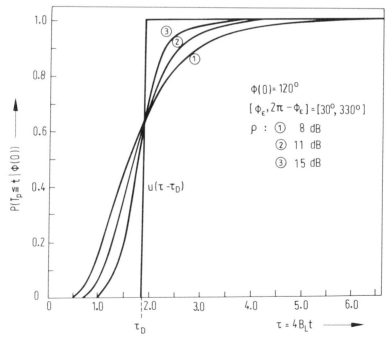

Figure 4.2-7. Conditional probability of acquiring lock $P[T_p \leq t \mid \phi(0) = 120°]$ (From Heinrich Meyr and Luitjens Popken, Phase Acquisition Statistics for Phase-Locked Loops, *IEEE Transactions on Communications*, Vol. 28, No. 8 © IEEE 1980).

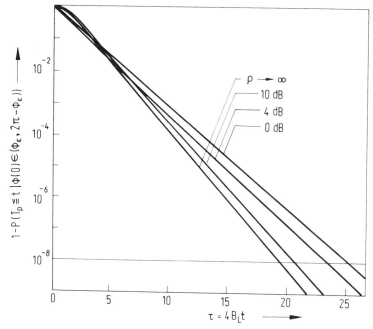

Figure 4.2-8. Probability of synchronization failure for a uniformly distributed initial phase error $\phi(0) \in [\phi_\varepsilon, 2\pi - \phi_\varepsilon]$ (From Heinrich Meyr and Luitjens Popken, Phase Acquisition Statistics for Phase-Locked Loops, *IEEE Transactions on Communications*, Vol. 28, No. 8 © IEEE 1980).

$$1 - P\{\tau_D \leq \tau \mid \phi(0) \in [\phi_\varepsilon, 2\pi - \phi_\varepsilon]\}$$

$$= 1 - \frac{1}{\pi - \phi_\varepsilon} \left\{ 2 \tan^{-1} \left[\exp(\tau) \tan \left(\frac{\phi_\varepsilon}{2} \right) \right] - \phi_\varepsilon \right\} \qquad (4.2\text{-}23)$$

Note that for reasonable SNRs ($\rho \geq 10\,\mathrm{dB}$) and small probabilities of failure to acquire lock, all curves are close to the asymptotic curve $\rho \to \infty$. In this case we can conclude that, on average, noise slightly prolongs acquisition.

Finally, we want to qualitatively discuss the case with loop detuning. In principle, no fundamental difference exists for the case with $\Delta\omega = 0$. The points of equilibrium are now

$$\phi_s = \begin{cases} \sin^{-1}(\Delta\omega/4B_L) & \text{stable} \\ \pi - \sin^{-1}(\Delta\omega/4B_L), & \text{unstable} \end{cases}$$

Hang-up occurs at $\phi_s = \pi - \sin^{-1}(\Delta\omega/4B_L)$ instead of $\phi_s = \pi$. (In other words, loop detuning is *no* remedy against hang-up.) Using the Fokker–Planck technique one can obtain the probability of acquisition $P(T_p \leq t)$ in the same way as for $\Delta\omega = 0$.

4.3. SECOND-ORDER LOOP

In a second-order loop, phase and frequency must be acquired before the loop can operate in its tracking mode. Usually, phase acquisition is not a problem. (There are, however, important applications where the unaided phase acquisition process is too slow.) Unaided acquisition of frequency is more difficult. It is a slow and unreliable process. As a consequence, the literature has concentrated on frequency acquisition to the extent that acquisition is almost synonymous with frequency acquisition of a second-order loop. Unaided acquisition of frequency is known as *pull-in* for reasons which will soon become clear.

4.3.1. Frequency Acquisition

We now assume a noise-free input signal; the influence of noise will be discussed later. The dynamic equations of a second-order loop with imperfect filter and with multiplier phase detector given a frequency offset $d\theta/d\tau = \Delta\omega/\omega_n$ are (see Chapter 2, equation (2.4-21))

$$\frac{d\phi(\tau)}{d\tau} = -2\zeta\left(1 - \frac{\beta}{2\zeta}\right)\sin\phi(\tau) - x_1(\tau) + \frac{\Delta\omega}{\omega_n}$$

$$\frac{dx_1(\tau)}{d\tau} = -\beta x_1(\tau) + (1 - 2\zeta\beta + \beta^2)\sin\phi(\tau)$$

(4.3-1)

For convenience, the equivalent mathematical model is repeated in Figure 4.3-1. Proceeding as for the first-order loop, we first determine the points of equilibrium. At equilibrium, the derivatives of the state variables with respect to time equal zero and they assume the values

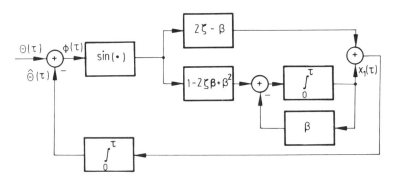

Figure 4.3-1. Equivalent baseband model of a second-order PLL with imperfect integrator.

$$\phi_s = \sin^{-1}\left(\frac{\Delta\omega}{\omega_n}\beta\right)$$

(4.3-2)

$$x_{1,s} = \frac{\Delta\omega}{\omega_n}(1 - 2\zeta\beta + \beta^2)$$

Because of the properties of the sine function, the first equation has two solutions

$$\phi_s = \begin{cases} \sin^{-1}\left(\dfrac{\Delta\omega}{\omega_n}\beta\right) + 2\pi k & \text{stable} \\[2ex] \pi - \sin^{-1}\left(\dfrac{\Delta\omega}{\omega_n}\beta\right) + 2\pi k & \text{unstable} \end{cases} , \quad k = 0, \pm 1, \ldots$$

(4.3-3)

Using the same reasoning as for the first-order loop, one can show that the first solution of (4.3-3) represents a stable point of equilibrium while the other corresponds to an unstable point of equilibrium.

The maximum frequency offset $\Delta\omega$ that the loop can track is the *hold-in range* $\Delta\omega_H$. From the first equation of (4.3-2)

$$\Delta\omega_H = \frac{\omega_n}{\beta} = K_0 K_D , \quad \text{hold-in range}$$

(4.3-4)

Equation (4.3-4) states that the hold-in range, in principle, can be made arbitrarily large by increasing the dc loop gain ($K_0 K_D$). Also, note that PLL with perfect integrator ($\beta = 0$) theoretically has an infinite hold-in range. (In a real loop the gain is always limited because of saturation of elements such as amplifiers within the loop.)

In order to understand the acquisition process it proves convenient to introduce *phase-plane techniques*. A phase plane is constructed by numerically solving the state equations (4.3-1) (no analytical solution exists). For every time τ, the state vector $x^T = [\phi(\tau), x_1(\tau)]$ is represented by a point in the plane with the coordinate axes ϕ (horizontal) and x_1 (vertical). A plot of a single solution $x(\tau)$, $\tau \geq 0$, of the state equation is known as a *trajectory*; it shows the dynamics of a loop as it settles (or fails to settle) towards a point of equilibrium. For clarity of presentation, the time history along the trajectory is usually omitted.

A typical phase plane portrait for a loop filter with perfect integrator ($\beta = 0$) is shown in Figure 4.3-2. Note that the vertical axis shows the difference $[x_1 - (\Delta\omega/\omega_n)]$ where $\Delta\omega/\omega_n$ would be the steady-state value of x_1 when $\beta = 0$. The shaded area is the *lock-in range* of the stable equilibrium point A_0. All trajectories that start within this area will terminate at point A_0; furthermore, the phase error of these trajectories will never exceed $\pm\pi$. Since the phase detector characteristic is periodic with period 2π, identical

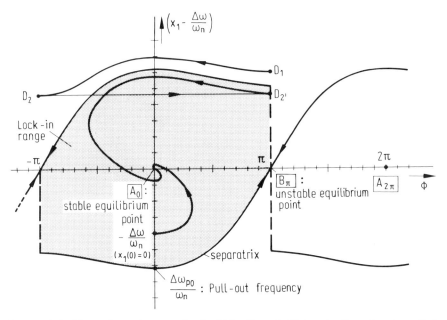

Figure 4.3-2. Phase plane of a second-order PLL with perfect integrator (loop damping ratio is $\zeta = 0.707$).

lock-in ranges are found at the other stable equilibrium points that are multiples of 2π away from point A_0.

Any trajectory that starts with an initial phase error $|\phi| < \pi$ but outside the shaded area will not terminate at the point A_0 but at another stable point $A_{k2\pi}$ (if at all). We then say that the PLL slips cycles before it settles at a stable equilibrium point. More precisely, a PLL is said to slip a cycle when the phase error increases by $+2\pi$ (positive cycle slip) or -2π (negative cycle slip).

As may be seen from Figure 4.3-2, the boundary (solid line) of the lock-in range goes through the unstable equilibrium point. This must be so; the boundary is the trajectory that separates the trajectories terminating at the stable equilibrium point from all other trajectories starting in a particular 2π interval. Since, by definition, it belongs to neither of the two classes, the trajectory must terminate at the unstable equilibrium point. Such a trajectory is called a *separatrix*. The trajectories are traversed only counterclockwise as shown by the flow arrows.

The phase detector is unable to distinguish between a phase error of ϕ and a phase error of $\phi + k2\pi$; therefore, it suffices to plot the trajectories in a single interval of 2π width without losing information. For example, the trajectory starting at D_1 will terminate at $A_{-2\pi}$; instead of continuing it into the interval to the left of $\phi = -\pi$ it is "restarted" at point $D_{2'}$.

While the lock-in range is a useful concept in understanding the acquisition process, we need a simpler though not as complete description. Let us assume that a frequency step is applied to the loop at time $\tau = 0$ which was initially in the null state, $x^T(0) = [\phi(0) = 0, x_1(0) = 0]$. There is a maximum frequency step for which the loop remains in lock without cycle slips and we denote this limit the *pull-out frequency* $\Delta\omega_{po}$. The pull-out frequency is simply found at the interception of the separatrix with the vertical axis, $\phi = 0$. Using the phase plane portrait Gardner [3: p. 57] has derived a very useful empirical formula for $\Delta\omega_{po}$.

$$\Delta\omega_{po} = 1.8\omega_n(\zeta + 1), \quad \text{pull-out frequency} \tag{4.3-5}$$

The formula is valid for $0.5 \leq \zeta \leq 1.4$ and a second-order loop with a perfect integrator and a multiplier type phase detector. It is also a good approximation for a high-gain loop with imperfect integrator ($\beta \ll 1$) and small loop detuning.

The larger the initial difference $|x_1(0) - (\Delta\omega/\omega_n)|$ the more cycles the loop will slip before it settles. Theoretically, a second-order loop with perfect integrator in the loop filter can pull in an arbitrarily large difference (shown later). This is not true for a second-order loop with a passive loop filter; since it can track only if the frequency difference $\Delta\omega$ is smaller than the hold-in range (4.3-4); it can certainly not pull in a larger frequency difference. In fact, the largest frequency difference which the loop pulls in is much smaller than the hold-in range. The maximum frequency offset $\Delta\omega$ for which the loop pulls in is called the *pull-in limit* and labeled $\Delta\omega_p$. The reader might be surprised that $\Delta\omega_p \ll \Delta\omega_H$. To see this, we have to consider the fundamental difference between tracking and acquisition. While the idea of the hold-in range presumes that the loop is already in lock and the input frequency is steadily increased until the loop loses lock, with the pull-in limit the loop is initially out of lock.

We now want to study the pull-in process. Firstly, we want to qualitatively understand the mechanism that pulls the loop into lock. Next we approximately compute the pull-in limit $\Delta\omega_p$, and, finally, we compute the pull-in limit T_f.

Let us assume that the initial difference $|x_1(0) - (\Delta\omega/\omega_n)|$ is much larger than the pull-out frequency, but smaller than the hold-in range. Then the loop will slip cycles. Looking at the phase plane (Figure 4.3-2) we see that the variation of x_1 over a cycle (difference in "heights" of the poins D_1 and D_2) is small when compared to the value of x_1. Thus, x_1 may be assumed constant during this period of time and the first equation in (4.3-1) looks like the equation of a first-order loop with normalized frequency detuning $(\Delta\omega_n - x_1) > 0$.

$$\frac{d\phi(\tau)}{d\tau} = -2\zeta\sin\phi(\tau) + \left(-x_1 + \frac{\Delta\omega}{\omega_n}\right), \quad 2\zeta\left(1 - \frac{\beta}{2\zeta}\right) \approx 2\zeta \tag{4.3-6}$$

with

$$\frac{\Delta\omega}{\omega_n} - x_1 \gg 2\zeta$$

From the previous section we know that a first-order loop with such conditions produces a periodic oscillation (beat note). The beat note is not symmetric but has a dc-component which is obtained by replacing $(\Delta\omega/4B_L)$ in equation (4.2-18) by $(1/2\zeta)[(\Delta\omega/\omega_n) - x_1]$

$$\overline{\sin\phi} = \frac{1}{2\zeta}\left(\frac{\Delta\omega}{\omega_n} - x_1\right) - \left[\left(\frac{1}{2\zeta}\right)^2\left(\frac{\Delta\omega}{\omega_n} - x_1\right)^2 - 1\right]^{1/2}$$

$$= \frac{1}{2\zeta}\left(\frac{\Delta\omega}{\omega_n} - x_1\right) - \frac{\pi}{\zeta\tau_{bn}} \qquad (4.3\text{-}7)$$

with

$$\tau_{bn} = \frac{\pi}{\zeta}\frac{1}{[[1/(2\zeta)^2][(\Delta\omega/\omega_n) - x_1]^2 - 1]^{1/2}}$$

It is exactly this dc-component that is integrated in the loop filter of a second-order loop and that reduces the initial difference $[\Delta\omega/\omega_n - x_1(0)]$ to its steady-state value. this follows easily from equation (4.3-7), which we slightly rearrange to obtain

$$\overline{\sin\phi} = \frac{1}{2\zeta}\left(\frac{\Delta\omega}{\omega_n} - x_1\right) - \frac{1}{2\zeta}\left(\frac{\Delta\omega}{\omega_n} - x_1\right)\left[1 - \frac{1}{(1/2\zeta)^2[(\Delta\omega/\omega_n) - x_1]^2}\right]^{1/2}$$

$$(4.3\text{-}8)$$

As an example, assume that x_1 is smaller than $\Delta\omega/\omega_n > 0$. In this case the average value of $\overline{\sin\phi}$ is positive and the accumulated voltage in the integrator of the loop filter during one cycle of the beat note is also positive.

$$\overline{\Delta x_1} = \int_0^{\tau_{bn}} \overline{\sin\phi}\, d\tau = \overline{\sin\phi}\,\tau_{bn} > 0 \qquad (4.3\text{-}9)$$

But this means that the difference $[(\Delta\omega/\omega_n) - x_1]$ steadily *decreases* with every cycle slip; the loop pulls in. For a perfect integrator in the loop filter the difference $[(\Delta\omega/\omega_n) - x_1]$ that the loop can pull in can be (theoretically) arbitrarily large. However, the average increment

$$\overline{\Delta x_1} = \overline{\sin\phi}\,\tau_{bn} = \frac{\pi}{2\zeta^2}\frac{(\Delta\omega/\omega_n) - x_1}{[(1/2\zeta)^2[(\Delta\omega/\omega_n) - x_1]^2 - 1]^{1/2}} - \frac{\pi}{\zeta}$$

$$(4.3\text{-}10)$$

becomes very small for a large difference $[(\Delta\omega/\omega_n) - x_1]$. Expanding the fraction in (4.3-10) into a Taylor series and retaining only the first term yields

$$\overline{\Delta x_1} \approx \frac{2\pi\zeta}{[(\Delta\omega/\omega_n) - x_1]^2} , \quad (\Delta\omega/\omega_n) - x_1 > 0 \text{ is assumed} \quad (4.3\text{-}11)$$

which shows that the pull-in effect is inversely proportional to the square of the frequency difference.

Next we want to study the pull-in process of a PLL with an imperfect integrator (see Figure 4.3-1). If we replace the input $\sin\phi$ by its average value, $\overline{\sin\phi}$ and assume that $1 - 2\zeta\beta + \beta^2 \approx 1$, the state variable x_1 obeys the differential equation

$$\frac{dx_1(\tau)}{d\tau} = -\beta x_1(\tau) + \overline{\sin\phi} \quad (4.3\text{-}12)$$

Again, assume that the frequency offset is positive and that x_1 is smaller than $\Delta\omega/\omega_n$ (negative $\Delta\omega$ are treated analogously)

$$\frac{\Delta\omega}{\omega_n} - x_1 > 0 , \quad \frac{\Delta\omega}{\omega_n} > 0 \quad (4.3\text{-}13)$$

Then the PLL can pull in *only* if the derivative $dx_1/d\tau$ is positive since otherwise the voltage Δx_1 accumulated during one cycle cannot be positive and the difference (4.3-13) will not decrease. We thus have found a necessary condition for pull-in

$$\frac{dx_1}{d\tau} = -\beta x_1 + \overline{\sin\phi} > 0 , \quad \text{for} \quad \frac{\Delta\omega}{\omega_n} - x_1 > 0 \quad (4.3\text{-}14)$$

Expanding (4.3-8) into a Taylor series and retaining only the first term of the series

$$\overline{\sin\phi} \approx \frac{\zeta}{(\Delta\omega/\omega_n) - x_1} \quad (4.3\text{-}15)$$

we obtain

$$-\beta x_1 + \frac{\zeta}{(\Delta\omega/\omega_n) - x_1} > 0 \quad (4.3\text{-}16)$$

as a necessary condition for pull-in. Since $[(\Delta\omega/\omega_n) - x_1]$ is positive we may multiply both sides of the inequality (4.3-16) by this factor to obtain

$$-\beta x_1\left(\frac{\Delta\omega}{\omega_n} - x_1\right) + \zeta > 0$$

or slightly rearranged

$$f(x_1) = x_1^2 - \left(\frac{\Delta\omega}{\omega_n}\right)x_1 + \frac{\zeta}{\beta} > 0 \qquad (4.3\text{-}17)$$

The quadratic form $f(x_1)$ (Figure 4.3-3) is strictly positive if the discriminant is negative

$$\left(\frac{\Delta\omega}{\omega_n}\right)^2 - \frac{4\zeta}{\beta} < 0$$

that is

$$\frac{\Delta\omega}{\omega_n} < 2\sqrt{\frac{\zeta}{\beta}} \qquad (4.3\text{-}18)$$

By definition, the right-hand side of the inequality (4.3-18) is the pull-in limit we are looking for

$$\Delta\omega_p = 2\omega_n\sqrt{\frac{\zeta}{\beta}}, \quad \text{pull-in limit} \qquad (4.3\text{-}19)$$

For $\Delta\omega > \Delta\omega_p$ the quadratic form has two real solutions

$$
\begin{aligned}
(x_1)_{\text{LIMIT}} &= \frac{1}{2}\frac{\Delta\omega}{\omega_n} - \frac{1}{2}\left[\left(\frac{\Delta\omega}{\omega_n}\right)^2 - \frac{4\zeta}{\beta}\right]^{1/2} \\
(x_1)_{\text{PULL-IN}} &= \frac{1}{2}\frac{\Delta\omega}{\omega_n} + \frac{1}{2}\left[\left(\frac{\Delta\omega}{\omega_n}\right)^2 - \frac{4\zeta}{\beta}\right]^{1/2}
\end{aligned}
\qquad (4.3\text{-}20)
$$

If the initial value $x_1(0)$ is larger than $(x_1)_{\text{PULL-IN}}$ the loop will converge towards its stable tracking point. For any other initial value the loop will not lock but produce a beat note (limit cycle) as we will explain.

If $x_1(0)$ is smaller than $(x_1)_{\text{LIMIT}}$ the quadratic form is positive (Figure 4.3-3), which implies by (4.3-14) that $dx_1/d\tau$ is also positive and the loop starts to pull in. When the value x_1 reaches $(x_1)_{\text{LIMIT}}$ the quadratic form and, hence, $dx_1/d\tau$ becomes zero. But since the quadratic form (and thus $dx_1/d\tau$) changes its sign for $x_1 > (x_1)_{\text{LIMIT}}$, the state variable decreases and the loop no longer pulls in but returns to the value $x_1 = (x_1)_{\text{LIMIT}}$. Since the value $(x_1)_{\text{LIMIT}}$ is different from $\Delta\omega/\omega_n$ the phase error $\phi(\tau)$ cannot be constant but produces a periodic oscillation with a frequency $\Delta\omega_{bn}$ equal to the difference of loop detuning $\Delta\omega/\omega_n$ and $(x_1)_{\text{LIMIT}}$

$$
\begin{aligned}
\frac{\Delta\omega_{bn}}{\omega_n} &= \frac{\Delta\omega}{\omega_n} - (x_1)_{\text{LIMIT}} \\
\frac{\Delta\omega_{bn}}{\omega_n} &= \frac{1}{2}\frac{\Delta\omega}{\omega_n} + \frac{1}{2}\left[\left(\frac{\Delta\omega}{\omega_n}\right)^2 - \frac{4\zeta}{\beta}\right]^{1/2}
\end{aligned}
\qquad (4.3\text{-}21)
$$

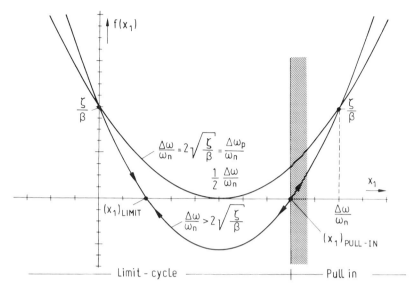

Figure 4.3-3. Quadratic form $f(x_1)$ with $(\Delta\omega/\omega_n)$ as parameter $(1 - 2\zeta\beta + \beta^2 \approx 1)$.

From (4.3-21) it can be seen that $\Delta\omega_{bn}/\omega_n$ is always smaller than the loop detuning $\Delta\omega/\omega_n$ but larger than half the pull-in limit

$$\frac{1}{2}\frac{\Delta\omega_p}{\omega_n} = \sqrt{\frac{\zeta}{\beta}} < \frac{\Delta\omega_{bn}}{\omega_n} < \frac{\Delta\omega}{\omega_n}$$

Example: Let us consider the case of a second-order loop with $\zeta = 0.707$ and $\beta = 0.031$. This means that $\Delta\omega_p/\omega_n = 2\sqrt{\zeta/\beta} = 9.6$. For a loop detuning of $\Delta\omega/\omega_n = 15$ which is larger than $\Delta\omega_p/\omega_n$, pull-in depends on the initial conditions. We have from (4.3-2) and (4.3-20)

$$x_{1,s} = 14.4$$

$$(x_1)_{\text{LIMIT}} = 1.72$$

$$(x_1)_{\text{PULL-IN}} = 13.3$$

By numerically solving the state equations (4.3-1), we obtain Figure 4.3-4 which shows exactly these results. For an initial value of the state variable x_1 of $x_1(0) = 13.5 > (x_1)_{\text{PULL-IN}}$, the loop locks into the tracking point $x_{1,s}$; if the initial value of x_1 is below $(x_1)_{\text{PULL-IN}}$ (in the figure we have chosen $x_1(0) = 13$), the loop cannot lock and produces a limit cycle.

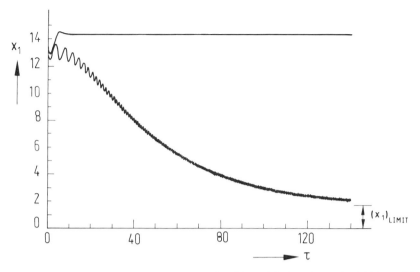

Figure 4.3-4. Example of the dependence of pull-in and limit cycle on the initial conditions of the state variables.

By solving the state equation, it can be seen that for values of x_1 in the close vicinity of the pull-in limit $(x_1)_{\text{PULL-IN}}$, the convergence of the system towards a stable tracking point depends on the initial phase error $\phi(0)$. This behavior is not predicted by our theory. However, coming back to the assumptions we have made to obtain $(x_1)_{\text{PULL-IN}}$ and $(x_1)_{\text{LIMIT}}$ one easily recognizes why this is so; we have approximated the second-order loop by a first-order loop and subsequently computed the *average* voltage Δx_1 accumulated during one beat note which is, of course, independent of the starting value $\phi(0)$. But the increment Δx_1 integrated over the first cycle depends on the initial conditions $[x_1(0), \phi(0)]$. If $x_1(0)$ lies very close to $(x_1)_{\text{PULL-IN}}$ then the value of x_1 after the first cycle is either slightly larger or smaller than $(x_1)_{\text{PULL-IN}}$, depending on $\phi(0)$. Thus, whether the loop pulls in depends on $\phi(0)$.

Figure 4.3-5 shows the phase plane for the PLL of the above example. We can read off $(x_1)_{\text{PULL-IN}} - x_{1,s} = -0.6$ which means $(x_1)_{\text{PULL-IN}} = 13.8$. This is slightly higher than the value predicted by (4.3-20) due to all the simplifying assumptions made in reaching that equation. The effect of $\phi(0)$ on pull-in can be seen clearly in this figure. For example, assume $x_1(0) - x_{1,s} = -0.7$. Then for the values of $\phi(0) - \phi_s = 0$ (point P_0) and 1.78 rad (point P_1), the loop will pull in to the points A_0 and $A_{2\pi}$, respectively; but for $\phi(0) - \phi_s = 2.22$ rad (point P_2) the loop cannot pull in at all and ends up with limit cycles.

We next consider the case where the initial value $x_1(0)$ is larger than $\Delta\omega/\omega_n > 0$. This is the area of the right-hand side of $x_1 = \Delta\omega/\omega_n$ in Figure 4.3-3. We claim that the loop will pull in for an arbitrary value

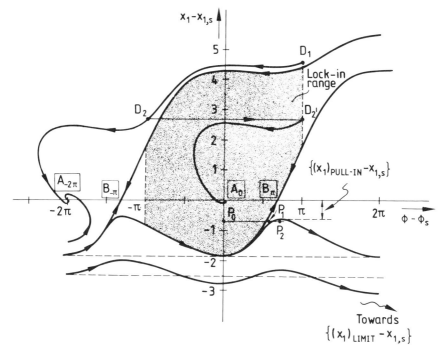

Figure 4.3-5. Phase plane of an imperfect second-order PLL with $\beta = 0.031$, $\zeta = 0.707$, and $\Delta\omega/\omega_n = 15 > (\Delta\omega_p/\omega_n)$.

$x_1(0) > \Delta\omega/\omega_n > 0$. Indeed, the dc-component $\overline{\sin\phi}$ is negative for $[(\Delta\omega/\omega_n) - x_1] < 0$ as follows from (4.3-8). Since $x_1 > 0$ it follows that the derivative

$$\frac{dx_1(\tau)}{d\tau} = -\beta x_1(\tau) + \overline{\sin\phi} < 0$$

is negative and, therefore, the value of x_1 decreases until it reaches the steady-state value.

The *pull-in time* τ_f can be obtained by inserting $\overline{\sin\phi}$ from (4.3-7) into (4.3-12). The resulting differential equation

$$\frac{dx_1(\tau)}{d\tau} = -\beta x_1(\tau) + \frac{1}{2\zeta}\left(\frac{\Delta\omega}{\omega_n} - x_1\right) - \left[\left(\frac{1}{2\zeta}\right)^2\left(\frac{\Delta\omega}{\omega_n} - x_1\right)^2 - 1\right]^{1/2}$$

can be solved by separation of the integration variables

$$d\tau = \frac{dx_1}{-\beta x_1 + (1/2\zeta)[(\Delta\omega/\omega_n) - x_1] - \{(1/2\zeta)^2[(\Delta\omega/\omega_n) - x_1]^2 - 1\}^{1/2}}$$

$$(4.3\text{-}22)$$

The pull-in time τ_f is defined as the time required for the frequency $x_1(\tau)$ to change from the initial value $x_1(0)$ to the lock limit $x_1(\tau_f)$ where the loop stops oscillating, i.e., $\tau_{bn} \to \infty$. From (4.3-7) we find that for $\tau_{bn} \to \infty$

$$\frac{1}{2\zeta} \left[\frac{\Delta\omega}{\omega_n} - x_1(\tau_f) \right] = 1 \tag{4.3-23}$$

Solving (4.3-23) for x_1 gives

$$x_1(\tau_f) = \left(\frac{\Delta\omega}{\omega_n} \right) - 2\zeta \tag{4.3-23a}$$

Integrating (4.3-22) between $x_1(0)$ and $x_1(\tau_f)$ yields

$$\tau_f = \int_0^{\frac{\Delta\omega}{\omega_n} - 2\zeta} \frac{dx_1}{-\beta x_1 + \frac{1}{2\zeta} \left[\frac{\Delta\omega}{\omega_n} - x_1 \right] - \left\{ \left(\frac{1}{2\zeta} \right)^2 \left[\frac{\Delta\omega}{\omega_n} - x_1 \right]^2 - 1 \right\}^{1/2}} \tag{4.3-24}$$

Lindsey [8] has carried out the integration. For large $\Delta\omega/\omega_n$ he gives the approximate formula for $T_f = \tau_f/\omega_n$

$$B_L T_f \approx \frac{\pi^2 (4\zeta^2 + 1)^3}{256\zeta^4} \left(\frac{\Delta f}{B_L} \right)^2, \qquad \frac{\Delta\omega}{\omega_n} < \frac{\Delta\omega_p}{\omega_n} = 2\sqrt{\frac{\zeta}{\beta}} \tag{4.3-25}$$

A narrowband loop can take a very long time to pull in. For example, the time required to pull the VCO 2 kHz will be greater than 16 s if $B_L = 100$ Hz, $\zeta = 0.707$.

Let us now consider the influence of noise on the pull-in effect. Qualitatively, the noise $m(\tau)$ causes a random fluctuation of the increment Δx_1. Its effect on pull-in can be roughly estimated by computing the variance of the integrated noise in the loop filter during one beat note. The approximate model during a beat note starting at $\tau = 0$ is depicted in Figure 4.3-6. If

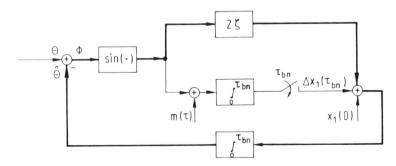

Figure 4.3-6. Approximate model of the PLL during a beat note ($0 \le \tau \le \tau_{bn}$).

$m(\tau) \equiv 0$ the model describes a first-order loop with detuning $[(\Delta\omega/\omega_n) - x_1(0)]$; this is exactly the model we used previously to compute the dc-component $\overline{\sin\phi}$.

In the model of Figure 4.3-6 the noise $m(\tau)$ only causes a random fluctuation of the increment Δx_1 and thus has no influence on the beat note. This is, of course, a simplifying assumption. Nevertheless, it serves our purpose well in demonstrating the disastrous effect of noise on the pull-in effect.

The noise $m(\tau)$ is modeled as a white gaussian noise. As shown in (3.4-12), the spectral density of $m(\tau)$ in normalized frequency $\Omega = \omega/\omega_n$ is given by

$$S_m(\Omega) = \omega_n \frac{N_0}{2A^2} \qquad (4.3\text{-}26)$$

It is well known that the Wiener process is the integral of a white process. The variance of the Wiener process Δx_1

$$\underbrace{\Delta x_1(\tau)}_{\substack{\text{Wiener} \\ \text{process}}} = \int_0^\tau m(\nu)\,d\nu \qquad (4.3\text{-}27)$$

is proportional to the time increment τ (see, e.g., Papoulis [6]). Hence, for the time interval of a beat note τ_{bn} we find

$$\sigma^2_{\Delta x_1} = \left(\omega_n \frac{N_0}{2A^2}\right)\tau_{bn} \qquad (4.3\text{-}28)$$

Inserting for τ_{bn}, the result of (4.3-7) yields for $(1/2\zeta)[(\Delta\omega/\omega_n) - x_1] \gg 1$ (basic assumption for a beat note in (4.3-6)) approximately

$$\sigma^2_{\Delta\omega_1} = \frac{\omega_n N_0}{2A^2} \frac{\pi}{\zeta} \frac{1}{\{(1/2\zeta)^2[(\Delta\omega/\omega_n) - x_1]^2 - 1\}^{1/2}} \approx \frac{\omega_n N_0}{2A^2} \frac{\pi}{\zeta} \frac{2\zeta}{(\Delta\omega/\omega_n) - x_1}$$

$$\approx \frac{4\zeta}{1 + 4\zeta^2} \frac{1}{\rho} \frac{2\pi}{(\Delta\omega/\omega_n) - x_1} \qquad (4.3\text{-}29)$$

with

$$\frac{1}{\rho} = \frac{N_0}{A^2} \frac{\omega_n \zeta}{2} \left(\frac{1 + 4\zeta^2}{4\zeta^2}\right)$$

For a reliable pull-in process, the RMS value $\sigma_{\Delta x_1}$ should be small compared to the useful signal, $\overline{\Delta x_1}$. Dividing $\sigma_{\Delta x_1}$ by $\overline{\Delta x_1}$ from (4.3-11) yields

$$\frac{\sigma_{\Delta\omega_1}}{\Delta x_1} = \left[\frac{2}{\pi\rho\zeta(1 + 4\zeta^2)} \left(\frac{\Delta\omega}{\omega_n} - x_1 \right)^3 \right]^{1/2} \qquad (4.3\text{-}30)$$

Example: For a PLL with a loop damping factor of $\zeta = \sqrt{2}/2$ and an initial frequency difference of ten times the normalized loop bandwidth

$$\frac{\Delta\omega}{\omega_n} - x_1(0) = 10\left(\frac{B_L}{\omega_n} \right)$$

a signal-to-noise ratio in the loop $\rho = 16.5$ dB is required to make the useful signal just equal in the RMS value of the random disturbance.

Putting together the various pieces of information of this section one comes to the simple conclusion that pull-in is *not* a practical solution for frequency acquisition because it is a weak and unreliable process. Some form of aided acquisition is needed as discussed in Chapter 5.

4.4. GENERALIZED STUDY OF FREQUENCY ACQUISITION FAILURE

In the previous section we have determined the condition for a limit cycle of an imperfect second-order loop with sinusoidal phase detector characteristic.

The general solution to the problem for an arbitrary phase detector and loop transfer function is of considerable interest. In a practical loop there always exist additional poles which may introduce additional points of false frequency lock. A practical loop may well have a dozen poles. Various approximate analytical solutions to the problem are known. Unfortunately, they are restricted in terms of either filter type, phase detector characteristic or both.

A completely general numerical solution to the problem has been given by Greenstein [7]. The idea behind this approach is to find conditions for a periodic solution of the loop equation which is a necessary condition for a limit cycle to exist. The following derivation closely follows Greenstein's work.

The block diagram of the generalized PLL with a fixed frequency offset $\Delta\omega$ as input is shown in Figure 4.4-1. The phase detector characteristic is assumed to have odd symmetry, a period of 2π and a unity peak value. The dc-gain of the loop filter transfer function is normalized to unity, $F(0) = 1$. Otherwise, the transfer function is arbitrary and may also include additional (perhaps unwanted) poles from various sources.

To begin the analysis, we write $a(t)$ as the complex Fourier series

$$a(t) = \sum_{k=-\infty}^{\infty} A_k \exp(j\omega_{bn}kt) \qquad (4.4\text{-}1)$$

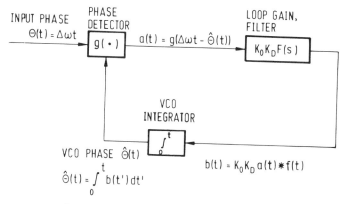

Figure 4.4-1. Generalized phase-locked loop.

where ω_{bn} is the frequency of the beat note (limit cycle) and

$$A_k = \frac{\omega_{bn}}{2\pi} \int_0^{2\pi/\omega_{bn}} a(t) \exp(-j\omega_{bn}kt) \, dt \qquad (4.4\text{-}2)$$

The filtered signal $b(t)$ is then found in frequency domain by its Fourier series coefficients as

$$B_k = A_k \cdot K_0 K_D F(s)\big|_{s=j\omega_{bn}k}$$

Therefore

$$b(t) = K_0 K_D \sum_{k=-\infty}^{\infty} A_k F(s = j\omega_{bn}k) \exp(j\omega_{bn}kt) \qquad (4.4\text{-}3)$$

and the VCO phase $\hat{\theta}(t)$ is

$$\hat{\theta}(t) = K_0 K_D A_0 \int_0^t dt' + \underbrace{\sum_{\substack{k=-\infty \\ k \neq 0}}^{\infty} K_0 K_D A_k \left[\frac{F(j\omega_{bn}k)}{j\omega_{bn}k} \right] \exp(j\omega_{bn}kt)}_{\theta_k} \qquad (4.4\text{-}4)$$

Notice that the infinite series in the last equation describes a periodic phase modulation of the VCO. Finally, the phase detector output can be written

$$a(t) = g\!\left((\Delta\omega - K_0 K_D A_0)t - K_0 K_D \sum_{\substack{k=-\infty \\ k \neq 0}}^{\infty} A_k \frac{F(j\omega_{bn}k)}{j\omega_{bn}k} \exp(j\omega_{bn}kt) \right) \qquad (4.4\text{-}5)$$

The fundamental frequency ω_{bn} of the beat note corresponds to the linearly increasing phase in (4.4-5). This leads to the relationship

$$\Delta\omega - K_0 K_D A_0 = \omega_{bn} \tag{4.4-6}$$

Using the normalization $\tau = K_0 K_D t$ and (4.4-1) and (4.4-5) we obtain the nonlinear equations describing the limit cycle

$$\sum_{k=-\infty}^{\infty} A_k \exp(j\Omega_{bn} k\tau) = g\left[\Omega_{bn}\tau - \sum_{\substack{k=-\infty \\ k \neq 0}}^{\infty} A_k \frac{F(j\Omega_{bn} K_0 K_D k)}{j\Omega_{bn} k} \exp(j\Omega_{bn} k\tau)\right] \tag{4.4-7}$$

with

$$\Omega_{bn} = \frac{\omega_{bn}}{K_0 K_D} \quad \text{and} \quad \Delta\Omega = \frac{\Delta\omega}{K_0 K_D}$$

Given the phase detector and filter function, the A_k that solve (4.4-7) are functions solely of Ω_{bn}. If these A_k exist and can be found, then the frequency offset $\Delta\omega$ that produces the beat note frequency ω_{bn} can be obtained from (4.4-6)

$$\Delta\Omega = \Omega_{bn} + A_0 \tag{4.4-8}$$

where A_0 is a function of Ω_{bn}.

In general, the variation of $\Delta\Omega$ with Ω_{bn} will have one of the two forms shown in Figure 4.4-2. In Figure 4.4-2a, $\Delta\Omega$ exhibits a minimum with respect to Ω_{bn} and there is also an Ω_{bn} below which no solution to (4.3-37) (i.e., no limit cycle) exists. The smallest $\Delta\Omega$ for which an Ω_{bn} exists equals the pull-in limit: for smaller values of $\Delta\Omega$ the loop cannot sustain a nonlinear oscillation and pulls in. In Figure 4.4-2b the minimum coincides with the value of $\Delta\Omega$ below which no solution to (4.4-7) exists. Computerized methods are necessary to solve equation (4.4-7). They are described in detail in Greenstein's paper [7].

Main Points of the Chapter

- The PLL has two distinct modes of operation. In the tracking mode all state variables have values near the steady-state values. The process of bringing the loop from its initial state to the tracking state is called acquisition. For an nth order loop n variables must be brought into near steady-state conditions; e.g., phase, frequency, frequency rate.

- *Ranges/Limits for Noise-Free Acquisition*
  The following definitions are useful in characterizing frequency acquisition

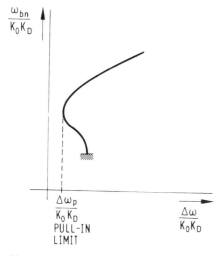

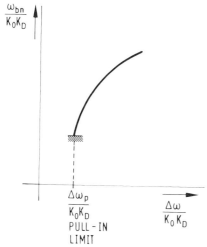

Figure 4.4-2. Two possible relations between frequency offset $\Delta\omega$ and limit cycle frequency.

Hold-in range $\Delta\omega_H$. The maximum frequency offset $\Delta\omega$ that the PLL can track (presuming the loop is already in lock)

$$\Delta\omega_H = \begin{cases} 4B_L = K_0K_D & \text{first-order loop} \\[2mm] \dfrac{\omega_n}{\beta} = K_0K_D & \text{imperfect second-order loop} \\[2mm] \infty & \text{perfect second-order loop} \end{cases}$$

Pull-in range $\Delta\omega_p$ (also called capture range). Maximum frequency offset $\Delta\omega$ between the input signal and the VCO signal so that acquisition is possible (after a finite number of cycle slips) for arbitrary PLL initial conditions $x(0)$.

$$
\Delta\omega_p = \begin{cases} \Delta\omega_H & \text{first-order loop} \\ 2\omega_n\sqrt{\dfrac{\zeta}{\beta}} = 2\sqrt{AK\zeta\omega_n} & \text{imperfect second-order loop} \\ \infty & \text{perfect second-order loop} \end{cases}
$$

Lock-in range. The range of initial conditions $x^T(0) = [\phi(0); x_1(0)]$ for which the phase error ϕ will never exceed $\pm\pi$ during acquisition. For a first-order loop the lock-in range equals the hold-in range. For a second-order loop the lock-in range can be seen in a phase-plane plot.
Pull-out frequency $\Delta\omega_{po}$. Maximum frequency step for which a second-order PLL returns to the stable equilibrium point without skipping cycles (initial condition $x(0) = 0$, i.e., $\Delta\omega = 0$ for $t < 0$).

$$
\Delta\omega_{po} = 1.8\omega_n(\zeta + 1) \quad \text{second order loop, multiplier phase detector}
$$

$$
0.5 < \zeta < 1.4
$$

The following relations between the various limits hold for an imperfect second-order loop

$$
\Delta\omega_{po} < \Delta\omega_p \ll \Delta\omega_H
$$

- *Acquisition in Noise*
  Phase acquisition ($\Delta\omega = 0$). Noise slightly prolongs phase acquisition
  Frequency acquisition. The frequency-difference sensitive control signal that pulls the loop in is small and can easily be ruined even by a small noise disturbance or unwanted dc-offset. As a rule of thumb, a loop SNR of 25 dB is needed for pull-in. Therefore, pull-in is *not* a practical solution for frequency acquisition.
- A second or higher order PLL may not pull-in to its steady state but produce a stable oscillation (limit cycle). The conditions for a limit cycle depend on the initial conditions and on the loop parameters.

REFERENCES

1. F. M. Gardner, Hang up in phase-lock loops, *IEEE Transactions on Communications*, Vol. 25, pp. 1210–1214, October 1977.
2. H. Meyr and W. C. Lindsey, Complete Statistical Description of the Phase-Error Process Generated by Correlative Tracking Systems, *IEEE Transactions on Information Theory*, Vol. 23, No. 2, pp. 194–202, March 1977.
3. F. M. Gardner, *Phaselock Techniques*, Wiley, New York, 1979.
4. D. Richman, Color Carrier Reference Phase Synchronization Accuracy in NTSC Color Television, *Proceedings of the IRE*, pp. 106–133, January 1954.

5. A. J. Viterbi, *Principles of Coherent Communications*, McGraw-Hill, New York, 1966.

6. A. Papoulis, *Probability, Random Variables, and Stochastic Processes*, Kogakusha, Tokyo and McGraw-Hill, New York, 1965.

7. L. L. Greenstein, Phase-Locked Loop Pull-In Frequency, *IEEE Transactions on Communications*, Vol. 22, pp. 1005–1013, August 1974.

8. W. C. Lindsey, *Synchronization Systems in Communication and Control*, Prentice Hall, Englewood Cliffs, NJ, 1972.

9. H. Meyr and L. Popken, Phase Acquisition Statistics for Phase-Locked Loops, *IEEE Transactions on Communications*, Vol. 28, No. 8, August 1980.

10. U. Mengali; Acquisition Behaviour of Generalized Tracking Systems in the Absence of Noise, *IEEE Transaction on Communications*, Vol. 21, pp. 820–826, July 1973.

APPENDIX 4.2A
PHASE ACQUISITION PROBABILITY OF A FIRST-ORDER PHASE-LOCKED LOOP WITH SINUSOIDAL PHASE DETECTOR

The first-order PLL is governed by a stochastic differential equation (in normalized time τ)

$$d\phi(\tau) = -\sin \phi(\tau) \, d\tau + \sqrt{\frac{2}{\rho}} \, d\beta(\tau) , \quad \Delta\omega = 0$$

$$\tau = 4B_L t$$

(4.2A-1)

where $d\beta$ is the increment of the Wiener process with unity variance parameter $E[d^2\beta] = d\tau$ and ρ the signal-to-noise ratio in the loop. The formal derivative of the Wiener process

$$\frac{d\beta(\tau)}{d\tau} = m(\tau)$$

(4.2A-2)

is a white gaussian noise process with spectral density

$$\frac{N_0}{2A^2} K_0 K_D = \frac{2}{\rho}$$

(see (3.3-3) and Table 3.4-1).

The conditional probability density function

$$q(\phi, \tau) \, d\phi = \Pr\left[\phi \in (\phi, \phi + d\phi); \tau \mid \phi(\tau = 0) = \phi_0\right] \quad (4.2A-3)$$

obeys a Fokker–Planck (F–P) equation of the form

$$\frac{\partial}{\partial \tau} q(\phi, \tau) + \frac{\partial}{\partial \phi} \left[-\sin(\phi)q(\phi, \tau) - \frac{1}{\rho} \frac{\partial}{\partial \phi} q(\phi, \tau) \right] = 0$$

(4.2A-4)

We define acquisition to be completed when the error drops below a given level ϕ_ε. Mathematically this can be modeled as follows. At time $\tau = 0$ the trajectory $\phi(\tau)$ starts at ϕ_0. The trajectories are stopped when they first reach one of the boundaries located at ϕ_ε or at $2\pi - \phi_\varepsilon$ (see Figure 4.2-5). It can be shown that by solving the F–P equation (4.2A-4) using the initial probability density

$$q(\phi, 0) = \delta(\phi - \phi_0)$$

(4.2A-5)

where $\delta(\phi)$ is the Dirac function, and the boundary conditions

$$q(\phi_\varepsilon, \tau) = q(2\pi - \phi_\varepsilon, \tau) = 0$$

(4.2A-6)

(absorbing barriers)

then $q(\phi, \tau)$ represents the probability that the trajectory at time τ is within the interval $(\phi, \phi + d\phi)$ where $\phi \in (\phi_\varepsilon, 2\pi - \phi_\varepsilon)$ without ever having reached the boundaries.

The probability that $\phi(\tau)$ is within $(\phi_\varepsilon, 2\pi - \phi_\varepsilon)$ at time τ is found by integrating $q(\phi, \tau)$ over this interval. But this probability is equivalent to the probability that the loop has not acquired lock, hence

$$P[T_p \le t \mid \phi_0, 0] = 1 - \int_{\phi_\varepsilon}^{2\pi - \phi_\varepsilon} q(\phi, \tau) \, d\phi$$

(4.2A-7)

The restricted PDF $q(\phi, \tau)$ needed in (4.2A-7) was obtained by Meyr and Lindsey by separation of variables and subsequently solving a boundary value problem of the Sturm–Liouville type. The reader interested in these mathematical details is referred to Chapter 11.

AIDED ACQUISITION

As was demonstrated in Chapter 4 unaided acquisition of a PLL is a long-lasting and unreliable process. For most applications auxillary circuits are employed to supply some sort of aided acquisition. These circuits are described in this chapter.

5.1. PHASE ACQUISITION

We saw in Section 4.2 that if the initial phase error of a PLL is close to the unstable null, the phase can dwell near the null for an extended time. This dwell phenomenon can be very troublesome in applications where rapid phase acquistion is essential, such as in time division multiple access (TDMA) systems to mention an important example.

TDMA communication is characterized by burst mode operation. The data from one station occur in a short burst, followed closely by a short burst from another station, etc. The difference in carrier frequencies between bursts is negligibly small, however, carrier phase is random from one burst to the next. A small number of symbols (preamble) at the start of each burst is used for synchronizing the carrier circuit. In order to use the communication channel efficiently, the preamble should be as short as possible. The hang-up problem associated with a standard PLL makes it unsuitable for such applications and some sort of aided acquisition is required.

A functional diagram of a phase-locked loop with phase acquisition aid is shown in Figure 5.1-1. The dashed line divides the diagram into the functions that are necessary for tracking and phase acquisition, i.e., a phase-locked loop for tracking, a detector to roughly estimate the initial phase error ϕ_0 at the start of a burst, and control logic to supervise all operations during acquisition.

The loop phase detector provides an output $x_I(t)$ proportional to the sine of the phase error $\phi = \theta - \hat{\theta}$ while the quadrature phase detector generates

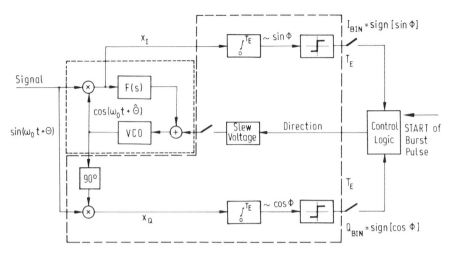

Figure 5.1-1. Functional diagram of a phase-locked loop with phase acquisition aid.

an output $x_Q(t)$ proportional to the cosine. Both the x_I and x_Q signals are low-pass filtered using integrate-and-dumps and subsequently binary quantized. The two binary signals I_{BIN} and Q_{BIN} are sampled T_E seconds after the start of burst pulse has been received and are subsequently used to estimate the quadrant in which the initial phase error is located.

If the initial phase error estimate lies in quadrant II or III (see Figure 5.1-2), a large pulse (slew) is applied to the VCO in such a way that the loop is pushed towards its stable equilibrium point. The polarity of the slew

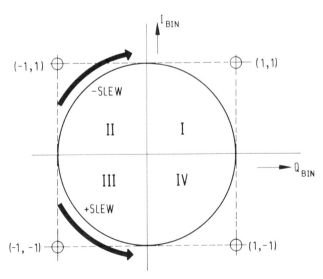

Figure 5.1-2. Definition of quadrants. Arrows indicate direction of slew provided estimate lies in respective quadrant.

voltage is determined by the sign of I_{BIN}. For $I_{BIN} = 1$ the polarity is such that the phase error is decreased by $\Delta\phi_a = 90°$ while $I_{BIN} = -1$ results in a phase shift of the same magnitude but opposite direction. No action is taken if the estimate lies in the right half-plane (quadrants I and IV).

Note that existence of an independent start of burst pulse is necessary for the proper operation of the circuit. This pulse is used in the control logic to start integration of the two signals $x_I(t)$ and $x_Q(t)$. After time T_E has elapsed, the binary signals (I_{BIN}, Q_{BIN}) are sampled and the integrators dumped.

The circuit has been analyzed by means of the Fokker–Planck technique [2]. The parameter of greatest practical interest is the probability of acquisition in a given interval of time.

Typical trajectories for the unaided as well as aided acquisition process are shown in Figure 5.1-3. As explained in Section 4.2.1 we consider the acquisition process to be completed when the phase error drops below the level ϕ_ε for the first time. This definition of lock is meaningful for a first-order loop and a highly damped second-order loop ($\zeta > 1$) where the other state variable does not change appreciably during the phase acquisition process. The acquisition time is now the sum of T_E, the time needed to estimate the initial phase error ϕ_0, plus the time it takes the trajectory to reach ϕ_ε starting from $\phi_A = \phi_0 + \Delta\phi_a$.

The probability that the loop has failed to acquire lock in a given time interval is shown in Figure 5.1-4. As an example of a signal-to-noise ratio in the loop $\rho = 10$ dB, an estimation time $T_E = 1/(2B_L)$ and a synchronization failure probability of 10^{-8}, only a modest improvement with respect to the

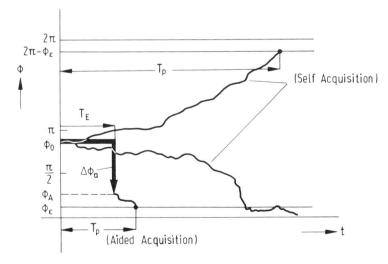

Figure 5.1-3. Typical phase trajectories for the unaided and aided aquisition process for the case of steady-state phase error $\phi_s = 0, 2\pi, \ldots$.

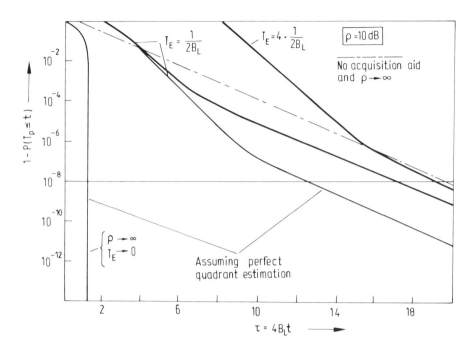

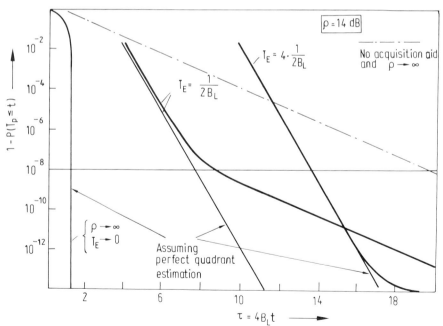

Figure 5.1-4. Probability of synchronization failure of a phase-locked loop with phase acquisition aid for $\phi_e = 30°$ and with quadrant estimation time T_E as parameter. (a) $\rho = 10$ dB; (b) $\rho = 14$ dB (from Heinrich Meyr and Luitjens Hopken, Phase Acquisition Statistics for Phase-Locked Loops, *IEEE Transactions on Communications*, Vol. 28, No. 8, © IEEE 1980).

unaided case is obtained. The influence of wrong decisions made by the quadrant estimator is clearly visible by comparing the result with the curve obtained for a perfect estimator. (A perfect estimator provides an error-free quadrant estimate for a given time T_E). Increasing T_E with $\rho = 10\,dB$ gives no improvement as shown for $T_E = 4/(2B_L)$. The nearly perfect quadrant estimator requires such a large T_E that the combined acquisition time is larger.

For $\rho = 14\,dB$ (Figure 5.1-4b) an estimation time of $T_E = 1/(2B_L)$ is nearly optimal. A perfect quadrant estimator would give only a slightly better result while a shorter T_E would result in a longer acquisition time due to wrong quadrant estimates. The improvement is a strongly increasing function of ρ assuming its maximum in the limiting case $\rho \to \infty$. Also, for large ρ and a synchronization failure probability of 10^{-8} the quadrant estimator operates nearly error-free for values of T_E negligible compared to the acquisition time.

The important conclusion that can be drawn from Figure 5.1-4 is that for moderately high SNRs ($\rho > 12\,dB$) the acquisition property of a phase-locked synchronizer is greatly improved using an acquisition aid. For values of ρ smaller than 12 dB, there exists no substantial difference between aided and unaided acquisition that would justify the inclusion of an acquisition aid. Use of a phase-locked synchronizer for these value of ρ is inevitably coupled with long acquisition times and therefore ruled out in application where rapid acquisition is necessary.

5.2. FREQUENCY ACQUISITION

5.2.1. Sweeping

Frequency acquisition can be accomplished by sweeping the frequency of the VCO thereby searching for the signal frequency. In the moment when the signal and VCO frequency coincide the loop will lock and remain in lock, provided the sweep rate is not too large (Figure 5.2-1). Notice that after the loop has locked, the VCO frequency remains constant despite the fact that the sweep voltage still increases. The price paid for this is a troublesome steady-state phase error as will be shown here.

Perfect Second-Order Loop. Although sweep can be applied as a ramp voltage directly to the VCO, for a perfect second-order loop a simpler approach is to insert a constant current I_{sw} at the junction of the R_2 and C (see Figure 5.2-2). Then a ramp voltage $e_{sw}(t)$ is superimposed on the output of the operational amplifier

$$e_{sw}(t) = -\frac{I_{sw}}{C}\,t$$

and the sweep rate is

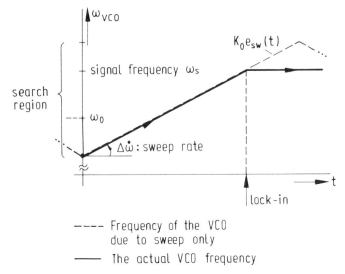

Figure 5.2-1. Frequency sweep process. – – – –, frequency of the VCO due to sweep only; ——, the actual VCO frequency.

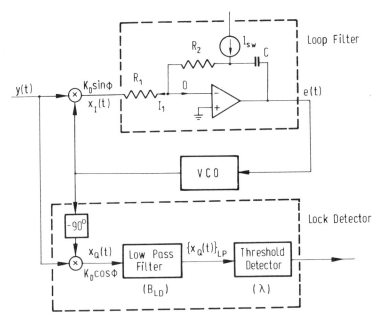

Figure 5.2-2. Frequency sweep circuit and lock detector.

$$\Delta\dot{\omega} = K_0 \frac{de_{sw}(t)}{dt} = -\frac{K_0 I_{sw}}{C} \tag{5.2-1}$$

In the locked mode, the normalized state variables take on the values (derived from (2.4-20) by adding the term $\Delta\dot{\omega}/\omega_n^2$ to the right-hand side of the second equation and setting both the derivatives equal to zero)

$$x_{1,s} = \frac{\Delta\dot{\omega}}{\omega_n^2} 2\zeta + \frac{d\theta(\tau)}{d\tau}$$

$$\phi_s = \sin^{-1}\left(-\frac{\Delta\dot{\omega}}{\omega_n^2}\right) \tag{5.2-2}$$

where $\Delta\dot{\omega}/\omega_n^2$ is the normalized sweep rate.

As will be shown in Chapter 6, a static phase error in the presence of noise greatly increases the cycle slip rate of a PLL. It is thus advisable to shut off the slew current once lock has been achieved. We are thus confronted with the problem of deciding when to stop slewing. This decision need not be particularly fast since the loop holds lock with the slew applied so that a sufficient amount of time can be taken for lock verification.

A lock indicator frequently used is shown in the lower part of Figure 5.2-2. The incoming signal $y(t)$ is multiplied in a second phase detector (known as quadrant or coherent amplitude detector) with a $(-90°)$ phase shifted version of the VCO signal. In the locked mode the dc-component of the quadrature phase detector is proportional to $\cos\phi$ while the main phase detector output voltage is proportional to $\sin\phi$. In the unlocked mode the outputs from both detectors are beat notes with very small dc-components. Thus, the filtered output of the quadrature phase detector provides an indication for lock. If the output voltage of the smoothing filter exceeds a level λ, the threshold detector indicates lock and the sweeping of the VCO is stopped. A narrowband smoothing filter is needed to avoid flicker of the threshold indicator that produces false indications of lock or loss of lock. On the other hand, too narrow a bandwidth unduly delays the indictions of lock from the actual occurrence of lock.

From a casual inspection of (5.2-2) one might infer that the maximum permissable sweep rate is $\Delta\dot{\omega}/\omega_n^2 = 1$. However, this is not the case. We explicitly had to assume that the loop was in the *locked state*. Therefore,

$$\frac{\Delta\dot{\omega}}{\omega_n^2} = 1 \tag{5.2-2a}$$

is the maximum permissible rate of change of frequency in the tracking mode and is otherwise considerably smaller. Viterbi [3] has discovered by phase-plane techniques that the sweep rate must be considerably smaller than 1 for certain (i.e., with probability one) acquisition. For example, for

$\zeta = \sqrt{2}/2$ we must have $\Delta\dot{\omega}/\omega_n^2 < 0.5$. For sweep rates larger than this value, but smaller than 1, there exists the possibility that the loop sweeps through the input frequency; whether the loop locks or not depends on the random initial phase and frequency conditions.

Noise greatly complicates the matter. Due to the noise, the phase error occasionally changes from a steady state value $|\phi_s|$ to $|\phi_s| + 2\pi$ or $|\phi_s| - 2\pi$; the loop slips a cycle. If a cycle slip occurs when sweep is applied the probability is high that the loop subsequently loses lock completely. The meantime between cycle slips is strongly dependent on the signal-to-noise ratio in the loop and on the static phase error $|\phi_s|$. Therefore, the stronger the noise the smaller the static phase error (5.2-2) and thus the smaller the sweep rate should be.

There is no known mathematical analysis for the determination of the maximum sweep rate. Several empirical design rules based on experiments and computer simulations have been published. One approach which leads to a practical design rule is to specify a minimum meantime between cycle slips. The meantime between slips could then be the time that the loop remains in lock if a sweep voltage is applied. It must be chosen such that a reliable lock detection for freezing the sweep voltage is possible.

The experimentally determined meantime between cycle slips as a function the signal-to-noise ratio ρ is displayed in Figure 5.2-3 for three values of

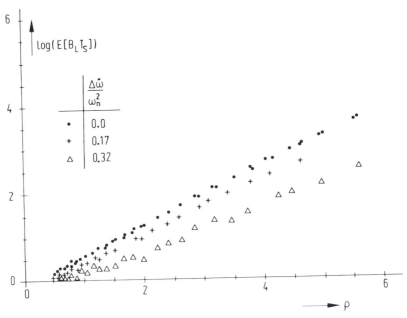

Figure 5.2-3. Normalized meantime between cycle slips $E[B_L T_s]$ of a second-order loop ($\zeta = 1.0$) for three values of the normalized sweep rate $\Delta\dot{\omega}/\omega_n^2$.

$\Delta\dot{\omega}/\omega_n^2$. For a specified normalized minimum meantime between cycle slips $E(T_sB_L)$ we obtain the maximum sweep rate $\Delta\dot{\omega}/\omega_n^2$ at the intersection of the horizontal line $E(T_sB_L) = \text{constant}$ with the function $E(T_sB_L)$. The results for three different values of $E(T_sB_L) = \text{constant}$ are displayed in Figure 5.2-4. Frazier and Page [4] have obtained an empirical formula that predicts the maximum sweep rate for an acquisition probability of 0.9

$$\left.\frac{\Delta\dot{\omega}}{\omega_n^2}\right|_{90\%} \approx 1 - \sqrt{\frac{2}{\rho}}, \quad \zeta > 0.7 \tag{5.2-3}$$

For comparison reasons the values for $\Delta\dot{\omega}/\omega_n^2$ given by (5.2-3) are also plotted in Figure 5.2-4. It is interesting that the resulting curve is quite similar to the curve obtained for $E(T_sB_L) = 100$. In most applications a probability of acquisition of 0.9 is too low. Choosing $E(T_sB_L) = 10^4$ a better preliminary design for the sweep rate would be

$$\frac{\Delta\dot{\omega}}{\omega_n^2} = \begin{cases} 1 - \left(\dfrac{2}{\rho - 4}\right)^{1/2}, & 6 < \rho < 9.5 \\ 0.4, & \rho > 9.5 \end{cases} \tag{5.2-4}$$

In order to provide a safety margin the maximum normalized sweep rate is limited to 0.4. Formula (5.2-4) implies that acquisition is impossible below $\rho = 6$ (≈ 8 dB).

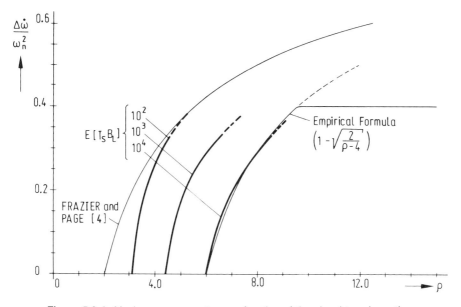

Figure 5.2-4. Maximum sweep rate as a function of the signal-to-noise ratio.

Lock Detector. Assume a noise corrupted received signal $y(t)$ as given by (3.2-1). Neglecting the double frequency terms, the outputs of the two phase detectors of Figure 5.2-2 are given by (see also Section 3.2)

Main phase detector: $x_I(t) = K_D[\sin \phi(t) + n'(t, \hat{\theta})]$

Quadrature phase detector: $x_Q(t) = K_D[\cos \phi(t) + n_Q'(t, \hat{\theta})]$

(5.2-5)

with the zero-mean processes

$$n'(t, \hat{\theta}) = \frac{n_c(t)}{A} \cos \hat{\theta} + \frac{n_s(t)}{A} \sin \hat{\theta}$$

$$n_Q'(t, \hat{\theta}) = \frac{n_c(t)}{A} \sin \hat{\theta} - \frac{n_s(t)}{A} \cos \hat{\theta}$$

(5.2-6)

Since the loop noise bandwidth B_L of the PLL is very small compared to the bandwidth of the noise process $n_c(t)$ we have argued in Chapter 3 that the statistics of the noise processes n' and n_Q' are approximately the same for the open and closed loop. Under the same conditions it is quite easy to show that $n'(t, \hat{\theta})$ and $n_Q'(t, \hat{\theta})$ are uncorrelated for a process $n(t)$ symmetric about ω_0 (and since gaussian, statistically independent)

$$E[n'(t, \hat{\theta})n_Q'(t + \tau, \hat{\theta})] = 0 , \quad \text{for all } \tau$$

$$E[n'(t, \hat{\theta})n'(t + \tau, \hat{\theta})] = E[n_Q'(t, \hat{\theta})n_Q'(t + \tau, \hat{\theta})] = \frac{R_{n_c}(\tau)}{A^2}$$

(5.2-7)

We next want to analyze the lock detector performance of Figure 5.2-2 as a function of the bandwidth B_{LD} of the smoothing filter and the threshold λ. In the locked mode, the output of the smoothing filter after the transients have died away is a stationary gaussian process with mean

$$E\{x_Q(t)\}_{LP} = E\left\{\int_{-\infty}^{t} h(t - t')K_D \cos \phi(t') \, dt'\right\}$$

$$\approx K_D H(0) \cos \phi_s \quad \text{(locked loop)}$$

(5.2-8)

with

$$H(0) = \int_0^{\infty} h(t') \, dt' \quad \text{(dc-gain of the low-pass filter)}$$

(5.2-9)

$$\cos \phi_s = (1 - \sin^2 \phi_s)^{1/2} = \left[1 - \left(\frac{\Delta\dot{\omega}}{\omega_n^2}\right)^2\right]^{1/2}$$

where $\{x_Q(t)\}_{LP}$ is the low-pass filter version of the quadrature phase detector output.

If the loop is unlocked the output of the smoothing filter has zero mean

$$E\{x_Q(t)\}_{LP} \approx 0 \quad \text{(unlocked loop)} \tag{5.2-10}$$

The variance of $\{x_Q(t)\}_{LP}$ in both cases equals

$$\sigma_{LP}^2 = \frac{1}{2\pi} \int_{-\infty}^{\infty} K_D^2 |H(\omega)|^2 S_{n_Q}(\omega) \, d\omega$$

$$= (H(0)K_D)^2 \frac{N_0 B_{LD}}{A^2} \tag{5.2-11}$$

where B_{LD} is the equivalent bandwidth of the smoothing filter

$$B_{LD} = \frac{1}{2\pi} \int_0^{\infty} \left| \frac{H(\omega)}{H(0)} \right|^2 d\omega \tag{5.2-12}$$

In solving the integral of (5.2-11), we have assumed an argument similar to that following (3.3-2). For the subsequent analysis, it proves helpful to relate the bandwidth B_{LD} to the loop bandwidth B_L.

$$\sigma_{LP}^2 = (H(0)K_D)^2 \frac{N_0 B_L}{A^2} \frac{B_{LD}}{B_L}$$

$$= (H(0)K_D)^2 \frac{1}{\rho} \left(\frac{B_{LD}}{B_L} \right) \tag{5.2-13}$$

Let us now compute the probability of a false alarm of the lock detector. A false alarm occurs when the output of the smoothing filter in the unlocked situation exceeds the threshold.

$$P_{FA} = \frac{1}{(2\pi\sigma_{LP}^2)^{1/2}} \int_{\lambda}^{\infty} \exp\left(-\frac{x^2}{2\sigma_{LP}^2} \right) dx$$

$$= \frac{1}{\sqrt{\pi}} \int_{\frac{\lambda}{\sqrt{2}\sigma_{LP}}}^{\infty} \exp(-y^2) \, dy = \frac{1}{2} \operatorname{erfc} \left(\frac{\lambda}{\sqrt{2}\sigma_{LP}} \right) \tag{5.2-14}$$

where $\operatorname{erfc}(x)$ is the complementary error function defined as

$$\operatorname{erfc}(x) = \frac{2}{\sqrt{\pi}} \int_x^{\infty} \exp(-y^2) \, dy \tag{5.2-15}$$

Since a false alarm stops the sweeping process, the probability P_{FA} should be small. From (5.2-14) one observes that P_{FA} becomes small if the variance σ_{LP}^2 is made small. This requires a small bandwidth B_{LD} with respect to the loop bandwidth B_L (see (5.2-13)). But since the duration of the transients in the smoothing filter is inversely proportional to the bandwidth B_{LD}, the bandwidth cannot be made too small in order not to unduly delay a lock indication after the actual occurrence of lock. Furthermore, in any case the transient period must be much smaller than the meantime between cycle slips. Equation (5.2-14) seems to indicate that the threshold λ should be chosen as large as possible. This conclusion is erroneous. As will be seen shortly, a large threshold λ increases the probability of a false indication of loss of lock in which case an unnecessary restart of the sweeping process would occur. This is a serious event which should be avoided.

The probability of a false indication of loss of lock, P_{FL}, equals the probability that the output of the smoothing filter drops below the threshold λ (see Figure 5.2-5)

$$P_{FL} = \frac{1}{(2\pi\sigma_{LP}^2)^{1/2}} \int_{-\infty}^{\lambda} \exp\left[-\frac{(x - K_D H(0)\cos\phi_s)^2}{2\sigma_{LP}^2} \right] dx \qquad (5.2\text{-}16)$$

or

$$P_{FL} = \frac{1}{\sqrt{2\pi}} \int_{-\infty}^{A} \exp\left(-\frac{y^2}{2} \right) dy \qquad (5.2\text{-}16a)$$

with

$$A = \frac{\lambda - K_D H(0)\cos\phi_s}{\sigma_{LP}}$$

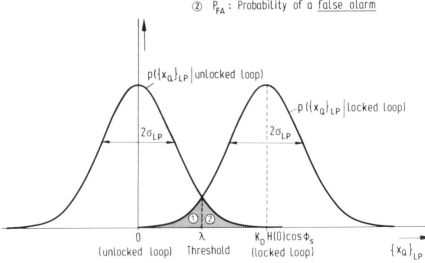

① P_{FL} : Probability of a false indication of Loss-of-Lock

② P_{FA} : Probability of a false alarm

Figure 5.2-5. Probability density function of the smoothing filter output $\{x_Q\}_{LP}$; probability of false alarm, P_{FA}, and probability of a false indication of loss-of-lock, P_{FL}, for a given threshold λ.

It is reasonable, although not necessary, to set $P_{FL} = P_{FA}$. From Figure 5.2-5 or the formulas (5.2-14) and (5.2-16), it immediately follows that the threshold λ must be placed in the middle between the mean value of the smoothing filter output in the locked state and in the unlocked state

$$\lambda = \frac{1}{2} K_D H(0) \cos \phi_s \qquad (5.2\text{-}17)$$

For this choice of threshold, the formulas (5.2-14) and (5.2-16) can be simplified. Replacing $\cos \phi_s$ and σ_{LP} in (5.2-16a) from (5.2-9) and (5.2-11), respectively, one obtains

$$P_{FA} = P_{FL} = \frac{1}{2} \operatorname{erfc}\left(\frac{\lambda}{\sqrt{2}\sigma_{LP}}\right) \qquad (5.2\text{-}18)$$

with

$$\frac{\lambda}{\sigma_{LP}} = \frac{1}{2} \left\{ \left[1 - \left(\frac{\Delta\dot{\omega}}{\omega_n^2}\right)^2\right] \rho\left(\frac{B_L}{B_{LD}}\right) \right\}^{1/2} \qquad (5.2\text{-}19)$$

Numerical Example:

Frequency uncertainty to be swept $\qquad\qquad \Delta\omega = 10^3 \text{ s}^{-1}$
Loop parameters (perfect second-order loop) $\quad \zeta = 1$
$\qquad\qquad\qquad\qquad\qquad\qquad\qquad\qquad\qquad \omega_n = 100 \text{ s}^{-1}$
$\qquad\qquad\qquad\qquad\qquad\qquad\qquad\qquad\qquad B_L = 62.5 \text{ Hz (Table 3.3-1)}$
Signal-to-noise ratio in the loop $\qquad\qquad\qquad \rho = \dfrac{A^2}{N_0 B_L} = 7$

From Figure 5.2-4 or the empirical formula (5.2-4) the sweep rate is found to be

$$\Delta\dot{\omega} = 0.18\omega_n^2 = 1800 \text{ s}^{-2}$$

To sweep the entire range of $\Delta\omega = 1000$ it takes the loop $t = \Delta\omega/\Delta\dot{\omega} = 5.5$ s. For a required probability of false indication $P_{FA} = P_{FL} = 10^{-4}$ the ratio λ/σ_{LP} equals 3.7. Solving (5.2-19) for the bandwidth of the smoothing filters yields

$$B_{LD} = \frac{[1 - (\Delta\dot{\omega}/\omega_n^2)^2]\rho}{(2\lambda/\sigma_{LP})^2} B_L = 0.124 B_L$$

$$B_{LD} = 7.7 \text{ Hz}$$

Finally, we want to check whether the time to indicate lock is indeed much smaller than the meantime between slips which is a necessary condition for the design rule (5.2-4) to apply. We arbitrarily define

$$\Delta t = 10/B_{LD}$$

as the time duration needed for the transients in the smoothing filter output to settle. Multiplying both sides by the loop bandwidth yields

$$\Delta t B_L = 10(B_L/B_{LD})$$
$$= 10/0.124 \approx 80$$

The sweep rate $\Delta\dot\omega/\omega_n^2 = 0.18$ was based on a (normalized) meantime between slips of $E(T_s B_L) = 10^4$ which is much larger than $\Delta t B_L = 80$.

The formula of Page and Frazier specifies a sweep rate of $\Delta\dot\omega/\omega_n^2 = 0.45$. Going back to Figure 5.2-4 we observe that for $\rho = 7$ the meantime between slips is only about 100 which seems to be the minimum for a successful acquisition by sweeping.

To supervise the in-lock state of the loop, the smoothing filter is often replaced by a narrower filter immediately after the in-lock decision has been made. Also, when the loop has acquired and indicated lock, an algorithm may be used to bring about an out-of-lock indication over an extended period of time before declaring the loop to be unlocked.

Imperfect Second-Order Loop. A second-order loop can be swept by either inserting a constant current into the loop filter capacitor, by adding a voltage ramp directly to the VCO control voltage or by adding a voltage step to the loop filter input. For a perfect second-order PLL the first two approaches are equivalent (the first being preferable because of simpler implementation). This is not true for an imperfect second-order loop.

We first consider the case where a constant current I_{sw} is inserted into the capacitor. The passive loop filter of Figure 5.2-6 obeys the equations

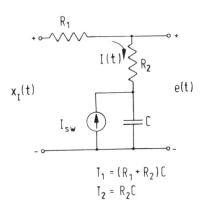

Figure 5.2-6. Injection of a constant current I_{sw} into the capacitor of a passive loop filter.

$$T_1 = (R_1 + R_2)C$$
$$T_2 = R_2 C$$

$$X_I(s) = (R_1 + R_2)I(s) + \frac{1}{sC}[I_{sw} + I(s)]$$

(5.2-20)

$$E(s) = R_2 I(s) + \frac{1}{sC}[I_{sw} + I(s)]$$

Solving for the VCO control voltage yields

$$E(s) = \frac{1 + sT_2}{1 + sT_1} X_I(s) + \frac{1}{1 + sT_1} R_1 I_{sw}$$

(5.2-21)

Using the normalized variables and parameters introduced in (2.4-21), we obtain the block diagram of an imperfect second-order loop with frequency acquisition aid as shown in Figure 5.2-7. Suppose now the loop starts sweeping through the frequency uncertainty region and eventually locks in. After lock, the values of the state variables are the solution of the system of equations

$$\frac{d\phi(\tau)}{d\tau} = 0 = -(2\zeta - \beta)\sin\phi_s - x_{1,s} + \frac{d\theta}{d\tau}$$

(5.2-22)

$$\frac{dx_1(\tau)}{d\tau} = 0 = (1 - 2\zeta\beta + \beta^2)\sin\phi_s + \frac{\Delta\dot{\omega}}{\omega_n^2} - \beta x_{1,s}$$

Solving these equations gives

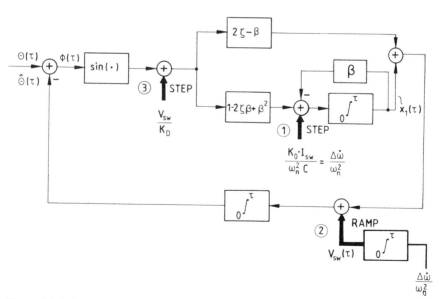

Figure 5.2-7. Three ways of providing frequency sweep to a second-order PLL. (1) (normalized) current step injected into filter capacitor; (2) (normalized) voltage ramp added to VCO control voltage; (3) (normalized) voltage step added to the loop filter input. (Note: For the perfect case, simply let $\beta = 0$. Then, (1) and (2) are equivalent for $V_{sw}(\tau) = (K_0 I_{sw}/\omega_n^2 C)\tau$.)

$$\sin \phi_s = -\frac{\Delta \dot{\omega}}{\omega_n^2} + \beta \frac{d\theta}{d\tau}$$

$$x_{1,s} = \frac{d\theta}{d\tau} (1 - 2\zeta\beta - \beta^2) + \frac{\Delta \dot{\omega}}{\omega_n^2} (2\zeta - \beta)$$

(5.2-23)

For a given frequency offset $d\theta/d\tau = \Delta\omega/\omega_n$, the loop stress $\sin \phi_s$ depends on whether the signal frequency $\omega_s = \omega_0 + \Delta\omega$ is approached from below ($\Delta\dot{\omega} > 0$) or above ($\Delta\dot{\omega} < 0$). Since the sign of the frequency offset $\Delta\omega$ is unknown the maximum allowable sweep rate is determined by the worst case. From (5.2-23) we learn that the largest loop stress occurs when the sweep rate and frequency offset have opposite signs

$$|\sin \phi_s| = \left|\frac{\Delta \dot{\omega}}{\omega_n^2}\right| + \beta \left|\frac{\Delta \omega}{\omega_n}\right| \quad \text{(maximum loop stress)} \qquad (5.2-24)$$

Compared to the perfect second order loop the loop stress is increased by $\beta|\Delta\omega/\omega_n|$. The maximum allowable sweep rate is therefore smaller than for a perfect second-order loop.

More important, however, is that the frequency uncertainty region one has to search with this approach is limited to the pull-in range (and *not* the hold-in range)

$$\left|\frac{\Delta \omega}{\omega_n}\right| < \frac{\Delta \omega_p}{\omega_n} = 2\sqrt{\frac{\zeta}{\beta}} \ll \Delta\omega_H \qquad (5.2-25)$$

Why is this so? Suppose that the loop has locked and the sweep current has been shut off. If $|\Delta\omega|$ is larger than $\Delta\omega_p$ and if x_1 after a cycle slip (caused by noise) is not in the suitable range for pull-in ((4.3-20) and Figure 4.3-3) the VCO frequency wanders off and the loop will completely lose lock.

The conclusion from the foregoing discussion is that for an imperfect second order loop (due to a limited search region) sweep by slewing current is not practical. In practice, if loop sweep is employed the loop filter is always realized as an active filter with dc gain arbitrarily high.

If a voltage ramp instead of the current source is added to the input of the VCO (Figure 5.2-7) the search region becomes independent of the loop parameters. However, the solution is more complicated to implement than the combination of an active filter and sweep by slewing current and is therefore of little practical interest.

Open Loop Frequency Sweeping. One can also perform an open loop frequency sweep. The sweep rate is now no longer restricted by the ramp tracking limits of the loop but solely a function of the response time of the lock indicator. It becomes necessary to detect frequency agreement very rapidly and then quickly stop sweeping and close the loop.

Sweeping of Higher Order Loops. In principle, sweep can also be applied to higher order loops. However, recognition must be given to the fact that a higher order loop must acquire several state variables besides frequency and phase. The problem is not well understood. Some preliminary results for third-order loops are given by Booth [6].

5.2.2. Frequency Discriminator Aided Acquisition

Another method to effect frequency acquisition is to add an additional frequency difference sensitive device to the conventional traditional phase detector in the PLL in the manner shown in Figure 5.2-8. With a large initial frequency error the dc-output of a multiplier type phase detector is practically zero while the frequency detector (FD) generates a voltage proportional to the frequency difference between signal and VCO, driving that difference to zero. The phase detector takes over when the frequency error is sufficiently small, thus completing acquisition. When the PLL is in-lock the output of the frequency detector should ideally be zero or have at least zero mean thus allowing the phase detector and the loop filter $F(s)$ to govern the loop dynamics.

The output of an ideal frequency detector is proportional to the difference between the signal frequency ω_s and the VCO frequency ω_{VCO}

$$X_f(s) = K_f[\omega_s(s) - \omega_{\mathrm{VCO}}(s)] \tag{5.2-26}$$

where K_f is the frequency detector gain. Writing ω_s and ω_{VCO} as usual as a sum

$$\omega_s(t) = \omega_0 + \frac{d\theta}{dt}$$

$$\omega_{\mathrm{VCO}}(t) = \omega_0 + \frac{d\hat{\theta}}{dt} \tag{5.2-27}$$

we can write (5.2-26) in the form

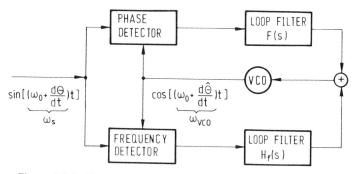

Figure 5.2-8. Phase-locked loop with phase and frequency detector.

$$X_f(s) = K_f[s\theta(s) - s\hat{\theta}(s)] \tag{5.2-28}$$

The VCO control law during frequency acquisition is

$$s\hat{\theta}(s) = K_0 H_f(s) K_f[s\theta(s) - s\hat{\theta}(s)] \tag{5.2-29}$$

Rearranging yields the transfer function of the frequency lock loop

$$H_{FD}(s) = \frac{s\hat{\theta}}{s\theta} = \frac{K_0 K_f H_f(s)}{1 + K_0 K_f H_f(s)} \tag{5.2-30}$$

The corresponding block diagram is shown in Figure 5.2-9. A simple integrator is sufficient as frequency loop filter $H_f(s)$. For an initial frequency offset $d\theta/dt = \Delta\omega$, $t \geq 0$ (i.e., a signal frequency step $s\theta = \Delta\omega/s$) we obtain with $H_f(s) = K_I/s$

$$s\hat{\theta}(s) = \frac{K_0 K_f K_I}{s + K_0 K_f K_I} \frac{\Delta\omega}{s} \tag{5.2-31}$$

The frequency error $s\theta - s\hat{\theta} = (\Delta\omega/s) - s\hat{\theta}$ is then found using (5.2-31) and we see that it decays exponentially with a time constant of $1/(K_0 K_f K_I)$

$$\frac{d\theta}{dt} - \frac{d\hat{\theta}}{dt} = \Delta\omega \exp(-K_0 K_f K_I t), \quad t > 0 \tag{5.2-32}$$

Equation (5.2-32) predicts that increasing $(K_f K_I)$ can result in arbitrarily fast acquisition. In reality, however, the fact that the frequency detector output in addition to the useful signal will have a randomly fluctuating voltage on its output places a limit on the size of the gain $(K_f K_I)$.

Practical realizations of the filters $H_f(s)$ and $F(s)$ do not require two different devices but could share the same operational amplifier in the manner shown in Figure 5.2-10 where the frequency detector output is used to charge the capacitor.

Frequency sensitive loops are not only employed for aided acquisition of a PLL but can be used as an independent unit for frequency control; the

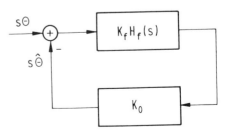

Figure 5.2-9. Equivalent linearized model during frequency acquisition mode.

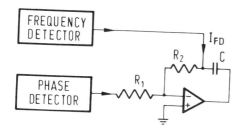

Figure 5.2-10. Combined loop filter for frequency discriminator aided acquisition.

so-called automatic frequency control (AFC) loops. They are treated in detail in Chapter 8.

Here we consider only the frequency-locked loop with a *rotational frequency detector* [7] because this type of detector is mainly used for aided acquisition of a PLL and not for an AFC loop. It is predominantly built in digital hardwave and has no filtering function.

The operation of this frequency detector can best be explained by using a phasor diagram (Figure 5.2-11). If the signal frequency ω_s is larger than the VCO frequency, the VCO phasor lags behind the signal phasor while for $\omega_s < \omega_{VCO}$, the VCO phasor advances with respect to the signal phasor. Thus, we need a circuit capable of determining whether the signal retards or advances relative to the VCO phasor in order to determine the sign of the frequency difference. Such a circuit can be built as follows.

We divide each cycle of the VCO phasor into four quadrants labeled A, B, C, D. This can be accomplished by actually running the VCO at four times the desired frequency and dividing it by four to obtain ω_{VCO} itself. We assume that the VCO waveform is a square. The output of the divider-by-four is shown in Figure 5.2-12 with the pertinent quadrant division. Further assume that the phase detector is designed such that in-lock, the PLL will maintain positive transitions of the signal in the vicinity of the positive transitions of the VCO (other phase detector designs can be handled analogously). Therefore, for the case of in-lock we will observe positive signal transitions predominantly or exclusively in quadrants A and D. Only during acquisition, should signal transitions occur in quadrants B and C. This fact is exploited by the frequency detector to generate an error signal.

The frequency detector monitors the signal transitions with respect to the four quadrants. Let the kth cycle be denoted by a subscript k. The situation $\omega_s > \omega_{VCO}$ can be recognized by observation of $C_k \rightarrow B_{k+1}$, in which case the frequency detector generates a positive pulse. By C_k we understand that at the kth transition of the signal phasor, the VCO phasor lies in quadrant C. Analogously B_{k+1} denotes the location of the VCO phasor at the $(k+1)$th transition of the signal phasor.

Similarly, if $B_k \rightarrow C_{k+1}$ is observed, the frequency detector generates a negative pulse in recognition of $\omega_s < \omega_{VCO}$. Note that the rotational frequency detector possesses the very desirable feature that during in-lock, no error signal is generated since the transition $B_k \rightarrow C_{k+1}$ or vice versa rarely occurs, if ever.

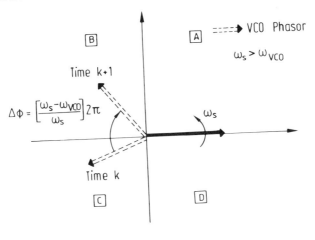

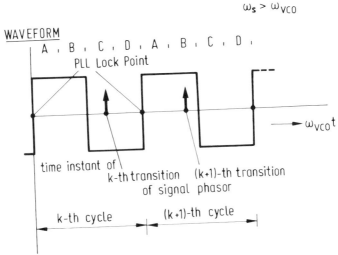

Figure 5.2-11. (a) Relative position of VCO phasor with respect to positive transitions of the signal phasor. (b) Division of the VCO cycle into four quadrants.

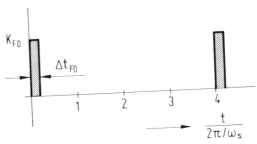

Figure 5.2-12. Waveform of the output of the frequency detector ($\Delta\omega/\omega_s = 1/4$).

The frequency detector can be characterized by its output voltage mean value since that mean value charges the integrating capacitor in the loop filter. This average value of the pulses depends on the phase increment per cycle of the signal

$$|\Delta\phi| = \left|2\pi - \omega_{VCO}\frac{2\pi}{\omega_s}\right| = \left|\frac{\omega_s - \omega_{VCO}}{\omega_s}\right|2\pi \qquad (5.2\text{-}33)$$

The phase increment $\Delta\phi$ is the amount of phase the VCO phasor advances or retards with respect to the signal phasor when the signal makes a full cycle. The mean value, assuming positive and negative pulses have equal area, is

$$g_f(\omega_s - \omega_{VCO}) = \overbrace{\frac{K_{FD}\,\Delta t_{FD}}{2\pi/\omega_s}}^{\substack{\text{Area of} \\ \text{FD pulse}}}[\text{Pr(positive pulse)} - \text{Pr(negative pulse)}]$$
$$(5.2\text{-}34)$$

where Δt_{FD} is the pulse width, $g_f(\omega_s - \omega_{VCO})$ is the mean value of pulses, $K_f = K_{FD}\,\Delta t_{FD}/(2\pi/\omega_s)$ is the frequency detector gain and Pr(positive pulse) = $\text{Pr}(C_k \rightarrow B_{k+1})$ = probability of generation of a positive pulse during one cycle of the signal phasor. The probability for a positive (or negative) pulse can be readily calculated by assuming that the phasor in cycle k is uniformly distributed. For example if the increment $|\Delta\phi|$ is less than $\pi/2$ then the pulse is generated only if the phasor lies within a sector of width $|\Delta\phi| = 2\pi|\omega_s - \omega_{VCO}|/\omega_s$, an event which has a probability of $|\Delta\phi|/2\pi = |\omega_s - \omega_{VCO}|/\omega_s$ (Figure 5.2-13). By a simple extension of this argument, the frequency detector characteristic of Figure 5.2-14 can be

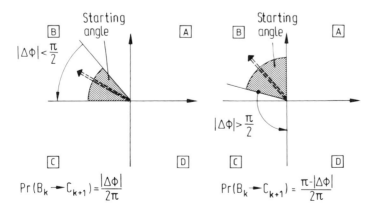

Figure 5.2-13. VCO phasor starting angle for $\beta_k \rightarrow C_{k+1}$ transition.

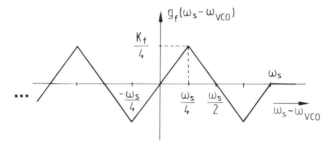

Figure 5.2-14. Frequency detector characteristic.

constructed. The characteristic is periodic for $\omega_s < \omega_{\mathrm{VCO}}$ since multiple cycles of the VCO phasor cannot be distinguished from a single cycle by the frequency detector as was described. It is not periodic for $\omega_s > \omega_{\mathrm{VCO}}$ since the VCO frequency must always be positive, hence

$$\omega_s - \omega_{\mathrm{VCO}} < \omega_s$$

As can be seen from Figure 5.2-14 the useful range of the frequency detector is

$$|\omega_s - \omega_{\mathrm{VCO}}| < \frac{\omega_s}{2}$$

that is, 50% offset in the initial VCO frequency can be tolerated. The linear range of the frequency detector over which the linear model in Figure 5.2-9 is applicable is $|\omega_s - \omega_{\mathrm{VCO}}| < \omega_s/4$.

An important question is how additive noise affects the application of the rotational frequency detector. The following arguments apply equally well to the sequential logic phase detector discussed in Section 2.6. To begin with, no theory exists which could predict the influence of additive noise on a sequential logic phase detector or a rotational frequency detector. The best we can offer are some qualitative remarks.

A frequently quoted statement is that additive noise causes extra transitions of the composite waveform of signal plus noise. If these extra transitions become excessive the detector fails to operate because it ignores the signal crossing if a noise crossing occurs first. This interpretation of the effect of noise is correct only if the bandwidth of the noise is very large. If, however, the noise spectrum is filtered, in the long term there are no extra transitions of the composite waveform. While in the unrestricted wideband case, noise can be viewed as "grass" it should rather be pictured in the latter case as blur. In this case, the effect of noise is to retard or advance the true zero crossings; it creates extra transitions only with low probability. Consequently, the PLL can work at a much lower SNR$_i$ compared with the wideband case.

After these general remarks we would like to analyze the effect of noise on the frequency detector in some detail. We model the effect of noise by a random additive phase term θ_k. The phase increment of the kth cycle is

$$\Delta \phi_k = \frac{\omega_s - \omega_{VCO}}{\omega_s} 2\pi + \theta_{k+1} - \theta_k \qquad (5.2\text{-}35)$$

To proceed, assume for the moment that θ_{k+1} and θ_k are known quantities. Mathematically we condition on θ_k and θ_{k+1}. Then the mean value of the frequency detector output voltage is exactly the same as previously determined but with $[(\omega_s - \omega_{VCO})/\omega_s]2\pi$ replaced by $[(\omega_s - \omega_{VCO})/\omega_s]2\pi + \theta_{k+1} - \theta_k$. Averaging over all possible values of θ_{k+1} and θ_k by taking expected values we have

$$\langle g_f(\omega_s - \omega_{VCO}) \rangle = E\left[g_f\left(\frac{\omega_s - \omega_{VCO}}{\omega_s} + \frac{\theta_{k+1}}{2\pi} - \frac{\theta_k}{2\pi} \right) \right] \qquad (5.2\text{-}36)$$

where g_f is the characteristic of Figure 5.2-17. If the argument of g_f is in the linear region of g_f with high probability, we can explicitly compute the expected value of (5.2-36)

$$\langle g_f(\omega_s - \omega_{VCO}) \rangle \approx K_f \frac{\omega_s - \omega_{VCO}}{\omega_s} + \frac{E(\theta_{k+1})}{2\pi} - \frac{E(\theta_k)}{2\pi}$$

$$\approx K_f \frac{\omega_s - \omega_{VCO}}{\omega_s} \qquad (5.2\text{-}37)$$

Here we made the further assumption that $E(\theta_{k+1}) = E(\theta_k)$. Notice we have not made any assumption on the independence of θ_{k+1} and θ_k, but have only required that the increment

$$\Delta\theta = \theta_{k+1} - \theta_k \qquad (5.2\text{-}38)$$

is small most of the time. For a wideband disturbance the increment $\Delta\theta$ is the difference of two independent random variables and has the variance

$$E[\theta_{k+1} - E(\theta_{k+1}) - \theta_k + E(\theta_k)]^2 = 2E[\theta_k - E(\theta_k)]^2 \qquad (5.2\text{-}39)$$

If the noise is filtered, the two random variables are statistically dependent and the variance is always smaller than the right-hand side of (5.2-39). Thus, for the same variance $E[\theta_k - E(\theta_k)]^2$, i.e., the same SNR_i, the probability of a small increment $\Delta\theta$ is larger when the noise is filtered.

It is interesting to observe from (5.2-37) that small phase jitter has no influence on the frequency detector. If, however, the variance of the phase jitter spans a significant portion of the period of g_f it is expected that the operation of the frequency detector will rapidly deteriorate above a certain threshold value.

Finally, we should keep in mind that the phase frequency detector discussed previously in Section 2.6 develops a frequency error signal when out-of-lock. The rotational frequency detector offers no advantage over the sequential phase frequency detector if the input signal has a periodic waveform. Its main application is in timing recovery when the input signal has randomly missing transitions and, therefore, the sequential logic phase detector fails to operate. Timing recovery will be discussed in detail in Volume 2.

5.2.3. Acquisition Aid Using a Nonlinearity

Pull-in of the PLL large frequency offset is a slow process because the average VCO control signal $\overline{\sin \phi}$ developed over a beat note is very small. While large phase errors normally occur only in the acquisition mode, the phase error fluctuations are small (around zero) in the tracking mode. Thus, incorporating a nonlinearity which produces a large additional control signal for a large phase error should accelerate the acquisiton process.

The schematic of such a modified phase-locked loop is shown in Figure 5.2-15. It differs from the conventional phase-locked loop by the addition of a circuit inserted between the phase detection and the loop filter. The upper signal path of this circuit contains a linear amplifier while the lower part contains an amplifier with a nonlinear amplification law. Ideally, the non-linear amplification law should have a dead zone around the tracking point. Then the tracking performance is identical to that of a conventional loop with phase detector gain $K_D' K_N$, while for large phase errors the loop dynamics are determined by the control signal produced in the lower signal path.

A practical realization of this concept is shown in Figure 5.2-16 [8]. The nonlinear network is made of a pair of antiparallel diodes shunted by a resistor R whose value is much smaller than the combined resistance of the diode pair when biased near zero voltage. The voltage is thus the sum of two terms; one is due to the current flowing through the resistance R and the other is caused by the diode currents. Its value varies with the error signal $K_D' \sin \phi$ as

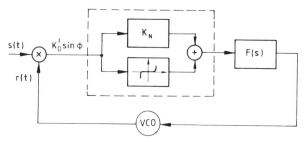

Figure 5.2-15. Phased-locked loop with additional nonlinear circuit to accelerate acquisition.

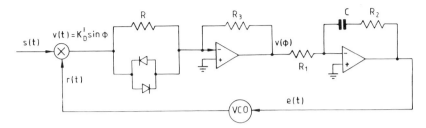

Figure 5.2.16. Practical realization of the configuration in Figure 5.2-15.

$$v(\phi) = R_3 \left[\frac{1}{R} K_D' \sin \phi + 2I_s \sinh (v_T^{-1} K_D' \sin \phi) \right]$$

The quantities v_T and I_s are parameters characterizing the diode current $i(t)$ with input signal $v(t)$

$$i(t) = I_s \left[\exp \left(+ \frac{v(t)}{v_T} \right) - 1 \right] \qquad (5.2\text{-}40)$$

Typically v_T is between 50 and 100 mV and I_s is in the order of 0.1 μA. The resistor R_3 determines the gain of the first operational amplifier.

In the tracking mode, the network current is mainly controlled by the resistor R because the fluctuations are small around zero where the resistance of the diode $(I_s/v_T)^{-1}$ is much larger than R. Consequently, $v(\phi)$ can be expressed as

$$v(\phi) \approx R_3 \left(\frac{1}{R} + \frac{2I_s}{v_T} \right) K_D' \sin \phi$$

$$\text{(tracking)}$$

$$\approx \frac{R_3}{R} K_D' \sin \phi =: K_D \sin \phi \qquad (5.2\text{-}41)$$

As a result, the loop configuration of Figure 5.2-16 provides the same performance as the conventional PLL in the tracking mode.

By contrast, in the unlocked mode an initial frequency offset gives rise to a large control signal when the phase error exceeds a given magnitude $|\phi_L|$. This is achieved by adjusting the phase detector gain K_D' to make $K_D' \sin |\phi_L|$ large enough to drive the diodes into large current conduction. In this mode the error signal at the input of the loop filter is orders of magnitude larger than $K_D := K_D' R_3/R$, the maximum value provided by the conventional loop. As a result, the capacitor in the loop filter is charged much faster.

The state equations of the modified loop in normalized time $\tau = \omega_n t$ are found by replacing $\sin \phi$ in (2.4-20) by

$$\sin \phi + \frac{2RI_s}{v_T} \frac{1}{\alpha_D} \sinh (\alpha_D \sin \phi) \qquad (5.2\text{-}42)$$

where α_D is the exponential factor

$$\alpha_D := \frac{K_D'}{v_T} \qquad (5.2\text{-}43)$$

The state equations read (see Figure 5.2-17)

$$\frac{d\phi}{d\tau} = -2\zeta \left[\sin \phi + \frac{2RI_s}{v_T} \frac{1}{\alpha_D} \sinh (\alpha_D \sin \phi) \right] - x_1 + \frac{\Delta\omega}{\omega_n}$$

$$\frac{dx_1}{d\tau} = \sin \phi(\tau) + \frac{2RI_s}{v_T} \frac{1}{\alpha_D} \sinh (\alpha_D \sin \phi) \qquad (5.2\text{-}44)$$

The state equations have been solved numerically [8]. The phase error $\phi(\tau)$ as a function of normalized time for $\zeta = 1$ and an initial phase error of zero is plotted in Figure 5.2-18. Large conduction in the diode occurs approximately for $\alpha_D \sin |\phi| > 2$. If we assume a linear tracking range of $|\phi_L|$ (where the diode current is negligible) then the value of α_D and of the phase detector gain K_D' are given by

$$\alpha_D = \frac{2}{\sin |\phi_L|} \approx \frac{2}{|\phi_L|}, \quad K_D' = v_T \alpha_D \qquad (5.2\text{-}45)$$

The results are presented in Figure 5.2-18 for $\alpha_D = 12$ corresponding to a linear tracking range of approximately $10°$ and for a wide range of normalized frequency offset $\Delta\omega/\omega_n$.

The results can be interpreted by the following approximate calculations. Since the input to the $\hat{\theta}(\tau)$ integrator is 2ζ-times larger than the input to the x_1-integrator (see Figure 5.2-17) the short-term transients are mainly determined by the proportional path of the loop filter and we may assume $x_1(\tau)$ to be constant during this period. Initially the phase error increases

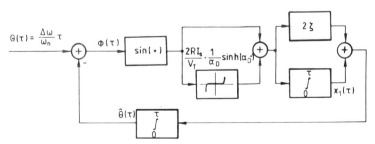

Figure 5.2-17. Baseband model of the modified perfect second-order loop in normalized time $(\tau = \omega_n t)$.

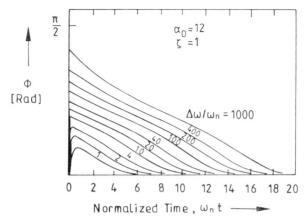

Figure 5.2-18. Variation of the phase error versus normalized time for $\zeta = 1$ and $\alpha_D = 12$ for various values of loop detuning $\Delta\omega/\omega_n$ (from Bernard Glance, New Phase-Lock Loop Circuit Providing Very Fast Acquisition Time, *IEEE Transactions on Microwave, Theory and Techniques*, Vol. 33, No. 9, © IEEE 1985).

linearly. Concurrently, the normalized control signal of the VCO increases as

$$\frac{d\hat{\theta}(\tau)}{d\tau} = 2\zeta \frac{2RI_s}{v_T} \frac{1}{\alpha_D} \sinh\left[\alpha_D \sin \phi(\tau)\right] \qquad (5.2\text{-}46)$$

until the slope $d\hat{\theta}/d\tau$ equals the slope $d\theta/d\tau = \Delta\omega/\omega_n$ of the incoming signal.

$$\frac{d\hat{\theta}}{d\tau} = 2\zeta \frac{2RI_s}{v_T} \frac{1}{\alpha_D} \sinh\left(\alpha_D \sin \phi_1\right)$$

$$= \frac{\Delta\omega}{\omega_n} \qquad (5.2\text{-}47)$$

This occurs during the first cycle of the variation of ϕ for a phase error $|\phi_1| < \pi/2$, in the very short time interval

$$\Delta\tau \approx \frac{\phi_1}{\Delta\omega/\omega_n} \qquad (5.2\text{-}48)$$

At this time the VCO is locked with a phase error ϕ_1. This duration decreases in inverse proportion to the initial frequency offset. For $\Delta\omega/\omega_n > 10$ this time is too short to be visible for the time scale used in the figures. Thereafter, $\phi(\tau)$ decreases to zero while the state variable $x_1(\tau)$ simultaneously increases until it reaches its steady-state final value $x_{1,s} = \Delta\omega/\omega_n$.

We define the acquisition time T_D as the time needed for $\phi(\tau)$ to decrease to one degree. From a curve fit of the numerical solution of the state equation, one obtains that the normalized acquisition time increases only logarithmically as a function of $\Delta\omega/\omega_n$

$$\omega_n T_D \approx 2\left[\ln\left(\frac{\Delta\omega}{\omega_n}\right) + 2.8\right], \quad \text{modified loop, } \zeta = 1 \qquad (5.2\text{-}49)$$

The effect of the damping ratio is illustrated by plotting the phase error ϕ versus τ/ζ. The curve thus obtained is almost the same for different values of ζ when the initial frequency offset is scaled by the damping ratio ζ. This effect is shown in Figure 5.2-19.

In comparison, for conventional phase-locked loops the pull-in time increases proportional to $(\Delta\omega/\omega_n)^2$ ((4.3-25) and Table 3.3-1)

$$\omega_n T_f \approx 0.5\left(\frac{\Delta\omega}{\omega_n}\right)^2, \quad \text{conventional loop, } \zeta = 1$$

$$\qquad (5.2\text{-}50)$$

$$T_D > T_f$$

Thus, the modified loop provides a great acceleration compared to the conventional loop.

The pull-out frequency which we defined in Section 4.3-1 as the maximum frequency step for which the loop reaches equilibrium without slipping a cycle is readily found from (5.2-47) by setting $\phi_1 = \pi/2$

$$\frac{\Delta\omega_{po}}{\omega_n} = 4\zeta \frac{RI_s}{v_T} \frac{1}{\alpha_D} \sinh(\alpha_D), \quad \text{modified loop} \qquad (5.2\text{-}51)$$

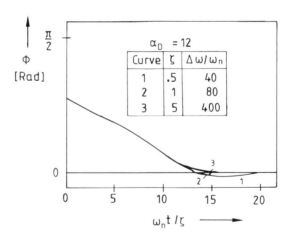

Figure 5.2-19. Effect of damping ratio on the variation of the phase error versus time normalized relative to loop natural frequency and damping ratio (from Bernard Glance, New Phase-Lock Loop Circuit Providing Very Fast Acquisition Time, *IEEE Transactions on Microwave, Theory and Techniques*, Vol. 33, No. 9, © IEEE 1985).

For typical parameter values of $\zeta = 1$, $2RI_s/v_T = 0.1$, and $\alpha_D = 12$ the normalized pull-out frequency becomes 1350 which is orders of magnitude larger than the corresponding value for the conventional loop with sinusoidal phase detector characteristic

$$\frac{\Delta\omega_{po}}{\omega_n} = 1.8(\zeta + 1) = 3.6, \quad \text{conventional loop}$$

The modified loop circuit also provides a much larger capacity to remain locked in the presence of fast frequency changes. The maximum rate of frequency change $\Delta\dot{\omega}/\omega_n^2$ which the conventional loop can sustain without losing lock was found in (5.2-2a)

$$\left(\frac{\Delta\dot{\omega}}{\omega_n^2}\right)_{max} = 1, \quad \text{conventional loop}$$

Using the same reasoning as for the above result we arrive, for the modified loop, at

$$\left(\frac{\Delta\dot{\omega}}{\omega_n^2}\right)_{max} = \frac{2RI_s}{v_T}\frac{1}{\alpha_D}\sinh(\alpha_D), \quad \text{modified loop} \qquad (5.2\text{-}52)$$

which is a very large improvement.

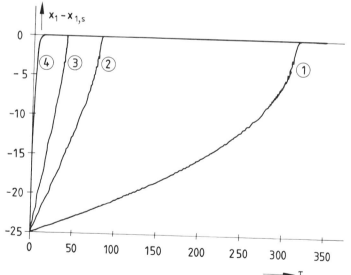

Figure 5.2-20. Frequency acquisition of a perfect second-order loop with $\zeta = 1.0$, $\Delta\omega/\omega_n = 100$, $\phi(0) = 0$, and $x_1(0) = 75$ for the case of: (1) unaided acquisition; (2) aided acquisition using current injection $\Delta\dot{\omega}/\omega_n = 0.2$; (3) aided acquisition using rotational frequency-discriminator applied as in Figure 5.2-13 with peak $I_{FD}K_0/\omega_n^2 C =: \Delta\dot{\omega}/\omega_n = 0.5$; (4) aided acquisition using two diodes as in Figure 5.2-16 with $\alpha_D = 12$ and $2RI_s/v_T = 0.1$.

An experimental circuit has been built to verify the performance predicted by the computer calculations. The results were found to be in good agreement with those predicted by the numerical calculations. For details the reader is referred to Glance [8]. No results are known at present for the performance of the modified loop in the presence of noise.

Figure 5.2-20 compares unaided acquisition of frequency with some of the aided versions discussed in this chapter. However, one has to keep in mind that in a realistic situation, i.e., in the presence of noise, the tracking behavior might place limitations on the parameters of the acquisition circuitry which otherwise would be chosen to speed up the acquisition arbitrarily.

Main Points of the Chapter

- *Aided Phase Acquisition*
  Aided phase acquisiton is needed only where rapid phase acquisition is necessary. For a significant reduction in phase acquisition time T_p the loop SNR must be larger than 12 dB. Below that value application of the PLL is ruled out when rapid phase acquisition is necessary.

- *Aided Frequency Acquisition*
  For most applications auxiliary circuits are employed to acquire frequency. The main aided frequency acquisition methods are

- Sweeping $\begin{cases} \text{closed loop} \\ \text{open loop} \end{cases}$

- Use of an additional detector to produce a frequency difference sensitive signal

- Incorporation of nonlinear elements into the loop

- Sweeping of a perfect second-order loop
  Insertion of a constant current into filter capacitor:

  (a) Maximum allowable (normalized) sweep rate in the presence of noise

$$\frac{\Delta\dot{\omega}}{\omega_n^2} = \begin{cases} 1 - \sqrt{\dfrac{2}{\rho - 4}}, & 6 < \rho < 9.5 \\ 0.4, & \rho > 9.5 \end{cases} \qquad (5.2\text{-}4)$$

  (b) The maximum sweep rate is smaller than the maximum rate of change of frequency in the tracking mode (noise-free case)

$$\frac{\Delta\dot{\omega}}{\omega_n^2} = 1 \qquad (5.2\text{-}2a)$$

  (c) Lock detector probabilities
     P_{FA}: probability of false alarm
     P_{FL}: probability of false indication of loss of lock
     A frequently used design rule is to set

$$P_{FA} = P_{FL} = \frac{1}{2}\,\mathrm{erfc}\left(\frac{\lambda}{\sqrt{2}\sigma_{LP}}\right) \qquad (5.2\text{-}18)$$

with

$$\frac{\lambda}{\sigma_{LP}} = \frac{1}{2}\left[\left[1 - \left(\frac{\Delta\dot{\omega}}{\omega_n^2}\right)^2\right]\rho\,\frac{B_L}{B_{LD}}\right]^{1/2}$$

where B_L is the loop noise bandwidth, B_{LD} is the noise bandwidth of lock detector, λ, the threshold of lock detector and σ_{LP}^2, the variance of lock detector output. The time to detect lock is inversely proportional to the bandwidth of the lock detector B_{LD}. As a rule of thumb we may take

$$\Delta\tau_{\mathrm{det}} = 10\,\frac{B_L}{B_{LD}}$$

Then we require that $\Delta\tau_{\mathrm{det}}$ be much smaller than the (normalized) mean time between cycle slips.

- Sweeping of an imperfect second-order loop

Insertion of a constant current into the filter capacitor allows searching only over the pull-in range

$$\left|\frac{\Delta\omega}{\omega_n}\right| < 2\sqrt{\frac{\zeta}{\beta}}$$

Maximum sweep must be reduced by $\beta\,\Delta\omega/\omega_n$ compared to the perfect second-order loop

$$\left(\frac{\Delta\dot{\omega}}{\omega_n^2}\right)_{\mathrm{imperfect}} = \left(\frac{\Delta\dot{\omega}}{\omega_n^2}\right)_{\mathrm{perfect}} - \beta\,\frac{\Delta\omega}{\omega_n}$$

- Sweeping of higher order loops is not well understood
- A modified loop with an additional nonlinearity inserted between phase detector and loop filter greatly accelerates the acquisition process. Acquisition time T_D increases with the logarithm of $\Delta\omega/\omega_n$ only

$$\omega_n T_D \approx 2\left[\ln\left(\frac{\Delta\omega}{\omega_n}\right) + 2.8\right], \quad \zeta = 1 \qquad (5.2\text{-}49)$$

Pull-out frequency is increased by orders of magnitude compared to the conventional loop

$$\frac{\Delta\omega_{po}}{\omega_n} = 4\zeta\,\frac{RI}{v_T}\,\frac{1}{\alpha_D}\,\sinh(\alpha_D) \qquad (5.2\text{-}51)$$

Also the ability to remain locked in the presence of fast frequency changes $\Delta\dot{\omega}/\omega_n^2$ is greatly improved

$$\left(\frac{\Delta\dot{\omega}}{\omega_n^2}\right)_{\max} = \frac{2RI_s}{v_T}\frac{1}{\alpha_D}\sinh(\alpha_D) \qquad (5.2\text{-}52)$$

REFERENCES

1. F. M. Gardner, Hang Up in Phase-Lock Loops, *IEEE Transactions on Communications*, Vol. 25, pp. 1210–1214, October 1977.
2. H. Meyr and L. Popken, Phase Acquisition Statistics for Phase-Locked Loops, *IEEE Transactions on Communications*, Vol. 28, No. 8, pp. 1365–1372, August 1980.
3. A. J. Viterbi, *Principles of Coherent Communications*, McGraw-Hill, New York, 1966.
4. J. P. Frazier and J. Page, Phase-Lock Loop Frequency, Acquisition Study, *IRE Transactions on Space Electronics and Telemetry*, Vol. 8, pp. 210–227, September, 1962.
5. D. Richman, Color Carrier Reference Phase Synchronization Accuracy in NTSC Color Television, *Proceedings of the IRE*, pp. 106–133, January 1954.
6. R. W. D. Booth, A note on the Design of Baseband AFC Discriminators, *Conference Record* National Telecommunications Conference, NTC'80, Vol. 2, paper 24.2.
7. D. G. Messerschmitt, Frequency Detection for PLL Acquisition in Timing and Carrier Recovery, *IEEE Transactions Communications*, Vol. 27, No. 9, September 1979.
8. B. S. Glance, New Phase-Lock Loop Circuit Providing Very Fast Acquisition Time, *IEEE Transactions on Microwave, Theory and Techniques*, Vol. 33, No. 9, September 1985.
9. U. Mengali, Acquisition Behaviour of Tracking Loops Operating in Frequency Search Modes, *IEEE Transactions on Aerospace and Electronic Systems*, Vol. 10, pp. 583–587, September 1974.

6

LOOP THRESHOLD

6.1. INTRODUCTION

The response of the PLL when the signal is disturbed by additive noise was studied in Chapter 3. The intensity of the noise was assumed such that the phase error remained small most of the time and, thus, linearization of the loop equations was permissible. We found that the noise causes a random angular phase disturbance $n'(t)$ (3.2-4)

$$x(t) = K_D[\phi(t) + n'(t)]$$

A good measure to verify the linearization approximation is the phase error variance σ_ϕ^2. When the variance becomes too large a phenomenon occurs that is inherent to the nonlinearity of the loop. At random points in time, the noise increases the phase error from the tracking value ϕ_s to $\phi_s + 2\pi$ or $\phi_s - 2\pi$. This means the loop has *slipped a cycle*. We speak of a positive cycle slip if the phase error increases by $+2\pi$ and of a negative cycle slip if the phase error increment equals -2π. Cycle slips are particularly destructive to operations in which every cycle counts, such as, for example, the recovery of digital clock timing.

The occurrence of a slip is an event with a very low probability for weak noise, but the probability rises steeply with increasing noise power, see Figure 6.1-1. It is clear that the loop ceases to function properly if cycle slips become frequent. Thus, the *loop threshold point* of a PLL can well be defined as the point at which the number of slips per unit time exceeds a certain level.

The definition given is subjective as is the whole notion of operating threshold. Generally speaking, we can say that the threshold is reached when a system ceases to fulfill its function or when it does so with severely degraded performance. Another widely accepted definition of loop threshold employs the variance of the VCO phase σ_ϕ^2 as a measure. The threshold of the device is reached when σ_ϕ^2, obtained by taking the

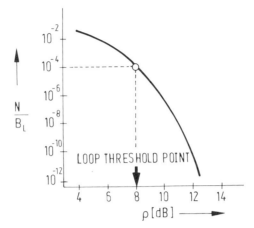

Figure 6.1-1. Normalized slip rate N/B_L (i.e., number of slips per unit of normalized time $\tau = B_L t$) as a function of the loop signal-to-noise ratio ρ. A second-order loop with $\zeta = 0.707$ is assumed.

nonlinearity into account, exceeds the value given by the linear equations by 1 dB (see Figure 6.1-2).

In comparing these two definitions of threshold, it can be seen that the departure of the variance σ_ϕ^2 from the value predicted by the linear theory is small whereas the cycle slip rate varies over orders of magnitude within a small interval of ρ. The cycle slip rate is thus a much more appropriate quantity to define loop threshold.

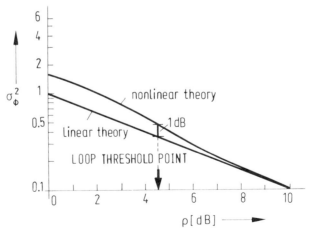

Figure 6.1-2. Comparison of the phase error variance as predicted by the linear or nonlinear theory.

Example: The 1 dB departure of σ_ϕ^2 sets the loop threshold point at $\rho \approx 5$ dB. If a value of 10^{-4} slips per unit of normalized time $\tau = B_L t$ is assumed, a signal-to-noise ratio $\rho \approx 8$ dB is required. The value $\rho = 5$ dB seems too optimistic since the corresponding slip rate of 10^{-3} is too large.

Cycle slips in a PLL are statistical nonlinear phenomena. The complicated interaction of nonlinearity and noise makes a mathematical analysis very difficult (see Chapters 9 to 12). By the very nature of cycle slips, matters cannot be simplified by merely making some sort of linearization.

Based on the experimental study [1], the underlying principles of the interaction between noise and nonlinearity are explained in Section 6.2. Avoiding advanced mathematical tools the experimental results are complemented by simple approximate analysis that agrees well with the experimental findings.

For the evaluation of loop thresholds, statistical data are needed. A set of diagrams on cycle slip statistics are given in Section 6.3 which should provide the reader with sufficient information for a large variety of design problems.

6.2. UNDERSTANDING CYCLE SLIPS

State variables are the only mathematical description that allows one to analyze the loop under nonlinear operating conditions. For convenience, the normalized state equations for a second-order loop with imperfect integrator in the loop filter are summarized below (see Chapter 2.4.3)

$$\frac{d\phi(\tau)}{d\tau} = -2\zeta\left[1 - \frac{\beta}{2\zeta}\right]\sin\phi(\tau) - x_1 - 2\zeta\left[1 - \frac{\beta}{2\zeta}\right]n'(\tau) + \frac{d\theta(\tau)}{d\tau}$$
$$\frac{dx_1(\tau)}{d\tau} = -\beta x_1(\tau) + [1 - 2\zeta\beta + \beta^2]\sin\phi(\tau) + [1 - 2\zeta\beta + \beta^2]n'(\tau)$$

(6.2-1)

with

$$\tau = \omega_n t, \quad x_1 = y_1/\omega_n, \quad \beta = \frac{\omega_n}{K_0 K_D}$$

(6.2-2)

The spectral density of the white noise process $n'(\tau)$ equals $\omega_n N_0/(2A^2)$ (Chapter 3.4, (3.4-12)). Throughout this chapter only high-gain loops ($\beta \ll 2\zeta$, see (2.4-26)) are considered. The equivalent model of the loop is depicted in Figure 6.2-1.

We already know from Figure 6.1-1 that the cycle slip rate is extremely dependent on the signal-to-noise ratio. As will be shown presently, the

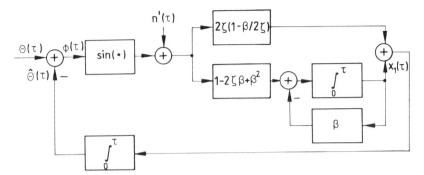

Figure 6.2-1. Equivalent model of a second-order phase-locked loop with imperfect integrator using normalized variables (for high gain loops: $\beta \ll 2\zeta$, (2.4-22)).

damping factor ζ also has a strong influence on the slip rate, as well as on the manner in which the slips occur.

In Figure 6.2-2 the probability that the time between two slips (T_s) is shorter than t is plotted for two values of ζ. For $\zeta = 0.24$ we observe a steep increase in $P(T_s < t)$ for small t, i.e., many slips have a very short duration compared to the mean time between slips. This means that the slips occur in

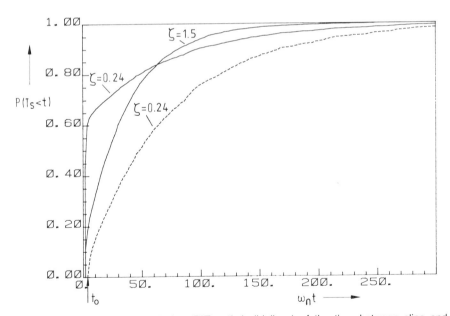

Figure 6.2-2. Probability distribution $P(T_s < t)$ (solid lines) of the time between slips and conditional distribution $P(T_s < t \mid T_s > t_0)$ (dashed line), conditioned on the fact that T_s exceeds a given time t_0. Signal-to-noise ratio $\rho = 2$ (numeric ratio) (from Gerd Ascheid and Heinrich Meyr, Cycle Slips in Phase-Locked Loops: A Tutorial Survey, *IEEE Transactions on Communications*, Vol. 30, No. 10, © IEEE 1982).

bursts; if viewed on an oscilloscope we see clusters of slips of very short duration separated from each other by long time intervals.

For large damping, the distribution $P(T_s < t)$ is virtually exponential. Probability theory states that the distribution of time periods between statistically independent events is exponential. Therefore we may conjecture that the cycle slips are independent, isolated events for large damping; an assumption which we will justify later.

Having recognized the burst-like appearance of slips for small damping factors and the apparently isolated appearance of slips for large damping, we need to explain this differing behavior.

6.2.1. Loops with Small Damping Factors

In Figure 6.2-3 we have plotted an example of phase error ϕ and state variable x_1 as a function of normalized time τ for $\zeta = 0.24$ and* a very small ρ. We clearly recognize the burst-like nature of the cycle slips. Before we turn our attention to the bursts of cycle slips we examine the behavior of the loop between two bursts, i.e., in its tracking mode. Then with a damping of

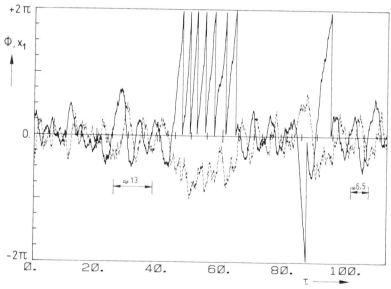

Figure 6.2-3. Trajectories of phase error $\phi(\tau)$ (solid line) and state variable $x_1(\tau)$ (dashed line) versus normalized time for $\beta = 0.03$, $\zeta = 0.24$, and $\rho = 2$ (numeric ratio) (from Gerd Ascheid and Heinrich Meyr, Cycle Slips in Phase-Locked Loops: A Tutorial Survey, *IEEE Transactions on Communications*, Vol. 30, No. 10, © IEEE 1982).

---

*Note that $\zeta = 0.24$ has been chosen to demonstrate more clearly the loop behavior for small damping factors. In practice such small values should never be used.

$\zeta = 0.24$ we expect to observe a weakly damped oscillation of $\phi(\tau)$ and $x_1(\tau)$ sustained by the noise process $n'(\tau)$. In the tracking mode we may linearize the state equation (6.2-1) to determine the properties of this oscillation. Its average period*, in normalized time, is

$$\tau_p = \frac{2\pi}{(1 - \zeta^2)^{1/2}} = 6.47 \qquad (6.2\text{-}3)$$

Oscillations with a period of approximately 6.5 are indeed found in Figure 6.2-3. Furthermore, we would expect an oscillation of $x_1(\tau)$ with the same period as $\phi(\tau)$, only shifted by $\pi/2$, which is again confirmed by Figure 6.2-3. To go one step further, $x_1(\tau)$ and $\phi(\tau)$ must have approximately the same amplitude, because of the normalizations that have been performed. For $x_1(\tau)$ and $\phi(\tau)$ we read off a value of 1 from Figure 6.2-3; converting the average amplitude of $\phi(\tau)$ into degrees gives approximately $60°$.

The weaker the noise the smaller the amplitude of $x_1(\tau)$ and $\phi(\tau)$ since weak noise can push the loop only slightly away from its stable equilibrium. From linear theory we know $\sigma_\phi^2 = 1/\rho$, hence the amplitude of the perturbed sinusoid would be $\sqrt{2/\rho}$. For strong noise, as in Figure 6.2-3, it is very probable that the amplitude of the fluctuations exceeds the $90°$ phase error (maximum restoring force), in which case a cycle slip is very likely to occur. Furthermore, in this case the $x_1(\tau)$ variable follows in taking a wrong value. After completion of the first cycle slip, $x_1(\tau)$, which is responsible for frequency correction, has a wrong initial value (loop stress) and consequently the loop is much more susceptible to another cycle slip.

The experimentally determined average value of x_1 taken at the completion of a cycle slip as a function of ρ is shown in Figure 6.2-4a. Indicated in this figure is the normalized pull-out frequency given by (4.3-5)

$$\frac{\Delta\omega_{po}}{\omega_n} = 1.8(1 + \zeta) \qquad (6.2\text{-}4)$$

and the pull-in frequency of the second-order loop with imperfect integrator (4.3-19)

$$\frac{\Delta\omega_p}{\omega_n} = 2\sqrt{\frac{\zeta}{\beta}} \qquad (6.2\text{-}5)$$

If the loop starts with zero phase error $\phi(0) = 0$ then $x_1(0) = \Delta\omega_{po}/\omega_n$ is the maximum initial frequency deviation for which the loop does not skip one or more cycles but remains in lock. The pull-in frequency is the maximum frequency difference for which the loop will lock, although this may require

---

*The average period equals the oscillation period of the homogeneous solution of the linearized state equation.

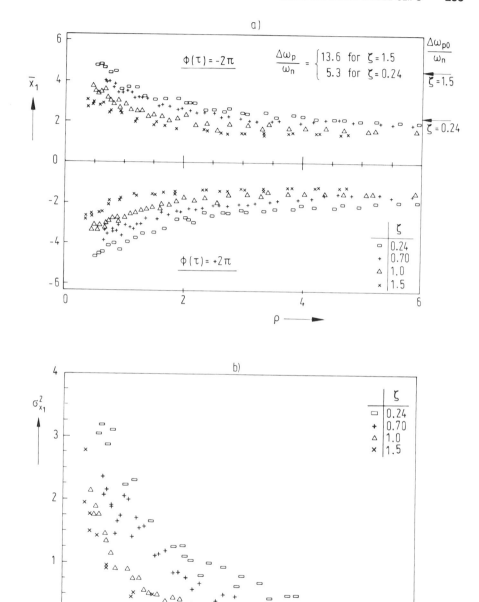

Figure 6.2-4. (a) Conditional experimental mean $\bar{x}_1$ of the state variable $x_1(\tau)$, taken at the instant $\phi(\tau) = \pm 2\pi$ (completion of a cycle slip) (b) Conditional experimental variance $\sigma_{x_1}^2 = \overline{(x_1 - \bar{x}_1)^2}$ taken at the completion of a cycle slip (from Gerd Ascheid and Heinrich Meyr, Cycle Slips in Phase-Locked Loops: A Tutorial Survey, *IEEE Transactions on Communications*, Vol. 30, No. 10, © IEEE 1982).

a long acquisition period. Both frequencies hold for the noise-free case only, but serve as a good indicator for the noisy case.

For a small damping factor the average value of x_1 taken at the completion of a slip is — even for large SNR — only slightly smaller than the pull-out frequency. This means that a burst of cycle slips rather than a single cycle slip is to be expected. For $\rho < 5$ (7.0 dB) and $\zeta = 0.24$ the average of x_1 is larger than the pull-out frequency and a single cycle slip is very unlikely. In addition, the larger the difference $|x_1 - \Delta\omega_{po}/\omega_n|$, the longer the mean duration of the burst will be.

The average of x_1 as a function of ρ increases rapidly with decreasing ρ, see Figure 6.2-4a. For $\rho = 1$ (0 dB) x_1 has come very close to the pull-in frequency of the loop. Since the experimental variance $\sigma_{x_1}^2$ (Figure 6.2-4b) taken at the completion of a cycle slip also increases with decreasing ρ, it is very likely that x_1 assumes a value larger than the pull-in frequency once the loop has started slipping cycles. Therefore, once the loop has lost lock it stays out of lock with high probability. Control of the VCO is lost and its frequency wanders off from the signal frequency. This phenomenon has been called drop-lock in the literature. Actually, there is no well-defined noise threshold below which the PLL falls out of lock and remains so. There is no fundamental difference between repeated cycle slips and drop of lock (but see the results on loop detuning to be discussed shortly).

That the VCO wanders off at extremely low SNR is easily explained if we recognize that the restoring force in the equivalent model of Figure 6.2-1 can be neglected where x_1 is larger than the pull-in frequency. But neglecting the restoring force $\sin \phi$ is equivalent to opening the feedback system (see Figure 6.2-1). In this case, the input to the integrator representing the VCO consists of the noise process plus $x_1(\tau)$. As is well known, the variance of an integrated noise process increases with τ, hence it is unbounded and the VCO wanders off.

It is very instructive to examine loop behavior during a burst of slips. We want to examine in detail the first and the last cycle as well as a cycle in the middle of a burst. Examples of bursts having the same parameters are shown in Figure 6.2-5 and in expanded scale in Figures 6.2-6 and 6.2-7.

Due to noise, the magnitude of both the phase error $\phi(\tau)$ and $x_1(\tau)$ increase at the beginning of the first cycle slip (Figure 6.2-6, region a). While the phase error $\phi(\tau)$ tends toward large negative values, $x_1(\tau)$ is positive. (Note that $-x_1(\tau)$ is shown in Figures 6.2-5 to 6.2-7.) But to correct a negative going phase error, $x_1(\tau)$ would have to be negative, also. The phase error passes through $-\pi/2$, the point of maximum restoring force, toward $-\pi$. In the interval of $-2\pi < \phi < -\pi$ the restoring force has the wrong polarity, as there is positive feedback in the loop. Since $x_1 > 0$, the phase error rapidly passes this region of positive feedback to reach $\phi = -2\pi$, which is of course equivalent to zero phase error. At the completion of the first slip, $x_1(\tau)$ assumes a random value of slightly more than the pull-out frequency. During the following three cycle slips the value

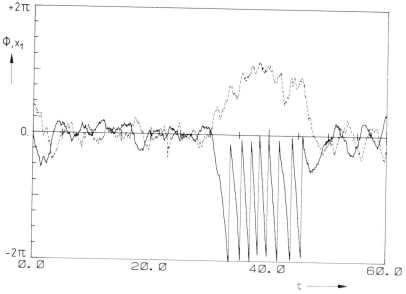

Figure 6.2-5. Typical burst of a weakly damped ($\zeta = 0.24$) second-order loop with signal-to-noise ratio $\rho = 3$ (numerical ratio) (from Gerd Ascheid and Heinrich Meyr, Cycle Slips in Phase-Locked Loops: A Tutorial Survey, *IEEE Transactions on Communications*, Vol. 30, No. 10, © IEEE 1982).

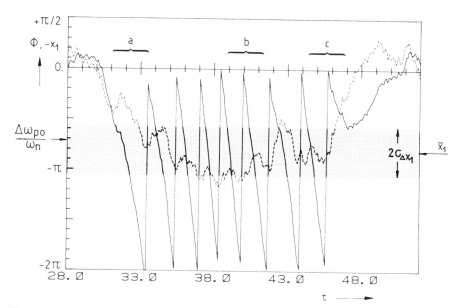

Figure 6.2-6. Part of the trajectory of Figure 6.2-5 in expanded scale. Note that $-x_1(\tau)$ is displayed (from Gerd Ascheid and Heinrich Meyr, Cycle Slips in Phase-Locked Loops: A Tutorial Survey, *IEEE Transactions on Communications*, Vol. 30, No. 10, © IEEE 1982).

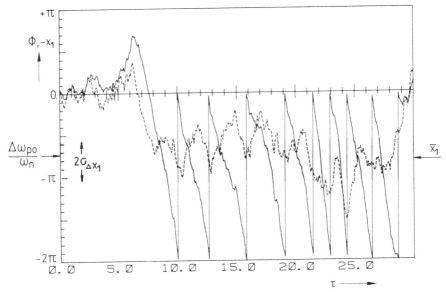

Figure 6.2-7. Part of another trajectory with the same parameters as in Figure 6.2-5. Note that $-x_1(\tau)$ is displayed (from Gerd Ascheid and Heinrich Meyr, Cycle Slips in Phase-Locked Loops: A Tutorial Survey, *IEEE Transactions on Communications*, Vol. 30, No. 10, © IEEE 1982).

of x_1 increases to its maximum before it slowly decreases to its correct average value of $x_1 = 0$. Another burst is shown in Figure 6.2-7. In contrast to the previous burst we do not observe a pumping-up of the x_1 variable during the first cycle slips. Rather the value of x_1, taken at the completion of a cycle slip, fluctuates around the experimental mean value $\bar{x}_1$ before it takes on a value $x_1 < \Delta\omega_{po}/\omega_n$ such that the loop can pull in.

Looking at the two (entirely different) bursts, the question arises whether we have observed examples of two different phenomena or whether a common mechanism exists in both cases. We will see that for both cases a common mechanism exists which consists of a systematic force driving the PPL towards its stable equilibrium and a random perturbation. We will first analyze the force and later on compute the variance of the random perturbation.

A typical cycle in the middle of a burst is shown in region b of Figure 6.2-6. It is typical in the sense that a value $x_1(\tau)$ exists that is erroneously interpreted as a frequency detuning by the PLL. Looking a little closer at this cycle slip, we notice that the phase error traverses the region $-2\pi < \phi < -\pi$ faster than the interval from $[0, -\pi]$. Since the restoring force $\sin \phi$ has the correct polarity within the interval $\tau_{bn}^+ = [0, -\pi]$, the absolute value of $x_1(\tau)$, as the integral of $\sin \phi$, is decreased. Within the interval $\tau_{bn}^- = [-\pi, -2\pi]$, however, the restoring force has the wrong polarity and instead of reducing $|x_1(\tau)|$ it moves it further away from the correct value $x_1(\tau) = 0$.

Since, however, the interval τ_{bn}^{-} is shorter than τ_{bn}^{+}, there is a net gain. To get a feeling for this net gain, we make an approximate computation.

Let us assume x_1 to be constant and neglect the noise $n(\tau)$. The loop is then governed by the differential equation

$$\frac{d\phi(\tau)}{d\tau} = -2\zeta \sin \phi(\tau) - x_1 \qquad (6.2\text{-}6)$$

Equation (6.2-6) is the differential equation of a first-order loop with frequency detuning $\Delta\omega = x_1$. If $x_1 > 2\zeta$, the loop will not lock but produce a periodic oscillation called beat note. In Figure 6.2-8 we have shown the phase error $\phi(\tau)$ and the phase detector output $\sin \phi(\tau)$.

The periodic waveform $\sin \phi(\tau)$ is not symmetric but has a dc-component that can be obtained by integrating the differential equation over a full cycle.

$$\phi(\tau_{bn}) - \phi(0) = -2\zeta \int_{0}^{\tau_{bn}} \sin \phi(\tau)\, d\tau - x_1 \tau_{bn} \qquad (6.2\text{-}7)$$

Since $\phi(\tau_{bn}) = -2\pi$ and $\phi(0) = 0$ we obtain

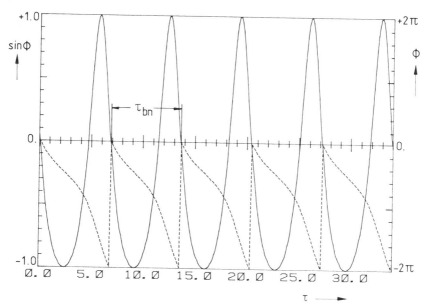

Figure 6.2-8. Phase error $\phi(\tau)$ (dashed line) and phase detector output $\sin \phi(\tau)$ (solid line) during a beat note (from Gerd Ascheid and Heinrich Meyr, Cycle Slips in Phase-Locked Loops: A Tutorial Survey, *IEEE Transactions on Communications*, Vol. 30, No. 10, © IEEE 1982).

$$\overline{\sin \phi} = \frac{1}{\tau_{bn}} \int_0^{\tau_{bn}} \sin \phi(\tau) \, d\tau = -\frac{x_1}{2\zeta} + \frac{2\pi}{2\zeta\tau_{bn}} , \quad x_1 > 0, \text{ negative cycle slip}$$

$$(6.2\text{-}8)$$

The beat note period was determined in Chapter 4.2. Replacing $\Delta\omega$ by x_1 and $4B_L$ by 2ζ in (4.2-15) we have*

$$\tau_{bn} = \frac{\pi}{\zeta} \frac{1}{[(x_1/2\zeta)^2 - 1]^{1/2}} \approx \frac{2\pi}{|x_1|} \frac{1}{1 - \frac{1}{2}(2\zeta/x_1)^2} , \quad \left(\frac{2\zeta}{x_1}\right)^2 \ll 1$$

$$(6.2\text{-}9)$$

Inserting (6.2-9) into (6.2-8) yields

$$\overline{\sin \phi} = -\frac{\zeta}{x_1} \tag{6.2-10}$$

Note that a positive value of x_1 produces a negative dc-term $\overline{\sin \phi}$ and vice versa. It is exactly this dc-term that is integrated in the loop filter of a second-order loop and reduces $x_1(\tau)$ to its correct value of $x_1(\tau) = 0$, thereby pulling the PLL back into its tracking range. The accumulated voltage during one cycle is

$$\overline{\Delta x_1} \approx \overline{\sin \phi} \, \tau_{bn} = -\frac{2\pi\zeta}{|x_1|x_1} , \quad |x_1| \gg 2\zeta \tag{6.2-11}$$

The noise process n' causes a random (mean free) fluctuation of the increment Δx_1. Under the same approximation as in (6.2-11), the random increment equals the integral of the white noise process $n'(\tau)$ over $[0, \tau_{bn}]$

$$\Delta x_1(\tau_{bn}) - \overline{\Delta x_1} = \int_0^{\tau_{bn}} n'(\tau) \, d\tau \tag{6.2-12}$$

The variance of the increment is

$$\sigma_{\Delta x_1}^2 = E[(\Delta x_1(\tau_{bn}) - \overline{\Delta x_1})^2]$$

$$= E\left[\int_0^{\tau_{bn}} \int_0^{\tau_{bn}} n'(\tau) n'(\nu) \, d\tau \, d\nu \right]$$

$$= \int_0^{\tau_{bn}} \int_0^{\tau_{bn}} E[n'(\tau) n'(\nu)] \, d\tau \, d\nu$$

The correlation function of a white noise process is a delta function

---

*Note that τ_{bn} in (6.2-9) corresponds to T_{bn} in (4.2-15) (see (4.2-1) and (6.2-6)).

$$E[n'(\tau)n'(\nu)] = \delta(\tau - \nu)\, \frac{1}{\rho\zeta}\, \frac{1}{1 + (1/2\zeta)^2}$$

where the weight is given by (3.4-13). Carrying out the integration and replacing τ_{bn} (6.2-9) yields

$$\sigma^2_{\Delta x_1} \approx \frac{4\zeta}{1 + (2\zeta)^2}\, \frac{1}{\rho}\, \frac{2\pi}{|x_1|} \qquad (6.2\text{-}13)$$

The mean value $\overline{\Delta x_1}$ in (6.2-11) can be interpreted as systematic drift force while the variance $\sigma^2_{\Delta x_1}$ describes a random diffusion. For the data given in Figure 6.2-5 (and Figure 6.2-6, respectively) and a value of $x_1 = 2.5$, we obtain for $\overline{\Delta x_1}$, $\sigma_{\Delta x_1}$, and for τ_{bn}

$$\tau_{bn} = 2.5\,, \quad \overline{\Delta x_1} = 0.24\,, \quad \sigma_{\Delta x_1} = 0.81 \qquad (6.2\text{-}14)$$

The random fluctuation is much larger than the systematic driving force $\overline{\Delta x_1}$. The different appearance of the two bursts is now easily understood. The systematic force $\overline{\Delta x_1}$ is covered by this random fluctuation. The values of x_1 taken at the completion of a cycle slip are within a band of $2\sigma_{\Delta x_1}$ width, which is centered around $\bar{x}_1$. What appears to be a systematic pumping-up effect in Figure 6.2-6 is nothing but a normal statistical fluctuation.

Our simple analysis is only approximate but predicts remarkably well the duration of a slip within a burst, as well as the statistical fluctuations of the increment Δx_1.

The distinctive features in region c are that the cycle slipping stops and x_1 rapidly converges toward zero. Immediately after the last cycle slip the phase error rapidly increases. The restoring force $\sin\phi$ is large and has the correct polarity during the convergence interval. Integration of the restoring force proceeds rapidly so that x_1 rapidly approaches zero.

In passing, formulas (6.2-11) and (6.2-13) give a hint why the loop permanently loses lock with decreasing ρ. The average pull-in effect $\overline{\Delta x_1}$ is inversely proportional to x_1^2, while the variance is inversely proportional to ρ and x_1. From Figure 6.2-4a we know that the average $\bar{x}_1$ increases with decreasing ρ. Therefore, the bursts tend to become longer until the loop eventually falls out of lock completely.

6.2.2. Overdamped Loops ($\zeta > 1$)

Having understood the rather complex structure of the cycle slips of an underdamped loop we turn our attention to loops having damping factors $\zeta > 1$. As already noted, cycle slips apparently occur in this case as isolated events, not in bursts. Why this difference?

The answer is again found by inspection of $\bar{x}_1$ in Figure 6.2-4a. We first note that for reasonably large ρ, $\bar{x}_1$ is less than $1/3$ of the pull-out frequency.

Secondly, the function $\bar{x}_1(\rho)$ is essentially flat for $\rho > 4$. It remains, for all ρ, much smaller than the pull-out frequency $\Delta\omega_{po}/\omega_n$ (not to mention the pull-in range $\Delta\omega_p/\omega_n$). From this and the variance $\sigma_{x_1}^2$ in Figure 6.2-4b it is clear that bursts of cycle slips are extremely unlikely events. The mean time between cycle slips converges for low ρ toward the values of a first-order loop, provided there is no frequency difference between VCO and signal.

We must still explain why the variable x_1 takes on such small values. As always, linear theory is instrumental in gaining an insight into nonlinear behavior.

A high damping factor implies a large time constant for the x_1-integrator; the response of such an integrator to a noise event is very sluggish and small in amplitude. Therefore, the proportional path of the filter determines the short time transients while the integral path compensates for a potential frequency difference between VCO and signal. Thus, for the computation of the short time transients we approximate the second-order PLL by a first-order PLL. This approach can be mathematically justified by the location of the poles of a highly damped second-order loop. The poles are

$$p = \frac{s}{\omega_n} = -\zeta \pm \zeta\left(1 - \frac{1}{\zeta^2}\right)^{1/2}$$

If $\zeta^2 \gg 1$ we may write for the square root

$$\left(1 - \frac{1}{\zeta^2}\right)^{1/2} \approx 1 - \frac{1}{2\zeta^2}$$

resulting in

$$p \approx -\zeta \pm \left(\zeta - \frac{1}{2\zeta}\right) \approx \begin{cases} -2\zeta \\ -\dfrac{1}{2\zeta} \end{cases}$$

The loop has two real poles which differ strongly. For the short time transients, the response of the pole $p_2 = -1/2\zeta$ remains essentially constant within a time interval of a few time constants of the first pole $p_1 = -2\zeta$.

The input to the integrator of the loop filter approximately equals the input to the $\hat{\theta}(\tau)$-integrator, if multiplied by $1/2\zeta$. Hence, the increments of x_1 and θ are approximately equal

$$x_1(\tau) - x_1(0) \approx \frac{1}{2\zeta}[\hat{\theta}(\tau) - \hat{\theta}(0)] \qquad (6.2\text{-}15)$$

since $\hat{\theta}(\tau) = -\phi(\tau)$ we may write

$$x_1(\tau) - x_1(0) \approx -\frac{1}{2\zeta}[\phi(\tau) - \phi(0)] \qquad (6.2\text{-}16)$$

for the short time transients.

A typical cycle slip is displayed in Figure 6.2-9. We observe a very similar shape of $x_1(\tau)$ and $\phi(\tau)$ up to the end of the cycle slip. Subsequently the value of $x_1(\tau)$ slowly decreases to its initial value of $x_1(\tau) = 0$ with a time constant of 2ζ, corresponding to the second pole. Note, however, that $\phi(\tau)$ after the slip is much less affected by $x_1(\tau)$.

Having recognized the similarity of $x_1(\tau)$ and $\phi(\tau)$ (as predicted by (6.2-16)), we are in a position to compute an estimate of $x_1(\tau)$ at the completion of a cycle slip. Let us denote by τ_0 and $\tau_{2\pi}$ the beginning and end of a cycle slip, respectively, and define a cycle slip as part of the trajectory $\phi(\tau)$ such that $\phi(\tau_0) = 0$ and $\phi(\tau \leq \tau_{2\pi}) \neq 0$. Furthermore, we assume $x_1(\tau_0) = 0$. Under these assumptions for $x_1(\tau_{2\pi})$ we obtain

$$x_1(\tau_{2\pi}) = \int_{\tau_0}^{\tau_{2\pi}} [\sin \phi(\tau) + n'(\tau)] \, d\tau \qquad (6.2\text{-}17)$$

where $\tau_{2\pi} - \tau_0$ is a random variable. Neglecting the integral path of the loop filter for the short time interval $(\tau_{2\pi} - \tau_0)$, the PLL is governed by a first-order differential equation

$$\frac{d\phi(\tau)}{d\tau} = -2\zeta[\sin \phi(\tau) + n'(\tau)] \qquad (6.2\text{-}18)$$

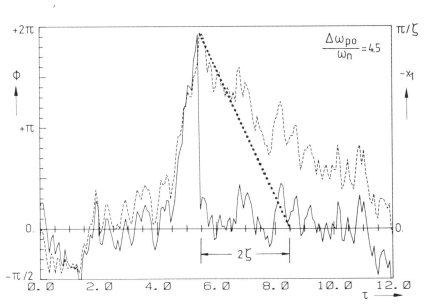

Figure 6.2-9. Cycle slip of an overdamped loop ($\zeta = 1.5$) for $\rho = 2.5$ (numeric ratio). Note that $-x_1(\tau)$ is displayed (from Gerd Ascheid and Heinrich Meyr, Cycle Slips in Phase-Locked Loops: A Tutorial Survey, *IEEE Transactions on Communications*, Vol. 30, No. 10, © IEEE 1982).

Note that the right-hand side of (6.2-18) multiplied by $-1/2\zeta$ equals the integrand in (6.2-17). Integration of the stochastic differential equation (6.2-18) from τ_0 to $\tau_{2\pi}$ yields

$$\phi(\tau_{2\pi}) - \phi(\tau_0) = -2\zeta \int_{\tau_0}^{\tau_{2\pi}} [\sin \phi(\tau) + n'(\tau)] \, d\tau \qquad (6.2\text{-}19)$$

but $\phi(\tau_0) = 0$ and $|\phi(\tau_{2\pi})| = 2\pi$, by definition. Replacing the right-hand side of (6.2-17) by $-\phi(\tau_{2\pi})/2\zeta$ yields

$$x_1(\tau_{2\pi}) = \begin{cases} \pi/\zeta & \text{negative cycle slip, } \phi(\tau_{2\pi}) = -2\pi \\ -\pi/\zeta & \text{positive cycle slip, } \phi(\tau_{2\pi}) = +2\pi \end{cases} \qquad (6.2\text{-}20)$$

As in the case of small damping factors ζ, $x_1(\tau_{2\pi})$ appears as a frequency detuning. However, the detuning is too small compared to the pull-out range to produce a burst. The inverse dependence of $x_1(\tau_{2\pi})$ has been well confirmed by experiments (see Figure 6.2-4). Due to the coupling of $x_1(\tau)$ and $\phi(\tau)$, the result (6.2-20) is, of course, only an approximation.

The fact that $x_1(\tau_{2\pi})$ is too small to significantly affect the phase error after a slip (as was also found in the example of Figure 6.2-9) justifies our assumption; large damping cycle slips are approximately isolated, independent events.

So far we have discussed two examples of loops with a damping factor of $\zeta = 0.24$ and $\zeta = 1.5$ and have found that there is a tendency for weakly damped loops to burst while overdamped loops do not. It would be interesting to identify a boundary between bursting and nonbursting loops. Of course, such a boundary cannot be a rigid one, but would merely provide information whether a loop is more likely to burst. On average, a loop will burst only if the mean value of x_1 taken at the completion of the slip is larger than the pull-out frequency. Using (6.2-20) and (6.2-4) yields the inequality

$$\pi/\zeta > 1.8(1 + \zeta) \quad \text{(condition for burst)} \qquad (6.2\text{-}21)$$

Solving (6.2-21) we find that loops with damping factors of $\zeta < 0.9$ will burst.

6.2.3. Frequency Detuning

In an imperfect second-order loop the frequency difference $\Delta\omega$, between transmitter and receiver, causes a static phase error; the x_1 variable assumes a value of $x_1 = \Delta\omega/\omega_n$ in order to compensate for the difference in frequency as shown by example in Figure 6.2-10. Because the difference between the maximum restoring force and the stable equilibrium point is reduced to $1 - \sin(\beta|\Delta\omega/\omega_n|)$, the loop slips much more often to this side than would

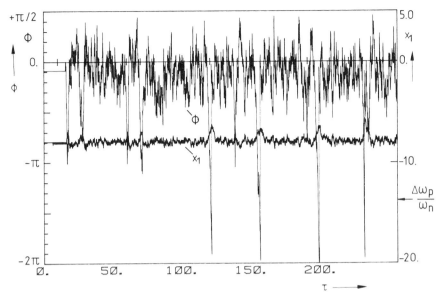

Figure 6.2-10. Trajectories $\phi(\tau)$ and $x_1(\tau)$ of a second-order loop with $\zeta = 1.0$ and $\rho = 2.8$ and a loop stress of $\Delta\omega/\omega_n = -8$ (from Gerd Ascheid and Heinrich Meyr, Cycle Slips in Phase-Locked Loops: A Tutorial Survey, *IEEE Transactions on Communications*, Vol. 30, No. 10, © IEEE 1982).

be the case for zero detuning. Note that $\Delta\omega > 0$ is assumed in the following analysis, the case $\Delta\omega < 0$ can be handled analogously.

In the experiment, it was observed that if $\Delta\omega/\omega_n$ was increased above the pull-in range, the loop, with very high probability, completely falls out of lock after once slipping a cycle; this behavior is not observed for the same signal-to-noise ratio with zero detuning.

In discussing this phenomenon a few preliminary remarks are helpful. In the noise-free case the steady-state value of x_1 remains constant, that is, $\dot{x}_1 = 0$. This is only possible if the signal $(-\beta x_1)$ at the input of the x_1 integrator is exactly compensated by a static phase error ϕ_s such that

$$-\beta x_1 + (1 - 2\zeta\beta + \beta^2)\sin\phi_s = 0 \qquad (6.2\text{-}22)$$

As long as $x_1 \le \Delta\omega/\omega_n < 1/\beta$ (hold-in range) the loop remains in lock.

If, due to noise, the loop slips a cycle there is a random difference $(\Delta\omega/\omega_n) - x_1$ after completion of the slip. A necessary condition for the PLL to resume lock is that, on average, it is capable of reducing this difference to zero. This might require a long pull-in period involving many cycle slips.

Mathematically, this condition requires that the average pull-in voltage $\overline{\sin\phi}$ for a given difference $(\Delta\omega/\omega_n) - x_1$ must be larger in amplitude than the decay βx_1 of the leaky integrator

$$\overline{\sin \phi} \geq \beta x_1 \qquad (6.2\text{-}23)$$

Otherwise the value of x_1 decays to zero and the PLL falls out of lock permanently, as is the case in Figure 6.2-11.

The worst case occurs for large frequency differences $(\Delta\omega/\omega_n) - x_1$. In this case we may use the result of (6.2-10) if we replace x_1 by $x_1 - (\Delta\omega/\omega_n)$

$$\overline{\sin \phi} = \frac{\zeta}{(\Delta\omega/\omega_n) - x_1} \geq \beta x_1 \qquad (6.2\text{-}24)$$

or slightly rearranged

$$x_1^2 - (\Delta\omega/\omega_n)x_1 + \zeta/\beta \geq 0 \qquad (6.2\text{-}25)$$

In general for a given $\Delta\omega/\omega_n$ and ζ/β, an interval of x_1 exists for which the inequality is not true. If the loop slips a cycle and x_1 accidentally assumes a value in this interval, resuming lock would be purely by chance. Such behavior is clearly unacceptable in a practical application. The question arises whether there are values of $\Delta\omega/\omega_n$ and ζ/β for which the quadratic form (6.2-25) is positive for all values of x_1. Then the loop could always reduce the difference $(\Delta\omega/\omega_n) - x_1$ to zero.

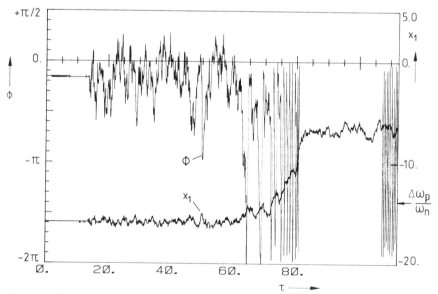

Figure 6.2-11. Trajectories $\phi(\tau)$ and $x_1(\tau)$ for the same parameters as in Figure 6.2-10, but for $\Delta\omega/\omega_n = -16$. The loop starts with correct initial conditions at $\tau = 0$ and drops lock after first slip (from Gerd Ascheid and Heinrich Meyr, Cycle Slips in Phase-Locked Loops: A Tutorial Survey, *IEEE Transactions on Communications*, Vol. 30, No. 10, © IEEE 1982).

Indeed the quadratic form is strictly positive if the discriminant is negative

$$(\Delta\omega/\omega_n)^2 - 4\zeta/\beta < 0 \qquad (6.2\text{-}26)$$

or

$$\Delta\omega/\omega_n < 2\sqrt{\frac{\zeta}{\beta}} \qquad (6.2\text{-}27)$$

But the right-hand side of (6.2-27) represents nothing but the pull-in frequency $\Delta\omega_p/\omega_n$. (The result is not surprising since we have used exactly the same argument as in Chapter 4.3 when we examined the pull-in behavior of a second-order loop.)

In conclusion, the loop must be designed such that $|\Delta\omega/\omega_n|$ is sufficiently smaller than the pull-in frequency $\Delta\omega_p$, this is particularly important for low SNRs. From (6.2-27) the conclusion may be drawn that a perfect integrator ($\beta = 0$) realized by means of an active loop filter is preferable. In practice, however, due to ever present drift currents there will always be a limit on the maximum permissible frequency difference.

Main Points of the Section

- On the completion of a cycle slip ($\tau = \tau_{2\pi}$) the state variable $x_1(\tau_{2\pi})$ in the mean has an incorrect value.
- For small damping factor $\zeta < 0.9$

  the mean of $x_1(\tau_{2\pi})$ is so large that additional slips are very likely at low signal-to-noise ratios (Figure 6.2-4) and cycle slips tend to appear in bursts.

  mean $\overline{\Delta x_1}$ and variance $\sigma^2_{\Delta x_1}$ of the increment Δx_1 of $x_1(\tau)$ over one cycle slip within a burst are given by

$$\Delta x_1 \approx -\frac{2\pi\zeta}{|x_1|x_1}, \qquad \sigma^2_{\Delta x_1} \approx \frac{1}{\rho}\frac{2\pi}{|x_1|}\frac{4\zeta}{1+(2\zeta)^2}$$

  Since the variance is inversely proportional to ρ the systematic drift term Δx_1 is increasingly covered by random fluctuations for decreasing signal-to-noise ratios and bursts tend to become longer.

- For loop damping factor $\zeta > 0.9$ the mean of $x_1(\tau_{2\pi})$ is small so that the influence of $x_1(\tau_{2\pi})$ on the phase error is negligible and cycle slips are isolated, independent events.
- With a frequency detuning above the pull-in range ($|\Delta\omega| > \Delta\omega_p$) the loop will drop lock after the first cycle slip (Figure 6.2-11). (Under noise-free operating conditions the loop is able to track frequency detuning within the hold-in range $|\Delta\omega| < \Delta\omega_H$).

6.3. CYCLE SLIP STATISTICS FOR A WIDEBAND NOISE DISTURBANCE

In the previous section we examined the physical phenomenon of a cycle slip. In a practical application one must make sure that the rate of the cycle slips is compatible with the function the loop is required to fulfill. For this purpose we need statistical data.

In this section we first give an overview of an experimental configuration to obtain cycle slip statistics. Then we discuss the experimental and theoretical results.

6.3.1. Measurement of Cycle Slips: Experimental Configuration

The experimental configuration is divided into two parts, the experiment itself in analog hardware and a microprocessor (μP) system to control the parameters and record the measured data.

A block diagram of the analog hardware is shown in Figure 6.3-1. An unmodulated carrier is provided by a crystal (X-tal) oscillator. The signal power can be set by a variable attenuator. Wideband gaussian noise from a random noise generator is added. The noise power can be varied by a second attenuator. Both signal and noise may also be switched off.

Filtered by the IF quartz filter, the noise becomes a narrowband gaussian

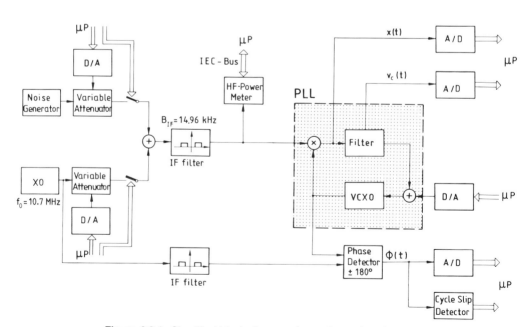

Figure 6.3-1. Simplified block diagram of experimental configuration.

process. At the output of the filter, the signal and the noise power are measured. The filter output is also the input to the phase detector of the PLL.

The phase detector is of the multiplier-type, the only applicable one at low SNR. The passive loop filters are exchangeable. The VCXO output is not only connected to the loop phase detector but also to the reference phase detector.

The undisturbed carrier is directed along a reference path to the other input of this linear $\pm 180°$ phase detector to determine the actual phase error $\phi(t)$. A second quartz filter, which is adjusted to compensate for the phase shift of the IF filter, has been inserted in the reference path.

The connections between the analog hardware and the μP system are marked by double lines in Figure 6.3-1. The μP controls the variable attenuators and the center frequency of the VCO.

Three analog-to-digital converters allow the μP system to record values of the phase detector output of the PLL $(x(t))$, the capacitor voltage of the loop filter of the second-order PLL $(v_c(t))$, and the actual phase error $\phi(t)$. Note that the capacitor voltage $v_c(t)$ is proportional to the state variable $y_1(t)$ and thus to $x_1(t)$ (2.4-22).

In addition, the output of the reference phase detector is connected to a cycle slip detector, which provides the μP with a signal whenever the phase error exceeds an absolute value of 2π. By means of a second signal, the direction of the cycle slips is transmitted to the μP system.

With this experimental configuration the trajectories of phase $\phi(t)$ and state variable $x_1(t)$ shown in Figures 6.2-3 and 6.2-5–6.2-11 have been measured as well as the statistical data of Figures 6.2-2, 6.2-4 and 6.3-4 and the experimental cycle slip rates in Figures 6.3-2, 6.3-3, and 6.3-6 to 6.3-8. To have sufficiently dense sampling of the trajectories the loop noise bandwidth had to be below 1 kHz. But then cycle slip rates above 10^5 cannot be measured since the duration of an experiment is too long, see Table 6.3-1.

For verification measurements of the slip rate of a first-order loop, where analytical results exist, were made. In the range $10^{0.5} < E(B_L T_S) < 10^5$ excellent agreement occurred (see Figure 6.3-2). At lower rates a degradation appears due to the limiting of the noise within the phase detector. At higher rates the duration of the measurement is too long.

TABLE 6.3-1   Mean Duration of an Experiment $E(T_m)$ ($B_L = 750$ Hz)

$E(B_L T_S)$	$E(T_s)$	$E(T_m)$/1600 events	$E(T_m)$/300 events
10^1	13.3 ms	21.3 s	
10^5	133.3 s	60 h	11 h
10^8	37 h		463 days

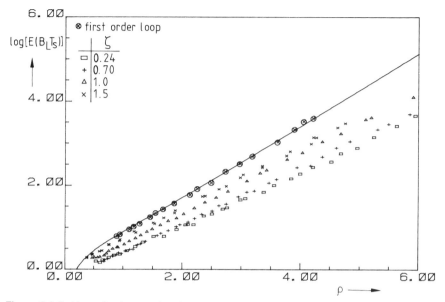

Figure 6.3-2. Normalized mean time between slips of a second-order phase-locked loop as a function of ρ with ζ as parameter. Zero loop detuning. Solid line shows analytical results for first-order PLL (from Gerd Ascheid and Heinrich Meyr, Cycle Slips in Phase-Locked Loops: A Tutorial Survey, *IEEE Transactions on Communications*, Vol. 30, No. 10, © IEEE 1982).

6.3.2. Experimental Results

The distribution of the mean time between slips, $P(T_s < t)$, was found to be very nearly exponential for sufficiently large damping ratios and signal-to-noise ratios

$$P(T_s < t) = 1 - \exp\left[-t/E(T_s)\right] \tag{6.3-1}$$

For small damping ratios a significant departure from the exponential distribution is visible for very small t; see Figure 6.2-2 for a typical example. This feature must be attributed to bursting of the slips.

If we exclude all the slips occurring very shortly after a predecessor, the conditional probability that $T_s < t$, under the condition that T_s exceeds t_0 results

$$P[T_s < t \mid T_s > t_0] \tag{6.3-2}$$

Using Bayes rule this yields

$$P[T_s < t \mid T_s > t_0] = \frac{P(T_s < t) - P(T_s < t_0)}{1 - P(T_s < t_0)} \tag{6.3-3}$$

Since bursts (at least approximately) are independent events we may expect that (6.3-3) is an exponential distribution function which is indeed found in Figure 6.2-2.

The mean time between slips versus ρ is shown in Figure 6.3-2 with the loop damping ratio ζ as parameter. It is interesting to observe that for zero loop detuning, the first-order loop always performs better than the second-order loop. This is explained by the fact that the integration in the loop filter is superfluous, only causing additional slips by feigning loop stress.

The effect of loop detuning on $E(B_L T_s)$ is shown in Figure 6.3-3. The frequency difference $\Delta\omega$ is normalized to the pull-in frequency $\Delta\omega_p$. From Figure 6.3-3 it may be concluded that the loop operates best for zero loop detuning.

The importance of the state variable x_1 has been discussed at length. A key finding of our discussion is the behavior of x_1 immediately following a slip; a consistent departure from the correct value has been identified. The statistics of the conditional mean $E(x_1 \mid \phi = \pm 2\pi)$ and variance $E[(x_1 - E(x_1))^2 \mid \phi = \pm 2\pi]$ can be found in Figure 6.2-4.

A typical distribution of x_1 immediately following a slip is shown in Figure 6.3-4a for $\zeta = 0.7$. As a consequence of the occurrence of bursts a significant skewness is visible. If one excludes the very short time interval

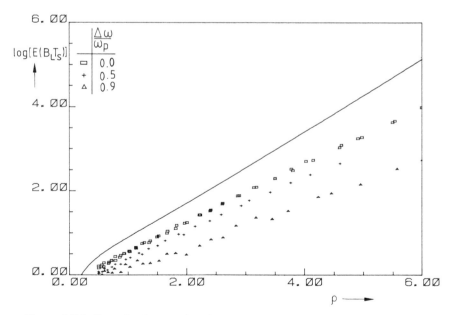

Figure 6.3-3. Normalized mean time between slips of a second-order phase-locked loop ($\zeta = 1.0$, $\beta = 0.03$) with loop detuning $\Delta\omega$ normalized to the pull-in frequency ω_p. Solid line shows analytical curve for first-order loop with zero detuning (from Gerd Ascheid and Heinrich Meyr, Cycle Slips in Phase-Locked Loops: A Tutorial Survey, *IEEE Transactions on Communications*, Vol. 30, No. 10, © IEEE 1982).

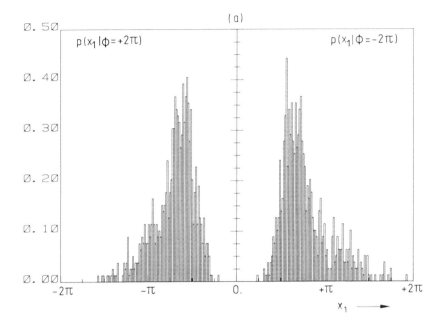

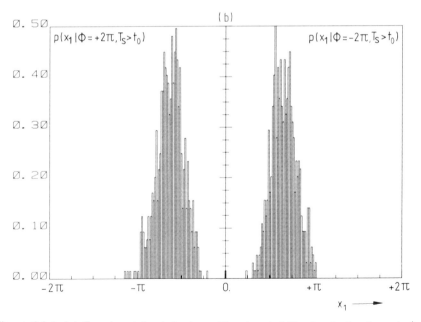

Figure 6.3-4. (a) Experimentally derived conditional probability density function $p(x_1 \mid \phi)$ for a second-order loop with $\zeta = 0.7$ and $\rho = 2.1$ (b) Conditional probability density function $p(x_1 \mid \phi, \; T_s > t_0)$ obtained if slips of duration T_s shorter than $t_0 \ll E(T_s)$ are excluded $(E(T_s)/t_0 = 43.5)$ (from Gerd Ascheid and Heinrich Meyr, Cycle Slips in Phase-Locked Loops: A Tutorial Survey, *IEEE Transactions on Communications*, Vol. 30, No. 10, © IEEE 1982).

between slips, the skewness disappears and both sides of the distribution assume a symmetrical shape, see Figure 6.3-4b.

6.3.3. Theoretical Results

For a first-order loop with sinusoidal nonlinearity an analytical formula for the mean time between slips exists (see Chapter 11 for mathematical details of the following results)

$$E(B_L T_s) = \frac{\rho}{2} \frac{\pi^2 |I_{j\gamma\rho}(\rho)|^2}{\cosh(\pi\gamma\rho)} \tag{6.3-4}$$

In this expression γ is the normalized loop detuning

$$\gamma = \frac{\Delta\omega}{K_0 K_D} = \frac{\Delta\omega}{4B_L}$$

and $I_{j\gamma\rho}(\rho)$ is the modified Bessel function of imaginary order. The function is plotted in Figure 6.3-2 for $\gamma = 0$. The curve for $\rho > 2$ is virtually a straight line on a logarithmic ordinate and is well approximated by

$$E(B_L T_s) \approx \frac{\pi}{4} \exp(2\rho) \tag{6.3-5}$$

With arbitrary loop detuning $|\gamma| < K_0 K_D$, the mean time between slips (6.3-4) for large signal-to-noise ratios ($\rho \gg 1$) is closely approximated by

$$E(B_L T_S) \approx \frac{\pi}{4} \frac{\exp(2\rho)}{\cosh(\pi\gamma\rho)} , \quad |\gamma| < K_0 K_D , \quad \rho \gg 1 \tag{6.3-6}$$

For $\Delta\omega \neq 0$ the stable tracking point is no longer the symmetry point of the nonlinearity. The loop will therefore slip a cycle much more often in the direction where it has to overcome the smaller difference between tracking point restoring force and maximum restoring force. The ratio between the rate of positive and negative cycle slips depends exponentially on the loop detuning and signal-to-noise ratio

$$\frac{N_+}{N_-} = \exp(2\pi\rho\gamma) \tag{6.3-7}$$

where $N_{\pm}$ is the rate of positive ($+$) and negative ($-$) slips. Knowing the total slip rate $N = E(T_s)^{-1}$ ((6.3-4), (6.3-6))

$$\frac{N}{B_L} = E(B_L T_s)^{-1}$$

and the ratio (6.3-7) it is possible to determine them independently

$$\frac{N_\pm}{B_L} \approx \frac{2}{\pi} \exp\left(\pm 2\pi\rho\gamma - 2\rho\right), \quad \rho \gg 1 \tag{6.3-8}$$

So far, the results presented were obtained for a sinusoidal phase detector characteristic. It is also interesting to see how $E(B_L T_s)$ is influenced by the phase detector characteristic. Table 6.3-2 summarizes the resulting expressions for three illustrative phase detector characteristics (sinusoidal, triangular, and sawtooth, all with unit slope at the origin). The mean time between cycle slips decreases exponentially with $a\rho$ where a is a factor depending on the phase detector type. It is apparent that both sawtooth and triangular phase detector outperform the sinusoidal phase detector. However, a word of caution is required. The effective phase detector characteristics* are triangular or sawtooth in nature only for very high input SNR_i conditions. The phase detector characteristic changes as a function of input SNR_i and deteriorates to sinusoidal at low SNR_i as was exemplified in Section 3.2. Hence, prior to applying the above formulas the efficient phase detector characteristic as a function of SNR_i must be evaluated to guarantee that it is sufficiently well approximated by a triangular or sawtooth characteristic.

No exact analytical formula for the mean time between cycle slips is known for a second-order loop. Based on a numerical approach (to be discussed in Chapter 12) the plots of Figures 6.3-5 and 6.3-6 were produced.

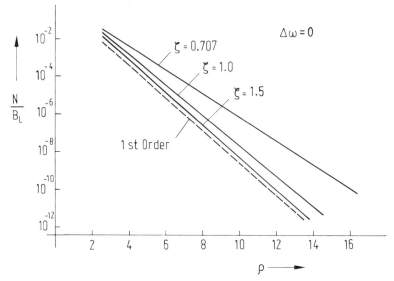

Figure 6.3-5. Normalized cycle slip rate $N/B_L = E(B_L T_s)^{-1}$ of a second-order phase-locked loop. Dashed line shows normalized slip rate for a first-order PLL. (Numerical computation by L. Popken).

*We recall from Section 3.2 that the effective phase detector characteristic is the mean output of the physical phase detector for a given ϕ.

TABLE 6.3-2 Mean Slip Time Approximations for Illustrative Equivalent Phase Detector Characteristics

	Equivalent Phase Detector Characteristic	Approximate Time Between Slips $E(B_L T_s)$
Sinusoidal	$g(\phi) = \sin \phi$	$\dfrac{\pi/2}{(2\cos\phi_s)\cosh\pi\rho\gamma} \exp\{2\rho\cos\phi_s + 2\rho\gamma\phi_s\}$ $\phi_s = \sin^{-1}(\gamma)$
Triangular	$g(\phi) = \begin{cases} \phi, & |\phi| < \pi/2 \\ \pi + \phi, & \phi < -\pi/2 \\ \pi - \phi, & \phi > \pi/2 \end{cases}$	$\dfrac{\pi/2}{2\cosh\pi\rho\gamma} \exp\{\rho(\pi^2/4 - \phi_s^2) + 2\rho\gamma\phi_s\}$ $\phi_s = \gamma$
Sawtooth	$g(\phi) = \phi$	$\left(\dfrac{\rho\pi}{2}\right)^{1/2} \dfrac{\rho\pi/2}{(\rho\pi)^2 - (\rho\gamma)^2} \cdot \dfrac{1}{\cosh\pi\rho\gamma} \exp\{\rho(\pi^2/2 - \phi_s^2/2) + \rho\gamma\phi_s\}$ $\phi_s = \gamma$

Source: From Chak Ming Chie, New Results on Mean Time-to-First-Slip for a First-Order Loop, *IEEE Transactions on Communications*, Vol. 33, No. 9, © IEEE 1985.

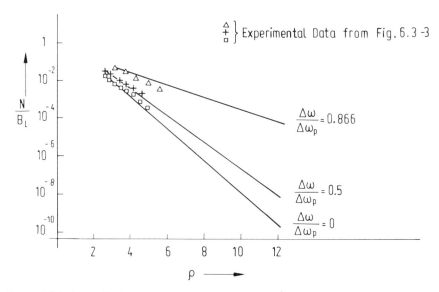

Figure 6.3-6. Normalized cycle slip rate $N/B_L = E(B_L T_s)^{-1}$ of a second-order phase-locked loop with normalized loop detuning $\Delta\omega/\omega_p$ as parameter ($\zeta = 1.0$, $\beta = 0.03$). (Numerical computation by L. Popken).

In Figure 6.3-5 the normalized cycle slip rate (N/B_L) for a second-order loop with a sinusoidal phase detector characteristic is shown with loop damping ratio ζ as a parameter. The effect of loop detuning on the cycle slipping rate is illustrated by Figure 6.3-6 where the frequency difference $\Delta\omega$ normalized to the pull-in frequency $\Delta\omega_p$ is the parameter. The results obtained by means of the numerical approach are in agreement with the experimental results shown in Figures 6.3-2 and 6.3-3.

From Figure 6.3-5 we may observe that the cycle slipping rate of a highly damped ($\zeta > 1.5$) second-order PLL disturbed by wideband noise approaches the rate of a first-order loop. To understand this behavior, the state equation (6.2-1) should be rewritten using a different normalization. We restrict the analysis to a second-order loop with perfect integrator ($\beta = 0$) here but the arguments also apply to a loop with imperfect integrator if $\beta\zeta \ll 1$. Dividing both sides of both state equations in (6.2-1) by $2/\zeta$ yields ($B_L = \zeta\omega_n/2$)

$$\frac{d\phi(t)}{d(B_L t)} = 4\sin\phi(t) - 4n'(t) - \frac{2}{\zeta}x_1(t) + \frac{d\theta(t)}{d(B_L t)}$$

$$\frac{dx_1(t)}{d(B_L t)} = \frac{2}{\zeta}\sin\phi(t) + \frac{2}{\zeta}n'(t)$$

(6.3-9)

Note that we now have the same time normalization ($\tau' = B_L t$) as for the mean time between cycle slips in Figure 6.3-5. It can be seen from the second equation that the fluctuations of $x_1(t)$ are slow (narrowband) for large damping ζ and that the influence of these fluctuations on the first equation are small for large ζ (which well agrees with the experimental findings, see Figures 6.2-4 and 6.2-9). We therefore may neglect these fluctuations in the first equation and replace x_1 by its steady state value x_{1s} in the noise free case. Considering a constant input offset frequency only

$$\frac{d\theta(t)}{d(B_L t)} = \frac{\Delta\omega}{B_L}$$

we find from (6.3-9)

$$\frac{2}{\zeta} x_{1s} = \frac{\Delta\omega}{B_L}$$

Thus, for large damping $\zeta \gg 1$ (6.3-9) reads

$$\frac{d\phi(t)}{d(B_L t)} = 4\sin\theta(t) - 4n'(t)$$

which is exactly the differential equation of a first-order loop with loop gain $AK = K_0 K_d = 4B_L$. Note that the frequency offset $\Delta\omega$ is completely eliminated which would not have been the case if actually a first-order loop would have been used.

6.3.4. Loop Parameters for Maximum Meantime between Cycle Slips

Optimization of loop parameters has already been discussed in Chapter 3. For the optimization we had assumed that linearization of the loop equations was permissible. It is also of interest to optimize the loop parameters to achieve maximum mean time between cycle slips. In general, minimization of the cycle slip rate is not the primary design goal but the optimization gives a deeper insight in how the various parameters affect the slip rate. Since linearization of the loop equation is not allowed, such an optimization is a formidable task which can be carried out only numerically on a digital computer or in the form of an experiment.

In a second-order loop there are essentially three loop parameters, namely bandwidth B_L, loop damping ζ, and β, to be optimized in a three-dimensional search. The loop parameters have to be clearly distinguished from the signal parameters $\{P_S, N_0, \Delta f\}$ which are given quantities.

In a first step we want to optimize the loop bandwidth B_L for a given damping ζ and a given β. For this purpose we seek a suitable normalized representation of the mean time between cycle slips $E(T_s)$ as a function of the bandwidth B_L. It is natural to modify the familiar plot of normalized mean time between slips $E(B_L T_s)$ versus ρ as depicted in Figure 6.3-3.

The signal-to-noise ratio is a function of the two signal parameters P_S, N_0, and the loop parameter B_L

$$\rho = \frac{P_s}{N_0} \frac{1}{B_L} \qquad (6.3\text{-}10)$$

If we multiply both numerator and denominator by Δf we obtain instead of (6.3-10)

$$\rho = \frac{P_s}{N_0 \, \Delta f} \frac{\Delta f}{B_L}$$

The three signal parameters can be grouped into a single quantity b

$$b = \frac{N_0 \, \Delta f}{P_S} \qquad (6.3\text{-}11)$$

which is related to the loop parameters γ_2 and ρ as follows

$$\rho = \left(\frac{1}{b}\right)\gamma_2 \qquad (6.3\text{-}12)$$

with

$$\gamma_2 = \frac{\Delta f}{B_L} \quad \text{(normalized loop stress)}$$

From (6.3-12) we learn that for plotting $E(B_L T_s)$ versus ρ, either the signal parameter b can be kept constant and γ_2 varied or vice versa. The two possibilities lead to different sets of curves; the case where a fixed normalized offset $\Delta f / B_L$ is maintained is shown in Figure 6.3-6 (with a different normalization), while in Figure 6.3-7 we have plotted the curves for a constant b on a double-logarithmic scale. The dashed curves in Figure 6.3-7 display $E(B_L T_s)$ for a first-order loop where an analytical formula exists (6.3-4). The curves labeled with $\square$ and $+$ are experimental results for a second-order loop with $\zeta = 1.0$ and $\beta = 0.03$. The maximum in these curves is qualitatively easily understood. A small ρ implies a large bandwidth B_L

$$\rho = \frac{1}{b} \frac{\Delta f}{B_L} \qquad (6.3\text{-}13)$$

resulting in a large number of slips due to poor noise suppression. For increasing ρ, the bandwidth is decreased and the loop slips less cycles. If, however, the bandwidth is decreased beyond the optimum, the static phase error caused by Δf starts to interfere and the mean time between slips starts to decrease again. The smaller b is, the smaller the optimum bandwidth; in the limiting case for $b = 0$ we obtain a strictly increasing curve. Note that a strictly increasing curve is also obtained for $\beta = 0$ (and arbitrary b).

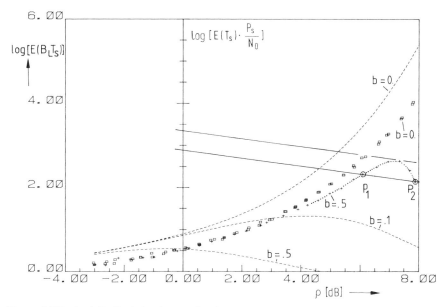

Figure 6.3-7. Analytically derived mean time between slips for first-order loop (dashed curves) and experimental results for second-order loop ($\zeta = 1.0$, $\beta = 0.03$) with b as parameter (from Gerd Ascheid and Heinrich Meyr, Cycle Slips in Phase-Locked Loops: A Tutorial Survey, *IEEE Transactions on Communications*, Vol. 30, No. 10, © IEEE 1982).

Because the normalized quantity $\log[E(B_L T_s)]$ is displayed in Figure 6.3-7 [and *not* $E(T_s)$] the optimum B_L is *not* found at the location of the maximum of these curves. We will show below how the optimization can be carried out by means of a graphic procedure.

Multiplying numerator and denominator of ρ by $E(T_s)$ and taking the logarithm yields

$$\log \rho = \log\left[\frac{P_s}{N_0} E(T_s)\right] - \log E(B_L T_s) \qquad (6.3\text{-}14)$$

or rearranged

$$\log E(B_L T_s) = -\log \rho + \log\left[\frac{P_s}{N_0} E(T_s)\right] \qquad (6.3\text{-}15)$$

For any given $E(T_s)$, (6.3-15) represents a straight line in Figure 6.3-7, and with increasing $E(T_s)$ the line moves upwards. Let us assume that such a line intersects a curve $\log[E(B_L T_s)]$ as illustrated for $b = 0.5$, and let us label the point of intersection P_1, and the corresponding ρ by ρ_1. Solving (6.3-13) for B_L yields

$$B_{L1} = \frac{\Delta f}{\rho_1 b} \qquad (6.3\text{-}16)$$

which determines the bandwidth of the loop for a given $E(T_s)$ and parameters ρ_1 and b. In our example, the line also intersects the same curve at the point P_2. This means that the same mean time between slips is obtained for two different loop bandwidths.

If we now increase $E(T_s)$, then according to (6.3-15) the straight line moves upwards until the two points P_1 and P_2 finally converge to one point defined as the tangent of the curve with the slope given by (6.3-15). Since for values of $E(T_s)$ above that point no intersection exists, we have found the maximum achievable $E(T_s)$ for any loop bandwidth. Therefore, the optimum bandwidth for a given signal parameter b can be found by constructing the tangent to the particular curve.

In Figure 6.3-8 we have plotted $\log[E(B_L T_s)]$ as a function of ρ for the two damping factors $\zeta = 0.7$ and $\zeta = 1.0$. We observe a slow increase up to the maximum and a steep descent beyond the optimum. As expected, the loop with the larger damping performs better. For $b = 0$ the function $\log[E(B_L T_s)]$ increases monotonically. In theory, any desired $E(T_s)$ can thus be obtained with a sufficiently narrow loop bandwidth. We also see that the first-order loop no longer performs better than the second-order loop when $b \neq 0$ (see also Figure 6.3-7).

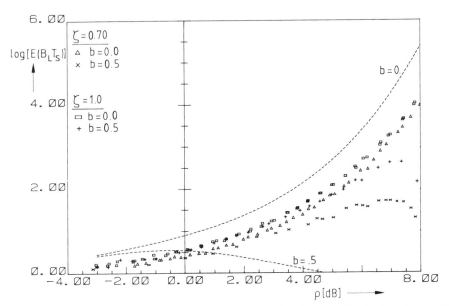

Figure 6.3-8. Experimental mean time between cycle slips for second-order loop for $\zeta = 1.0$ and $\zeta = 0.7$ ($\beta = 0.03$) with b as parameter; analytically derived mean time between slips for first-order loop shown for comparison (dashed curves) (from Gerd Ascheid and Heinrich Meyr, Cycle Slips in Phase-Locked Loops: A Tutorial Survey, *IEEE Transactions on Communications*, Vol. 30, No. 10, © IEEE 1982).

Main Points of the Section

- An analytical formula for the mean time between cycle slips is known for first-order loops only

$$E[B_L T_S] = \frac{1}{\rho} \frac{\pi^2 |I_{j\gamma\rho}(\rho)|^2}{\cosh(\pi\gamma\rho)} \tag{6.3-4}$$

(For a numerical approach to evaluate the mean time between cycle slips for second-order loops see Chapter 12.)

- Without frequency detuning, the mean time between cycle slips is maximum for a first-order loop and decreasing with decreasing damping factor ζ for second-order loops (Figure 6.3-2).
- With frequency detuning, second-order loops perform better than first-order loops.

REFERENCES

1. G. Ascheid and H. Meyr, Cycle Slips in Phase-Locked Loops: A Tutorial Survey, *IEEE Transactions on Communications*, Vol. 30, No. 10, pp. 2228–2241, October 1982.
2. R. W. Sannemann and J. R. Rowbotham, Unlock Characteristics of the Optimum Type II Phase-Locked Loop, *IEEE Transactions on Aerospace and Navigational Electronics*, Vol. 11, pp. 14–24, March 1964.
3. R. C. Tausworthe, Cycle Slipping in Phase-Locked Loops, *IEEE Transactions on Communication Technology*, Vol. 15, pp. 417–421, June 1967.
4. F. J. Charles and W. C. Lindsey, Some Analytical and Experimental Phaselocked Loop Results for Low Signal-to-Noise Ratios, *Proceedings of the IEEE*, Vol. 54, pp. 1152–1166, September 1966.
5. C. M. Chie, New Results on Mean Time-to-First-Slip for a First-Order Loop, *IEEE Transactions on Communications*, Vol. 33, No. 9, pp. 897–903, 1985.

PART 3

7

AMPLITUDE CONTROL

In many applications of synchronizers the signal and, thus, the signal-to-noise ratio vary over a wide range (up to 100 dB). With multiplier type phase detectors the phase detector gain is proportional to the signal amplitude (see Section 2.6). As a consequence, most of the phase-locked loop parameters depend on the input signal level. To keep a PLL with multiplier type phase detector operating in the vicinity of its "design point" some kind of input amplitude control is inevitable. In this chapter both types of amplitude control are considered, namely limiter, and automatic gain control (AGC).

Synchronizer performance is not the only reason for the use of amplitude control devices, therefore we present a more general viewpoint. For instance, amplitude control is also important to keep the signal within the dynamic range of the devices and to provide the detection circuits with a predefined signal level. The latter, in particular, is inevitable with (quadrature) amplitude modulation.

7.1. LIMITERS

We first study the influence of a limiter on a noisy bandpass signal. The results are not relevant, however, when the limiter is followed by a phase detector. This case is studied in Section 7.1.2.

7.1.1. Bandpass Limiters

Figure 7.1-1 shows a typical bandpass limiter configuration for amplitude control. The bandlimited input signal $z(t)$ to the limiter consists of the signal $s(t)$ corrupted by additive gaussian noise $n(t)$

$$z(t) = s(t) + n(t)$$
$$= \sqrt{2} A \sin\left[\omega_0 t + \theta(t)\right] + \sqrt{2} n_c(t) \cos \omega_0 t - \sqrt{2} n_s(t) \sin \omega_0 t \quad (7.1\text{-}1)$$

Figure 7.1-1. Mechanization of a bandpass limiter.

Here $\theta(t)$ is an arbitrary phase modulation statistically independent of $n(t)$. We restrict our attention to hard limiters where the output signal $y(t)$ is given by

$$y(t) = \text{sgn } \{z(t)\} = \begin{cases} +1, & z > 0 \\ -1, & z < 0 \end{cases}$$

The limiter may be followed by a zonal filter which removes all harmonics and only passes the fundamental band centered around ω_0.

The basic performance analysis of bandpass limiters is mainly due to Davenport [1] and Springett and Simon [2]. Following their approach we write the limiter output signal

$$y(t) = \text{sgn } \{\sin [\omega_0 t + \theta(t) + \theta_n(t)]\} \qquad (7.1\text{-}2)$$

where (see (3.2-14))

$$\theta_n(t) = \tan^{-1}\left[\frac{N_c(t)}{A - N_s(t)}\right]$$

and

$$N_c(t) = n_s(t) \sin \theta(t) + n_c(t) \cos \theta(t)$$

$$N_s(t) = n_s(t) \cos \theta(t) - n_c(t) \sin \theta(t)$$

For the following analysis it is more convenient to represent $y(t)$ as a series of sinusoids (see Appendix 7.1A)

$$y(t) = \frac{4}{\pi} \sum_{k=0}^{\infty} \frac{1}{2k+1} \sin \{(2k+1)[\omega_0 t + \theta(t) + \theta_n(t)]\}$$

The output of the zonal filter $y_1(t)$ now equals the first term of the sum ($k = 0$)

$$y_1(t) = \frac{4}{\pi} \sin [\omega_0 t + \theta(t) + \theta_n(t)] \qquad (7.1\text{-}3)$$

One approach to the analysis of the bandpass limiter performance is to split up $y_1(t)$ into a useful signal $s_1(t)$ and a noise term $n_1(t)$. At this point we are faced with the problem of which portion of $y_1(t)$ is to be considered as the useful signal. A reasonable definition from a practical, as well as from a theoretical point of view, is the conditional expectation

$$s_1(t) = E[y_1 \mid s]$$

Note that with this definition $n_1(t)$ is orthogonal to $s_1(t)$ and $s(t)$. Evaluating the expectation we obtain (see (7.1B-3), (7.1B-5) for $E[\cos \theta_n]$ and $E[\sin \theta_n]$)

$$y_1(t) = s_1(t) + n_1(t) = \frac{4}{\pi} \alpha(\rho_i) \sin [\omega_0 t + \theta(t)] + n_1(t) \qquad (7.1\text{-}4)$$

where

$$\alpha(\rho_i) = \frac{1}{2} \sqrt{2\pi} \sqrt{\rho_i/2} \exp (-\rho_i/2)[I_0(\rho_i/2) + I_1(\rho_i/2)] \qquad (7.1\text{-}5)$$

is referred to as the *limiter signal suppression factor* and ρ_i is the input signal-to-noise ratio (SNR_i)

$$\text{SNR}_i = \rho_i = \frac{A^2}{E[n^2]} = \frac{A^2}{\sigma_n^2} \qquad (7.1\text{-}6)$$

(for a definition of the modified Bessel functions $I_0(x)$, $I_1(x)$ see Appendix 7.1C). Since $s_1(t)$ and $n_1(t)$ are orthogonal, the output noise power P_{n_1} is given by

$$P_{n_1} = E[n_1^2] = E[y_1^2] - E[s_1^2]$$

With (from (7.1-3))

$$E[y_1^2] = \frac{1}{2} (4/\pi)^2 \quad \text{and} \quad E[s_1^2] = \frac{1}{2} (4/\pi)^2 \alpha^2(\rho_i)$$

the output signal-to-noise ratio (SNR_o) is found to be

$$\text{SNR}_o = \rho_o = \frac{E[s_1^2]}{E[n_1^2]} = \frac{\alpha^2(\rho_i)}{1 - \alpha^2(\rho_i)} \qquad (7.1\text{-}7)$$

One performance measure of interest is the amplitude of the output signal $s_1(t)$ as a function of the SNR_i. From Figure 7.1-2 it can be seen that the limiter suppression factor, which is proportional to the amplitude of $s_1(t)$, is constant over a wide range of the SNR_i. But for low signal-to-noise ratios (<10 dB) the amplitude of $s_1(t)$ decreases with decreasing SNR_i. Although the total output power still remains constant (see (7.1-3)), the limiter no longer provides a constant signal level because of the increasing noise

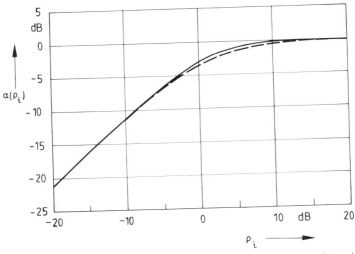

Figure 7.1-2. Limiter suppression factor $\alpha(\rho_i)$ as a function of the input signal-to-noise ratio ρ_i (solid line) and approximation (7.1-8) (dashed line).

power. Figure 7.1-2 also shows a very simple approximation for the limiter suppression factor $\alpha(\rho_i)$

$$\alpha(\rho_i) \approx \left(\frac{\rho_i}{4/\pi + \rho_i} \right)^{1/2} \tag{7.1-8}$$

which is sufficiently precise for most calculations.

A second performance measure of interest is the ratio of input to output signal-to-noise ratio $(\text{SNR}_i/\text{SNR}_o)$ shown in Figure 7.1-3 as a function of

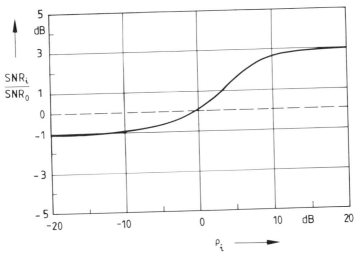

Figure 7.1-3. Ratio of input signal-to-noise ratio (SNR_i) to output signal-to-noise ratio (SNR_o) as a function of the input signal-to-noise ratio $(\text{SNR}_i = \rho_i)$.

the input signal-to-noise ratio ρ_i. It is found that the SNR_o is degraded by no more than $-1.05\,\text{dB}$ at low SNR_i and even improved by up to $3\,\text{dB}$ at very high SNR_i (but see Section 7.1.2). This improvement is easily understood when we rewrite the zonal filter output (7.1-3)

$$y_1(t) = \frac{4}{\pi} \{ \sin[\omega_0 t + \theta(t)] \cos \theta_n(t) + \cos[\omega_0 t + \theta(t)] \sin \theta_n(t) \}$$

For high signal-to-noise ratios $\theta_n(t) \approx N_c(t)/A \ll 1$ and $y_1(t)$ approaches

$$y_1(t) = \frac{4}{\pi} \left\{ \sin[\omega_0 t + \theta(t)] + \frac{N_c(t)}{A} \cos[\omega_0 t + \theta(t)] \right\} \qquad (7.1\text{-}9)$$

On the other hand we have for the limiter input signal (7.1-1) when we introduce $N_c(t)$ and $N_s(t)$ as defined by (7.1-2)

$$z(t) = \sqrt{2}A\left\{ \sin[\omega_0 t + \theta(t)] + \frac{N_c(t)}{A} \cos[\omega_0 t + \theta(t)] \right.$$
$$\left. - \frac{N_s(t)}{A} \sin[\omega_0 t + \theta(t)] \right\} \qquad (7.1\text{-}10)$$

Obviously the limiter suppresses the in-phase component of the noise.

7.1.2. Limiter Followed by a Phase Detector

Now let us consider that a phase-locked loop follows the bandpass limiter. In this case the performance parameters discussed so far are no longer applicable. We rather are interested in the influence of the limiter on loop parameters and loop performance. The first step in the analysis is the determination of the phase detector output when a bandpass limiter is used as shown in Figure 7.1-4.

Input to the phase detector is now either the limiter output signal $y(t)$ (7.1-2) or, with a zonal filter, the filter output $y_1(t)$ (7.1-3). These signals have two important properties in common. First, both have an amplitude independent of the input signal or noise power (1 for $y(t)$, $4/\pi$ for $y_1(t)$). Second, the additive noise at the limiter input causes a random phase modulation $\theta_n(t)$.

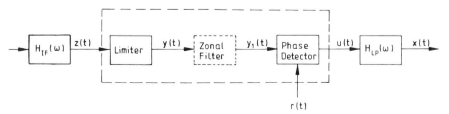

Figure 7.1-4. Bandpass limiter followed by a phase detector.

For all phase detectors implemented with digital circuits a rectangular waveform like the limiter output is the "natural" input signal waveform. For any memoryless phase detector* of this type, the low-pass filter output $x(t)$ is, therefore, given by

$$x(t) = K_D g_0[\phi(t) + \theta_n(t)]$$

$$\phi(t) = \theta(t) - \hat{\theta}(t)$$

(7.1-11)

as long as the variations of $\theta_n(t)$ are not too fast in comparison to the carrier frequency (for details on the restriction see Section 3.2.2). Here K_D is the phase detector gain, depending only on the implementation, and $g_0(\cdot)$ is the normalized noise-free phase detector characteristic with unit slope at the origin, i.e., $g'_0(0) = 1$.

With a zonal filter the input signal to the phase detector $y_1(t)$ is a sinusoid (7.1-3) of amplitude $4/\pi$ and phase $\phi(t) + \theta_n(t)$. Therefore, for a multiplier type phase detector, $x(t)$ is also given by (7.1-11). When the zonal filter is omitted, (7.1-11) can still be used for $x(t)$ but K_D and $g_0(\cdot)$ are no longer gain and characteristic of the phase detector itself and, therefore, have to be determined for each case as described in Section 3.2.2. The results for the most interesting practical combinations using sinusoidal and rectangular VCO signals $r(t)$ are summarized in Table 7.1-1.

The phase detector output (7.1-11) can be split up into a useful control signal given by the expected value of $x(t)$ and a random disturbance. Introducing the effective phase detector characteristic

$$g(\phi) = E[g_0(\phi + \theta_n)]$$

(7.1-12)

(7.1-11) can be written

TABLE 7.1-1   Gain K_D and Normalized Noise-Free Characteristic of a Multiplier Type Phase Detector with Limiter

VCO Signal	Input Signal	K_D/K_m	$g_0(\phi)$
$\sqrt{2}\cos(\omega_0 t + \hat{\theta})$	$y(t), y_1(t)$	$2\sqrt{2}/\pi$	$\sin(\phi)$
sgn $\{\cos(\omega_0 t + \hat{\theta})$	$y_1(t)$	$8/\pi^2$	$\sin(\phi)$
	$y(t)$	$2/\pi$	$\Lambda_0(\phi)$

K_m multiplier gain.

$\Lambda_0(\phi)$ triangular phase detector characteristic (type III, Figure 3.2-3c).

$$\Lambda_0(\phi) = \begin{cases} \pi - \phi & \pi/2 \le \phi \le \pi \\ \phi & -\pi/2 \le \phi \le \pi/2 \\ -\pi - \phi & -\pi \le \phi \le \pi \le \pi/2 \end{cases} \quad \text{periodic with } \phi \text{ modulo } 2\pi.$$

$y(t)$ limiter output (7.1-1).

$y_1(t)$ zonal filter output (7.1-3)

*A sequential phase detector for instance has memory and cannot be described so compactly.

$$x(t) = K_D\{g[\phi(t)] + n'(t,\ \phi)\} \tag{7.1-13}$$

The random disturbance $n'(t,\ \phi)$ is given by (see also Section 3.2.2)

$$n'(t,\ \phi) := g_0[\phi(t) + \theta_n(t)] - g[\phi(t)]$$

and has variance

$$\sigma_{n'}^2(t) = E\{[n'(t,\ \phi)]^2\} = E\{g_0^2[\phi(t) + \theta_n(t)]\} - g^2[\phi(t)] \tag{7.1-14}$$

Note that we assume $\phi(t)$ to be (at least approximately) uncorrelated with $n'(t,\ \phi)$ for the analysis, i.e., the synchronizer bandwidth is assumed to be much smaller than the noise bandwidth.

Linear Approximation. For a linear model of a synchronizer the effective phase detector characteristic is linearized around $\phi = 0$

$$g'(0) := \frac{dg(\phi)}{d\phi}\bigg|_{\phi=0}$$

Then, the phase detector output signal in the linear model is (assuming $g(0) = 0$)

$$x(t) = K_D g'(0)\left[\phi(t) + \frac{n'(t,\ \phi)}{g'(0)}\right] \tag{7.1-15}$$

Since K_D is a constant in the considered cases, $g'(0)$ determines the amplitude control capability of the limiter. Another point of primary interest is the influence of the limiter on the noise in the synchronizer. Neglecting the influence of the limiter on the noise power spectrum in a first approach, the relevant parameter is the variance of $n'(t,\ \phi)/g'(0)$

$$\sigma_l^2 = \sigma_{n'}^2/[g'(0)]^2 \tag{7.1-16}$$

Sinusoidal Phase Detector Characteristic. To evaluate the performance parameters of a given limiter phase detector combination, we first have to determine $g(\phi)$ and $E[g_0^2(\phi + \theta_n)]$. With $g_0(\phi) = \sin\phi$ we have

$$g(\phi) = E[\sin(\phi + \theta_n)] = E[\sin\theta_n]\cos\phi + E[\cos\theta_n]\sin\phi$$

Inserting the expected values as evaluated in Appendix 7.1B ((7.1B-3) and (7.1B-6)) yields

$$g(\phi) = \alpha(\rho_i)\sin\phi \tag{7.1-17}$$

We find that the amplitude control capability is given by the limiter signal suppression factor $\alpha(\rho_i)$ shown in Figure 7.1-2. For high input SNR ($\rho_i > 5$ dB), the amplitude control is perfect but for low input SNR ($\rho_i < 0$ dB), according to approximation (7.1-8)

$$\alpha(\rho_i) \approx \sqrt{\frac{\pi}{4}\,\rho_i} = \sqrt{\frac{\pi}{4}}\,\frac{A}{\sigma_n}$$

i.e., the phase detector gain is proportional to the signal amplitude (assuming constant noise variance).

To find the variance $\sigma_{n'}^2$ we first have to determine

$$E[g_0^2(\phi + \theta_n)] = E[\sin^2(\phi + \theta_n)]$$

$$= \frac{1}{2}[1 + E(\sin 2\theta_n \sin 2\phi) - E(\cos 2\theta_n \cos 2\phi)]$$

Again inserting the expectations evaluated in Appendix 7.1B yields

$$E[g_0^2(\phi + \theta_n)] = \frac{1}{2}\left[1 - \left(\frac{\pi}{4}\,\rho_i\right)^{1/2}\exp\left(-\frac{\rho_i}{2}\right)\left[I_{3/2}\left(\frac{\rho_i}{2}\right) + I_{1/2}\left(\frac{\rho_i}{2}\right)\right]\cos 2\phi\right]$$

The modified Bessel functions $I_{3/2}(x)$, $I_{1/2}(x)$ can be represented by a finite sum ((7.1C-7) and (7.1C-8))

$$E[g_0^2(\phi + \theta_n)] = \frac{1}{2}\left\{1 - \left[1 - \frac{1 - \exp(-\rho_i)}{\rho_i}\right]\cos 2\phi\right\} \qquad (7.1\text{-}18)$$

Finally, we have to insert (7.1-17) and (7.1-18) in (7.1-14) to obtain the variance

$$\sigma_{n'}^2(\phi) = \frac{1}{2}\left\{1 - \alpha^2(\rho_i) - \left[1 - \alpha^2(\rho_i) - \frac{1 - \exp(-\rho_i)}{\rho_i}\right]\cos 2\phi\right\}$$
$$(7.1\text{-}19)$$

shown in Figure 7.1-5. In the case of high input SNR, the variance depends on ϕ but in this operating range a synchronizer produces only small phase errors. As a result, in analyzing the loop behavior, it is generally sufficient to use

$$\sigma_{n'}^2(0) = \frac{1}{2\rho_i}(1 - \exp(-\rho_i))$$

To make the influence of the limiter on the noise more apparent, Figure 7.1-6 shows the variance $\sigma_l^2(0) = \sigma_{n'}^2(0)/[g'(0)]^2$ normalized to $1/2\rho_i$, which is the variance of the noise for a sinusoidal phase detector without limiter.

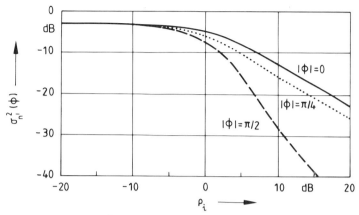

Figure 7.1-5. The variance $\sigma_n^2(\phi)$ of the noise $n'(t, \phi)$ for a limiter phase detector with sinusoidal characteristic (ρ_i, input signal-to-noise ratio).

It is found that the deterioration due to the limiter is minimal at low signal-to-noise ratios (below $-1.05\,\text{dB} = 10\log_{10}(\pi/4)$). The influence on the synchronizer performance is even less. The synchronizer bandwidth is usually much narrower than the input filter bandwidth. Therefore, the loop signal-to-noise ratio depends on the noise power spectral density at low frequencies. As was shown by Springett and Simon [2] the noise power spectrum is spread over a wider frequency band by the limiter, thus reducing $S_{n'}(0)$. The actual deterioration due to the limiter was found to be about $-0.2, \ldots, -0.7\,\text{dB}$ depending on the bandpass filter type, which is negligibly small.

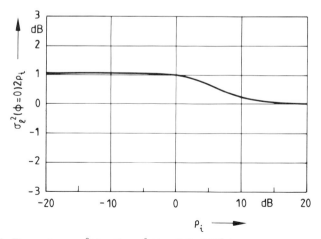

Figure 7.1-6. The variance $\sigma_l^2(\phi = 0) = \sigma_n^2(\phi = 0)/[g'(0)]^2$ for a limiter phase detector with sinusoidal characteristic, normalized to the noise variance $(1/2\rho_i)$ for a sinusoidal phase detector without limiter.

For high input signal-to-noise ratios the noise variance is the same with or without limiter. The 3 dB gain in signal-to-noise ratio found for the limiter output signal (see Section 7.1.1) is no longer in effect. The reason is that the gain is accomplished by suppression of the in-phase component of the noise (see (7.1-9)). But with a coherent detector the in-phase component is also suppressed as was shown in Section 3.2 and, therefore, the signal-to-noise ratio improvement by the limiter is irrelevant.

Nonsinusoidal Phase Detector Characteristic. When we want to analyze the performance of an arbitrary nonsinusoidal limiter phase detector, $g(\phi)$ and $E[g_0^2(\phi + \theta_n)]$ may be determined using Fourier series representations as described in Appendix 7.1B. The resulting expressions are quite cumbersome. Since it is sufficient in most cases an approximate analysis is given instead. In critical cases the sinusoidal limiter phase detector should be used anyway because it has been proved to have the best performance at low input SNR.

For high input SNR ($\rho_i \gg 1$), the probability density function $p(\theta_n)$ is approximately gaussian with variance $\sigma_{\theta_n}^2 = 1/(2\rho_i)$ and zero mean given by (7.1B-10). When the variance is small $p(\theta_n)$ is a small peak. With

$$g_0(\phi + \theta_n) \approx g_0(\phi) + g_0'(\phi)\theta_n$$

the effective phase detector characteristic is

$$g(\phi) = E[g_0(\phi + \theta_n)] \approx g_0(\phi)$$

and the variance

$$\sigma_{n'}^2(\phi) \approx \frac{1}{2\rho_i}[g_0'(\phi)]^2$$

In the case of small phase errors (ϕ about 0) we obtain $\sigma_{n'}^2(0) = 1/2\rho_i$ as for a sinusoidal phase detector characteristic.

For low SNR ($\rho_i \ll 1$), the term with the smallest value of k for which $b_k \neq 0$ becomes dominating in the Fourier series representation (7.1B-4). This is due to the fact that the Bessel functions decrease significantly with increasing order for small arguments ($\ll 1$) (see (7.1C-2)). When the minimum period is 2π the effective phase detector characteristic, therefore, approaches

$$g(\phi) \approx b_1 \alpha(\rho_i) \sin \phi \qquad (7.1-20)$$

Since the probability density function $p(\theta_n)$ approaches uniform distribution we have

$$\sigma_{n'}^2 \approx \frac{1}{2\pi} \int_{-\pi}^{+\pi} g_0^2(\phi) \, d\phi = \text{constant}$$

The variance of interest in the linear model is

$$\sigma_l^2 = \frac{\sigma_{n'}^2}{b_1^2 \alpha(\rho_i)^2}$$

$\sigma_{n'}^2$ may be considered as the average power of the signal $g_0(t)$ and $b_1^2/2$ as the power of its fundamental harmonic. Therefore, $\sigma_{n'}^2/b_1^2 \geq 1/2$ and the minimum is achieved for $g_0(\phi) = \sin(\phi)$, i.e., the minimum noise variance is obtained for a sinusoidal phase detector. Since the effective phase detector characteristic $g(\phi)$ is proportional to $\alpha(\rho_i)$ for all types of phase detectors when $\rho_i \ll 1$, sinusoidal detectors are optimum at low input SNR.

Main Points of the Section

- Bandpass limiter (incoherent case).
  Useful signal amplitude: $\alpha(\rho_i) \approx \{\rho_i/[\rho_i + (\pi/4)]\}^{1/2}$ ($\alpha(\rho_i)$ see Figure 7.1-2).
  Input signal-to-noise to output signal-to-noise ratio between $+3$ dB at high signal-to-noise ratios and -1.05 dB at low signal-to-noise ratios.
- Limiter followed by sinusoidal phase detector.
  Phase detector gain: $K_D \alpha(\rho_i)$.
  Noise variance $\sigma_{n'}^2/[g'(0)]^2 \approx 1/2\rho_i$, approximately the same as without limiter.
- Limiter followed by nonsinusoidal phase detector.
  Same noise variance $\sigma_{n'}^2/[g'(0)]^2 \approx 1/2\rho_i$ as with sinusoidal phase detector for high signal-to-noise ratios.
  Approaches sinusoidal characteristic but with higher noise variance for low signal-to-noise ratios ($\rightarrow$ sinusoidal phase detector superior at low signal-to-noise ratios).
- In all cases, amplitude control is perfect at high signal-to-noise ratios (>5 dB) but poor at low signal-to-noise ratios.

7.2. AUTOMATIC GAIN CONTROL CIRCUITS

Instead of limiting the input signal, the amplitude may also be kept close to a desired value by appropriately controlling the gain of an amplifier. This leads to the second method of amplitude control, the so-called *automatic gain control* (AGC) *circuits*. The control voltage of the amplifier may be derived in two different ways as shown in Figures 7.2-1 and 7.2-2.

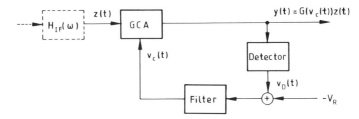

Figure 7.2-1. Block diagram of a feedback AGC (GCA, gain control amplifier).

In the first case, the amplitude of the amplifier output signal is detected, compared to a reference level, filtered, and fed back to the amplifier. This is the conventional *feedback AGC*. As shown in Figure 7.2-2 the input signal amplitude may also be measured, filtered, and fed forward to the amplifier. To compensate for delays in the control path, a properly chosen delay is introduced in the signal path of this *feedforward AGC*. Although a faster response to input signal amplitude variations may be achieved with the feedforward AGC it has not been used frequently up to now since it is much more sensitive to parameter variations in the circuits. Therefore, in this section more emphasis is placed on the analysis of feedback AGCs.

Note that we follow the common convention to denote feedback AGCs simply as AGCs wherever the type referred to is clear from the context.

7.2.1. Gain Controlled Amplifiers

Amplifiers with controllable gain may be implemented in many different ways. To avoid overload and to keep the influence of the intrinsic noise low the gain control is often distributed over several amplifier stages, in particular when the dynamic range of the input signal is large. Therefore, a large variety of gain control characteristics exists. In general, we can assume for the analysis that the amplifier instantaneously responds to a change of the control signal $v_c(t)$ since the time constants of the amplifier are much smaller than the AGC loop time constants. Then we obtain

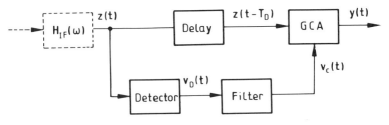

Figure 7.2-2. Block diagram of a feedforward AGC.

$$y(t) = G[v_c(t)]z(t) \qquad (7.2\text{-}1)$$

for the amplifier output signal where $G(v_c)$ is the overall *gain control characteristic* of the gain controlled amplifier (GCA). Note that in contrast to a limiter the GCA acts as a linear device with respect to the input signal $z(t)$ for a given gain $G[v_c(t)]$. In an AGC loop $v_c(t)$ is a function of $y(t)$ and therefore (7.2-1) is no longer linear with respect to $z(t)$. But when $v_c(t)$ is constant or varies slowly in comparison to $z(t)$ the GCA acts approximately linear even in the closed AGC loop.

A large variety of gain control characteristics is used. Some correspond to well-known implementations of GCAs (e.g., balanced mixers, field-effect transistor amplifiers) others were found analytically as yielding a desirable AGC performance and therefore implemented. Frequently used are, for example:

1. A linear gain control characteristic

$$G(v_c) = G_0 v_c \qquad (7.2\text{-}2)$$

   where G_0 is a design constant. This is the gain control characteristic of multipliers.

2. An exponential gain control characteristic

$$G(v_c) = G_0 \exp(G_1 v_c) \qquad (7.2\text{-}3)$$

   where G_0, G_1 are design constants. Note that the exponential characteristic is linear when the gain is expressed in decibels (dB). With this characteristic an upper limit of the gain G_{max}, which always exists in implementations, may easily be accounted for by using the gain control characteristic

$$G(v_c) = [G_{max}^{-1} + G_0^{-1} \exp(G_1 v_c)]^{-1} \qquad (7.2\text{-}4)$$

3. An approximately exponential gain control characteristic

$$G(v_c) = G_0[\exp(G_1 v_c) - 1] \qquad (7.2\text{-}5)$$

4. A hyperbolic gain control characteristic

$$G(v_c) = \frac{G_0}{v_c} \qquad (7.2\text{-}6)$$

   which corresponds to a divider. This characteristic is of particular interest for the feedforward AGC (see Section 7.2.7).

7.2.2. Detectors

The detector is a critical part of the AGC. When the signal amplitude (or power) is measured incorrectly, the output signal of the AGC is not adjusted accurately. Therefore a careful circuit design is important. Also, when the input is corrupted by strong noise the measurement may be biased depending on the detection principle as will be seen later.

Assuming a sinusoidal AGC input signal $s(t)$ corrupted by narrowband gaussian noise, the input $y(t)$ to the detector is (see Figure 7.2-1)

$$y(t) = G[v_c(t)]z(t)$$
$$= G[v_c(t)]\{\sqrt{2}A(t)\sin[\omega_0 t + \theta(t)]$$
$$+ \sqrt{2}n_c(t)\cos\omega_0 t - \sqrt{2}n_s(t)\sin\omega_0 t\} \qquad (7.2\text{-}7)$$

For the analysis of the different types of detectors we further assume that $G[v_c(t)]$ is a given function independent of the detector output, i.e., that the AGC loop is opened after the detector.

It is now the task of the detector to provide an estimate of the AGC output signal amplitude $G[v_c(t)]A(t)$. Two classes of detectors are distinguished. When an estimate $\hat{\theta}(t)$ of the phase $\theta(t)$ is available to the detector we may use coherent detection, otherwise we are restricted to noncoherent detection. Note that according to the detector type used, the AGC is either referred to as coherent or noncoherent.

Coherent Detection. In this case the input signal $y(t)$ is multiplied by a reference signal $r(t) = \sqrt{2}\sin[\omega_0 t + \hat{\theta}(t)]$ which may be provided by the oscillator of a synchronizer. Neglecting double frequency terms which are assumed to be suppressed by an appropriate low-pass filter, the output of the detector $v_D(t)$ is

$$v_D(t) = K_D G[v_c(t)][A(t)\cos\phi(t) + n_c(t)\sin\hat{\theta}(t) - n_s(t)\cos\hat{\theta}(t)]$$
$$= K_D G[v_c(t)]\{A(t)\cos\phi(t) + n''[t, \hat{\theta}(t)]\} \qquad (7.2\text{-}8)$$

where K_D is the detector gain. With the same arguments as in Appendix 3.2A (see the definition of $n'[t, \hat{\theta}(t)]$ in (3.2A-3)), we may assume that $n''[t, \hat{\theta}(t)]$ is independent of $\hat{\theta}(t)$ when the synchronizer bandwidth is small compared to the IF filter bandwidth B_{IF}. Then, the autocorrelation function $R_{n''}(\tau)$ is obtained from Appendix 3.2A by analogy as

$$R_{n''}(\tau) = R_{n_c}(\tau) = R_{n_s}(\tau) \qquad (7.2\text{-}9)$$

$$S_{n''}(\omega) = S_{n_c}(\omega) = S_{n_s}(\omega)$$

and the variance of $n''[t, \hat{\theta}(t)]$ as

$$\sigma_{n''}^2 = \frac{N_0}{4} 2B_{IF} = \frac{N_0 B_{IF}}{2} \tag{7.2-10}$$

For small phase errors $\phi \approx 0$ we have the desired result that the expected value of the detector output is proportional to the amplitude of the AGC output signal

$$E[v_D(t)] = K_D G[v_c(t)] A(t)$$

Neglecting the influence of $\hat{\theta}(t)$ on $n''[t, \hat{\theta}(t)]$ (with the arguments given above) the detector output can be expressed as

$$v_D(t) = K_D G[v_c(t)][\hat{A}(t) + n''(t)] \tag{7.2-11}$$

with

$$\hat{A}(t) = A(t)$$

i.e., the estimate $\hat{A}$ of the input signal amplitude equals A.

Noncoherent Detection. Without a phase coherent reference signal an *envelope detector* may be used to estimate the amplitude of the GCA output signal

$$v_D(t) = K_{en} \left| G[v_c(t)] z(t) \right| * h_L(t)$$

where K_{en} is a possible gain. The convolution with $h_L(t)$ refers to the assumption that a low-pass filter follows the detector to remove all components of $v_D(t)$ at harmonics of ω_0. Inserting (7.2-7) and following (3.2-13) in Section 3.2.2, yields

$$v_D(t) = \sqrt{2} K_{en} G[v_c(t)]$$
$$\{[A(t) - N_s(t)]^2 + N_c^2(t)\}^{1/2} \left| \sin [\omega_0 t + \theta(t) + \theta_n(t)] \right| * h_L(t)$$
$$\approx K_D G[v_c(t)]\{[A(t) - N_s(t)]^2 + N_c^2(t)\}^{1/2} \tag{7.2-12}$$

with

$$K_D = \frac{2\sqrt{2}}{\pi} K_{en}$$

It is well known that for a given value of $A(t)$ the envelope

$$V(t) = \{[A(t) - N_s(t)]^2 + N_c^2(t)\}^{1/2} = \frac{v_D(t)}{K_D G[v_c(t)]}$$

has a Rician probability density function

$$p(V \mid A) = 2 \frac{V}{\sigma_n^2} \exp\left(-\frac{V^2 + A^2}{\sigma_n^2}\right) I_0\left(2\frac{VA}{\sigma_n^2}\right) \qquad (7.2\text{-}13)$$

where

$$\frac{\sigma_n^2}{2} = \frac{E(n^2)}{2} = E(N_s^2) = E(N_c^2) = \frac{N_0 B_{IF}}{2}$$

Splitting up $v_D(t)$ into a useful signal

$$E[v_D(t) \mid A(t)] = K_D G[v_c(t)] \hat{A}(t)$$

and a noise term $n''(t)$ so that

$$v_D(t) = K_D G[v_c(t)][\hat{A}(t) + n''(t)] \qquad (7.2\text{-}14)$$

we find that $\hat{A}(t)$ is a biased estimate

$$\hat{A}(t) = \frac{E[v_D(t) \mid A(t)]}{K_D G[v_c(t)]} = E[V(t) \mid A(t)] \qquad (7.2\text{-}15)$$

since the conditioned mean from (7.2-13) is obtained as

$$\hat{A} = \frac{(\pi\sigma_n^2)^{1/2}}{2} \exp\left(-\frac{A^2}{2\sigma_n^2}\right)\left[\left(1 + \frac{A^2}{\sigma_n^2}\right)I_0\left(\frac{A^2}{2\sigma_n^2}\right) + \frac{A^2}{\sigma_n^2} I_1\left(\frac{A^2}{2\sigma_n^2}\right)\right]$$

Introducing the input signal-to-noise ratio $\rho_i(t) = A^2(t)/\sigma_n^2$ the ratio of estimate $\hat{A}$ to signal amplitude A is

$$\beta_{en}[\rho_i(t)] = \frac{\hat{A}(t)}{|A(t)|}$$

$$= \left[\frac{\pi}{4\rho_i(t)}\right]^{1/2} \exp\left[-\frac{\rho_i(t)}{2}\right]$$

$$\times \left\{[1 + \rho_i(t)]I_0\left(\frac{\rho_i(t)}{2}\right) + \rho_i(t)I_1\left(\frac{\rho_i(t)}{2}\right)\right\}$$

where we refer to $\beta_{en}(\rho_i)$ as the signal suppression factor of the envelope detector.

The variance of the noise $n''(t)$ is

$$\sigma_{n''}^2(t) = E[V^2(t) \mid A(t)] - \hat{A}^2(t) = E\{[A(t) - N_s(t)]^2 + N_c^2(t)\} - \hat{A}^2(t)$$

$$= A^2(t) + \sigma_n^2 - \hat{A}^2(t) \qquad (7.2\text{-}17)$$

Using the asymptotic expansions (7.1C5) of the modified Bessel functions for high input signal-to-noise ratios we obtain

$$\hat{A}(t) \approx |A(t)| \frac{1 + 4\rho_i(t)}{4\rho_i(t)} \approx |A(t)|$$

$$\text{Approximation for } \rho_i \gg 1 \qquad (7.2\text{-}18)$$

$$\sigma_{n''}^2 \approx \sigma_n^2/2$$

The 3 dB reduction in noise variance corresponds to the fact that $n''(t)$ approaches $-N_S(t)$ (see (7.2-14) and (7.2-12)) for high signal-to-noise ratios. For $\rho_i \ll 1$ we have $I_0(\rho_i/2) \approx 1$, and $I_1(\rho_i/2) \approx \rho_i/4$. Thus, at low signal-to-noise ratios we have

$$\hat{A} \approx \sqrt{\frac{\pi}{4}} \, \sigma_n$$

$$\text{Approximation for } \rho_i \ll 1 \qquad (7.2\text{-}19)$$

$$\sigma_{n''}^2 \approx \left(1 - \frac{\pi}{4}\right)\sigma_n^2$$

The exact curves for the ratio of estimate $\hat{A}$ to signal amplitude A and the ratio $\sigma_{n''}^2/\sigma_n^2$ are shown in Figure 7.2-3 as a function of the input signal-to-

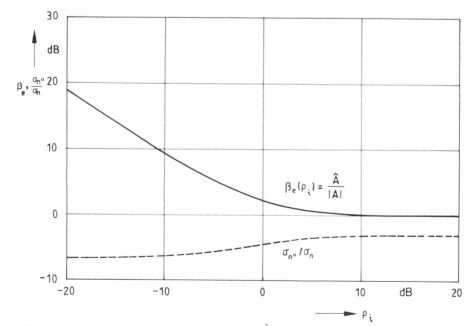

Figure 7.2-3. Ratio of estimated signal amplitude $\hat{A}$ to actual signal amplitude A (solid line) and ratio of detector output noise variance $\sigma_{n''}$ to input noise variance σ_n (dashed line) as a function of the input signal-to-noise ratio ρ_i.

noise ratio ρ_i. While we have the desired performance at high signal-to-noise ratios, i.e., $\hat{A} \approx A$ and the noise is reduced by 3 dB, at low signal-to-noise ratios the detector responds to the noise only. This bias cannot be corrected by the AGC loop and therefore causes a steady-state error in the AGC output signal amplitude.

For the analysis of the closed AGC loop the power spectral density $S_{n''}(\omega)$ of the noise term $n''(t)$ is often more important than the variance $\sigma_{n''}$. Assuming that the variation of the amplitude $A(t)$ is much slower than that of the noise components $N_s(t)$, $N_c(t)$, from (7.2-12) we find that $S_{n''}(\omega)$ approaches the power spectral density $S_{N_s}(\omega)$ of $N_s(t)$ for high input signal-to-noise ratios. Assuming that the power spectral densities of $N_s(t)$, $N_c(t)$ are approximately flat in $[-B_{IF}/2, +B_{IF}/2]$, $S_{n''}(\omega)$ is approximately triangular [11] in $[-B_{IF}, B_{IF}]$ with $S_{n''}(0) \approx \sigma_{n''}^2/B_{IF}$ at low input signal-to-noise ratios. When the detector output passes a narrowband filter with single-sided noise bandwidth $B_F \ll B_{IF}$ the variance of the filtered noise may in this case be approximated by

$$\sigma_F^2 \approx 2B_F S_{n''}(0) = \sigma_{n''}^2 \frac{2B_F}{B_{IF}} \tag{7.2-20}$$

Another noncoherent detector of interest is the *squaring detector* with

$$\begin{aligned} v_D(t) &= K_D \{ G^2[v_c(t)] z^2(t) \} * h_L(t) \\ &\approx K_D G^2[v_c(t)] \{ [A(t) - N_s(t)]^2 + N_c^2(t) \} \end{aligned} \tag{7.2-21}$$

Again splitting up the output signal

$$v_D(t) = K_D G^2[v_c(t)][\hat{A}^2(t) + n''(t)] \tag{7.2-22}$$

The useful signal is now an estimate of the input signal power

$$\hat{A}^2(t) = E[v_D(t) \mid A^2(t)] = A^2(t) + \sigma_n^2 \tag{7.2-23}$$

The signal suppression factor $\beta_{sq}(\rho_i)$ of the squaring detector is given by

$$\beta[\rho_i(t)] = \frac{\hat{A}(t)}{A(t)} = \left[1 + \frac{1}{\rho_i(t)} \right]^{1/2} \tag{7.2-24}$$

The variance of the noise term is easily found to be

$$\sigma_{n''}^2(t) = 2A^2(t)\sigma_n^2 + \sigma_n^4 \tag{7.2-25}$$

In this case the noise power spectral density can be given explicitly if $A(t)$ varies much slower than the noise components $N_s(t)$, $N_c(t)$

$$S_{n''}(\omega) = 4A^2 S_{N_c}(\omega) + 4S_{N_c}(\omega) * S_{N_c}(\omega)$$

As in the case of the envelope detector, the noise variance after a filter of equivalent noise bandwidth $B_F \ll B_{IF}$ is sufficiently accurately given by

$$\sigma_F^2 = 2B_F S_{n''}(0) = (2A^2 + \sigma_n^2)\sigma_n^2 \frac{2B_F}{B_{IF}} \tag{7.2-26}$$

When the input signal-to-noise ratio is unknown, it is impossible to provide an unbiased estimate with noncoherent detection. The remedies out of this situation are either to increase ρ_i by reduction of the IF filter bandwidth (if possible) or to employ coherent detection.

7.2.3. The Automatic Gain Control Loop

We may face two basically different applications of AGC. Either the fluctuations of the input signal amplitude $A(t)$ only reflect the time-varying nature of the transmission channel to be suppressed by the AGC, or we additionally have an information-carrying amplitude modulation to be passed without alteration. Here we restrict the analysis to the first case where we have $A(t) > 0$ and a power spectral density of $A(t)$ much narrower than $S_{n''}(\omega)$.

The general model shown in Figure 7.2-4 can be used for the analysis of the various noncoherent and coherent AGCs obtained when combining any of the detectors with any of the GCA characteristics. The output of the detector is

$$v_D(t) = K_D G[v_c(t)][\hat{A}(t) + n''(t)] = K_D G[v_c(t)]z_D(t) \tag{7.2-27}$$

where

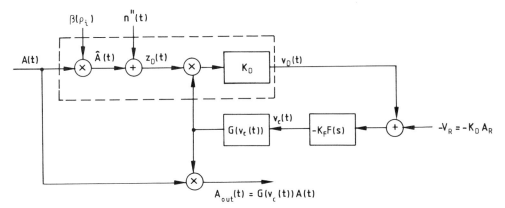

Figure 7.2-4. General loop model for coherent and noncoherent AGCs.

$$\hat{A}(t) = \beta[\rho_i(t)]A(t)$$

is the expected value of the noisy amplitude estimate

$$z_D(t) = \hat{A}(t) + n''(t)$$

The AGC output signal amplitude $A_{out}(t)$ is

$$A_{out}(t) = G[v_c(t)]A(t) \tag{7.2-28}$$

With coherent detection $\beta(\rho_i) = \cos \phi$ and $n''(t)$ is given by (7.2-8). With envelope detection $\beta(\rho_i) = \beta_{en}(\rho_i)$ (7.2-16) and $n''(t)$ is given by (7.2-14). When we make the following replacement in the signal model of Figure 7.2-4

$$A(t) \rightarrow A^2(t) = P_s$$

$$A_{out}(t) \rightarrow A^2_{out}(t)$$

$$G[v_c(t)] \rightarrow G^2[v_c(t)]$$

we find that with (see (7.2-24))

$$\beta(\rho_i) = \beta^2_{sq}(\rho_i) = \frac{1 + \rho_i}{\rho_i}$$

the model also applies to an AGC with square law detector.

After subtracting the reference value $V_R = K_D A_R$ the detector output is filtered and, possibly, amplified. Denoting the impulse response of the filter by $f(t)$, and the amplifier gain by $-K_F$ the input signal to the GCA can be written

$$\begin{aligned} v_c(t) &= -K_F K_D \{ G[v_c(t)]z_D(t) - A_R \} * f(t) \\ &= -K_F K_D \{ G[v_c(t)][\hat{A}(t) + n''(t)] - A_R \} * f(t) \end{aligned} \tag{7.2-29}$$

The amplifier gain is assumed to be negative for stability reasons, since with the GCA characteristics (7.2-2)–(7.2-5) the gain increases with increasing control signal $v_c(t)$ (the hyperbolic gain characteristic (7.2-6) is not considered here).

With the chosen gain control characteristic inserted, (7.2-29) completely describes the dynamics of the AGC loop. But $v_c(t)$ can be evaluated from (7.2-29) only in certain special cases. Therefore we start with the much simpler steady-state analysis followed by linear approximations which may be used when the variations around the steady state are small. Finally, the exact solution of (7.2-29) in special cases will be discussed in Section 7.2.6.

7.2.4. Steady-State Analysis

For the steady-state analysis we assume that the input signal amplitude $A(t)$ is constant or varying so slowly that the dynamic response of the AGC to the amplitude fluctuation is negligible. Strictly speaking there is no steady state when the input signal is corrupted by noise, rather we have expected values and random fluctuations around these values. We may, however, approximately determine these expected values if the filter bandwidth is sufficiently small by letting $n''(t) \approx 0$. Then we have to distinguish two cases, namely filters with a pole at zero, i.e., truly integrating filters, and low-pass filters with $F(0) = 1$ (a potential gain is assumed to be absorbed in K_F).

With an integrating filter in the steady state the integrator input must be zero

$$G(v_{c,s})\hat{A} - A_R = 0 \qquad (7.2\text{-}30)$$

Here the steady-state value of $v_c(t)$ is denoted by $v_{c,s}$. With this condition, (7.2-28) yields for the steady-state value of the AGC output signal amplitude $A_{out,s}$

$$A_{out,s} = G(v_{c,s})A = A_R/\beta(\rho_i) \qquad (7.2\text{-}31)$$

Thus, the output signal amplitude depends on the signal suppression factor $\beta(\rho_i)$ of the detector only. When the phase error ϕ is small, using coherent detection the output signal amplitude equals the reference value A_R for arbitrary input signal-to-noise ratios. In Figure 7.2-5 the AGC signal suppression factor $\alpha_{AGC}(\rho_i) = 1/\beta(\rho_i)$ is shown in comparison to the limiter suppression factor $\alpha(\rho_i)$ for the three types of detectors considered here.

With a low-pass filter, loop equation (7.2-29) in the steady state reads

$$v_{c,s} = -K_F K_D[G(v_{c,s})\hat{A} - A_R] \qquad (7.2\text{-}32)$$

Since in general (7.2-32) cannot be solved for $v_{c,s}$ explicitly, we will analyze the steady-state performance by means of a graphic procedure. Rewriting (7.2-32)

$$AG(v_{c,s}) = \frac{A_R}{\beta(\rho_i)} - \frac{v_{c,s}}{K_D K_F \beta(\rho_i)}$$

it is found that $v_{c,s}$ is given by the intersection of the curve $AG(v_c)$ with the straight line $[A_R/\beta(\rho_i)] - [v_c/K_D K_F \beta(\rho_i)]$ as shown in Figure 7.2-6 in principle. Note that the $AG(v_{c,s})$ equals the steady-state AGC output signal amplitude $A_{out,s}$. When

$$v_{c,0} = G^{-1}\left(\frac{A_R}{\beta(\rho_i)A}\right) \ll K_F K_D A_R$$

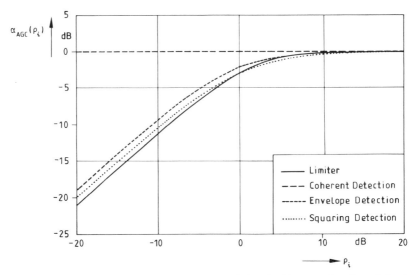

Figure 7.2-5. AGC signal suppression factor $\alpha_{AGC}(\rho_i) = A_{out}/A_R$ as a function of the input signal-to-noise ratio ρ_i for various types of detectors and integrating loop filter.

it follows from Figure 7.2-6 that $v_{c,s}$ is close to $v_{c,0}$ and we may linearize $G(v_c)$ about $v_c = v_{c,0}$

$$G(v_{c,s}) \approx G(v_{c,0}) + (v_{c,s} - v_{c,0})g_0 \qquad (7.2\text{-}33)$$

where

$$g_0 = \frac{\partial G(v_c)}{\partial v_c}\bigg|_{v_c = v_{c,0}}$$

Inserting the linearization into (7.2-32) and solving for $v_{c,s}$ yields

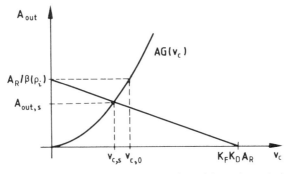

Figure 7.2-6. Output signal amplitude A_{out} as a function of the gain control voltage v_c for an AGC with low-pass loop filter.

$$v_{c,s} \approx \frac{K_F K_D \beta(\rho_i) A g_0}{1 + K_F K_D \beta(\rho_i) A g_0} v_{c,0}$$

From (7.2-28) we obtain for the steady-state AGC output signal amplitude

$$A_{\text{out},s} = G(v_{c,s}) A = G\left(\frac{K_F K_D \beta(\rho_i) A g_0}{1 + K_F K_D \beta(\rho_i) A g_0} v_{c,0} \right) A \qquad (7.2\text{-}34)$$

When $K_F K_D \beta(\rho_i) A g_0 \gg 1$ we approximately have the same steady-state AGC characteristic as for an integrating filter (7.2-31)

$$A_{\text{out},s} \approx G(v_{c,0}) A = A_R / \beta(\rho_i) \qquad (7.2\text{-}35)$$

Both conditions, $G^{-1}(A_R / \beta(\rho_i) A) \ll K_F K_D A_R$ and $K_F K_D \beta(\rho_i) A g_0 \gg 1$, are satisfied for an arbitrary (monotonically increasing) gain control characteristic when the input signal amplitude A is sufficiently large, i.e., (7.2-35) describes the asymptotic behavior of the AGC for $A \to \infty$.

When the input signal amplitude A approaches zero from Figure 7.2-6 (and from (7.2-31)) we find $v_{c,s} \approx K_F K_D A_R$. The steady-state AGC output signal amplitude then is proportional to A

$$A_{\text{out},s} \approx A G(K_F K_D A_R)$$

Combining the results for $A \to \infty$ and for $A \to 0$ yields a steady-state AGC control characteristic as shown in Figure 7.2-7 in principle. For $A G(K_F K_D A_R) \gg A_R / \beta(\rho_i)$ we have the desired effect that the output signal amplitude is independent of the input signal amplitude, as with an integrating filter. When A is too small the AGC with low-pass loop filter is unable to keep the output signal amplitude at a given level.

Note that AGCs with low-pass loop filter are sometimes operated without a reference signal, i.e., $A_R = 0$. As was seen, amplitude control is poor for

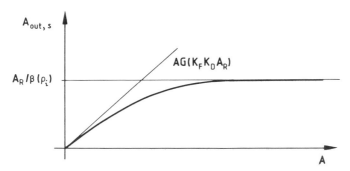

Figure 7.2-7. Steady-state output signal amplitude $A_{\text{out},s}$ as a function of the input signal amplitude A for an AGC with low-pass loop filter.

$\beta(\rho_i)AG(v_{c,s}) \ll A_R$. Without reference, amplitude control starts at arbitrarily low input signal amplitudes. Therefore AGCs with $A_R = 0$ are sometimes called "undelayed" AGCs. The advantage of AGCs with $A_R \neq 0$ is that above the threshold, amplitude control is much better than with an "undelayed" AGC. Note that some gain control characteristics, which includes the linear characteristic, must not be used in undelayed AGCs.

So far we have assumed that $A(t)$ is slowly varying. With a digital amplitude modulation, $A(t)$ varies rapidly. When the AGC loop filter bandwidth is sufficiently narrow we can neglect these fluctuations. Then, we may replace the signal amplitude by the appropriate expected value in the steady-state analysis. For instance, with envelope detection $E[\|A\|]$ replaces A, where $\|A\|$ stands for the square root of the sum of the squared quadrature component amplitudes ($\|A\| := (A_s^2 + A_c^2)^{1/2}$).

7.2.5. Linear Approximations

When the gain control voltage exhibits only small fluctuations around its steady-state value $v_{c,s}$ we again may omit all terms of order ≥ 2 of the Taylor series expansion of $G[v_c(t)]$

$$G[v_c(t)] \approx G(v_{c,s}) + [v_c(t) - v_{c,s}]g \qquad (7.2\text{-}36)$$

with

$$g = \left. \frac{\partial G(v_c)}{\partial v_c} \right|_{v_c = v_{c,s}}$$

Then (7.2-29) reads

$$c(t) \approx -K_F K_D A_0 g[c(t) * f(t)] - K_F K_D G(v_{c,s})\left[\left(1 + \frac{gc(t)}{G(v_{c,s})}\right)a(t)\right] * f(t) \qquad (7.2\text{-}37)$$

with

$$c(t) = v_c(t) - v_{c,s}$$

$$a(t) = \hat{A}(t) + n''(t) - A_0$$

where A_0 is chosen so that $a(t)$ has zero mean. Note that in (7.2-37) equality holds for a linear gain control characteristic. We still have a nonlinear equation but when the gain fluctuations around the steady-state value are small as assumed, $|gc(t)|/G(v_{c,s}) \ll 1$. Thus, we may neglect this term in (7.2-37). We finally obtain a linear differential equation for $c(t)$

$$c(t) \approx -K_F K_D A_0 g[c(t) * f(t)] - K_F K_D G(v_{c,s})[a(t) * f(t)] \tag{7.2-38}$$

Defining a closed-loop AGC transfer function

$$H_{\mathrm{AGC}}(s) = \frac{K_F K_D A_0 g F(s)}{1 + K_F K_D A_0 g F(s)} \quad \bullet\!\!-\!\!\circ\, h_{\mathrm{AGC}}(t) \tag{7.2-39}$$

we have for (7.2-37) in the Laplace domain

$$C(s) \approx -\frac{G(v_{c,s})}{A_0 g} H_{\mathrm{AGC}}(s) A(s) \tag{7.2-40}$$

$$c(t) \approx -\frac{G(v_{c,s})}{A_0 g} h_{\mathrm{AGC}}(t) * a(t)$$

(Corresponding transform pairs are denoted by the same letter with a capital letter for the spectrum.) The gain fluctuations around $G(v_{c,s})$ are given approximately by $gc(t)$. Thus, for the output signal of the AGC we have

$$y(t) \approx [G(v_{c,s}) + gc(t)]z(t)$$

Multiplying both sides of (7.2-40) by $g/G(v_{c,s})$ we find that the relative gain fluctuations equal the relative amplitude fluctuations $a(t)/A_0$ filtered by $H_{\mathrm{AGC}}(j\omega)$. The remarkable point is that for the linearization to be accurate it is sufficient that the filtered amplitude fluctuations $h_{\mathrm{AGC}}(t) * a(t)/A_0$ are small but it is not required that $a(t)/A_0$ itself is small. Therefore we can approximately determine the gain fluctuations of the AGC from $gc(t)$ even at very low input signal-to-noise ratios provided that the AGC loop bandwidth is so small that $|h_{\mathrm{AGC}}(t) * a(t)(1/A_0)| \ll 1$.

Example: With an integrating filter

$$F(s) = \frac{1}{sT_F} \tag{7.2-41}$$

we have

$$C(s) \approx -\frac{G(v_{c,s})}{gA_0} \frac{1}{1 + s(T_F/K_F K_D A_0 g)} A(s) \tag{7.2-42}$$

The variance of the gain fluctuations is then approximately (see (7.2-20))

$$E[(gc)^2] \approx \left[\frac{G(v_{c,s})}{A_0}\right]^2 \sigma_{n''}^2 \frac{K_F K_D A_0 g}{2 B_{IF} T_F} \tag{7.2-43}$$

Example: With an exponential gain characteristic (7.2-3) and an integrating filter (7.2-41)

$$g = g_0 = G_1 G(v_{c,0}) = G_1 A_R / A_0$$

and, thus, we have the important result

$$H_{\text{AGC}}(s) = \frac{1}{1 + s(T_F / K_F K_D G_1 A_R)} \qquad (7.2\text{-}44)$$

i.e., the small signal response does not depend on the operating point ($\neq f(A)$). Within the dynamic range of the input signal amplitude A where approximation (7.2-35) applies, this also holds for a low-pass filter. For this reason the exponential gain characteristic is often chosen when the dynamic range of the input signal amplitude is very large and loop design is critical (strong noise, wanted amplitude modulation, etc.). Also note that the time constant of the AGC transfer function is much smaller than the filter time constant T_F since $K_F K_D A_0 g \gg 1$.

While the linearization described so far applies for AGCs with arbitrary gain control characteristics, for an exponential gain control characteristic with a different approach (and under different restrictions) a linearized model can be derived. In this case we have from (7.2-29)

$$G[v_c(t)] = G_0 \exp\{-G_1 K_F K_D [G(v_c(t))(\hat{A}(t) + n''(t)) - A_R] * f(t)\} \qquad (7.2\text{-}45)$$

Taking $20 \log_{10}\{\cdot\}$ of both sides to arrive at a decibel representation yields

$$20 \log_{10}\{G(v_c(t))\} = 20 \log_{10}\{G_0\}$$
$$- G_1 K_F K_D K_e\{G[v_c(t)][\hat{A}(t) + n''(t)] - A_R\} * f(t) \qquad (7.2\text{-}46)$$

with the constant

$$K_e = 20 \log_{10}\{e\} \approx 8.686$$

Using Taylor series expansion, for instance, one can show that $\ln(x) \approx x - 1$ for $|x - 1| \ll 1$, or equivalently

$$20 \log_{10}(x) \approx K_e(x - 1)$$

If the loop is not too far from the design point $G[v_c(t)]\hat{A} \approx A_R$ we may approximate

$$\frac{G[v_c(t)]\hat{A}(t) - A_R}{A_R} \approx \frac{1}{K_e} 20 \log_{10}\left\{\frac{G[v_c(t)]\hat{A}(t)}{A_R}\right\} \qquad (7.2\text{-}47)$$

Denoting variables expressed in decibels by the subscript dB, i.e.,

$$G_{dB}(t) = 20 \log\{G[v_c(t)]\}$$

$$G_{0,dB} = 20 \log\{G_0\}$$

$$\hat{a}_{dB}(t) = 20 \log\{\hat{A}(t)/A_R\}$$

and inserting the approximation (7.2-47) in (7.2-46) yields

$$\begin{aligned}
G_{dB}(t) = G_{0,dB} &- G_1 K_F K_D A_R[G_{dB}(t) + \hat{a}_{dB}(t)] * f(t) \\
&- G_1 K_F K_D K_e\{G[v_c(t)]n''(t)\} * f(t) \qquad (7.2\text{-}48)
\end{aligned}$$

The corresponding block diagram is shown in Figure 7.2-8. With

$$a_{dB}(t) = 20 \log_{10}[A(t)/A_R]$$

and

$$a_{out,dB}(t) = 20 \log_{10}[A_{out}(t)/A_R]$$

the output signal amplitude in dB is

$$a_{out,dB}(t) = G_{dB}(t) + a_{dB}(t)$$

From (7.2-48) we can also derive a closed-loop transfer function

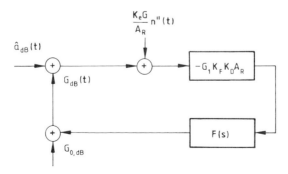

Figure 7.2-8. Linearized model of an AGC with exponential gain control characteristic, all variables given in decibels (dB).

$$H_{\mathrm{AGC}}(s) = -\frac{G_1 K_F K_D A_R F(s)}{1 + G_1 K_F K_D A_R F(s)} \tag{7.2-49}$$

When the approximation (7.2.47) is valid we have $G(v_{c,s})A_0 \approx A_R$. Since $G_1 G(v_{c,s}) = g$ for an exponential gain control characteristic, the transfer function (7.2-49) for the variables expressed in dB is approximately the same as for the linear approximation (7.2-39). Actually, if deviations from the steady state are small, results from both approaches converge. On the other hand, if deviations are larger, the results of the above approach are more precise.

7.2.6. Exact Dynamics

Exact solutions of (7.2-29) have been found for AGCs with linear and hyperbolic (see, e.g., Davis [10]) gain control characteristics. When we have an integrating loop filter an explicit solution can also be given for an AGC with exponential gain control characteristic.

Multiplying both sides of (7.2-29) by G_0 yields for an *AGC with linear gain control characteristic*

$$G[v_c(t)] = -K_F K_D G_0 \{G[v_c(t)]z_D(t) - A_R\} * f(t) \tag{7.2-50}$$

For an nth order loop filter $F(s)$, this corresponds to an nth order linear differential equation for the gain $G[v_c(t)]$. With the loop filter

$$F(s) = \frac{1}{F_0 + sT_F}$$

which corresponds to a simple resistor-capacitor (RC) filter for $F_0 = 1$ and to an integrating filter for $F_0 = 0$, we have a linear first-order differential equation

$$[F_0 + K_F K_D G_0 z_D(t)]G(t) + T_F \dot{G}(t) = K_F K_D G_0 A_R$$

Note that to simplify the notation we write $G(t)$ for $G[v_c(t)]$ here. The solution of such an equation is well known. In our case it reads

$$
\begin{aligned}
G(t) = \exp\left(-\frac{t}{T_F}F_0\right)\exp\left(-\frac{K_F K_D G_0}{T_F}\int_0^t z_D(x)\,dx\right) \\
\times\left[\frac{K_F K_D G_0 A_R}{T_F}\int_0^t \exp\left(\frac{x}{T_F}F_0\right)\right. \\
\left.\times \exp\left(\frac{K_F K_D G_0}{T_F}\int_0^x z_D(y)\,dy\right)dx + G(0)\right]
\end{aligned}
\tag{7.2-51}
$$

This result is not very helpful in noise analysis (for the case where the input to the AGC is pure gaussian noise see, e.g., Schachter and Bergstein [7]) but the response of the AGC to large deterministic variations of the input signal amplitude can be evaluated easily from (7.2-51). For instance, to analyze acquisition without noise $(n''(t) = 0)$ we may assume

$$z_D(t) = \hat{A}(t) = A_0, \quad t \geq 0$$

Solving (7.2-51) in this case yields

$$G(t) = \left(G(0) - \frac{A_R/A_0}{1 + (F_0/K_F K_D G_0 A_0)} \right) \exp\left[-\frac{F_0 + K_F K_D G_0 A_0}{T_F} t \right]$$
$$+ \frac{A_R/A_0}{1 + (F_0/K_F K_D G_0 A_0)}$$

From this result it is clear that (see also Section 7.2.4)

$$G(t \rightarrow \infty) = \frac{A_R}{A_0} \frac{1}{1 + (F_0/K_F K_D G_0 A_0)}$$

Defining the time constant

$$\tau = \frac{T_F}{F_0 + K_F K_D G_0 A_0}$$

we have the simple solution

$$G(t) = G(\infty) + [G(0) - G(\infty)] \exp(-t/\tau) \tag{7.2-52}$$

In the operating region where the loop gain is large, i.e., $K_F K_D G_0 A_0 \gg 1$ the behavior is approximately the same for the RC filter and the integrating filter. Since $g = G_0$ (from (7.2-2) and (7.2-36)) we have the same time constant τ, depending on the detected amplitude A_0, as in the linear approximation.

The second case to be considered here where a closed form solution can be given is an *AGC with exponential gain control characteristic and an integrating filter* $F(s) = 1/sT_F$. In this case the loop equation (7.2-29) reads

$$\frac{dv_c(t)}{dt} = \dot{v}_c(t) = -\frac{1}{T_F} K_F K_D \{ z_D(t) G[v_c(t)] - A_R \}$$

Since

$$\frac{dG[v_c(t)]}{dt} = \dot{G}[v_c(t)] = G[v_c(t)] G_1 \dot{v}_c(t)$$

again using the simplified notation $G(t)$ for $G[v_c(t)]$ we have the differential equation for the gain

$$\dot{G}(t) + \frac{1}{T_F} K_F K_D G_1 z_D(t) G^2(t) - \frac{1}{T_F} \cdot K_F K_D G_1 A_R G(t) = 0$$

which is a form of Bernoulli's equation that is known to have the solution

$$G(t) = \left[\frac{\exp(-t/\tau)}{G(0)} + \frac{1}{A_R} \int_0^t \frac{1}{\tau} \exp\left(-\frac{t-x}{\tau}\right) z_D(x) \, dx \right]^{-1}$$

with

$$\tau = \frac{T_F}{K_F K_D G_1 A_R}$$

Defining an impulse response of the loop

$$h_{\mathrm{AGC}}(t) = \frac{1}{\tau} \exp\left(-\frac{t}{\tau}\right), \quad t \geq 0$$

we finally obtain for the gain

$$G(t) = \left[\frac{\exp(-t/\tau)}{G(0)} + \frac{1}{A_R} z_D(t) * h_{\mathrm{AGC}}(t) \right]^{-1} \tag{7.2-53}$$

From this equation we may again determine the response of the AGC to a step of the input signal amplitude to analyze the acquisition performance. Neglecting noise ($n''(t) = 0$) we assume

$$z_D(t) = \hat{A}(t) = A_0, \quad t \geq 0$$

Solving (7.2-53) yields

$$G(t) = \left[\frac{\exp(-t/\tau)}{G(0)} + \frac{A_0}{A_R} \left(1 - \exp\left(-\frac{t}{\tau}\right)\right) \right]^{-1} \tag{7.2-54}$$

Since

$$\frac{A_0}{A_R} = G(\infty)$$

We can rewrite the result (7.2-54)

$$G(t) = \frac{G(\infty)}{1 + \{[G(\infty)/G(0)] - 1\} \exp(-t/\tau)}$$

In contrast to the AGC with linear gain control characteristic, here the time constant τ is independent of the input signal amplitude. Because of the simple form of (7.2-53) in this case, we can go even one step further in the analysis. The first term corresponds to an initial condition which decays to zero after a while. After the system has operated for some time we may neglect this term. Then the exact gain is given by

$$G(t) = \frac{1}{\frac{1}{A_R}[\hat{A}(t) + n''(t)] * h_{AGC}(t)} \tag{7.2-55}$$

and the output signal amplitude is

$$A_{out}(t) = A(t)G(t) = \frac{A(t)}{(1/A_R)[\hat{A}(t) + n''(t)] * h_{AGC}(t)} \tag{7.2-56}$$

An equivalent block diagram is shown in Figure 7.2-9. Note that $h_{AGC}(t)$ corresponds exactly to the transfer function (7.2-44) found in the linearized model. Before we proceed in the analysis, a word of caution must be given. Equation (7.2-55) only applies when the denominator is positive. This is always the case with noncoherent detection. But with coherent detection $\hat{A}(t) + n''(t) = A(t)\cos(\phi) + n''(t)$ may remain negative long enough for the denominator of (7.2-55) to approach zero

$$\frac{1}{A_R}[A(t)\cos(\phi) + n''(t)] * h_{AGC}(t) \to 0$$

Then the AGC gets stuck at $G(t) \to \infty$ and (7.2-55) no longer applies. Since the limited validity of this equation has been ignored, the model of Figure 7.2-9 does not show this effect. In a real system the gain cannot become infinite. In fact, the integrating loop filter stops integrating when its output voltage (which controls the amplifier) has reached the supply voltage thus setting the gain to its maximum value. As soon as the detector output $A(t) + n''(t)$ becomes positive, the AGC starts to recover.

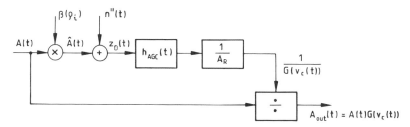

Figure 7.2-9. Exact model of an AGC with exponential gain control characteristic and integrating filter (the model is valid for $G[v_c(t)] > 0$ only).

Although the exact AGC response to large variations of the input signal amplitude is found easily from (7.2-55), in general the response to random processes (noise) cannot be determined exactly. With coherent detection the noise is gaussian and therefore has a finite probability of being negative to such an extent that (7.2-55) no longer applies. For weak noise an approximate analysis may be given by neglecting this probability. With noncoherent detection the second-order statistics of the noise are mathematically not tractable.

The dynamics of an AGC with hyperbolic gain control characteristic which is not considered here is described by Davis [10].

Another word of caution must be given here finally. The exact solutions are correct only in the gain range where the assumed gain control characteristic applies. For instance, when a strong signal is applied suddenly to a receiver the gain control characteristic breaks down because the receiver overloads. The resulting AGC dynamic characteristic is complicated by this extra, often unpredictable, nonlinearity.

7.2.7. Acquisition of Coherent Automatic Gain Control and Phase-Locked Loop

For coherent detection in the AGC a phase reference $\hat{\theta}$ is required. This reference is usually generated from a PLL following the AGC as shown in Figure 7.2-10. When the PLL is in the tracking mode the phase error is small and $\cos \phi \approx 1$. But during acquisition it is necessary to observe the interaction between AGC and PLL. For a typical configuration the combined differential equations are summarized in Table 7.2-1. We have a system of three coupled nonlinear differential equations which in general cannot be solved. Acquisition under noise-free operating conditions has been studied by Green [13]. We will not give a full analysis here but discuss the interaction of PLL and coherent AGC qualitatively.

For $\phi = 0$ the output of the coherent detector is $A(t)$ but when the PLL is out of lock the output is smaller than $A(t)$ and approaches $A(t)$ only for brief moments. If the AGC response is not too fast the loop mainly responds to a short term average of the detector output, thus setting the

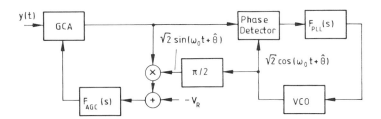

Figure 7.2-10. Coherent automatic gain control and phase-locked loop.

TABLE 7.2-1 Combined Differential Equations of a Coherent AGC with Exponential Gain Control Characteristic and an Integrating Filter Followed by a PLL with Sinusoidal Phase Detector Characteristic and an Imperfect Integrator

$$\dot{G}(t) = -K_{AGC} G^2(t)A(t)\{\cos[\phi(t)] + n'_Q(t)\} + K_{AGC} G(t)A_R$$

$$\dot{\phi}(t) = -K_{PLL} G(t)A(t)\frac{T_2}{T_1}\sin[\phi(t)] + n'_I(t)\} - y_1(t) + \Omega$$

$$\dot{y}_1(t) = -\frac{1}{T_1}y_1(t) + \frac{1 - (T_2/T_1)}{T_1}K_{PLL}G(t)A(t)\{\sin[\phi(t)] + n'_I(t)\}$$

AGC*

$$F(s) = \frac{1}{sT_I}, \quad K_{AGC} = \frac{K_F K_{D,AGC} G_1}{T_I}, \quad n'_Q(t) = \frac{n''(t)}{A}$$

$$n'_Q(t) = \frac{n''(t)}{A} = \frac{n_c(t)}{A}\sin(\hat{\theta}) - \frac{n_s(t)}{A}\cos(\hat{\theta})$$

PLL*

$$F(s) = \frac{1 + sT_2}{1 + sT_1}, \quad K_{PLL} = \frac{K_0 K_{D,PLL}}{GA}$$

$$n'_I(t) = n'(t) = \frac{n_c(t)}{A}\cos(\hat{\theta}) + \frac{n_s(t)}{A}\sin(\hat{\theta})$$

*Note that K_D is used in this book for both the detector gain of the AGC amplitude detector and the gain of the PLL phase detector. For proper distinction $K_{D,PLL}$ and $K_{D,AGC}$ are used here instead of K_D.

gain significantly higher than required for $GA(t) \approx A_R$. However, with a larger signal amplitude the phase-locked loop gain is above the design point and as a consequence the pull-in range is increased (at the cost of an increased noise bandwidth). To summarize, the coherent AGC in addition to having an amplitude control function improves the PLL acquisition. More details are given by Green [13].

7.2.8. Miscellaneous Modifications of Automatic Gain Control

Several modifications of feedback AGCs are discussed in the literature. For instance, to avoid gain fluctuations due to long strings of equal symbols in M-ary Amplitude Phase Shift Keying (MAPSK) transmission a decision-directed amplitude detector may be used [14].

Digital (discrete-time) realizations of parts of the AGC loop have drawn a lot of attention (see, e.g., [15]). With digital signal processing more complex algorithms may be used to enhance AGC performance. This topic will be covered in Volume 2.

The use of feedforward AGCs (see Figure 7.2-2) on mobile radio channels is discussed by Sladen and McGeehan [16]. Parameter variations of the hardware are not suppressed in a feedforward AGC. Therefore, the application is restricted to cases where gain control is extremely difficult and this drawback does not count. The situation is different with an all-digital AGC which does not suffer from parameter variations.

Main Points of the Section

- With noncoherent detection the amplitude estimate $\hat{A}$ is biased at low signal-to-noise ratios causing signal suppression by the AGC loop (see Figure 7.2-5).
- Coherent detection requires a phase reference but yields an unbiased amplitude estimate $\hat{A} = A$ at all signal-to-noise ratios.
- The operation of all types of feedback AGCs is described by the unified model (Figure 7.2-4)

$$v_c(t) = -K_F K_D \{ G[v_C(t)] z_D(t) - A_R \} * f(t) \qquad (7.2\text{-}29)$$

- Two different linear approximations
Equation (7.2-41)

$$C(s) \approx -\frac{G(v_c)}{A_0 g} H_{\mathrm{AGC}}(s) A(s)$$

applies to all types of AGCs when gain variations $gc(t)$ around the design point $G(v_c)$ are small.

Approximation (7.2-48) for the variables given in decibels applies to AGCs with exponential gain control characteristics only.

- The transfer function of the linear approximation is independent of the input signal amplitude for an AGC with exponential gain control characteristic only ((7.2-44) and (7.2-49)).
- An "exact" model for an AGC with exponential gain control characteristic and an integrating filter is given by (Figure 7.2-9 and (7.2-54))

$$G(t) = \left\{ \frac{1}{A_R} [\hat{A}(t) + n''(t)] * h_{\mathrm{AGC}}(t) \right\}^{-1}$$

- A coherent AGC improves the acquisition of a subsequent PLL.

REFERENCES

1. W. B. Davenport, Jr., Signal-to-Noise Ratios in Band-Pass Limiters, *Journal of Applied Physics*, Vol. 24, No. 6, pp. 720–727, June 1953.
2. J. C. Springett and M. K. Simon, An Analysis of the Phase Coherent-Incoherent Output of the Bandpass Limiter, *IEEE Transactions on Communication Technology*, Vol. 19, No. 1, pp. 42–49, February 1971.

3. W. C. Lindsey, Synchronization Systems in Communication and Control, Prentice Hall, Englewood Cliffs, NJ, 1972.

4. M. Abramowitz and I. Stegun, *Handbook of Mathematical Functions*, National Bureau of Standards, Applied Mathematics Series, Vol. 55, June 1964.

5. D. V. Mercy, A Review of Automatic Gain Control Theory, *The Radio and Electronic Engineer*, Vol. 51, No. 51, No. 11/12, pp. 579–590, November/December 1981.

6. E. D. Banta, Analysis of an Automatic Gain Control (AGC), *IEEE Transactions on Automatic Control*, Vol. 9, No. 2, pp. 181–182, April 1964.

7. H. Schachter and L. Bergstein, Noise Analysis of an Automatic Gain Control System, *IEEE Transactions on Automatic Control*, Vol. 9, No. 3, pp. 249–255, July 1964.

8. W. K. Victor and M. H. Brockman, The Application of Linear Servo Theory to the Design of AGC Loops, *Proceedings of the IRE*, Vol. 48, No. 2, pp. 234–238, February 1960.

9. J. E. Ohlson, Exact Dynamics of Automatic Gain Control, *IEEE Transactions on Communications*, Vol. 22, No. 1, pp. 72–75, January 1974.

10. R. C. Davis, Mathematical Analysis of Automatic Gain Control Circuits, IRE WESCON Convention 1961, *WESCON Convention Record*, part 28/2, 1961.

11. W. R. Braun, Analysis and Synthesis of Automatic Gain Control Systems, Ph.D. Dissertation, University of Southern California, Los Angeles, June 1976.

12. B. M. Oliver, Automatic Volume Control as a Feedback Problem, *Proceedings of the IRE*, Vol. 36, No. 4, pp. 466–473, April 1948.

13. D. N. Green, Lock-In, Tracking, and Acquisition of AGC-Aided Phase-Locked Loops, *IEEE Transactions on Circuits and Systems*, Vol. 32, No. 6, pp. 559–568, June 1985.

14. W. J. Weber, III, Decision-Directed Automatic Gain Control for MAPSK Systems, *IEEE Transactions on Communications*, Vol. 23, No. 5, pp. 510–517, May 1975.

15. D. R. Morgan, On Discrete-Time AGC Amplifiers, *IEEE Transactions on Circuits and Systems*, Vol. 22, No. 2, pp. 135–146, February 1975.

16. J. P. H. Sladen and J. P. McGeehan, The Performance of Combined Feedforward AGC and Phase-Locked AFC in a Single Sideband Mobile Radio Receiver, *Proceedings of the IEE*, Vol. 131, Part F, No. 5, pp. 437–441, August 1984.

APPENDIX 7.1A
SERIES REPRESENTATION OF THE HARD LIMITER OUTPUT SIGNAL

The hard limiter nonlinearity $f(z)$ is given by

$$f(z) = \text{sgn}\{z\}$$

Rewriting $z(t)$ (see also (3.2-13))

$$z(t) = \sqrt{2}A \sin[\omega_0 t + \theta(t)] - \sqrt{2}n_s(t) \sin(\omega_0 t) + \sqrt{2}n_c(t) \cos(\omega_0 t)$$
$$= \sqrt{2}[A - N_s(t)] \sin[\omega_0 t + \theta(t)] + \sqrt{2}N_c(t) \cos[\omega_0 t + \theta(t)]$$
$$= \sqrt{2}V(t) \sin[\omega_0 t + \theta(t) + \theta_n(t)]$$

with

$$N_c(t) = n_s(t) \sin\theta(t) + n_c(t) \cos\theta(t)$$
$$N_s(t) = n_s(t) \cos\theta(t) - n_c(t) \sin\theta(t)$$

and

$$V(t) = \{[A - N_s(t)]^2 + N_c^2(t)\}^{1/2}$$

The limiter output is obtained as

$$y(t) = \text{sgn}\{\sin[\omega_0 t + \theta(t) + \theta_n(t)]\}$$

But $\text{sgn}\{\sin(x)\}$ is a square wave with period 2π for which the Fourier series representation is known to be

$$\text{sgn}\{\sin(x)\} = \frac{4}{\pi} \sum_{k=0}^{\infty} \frac{1}{2k+1} \sin[(2k+1)x]$$

Thus, the hard limiter output signal can be represented by a series of sinusoids

$$y(t) = \frac{4}{\pi} \sum_{k=0}^{\infty} \frac{1}{2k+1} \sin\{(2k+1)[\omega_0 t + \theta(t) + \theta_n(t)]\} \quad (7.1A\text{-}1)$$

APPENDIX 7.1B
EXPECTED VALUES $E[g_0(\phi + \theta_n)]$, $E[g_0^2(\phi + \theta_n)]$

The signal $z(t)$ assumed in Section 7.1

$$z(t) = \sqrt{2}[A - N_s(t)] \sin[\omega_0 t + \theta(t)] + N_c(t) \cos[\omega_0 t + \theta(t)]$$

can also be expressed

$$z(t) = \sqrt{2}V(t) \sin[\omega_0 t + \theta(t) + \theta_n(t)]$$

where

$$V(t) = \{[A - N_s(t)]^2 - N_c^2(t)\}^{1/2}$$

and

$$\theta_n(t) = \tan^{-1}\left[\frac{N_c(t)}{A - N_s(t)}\right], \quad -\pi < \theta_n(t) \leq +\pi$$

From the joint probability density function of N_s, N_c

$$p(N_s, N_c) = \frac{1}{2\pi(\sigma_n^2/2)} \exp\left(-\frac{N_s^2}{\sigma_n^2}\right) \exp\left(-\frac{N_c^2}{\sigma_n^2}\right)$$

with

$$E[N_s^2] = E[N_c^2] = \sigma_n^2/2$$

The probability density function of the phase noise $\theta_n(t)$

$$p(\theta_n) = \int_0^{+\infty} p(V, \theta_n) \, dV$$

is obtained as

$$p(\theta_n) = \frac{1}{2\pi} \exp(-\rho_i) + \frac{\cos\theta_n}{(\pi/\rho_i)^{1/2}} \frac{1 + \text{erf}\left[\sqrt{\rho_i}\cos\theta_n\right]}{2}$$

$$\times \exp(-\rho_i \sin^2\theta_n), \quad -\pi < \theta_n \leq +\pi \qquad (7.1B\text{-}1)$$

where $\rho_i = A^2/\sigma_n^2$ is the input signal-to-noise ratio. Here $\text{erf}(x)$ denotes the error function

$$\text{erf}(x) := \frac{2}{\sqrt{\pi}} \int_0^x \exp(-t^2) \, dt$$

Since memoryless phase detector chacteristics are periodic with 2π they can be represented by Fourier series. Restricting our attention to odd phase detector characteristics $g_0(\phi) = -g_0(-\phi)$ (which includes the multiplier type phase detectors), only sine terms are present

$$g_0(\phi) = \sum_{k=1}^{\infty} b_k \sin(k\phi)$$

$$(7.1B\text{-}2)$$

$$b_k = \frac{1}{\pi} \int_0^{2\pi} g_0(\phi) \sin(k\phi) \, d\phi = \frac{2}{\pi} \int_0^{\pi} g_0(\phi) \sin(k\phi) \, d\phi$$

Note that for the multiplier type phase detectors $b_k = 0$ for even k, due to $g_0(\phi + \pi) = -g_0(\phi)$. Using this series expansion, the effective phase detector characteristic can also be written

$$g(\phi) = E[g_0(\phi + \theta_n)] = \sum_{k=1}^{\infty} b_k \int_{-\pi}^{\pi} \sin[k(\phi + \theta_n)] p(\theta_n) \, d\theta_n$$

$$= \sum_{k=1}^{\infty} b_k \{\sin(k\phi) E[\cos(k\theta_n)] + \cos(k\phi) E[\sin(k\theta_n)]\}$$

Since $p(\theta_n)$ is an even function

$$E[\sin (k\theta_n)] = 0 \tag{7.1B-3}$$

and

$$g(\phi) = \sum_{k=1}^{\infty} b_k \sin (k\phi) E[\cos (k\theta_n)] \tag{7.1B-4}$$

Thus, we only have to determine

$$E[\cos (k\theta_n)] = \int_{-\pi}^{\pi} \cos (k\theta_n) p(\theta_n) \, d\theta_n = 2 \int_0^{\pi} \cos (k\theta_n) p(\theta_n) \, d\theta_n$$

Inserting (7.1B-1) we obtain

$$E[\cos (k\theta_n)] = 2 \int_0^{\pi} \frac{1}{2\pi} \exp (-\rho_i) \cos (k\theta_n) \, d\theta_n$$

$$+ 2 \frac{\sqrt{\rho_i}}{\pi} \int_0^{\pi} \exp (-\rho_i \sin^2 \theta_n)$$

$$\times \cos (k\theta_n) \cos \theta_n \int_0^{\infty} \exp [-(x - \sqrt{\rho_i} \cos \theta_n)^2] \, dx \, d\theta_n$$

The first integral equals zero. Interchanging the order of integration and replacing

$$\cos (k\theta_n) \cos \theta_n = \frac{1}{2} \{\cos [(k + 1)\theta_n] + \cos [(k - 1)\theta_n]\}$$

we arrive after some straightforward manipulations at

$$E[\cos (k\theta_n)] = \sqrt{\rho_i} \exp (-\rho_i) \int_0^{\infty} \exp (-x^2) \frac{1}{\pi} \int_0^{\pi} \{\cos [(k + 1)\theta_n]$$

$$+ \cos [(k - 1)\theta_n]\} \exp [2x\sqrt{\rho_i} \cos \theta_n] \, d\theta_n \, dx$$

The inner integral represents the sum of modified Bessel functions of order $k + 1$ and $k - 1$ (see (7.1C-1))

$$E[\cos (k\theta_n)] = \sqrt{\rho_i} \exp (-\rho_i) \int_0^{\infty} \exp (-x^2)$$

$$\times \{I_{k+1}[2x\sqrt{\rho_i}] + I_{k-1}[2x\sqrt{\rho_i}]\} \, dx$$

This integral is found in mathematical handbooks (see, e.g., Abramowitz and Stegun [4]). Finally we obtain

$$E[\cos (k\theta_n)] = \left(\frac{\pi}{4} \rho_i\right)^{1/2} \exp \left(-\frac{\rho_i}{2}\right) \left[I_{(k+1)/2}\left(\frac{\rho_i}{2}\right) + I_{(k-1)/2}\left(\frac{\rho_i}{2}\right)\right] \tag{7.1B-5}$$

Note that for $k = 1$ the limiter suppression factor results (see (7.1-5)).

$$E[\cos \theta_n] = \alpha(\rho_i) \tag{7.1B-6}$$

Similarly $E[g_0^2(\phi + \theta_n)]$ may be determined using Fourier series expansion of $g_0^2(\phi)$. Since $g_0^2(\phi)$ is an even function when $g_0(\phi)$ is an odd function of ϕ

$$g_0^2(\phi) = \sum_{k=0}^{\infty} s_k \cos(k\phi) \tag{7.1B-7}$$

with

$$s_k = \frac{2}{\pi} \int_0^{\pi} g_0^2(\phi) \cos(k\phi) \, d\phi$$

Thus, the expected value is

$$E[g_0^2(\phi + \theta_n)] = \sum_{k=0}^{\infty} s_k \cos(k\phi) \, E[\cos(k\theta_n)] \tag{7.1B-8}$$

For most phase detector characteristics (including multiplier type phase detectors) the symmetries $-g_0(\phi + \pi) = g_0(\phi)$, and as a consequence $g_0^2(\phi + \pi) = g_0^2(\phi)$, hold. In this case $s_{2k+1} = 0$ and (7.1B-8) reduces to

$$E[g_0^2(\phi + \theta_n)] = \sum_{k=0}^{\infty} s_{2k} \cos(2k\phi) \, E[\cos(2k\theta_n)] \tag{7.1B-9}$$

The expectations now only consist of modified Bessel functions of order $n + 1/2$ ($n = k, k - 1$) which can be represented by a finite sum (7.1C-6). For nonsinusoidal phase detector characteristics the results are still rather impractical. Therefore, approximations of the probability density function $p(\theta_n)$ for high and low input SNR_i are very helpful in these cases.

For high input signal-to-noise ratios ($\rho_i \gg 1$) the first term in (7.1B-1) is small. The exponential function in the second term takes on significant values only for θ_n close to 0 and $\pm \pi$. Since $\mathrm{erf}(x) \to 1$ and $\mathrm{erf}(-x) \to -1$ for $x \gg 1$, the second term is small for $|\theta_n| \approx \pi$. Thus, the probability density function $p(\theta_n)$ is closely approximated by

$$p(\theta_n) = \frac{\cos \theta_n}{(\pi/\rho_i)^{1/2}} \exp(-\rho_i \sin^2 \theta_n)$$

$$\approx \frac{1}{[2\pi(1/2\rho_i)]^{1/2}} \exp\left(-\frac{\theta_n^2}{2\frac{1}{2\rho_i}}\right), \quad -\pi < \theta_n < +\pi, \ \rho_i \gg 1 \tag{7.1B-10}$$

The phase noise has an approximately gaussian distribution with zero mean and variance $1/2\rho_i$.

For low input SNR ($\rho_i \ll 1$) the exponential function approaches 1. Thus, the probability density function is approximately

$$p(\theta_n) \approx \frac{1}{2\pi} + \left(\frac{\rho_i}{4\pi}\right)^{1/2} \cos \theta_n \approx \frac{1}{2\pi}, \quad -\pi < \theta_n < +\pi, \; \rho_i \ll 1$$

$$\text{(7.1B-11)}$$

APPENDIX 7.1C
THE MODIFIED BESSEL FUNCTIONS $I_\nu(x)$*

An integral representation of the modified Bessel functions is given by Abramowitz and Stegun [4]

$$I_\nu(x) = \frac{1}{\pi} \int_0^\pi \exp(x \cos \psi) \cos(\nu\psi) \, d\psi$$

$$- \frac{\sin(\nu\pi)}{\pi} \int_0^\infty \exp(-x \cosh(t) - \nu t) \, dt \quad \text{(7.1C-1)}$$

For integer order $\nu = n$, the second term equals 0 and the integral representation reduces to

$$I_n(x) = \frac{1}{\pi} \int_0^\pi \exp(x \cos \psi) \cos(n\psi) \, d\psi$$

The following series representation of the modified Bessel functions is very useful in particular for small arguments ($|x| \ll 1$)

$$I_n(x) = \sum_{k=0}^\infty \frac{1}{k!(n+k)!} \left(\frac{x}{2}\right)^{n+2k} \quad \text{(7.1C-2)}$$

For large arguments ($|x| \gg 1$) the following asymptotic expansions may be used

$$I_n(x) \sim \frac{e^x}{\sqrt{2\pi x}} \sum_{k=0}^\infty \frac{(-1)^k}{(2x)^k} \frac{\Gamma\left(n+k+\frac{1}{2}\right)}{k!\,\Gamma\left(n-k+\frac{1}{2}\right)} \quad \text{(7.1C-3)}$$

where $\Gamma(x)$ is the gamma function which for integer argument $x = n$ is given by

$$\Gamma(n) = (n-1)! = 1 \times 2 \times 3 \times \cdots \times (n-1) \quad \text{for } n \text{ integer}$$

and for arguments $z = n + 1/2$ by

*Note that only real arguments x are considered here.

$$\Gamma\left(n + \frac{1}{2}\right) = 1 \times 3 \times 5 \cdots \times (2n - 1) \times \frac{\sqrt{\pi}}{2^n} \quad \text{for } n \text{ integer}$$

in the special cases of $n = 0$, and $n = 1$ we have

$$I_0(x) \approx 1 + \frac{x^2}{4} + \cdots$$

$$|x| \ll 1 \qquad (7.1\text{C-}4)$$

$$I_1(x) \approx \frac{x}{2} + \cdots$$

and

$$I_0(x) \sim \frac{e^x}{\sqrt{2\pi x}}\left(1 + \frac{1}{8x} + \cdots\right)$$

$$|x| \gg 1 \qquad (7.1\text{C-}5)$$

$$I_1(x) \sim \frac{e^x}{\sqrt{2\pi x}}\left(1 - \frac{3}{8x} + \cdots\right)$$

Another case of particular interest are Bessel functions of order $n + 1/2$ which can be represented by a finite sum

$$I_{\pm(n+1/2)}(x) = \frac{1}{\sqrt{2\pi x}}\left[e^x \sum_{k=0}^{n} \frac{(-1)^k (n + k)!}{k!(n - k)!(2x)^k} \right.$$

$$\left. \pm\, e^{-x}(-1)^{n+1} \sum_{k=0}^{n} \frac{(n + k)!}{k!(n - k)!(2x)^k} \right] \qquad (7.1\text{C-}6)$$

In the special cases of $n = 0$ and $n = 1$ we have

$$I_{\pm 1/2}(x) = \frac{1}{\sqrt{2\pi x}}\left[e^x \mp e^{-x} \right] \qquad (7.1\text{C-}7)$$

and

$$I_{\pm 3/2}(x) = \frac{1}{\sqrt{2\pi x}}\left[e^x\left(1 - \frac{1}{x}\right) \pm e^{-x}\left(1 + \frac{1}{x}\right) \right] \qquad (7.1\text{C-}8)$$

8

AUTOMATIC FREQUENCY CONTROL

8.1. INTRODUCTION

The automatic frequency control (AFC) loop is used for controlling the frequency of a received signal that is disturbed by noise. It is used in various digital data links.

Digital data transmission over satellite links is dominated by coherent phase shift keying as, for example, quadrature phase shift keying (QPSK) or coded octal phase shift keying (8PSK). Coherent demodulation is performed with the help of a phase-locked loop. However, as was explained in Chapter 4, during acquisiton the performance of the PLL is very poor, especially if the incoming signal frequency is unknown (e.g., as a result of an unknown Doppler shift). For this reason an AFC is a popular acquisition aid for a phase-locked loop (see Chapter 5).

In the case of severely disturbed channels (e.g., fading channels) it might not be possible to carry out coherent demodulation due to synchronization problems. As a result noncoherent transmission systems may be preferable (e.g., frequency shift keying). Then the use of an AFC loop for controlling the frequency is a necessity.

8.2. STRUCTURES OF FREQUENCY DETECTORS

8.2.1. Optimal Frequency Estimator

In this section we are concerned with the problem of estimating the frequency of a signal $s(t)$ when the noisy received signal

$$y(t) = s(t) + n(t) = \sqrt{2}A \sin(\omega_{i,0}t + \theta_0) + n(t) \qquad (8.2\text{-}1)$$

is given in the time interval $0 \le t \le T_E$. In this case $n(t)$ represents white gaussian noise. The phase angle θ_0 has an unknown value and is assumed to be constant in $0 \le t \le T_E$. The only thing we know about the frequency is that its value lies in the range $[\omega_L, \omega_L + 2\pi W]$.

What we are looking for is the maximum likelihood (ML) estimator for the unknown signal frequency. The ML-estimate $\hat{\omega}_i$ is that value of the trial parameter ω_i which most likely has produced the received signal $y(t)$ in the time interval $[0, T_E]$. Therefore, to obtain $\hat{\omega}_i$ we have to proceed as follows: we must try out values for ω_i in (8.2-1) and compute their effect on the statistics of the received signal. By comparison with the actually received signal $y(t)$ we can decide which value of ω_i has most likely produced $y(t)$. ML techniques will be explained in detail in Volume 2. A short summary is given in Appendix 8.2A.

Here we are facing the problem that the signal $s(t)$ not only depends on the parameter ω_i but also on a random variable θ (the unknown carrier phase). According to Appendix 8.2A we first calculate the function $p(y \mid \omega_i, \theta)$ and then obtain the likelihood function $p(y \mid \omega_i)$ by averaging over θ.

Since the received signal $y(t)$ is corrupted by white gaussian noise we have to make use of the concept of a "sufficient statistics" and applying the results of Appendix 8.2A we obtain (instead of $p(y(t) \mid \omega_i, \theta)$)

$$\Lambda(y \mid \omega_i, \theta) = \exp\left\{ -\frac{1}{N_0} \int_0^{T_E} [y(t) - s(t, \omega_i, \theta)]^2 \, dt \right\} \qquad (8.2\text{-}2)$$

where in the notation

$$s(t, \omega_i, \theta) = \sqrt{2} A \sin(\omega_i t + \theta)$$

we have stressed that the signal $s(t)$ depends on ω_i and θ. The ML principle demands that the likelihood function

$$\Lambda(y \mid \omega_i) = \int_{-\infty}^{+\infty} \Lambda(y \mid \omega_i, \theta) p(\theta) \, d\theta \qquad (8.2\text{-}3)$$

is to be maximized, i.e., the ML estimate $\hat{\omega}_i$ is that value of the trial parameter ω_i for which Λ assumes its maximum. Before averaging over θ we consider the exponent in (8.2-2) in more detail. The exponent can be subdivided into three terms

$$-\frac{1}{N_0} \int_0^{T_E} [y(t) - s(t)]^2 \, dt = -\frac{1}{N_0} \int_0^{T_E} y^2(t) \, dt - \frac{1}{N_0} \int_0^{T_E} s^2(t) \, dt$$

$$+ \frac{2}{N_0} \int_0^{T_E} y(t)s(t) \, dt \qquad (8.2\text{-}4)$$

The first term in (8.2-4) is the received signal itself. It cannot be maximized and may therefore be discarded. The second term is the integral over a fast oscillation ($\omega_i T_E \gg 1$). It is therefore approximately independent of ω_i and for this reason is also not included in the maximization. Thus only the last term remains

$$\Lambda(y \mid \omega_i, \theta) = \exp \left\{ \frac{2\sqrt{2}A}{N_0} \int_0^{T_E} y(t) \sin (\omega_i t + \theta) \, dt \right\} \qquad (8.2\text{-}5)$$

Now the condition on θ is removed using (8.2-3). Assuming uniform distribution of the phase angle we obtain

$$\Lambda(y \mid \omega_i) = \frac{1}{2\pi} \int_{-\pi}^{\pi} \exp \left\{ \frac{2\sqrt{2}A}{N_0} \left[\left(\int_0^{T_E} y(t) \sin \omega_i t \, dt \right) \cos \theta \right. \right.$$
$$\left. \left. + \left(\int_0^{T_E} y(t) \cos \omega_i t \, dt \right) \sin \theta \right] \right\} d\theta \qquad (8.2\text{-}6)$$

With the abbreviations

$$X(\omega_i) = \int_0^{T_E} y(t)\sqrt{2} \cos \omega_i t \, dt$$
$$\qquad\qquad\qquad\qquad\qquad (8.2\text{-}7)$$
$$Z(\omega_i) = \int_0^{T_E} y(t)\sqrt{2} \sin \omega_i t \, dt$$

one eventually arrives at

$$\Lambda(y \mid \omega_i) = I_0 \left\{ \frac{2A}{N_0} \left[X^2(\omega_i) + Z^2(\omega_i) \right]^{1/2} \right\} \qquad (8.2\text{-}8)$$

where $I_0(\cdot)$ is the modified Bessel function of first kind and zero order [1].
$I_0(x)$ is a monotonically increasing function for the positive arguments present in (8.2-8). Therefore we can equivalently maximize

$$M^2(\omega_i) = X^2(\omega_i) + Z^2(\omega_i) \qquad (8.2\text{-}9)$$

The ML estimate $\hat{\omega}_i$ is, by definition, the value of ω_i which maximizes $M^2(\omega_i)$. Since ω_i can assume an infinite number of values the optimal ML estimator requires an uncountable number of elements as illustrated in Figure 8.2-1. The ML estimate $\hat{\omega}_i$ is obtained by selecting that element of the estimator which yields the largest output.

Since the optimum estimator requires an uncountable number of these elements, a physical realization is clearly impossible. However, a reasonable approximation to this estimator can be achieved by using no more than $2WT_E + 1$ elements, each tuned to frequencies spaced $\Delta\omega = \pi/T_E$ rad/s. The reason for this is explained later in this section. The element that gives rise to the largest output is then chosen as the one most likely to match the

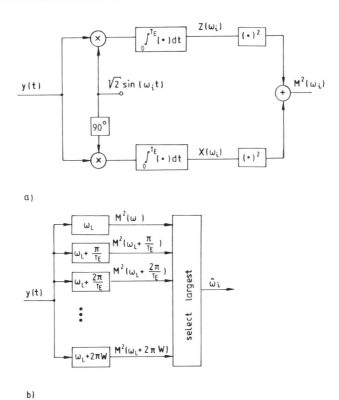

Figure 8.2-1. Maximum-likelihood frequency estimation. (*a*) An element of the noncoherent frequency estimator. (*b*) ML estimator for ω_i realized as a bank of $(2WT_E + 1)$ elements. The trial frequency ω_i assumes the value $\omega_L + k(\pi/T_E)$ in the $(k + 1)$st interval.

signal frequency. The frequency of the local oscillator of the receiver is then set to this frequency.

Our next step is to analyze the estimator structure, and in doing so we encounter two interesting interpretations of the estimator. The first interpretation is that each element of the frequency estimator can be viewed as an *correlation detector*. Let us assume that we are observing the $(k + 1)$st device. In this device the received signal is cross-correlated by sine and cosine of the frequency $\omega_i = \omega_L + k\Delta\omega$. Therefore this device can be interpreted as a filter matched to the single tone $\omega_i = \omega_L + k\Delta\omega$. For the sake of simplicity we assume noise-free operation at this point. Then the output of the correlators equals

$$X(\omega_i) = \int_0^{T_E} \sqrt{2}\,A \sin(\omega_{i,0}t + \theta_0)\sqrt{2} \cos \omega_i t\, dt$$

$$= AT_E\left\{\frac{\cos\theta_0 - \cos\left[(\omega_{i,0} - \omega_i)T_E + \theta_0\right]}{(\omega_{i,0} - \omega_i)T_E}\right\} \qquad (8.2\text{-}10a)$$

and
$$Z(\omega_1) = AT_E\left\{ \frac{\sin\left[(\omega_{i,0} - \omega_i)T_E + \theta_0\right] - \sin\theta_0}{(\omega_{i0} - \omega_i)T_E} \right\} \tag{8.2-10b}$$

In the calculation of (8.2-10) we have omitted the terms with frequency $(\omega_{i,0} + \omega_i)$. Since $(\omega_{i,0} + \omega_i)T_E$ is very large compared to unity the terms represent an integration over a fast oscillating function which yields approximately zero (the integration following the multiplication, see Figure 8.2-1a, can be viewed as low-pass filtering).

The output of the two correlators still depends on the phase θ_0. By squaring and adding $X(\omega_i)$ and $Z(\omega_i)$ this dependency can be eliminated. Thereby we obtain

$$M^2(\omega_i) = A^2 T_E^2 \left[\frac{\sin\left(\omega_{i,0} - \omega_i\right)T_E/2}{(\omega_{i,0} - \omega_i)T_E/2} \right]^2 \tag{8.2-11}$$

The cross-correlator most closely matched to $\omega_{i,0}$ yields the greatest output. As can be seen from (8.2-11) the frequency response of the filter matched to a single tone is the well-known Dirichlet kernel with its $\sin(x)/x$ behavior [2].

With a spacing of the frequency estimator elements of π/T_E, the received frequency can be no more than $\pi/2T_E$ rad/s from the frequency of some element ω_i. From (8.2-11) it follows that when there is a difference of $\pi/2T_E$, the output of the ith element yields $M^2(\omega_i) = A^2 T_E^2(8/\pi^2)$. Thus with the use of estimator elements spaced π/T_E, we reduce the signal power by $(8/\pi^2)$ at the most, which seems to be tolerable. Thus, $2WT_E + 1$ elements are sufficient for the frequency estimator.

The second interpretation views the estimator as a *frequency analyzer*. The Fourier transform of the signal segment

$$y_T(t) = \begin{cases} y(t), & 0 \le t \le T_E \\ 0, & \text{otherwise} \end{cases} \tag{8.2-12}$$

is given by

$$Y_T(\omega) = \int_0^{T_E} y(t) \exp(-j\omega t)\, dt$$
$$= \int_0^{T_E} y(t) \cos\omega t\, dt - j\int_0^{T_E} y(t) \sin\omega t\, dt \tag{8.2-13}$$

In every element of the frequency detector the real and imaginary parts of the spectrum $Y_T(\omega)$ are computed for $\omega = \omega_i$

$$Y_T(\omega_i) = \frac{1}{\sqrt{2}}[X(\omega_i) - jZ(\omega_i)] \tag{8.2-14}$$

The output $M^2(\omega_i)$ is, up to a constant, nothing else but the power spectral density (PSD)

$$S_{y_T}(\omega_i) = |Y_T(\omega_i)|^2 = \frac{1}{2}[X^2(\omega_i) + Z^2(\omega_i)] = \frac{1}{2}M^2(\omega_i) \quad (8.2\text{-}15)$$

We may therefore view the ML frequency estimator as a device that estimates the PSD from a segment of the received signal $y(t)$. The location of the maximum of the PSD is the ML estimate $\hat{\omega}_i$. Since $M^2(\omega_i)$ is the value of the PSD at location ω_i it can also be considered as the Fourier transform of the autocorrelation function of $y(t)$. The output functions $M^2(\omega_i)$ are related to the correlation function of $y(t)$ through

$$M^2(\omega_i) = 4\int_0^{T_E} R_{y_T}(\tau)\cos\omega_i\tau\, d\tau \quad (8.2\text{-}16)$$

where

$$R_{y_T}(\tau) = \int_0^{T_E - |\tau|} y(t)y(t+\tau)\, dt$$

so that $M^2(\omega_i)$ is proportional to the transform of the finite time autocorrelation function $R_{y_T}(\tau)$ of $y(t)$, the so-called periodogram. The verification of (8.2-16) is left as an exercise for the reader.

8.2.2. Suboptimal Frequency Estimation Methods

The optimal frequency estimator presented above requires the simultaneous computation of $(2WT_E + 1)$ values of the power spectral density of the received signal. The frequency estimate is then ascertained by a maximum selection process. In contrast to such an open-loop structure, this section deals with closed-loop structures which at the expense of a suboptimal performance require far less computational effort.

At this stage a change in notation seems favorable. In order to comply with the notation used in this book for closed-loop estimators we denote the estimate as $\hat{\omega}_i$ and the true (unknown) value as ω_i. This is in contrast to the previous section where ω_i was a trial parameter and $\omega_{i,0}$ was the true value.

A necessary condition to obtain a maximum of $M^2(\hat{\omega}_i)$ is that the derivative of this function with respect to $\hat{\omega}_i$ is zero. Starting with this statement it is possible to derive a closed-loop structure based on the ML principle. A possible error signal that can be used for adjusting the frequency of a local oscillator is then

$$\frac{\partial M^2(\hat{\omega}_i)}{\partial\hat{\omega}_i} \approx \frac{M^2(\hat{\omega}_i + \Delta\omega) - M^2(\hat{\omega}_i - \Delta\omega)}{2\Delta\omega} =: D(\omega_i - \hat{\omega}_i)$$

$$(8.2\text{-}17)$$

with $\Delta\omega = \pi/T_E$, i.e., the difference between the output signals of two devices of Figure 8.2-1 is used as an error signal. Equation (8.2-17) can be further manipulated. The first term yields

$$M^2(\hat{\omega}_i + \Delta\omega) = \left[\int_0^{T_E} y(t)\sqrt{2}\sin(\hat{\omega}_i t + \Delta\omega t)\,dt\right]^2$$

$$+ \left[\int_0^{T_E} y(t)\sqrt{2}\cos(\hat{\omega}_i t + \Delta\omega t)\,dt\right]^2$$

$$= \left[\int_0^{T_E} \tilde{y}_s(t)\cos\Delta\omega t + \tilde{y}_c(t)\sin\Delta\omega t)\,dt\right]^2$$

$$+ \left[\int_0^{T_E} (\tilde{y}_c(t)\cos\Delta\omega t - \tilde{y}_s(t)\sin\Delta\omega t)\,dt\right]^2 \qquad (8.2\text{-}18)$$

with $\tilde{y}_c(t) = y(t)\sqrt{2}\cos\hat{\omega}_i t$ and $\tilde{y}_s(t) = y(t)\sqrt{2}\sin\hat{\omega}_i t$.

According to the second interpretation of the ML estimator, (8.2-18) represents the power spectral density of the quadrature components $\tilde{y}_s(t)$ and $\tilde{y}_c(t)$. These components are obtained as a result of shifting the received signal $y(t)$ to baseband by multiplying it by $\sin\hat{\omega}_i t$ and $\cos\hat{\omega}_i t$, respectively. The difference signal, (8.2-17), is zero whenever the received signal frequency ω_i coincides with the local frequency $\hat{\omega}_i$ (in the absence of noise). If $\omega_i \neq \hat{\omega}_i$, then the error signal $D(\omega_i - \hat{\omega}_i)$, after appropriate filtering, can be used to drive the frequency of the local oscillator towards the incoming frequency ω_i.

An AFC configuration results as is illustrated in Figure 8.2-2. This is known as the *discrete Fourier transform AFC* (DFTAFC) [3], since the frequency discriminator carries out a discrete Fourier transformation (if the circuit is built up digitally as is usual). The low-pass filters (LPF) have the function of suppressing the double frequency terms which arise from mixing.

It can be shown from (8.2-17) that the error signal has the following characteristics

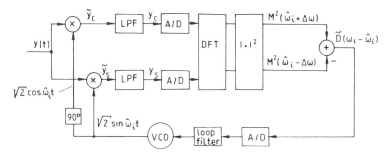

Figure 8.2-2. DFTAFC configuration.

$$D(\omega_i - \hat{\omega}_i) = \frac{8A^2 T_E^4}{\pi^4} \frac{(\omega_i - \hat{\omega}_i) \cos^2 [T_E(\omega_i - \hat{\omega}_i)/2]}{\{[T_E(\omega_i - \hat{\omega}_i)/\pi] + 1\}^2 \{[T_E(\omega_i - \hat{\omega}_i)/2] - 1\}^2}$$

(8.2-19)

The error signal is presented in Figure 8.2-3.

In the closed-loop structure of Figure 8.2-2 the values of the power spectral density are only evaluated for two frequencies. The advantage of the less complex hardware must be traded off against the disadvantage that the frequency difference between the received signal and the local oscillator must not be too large to ensure that the error signal produced by the frequency discriminator is still proportional to the frequency difference (see Figure 8.2-3). If the frequency difference is substantially larger than $\Delta\omega = \pi/T_E$ additional hardware is needed to bring $\Delta\omega$ into the tracking range of the AFC loop. (Compare with similar observations for the PLL in Chapter 4.)

So far we have discussed structures that use the maximum likelihood principle to obtain an estimate for the frequency of a noisy input signal. Another method is to first determine the optimal estimate for the phase angle and then obtain a frequency estimate through differentiation. This leads directly to the well-known balanced quadricorrelator. For this purpose we first assume that in (8.2-1) the frequency ω_i is now known and the carrier phase θ is the parameter we are looking for.

By making an analogous derivation as in Section 8.2.1, the structure shown in Figure 8.2-4 is obtained to produce the optimal phase estimate (a detailed derivation is given in Appendix 8.2B). This structure is now to be extended in order to estimate a frequency. Fot this purpose we replace the integrate and dump circuits in Fig. 8.2-4 by low-pass filters. The LPFs have the sole function of suppressing the double frequency terms. Equation (8.2-1) can be written as

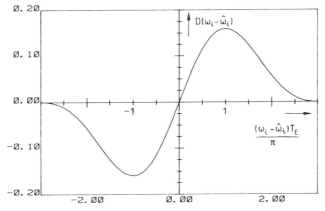

Figure 8.2-3. Discriminator curve of DFTAFC.

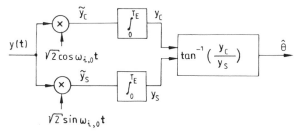

Figure 8.2-4. Optimal phase estimator.

$$y(t) = \sqrt{2}A \sin\left[\hat{\omega}_1 t + (\omega_i - \hat{\omega}_i)t + \theta\right] + n(t)$$

$$= \sqrt{2}A \sin\left[\hat{\omega}_1 t + \theta(t)\right] + n(t) \tag{8.2-20}$$

where $\hat{\omega}_i$ is the known local frequency. The deviation $\omega_i - \hat{\omega}_i$ from this frequency can be represented together with the unknown carrier phase by a time variant phase $\theta(t)$.

The output of the LPF equals (noise free case)

$$y_c(t) = A \sin \theta(t)$$

$$y_s(t) = A \cos \theta(t) \tag{8.2-21}$$

Since

$$\omega_i - \hat{\omega}_i = \frac{d\theta(t)}{dt} \tag{8.2-22}$$

we arrive at an estimate for the difference between the received signal frequency and the local frequency of the VCO through differentiating the phase estimate. We have

$$\frac{d}{dt} \tan^{-1}\left(\frac{y_c}{y_s}\right) = \frac{\dot{y}_c y_s - y_c \dot{y}_s}{y_s^2 + y_c^2} = (\omega_i - \hat{\omega}_i) \tag{8.2-23}$$

Thereby we arrive at a further estimator structure called the *differentiator AFC*. The discriminator, i.e., the part that generates the error signal $D(\omega_i - \hat{\omega}_i)$, is known as the *balanced quadricorrelator* (Figure 8.2-5).

The error signal $D(\omega_i - \hat{\omega}_i)$ is directly proportional to the frequency error. The dependency is no longer linear when the frequency difference becomes so large that the influence of the low-pass filters becomes apparent.

In the discrete-time domain we can find an analogy to the differentiator AFC (see Figure 8.2-6) which is easily derived by replacing the differential quotient with the difference quotient; $\dot{y}_c$ becomes, for example

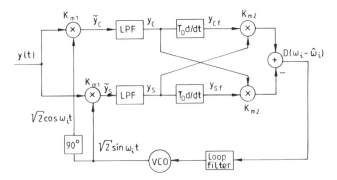

Figure 8.2-5. Differentiator automatic frequency control.

$$\dot{y}_c(t + k\,\Delta T) \approx \frac{y_c(t = k\,\Delta T) - y_c(t = (k-1)\,\Delta T)}{\Delta T} \qquad (8.2\text{-}24)$$

which results in

$$\dot{y}_c y_s - \dot{y}_s y_c \approx \frac{1}{\Delta T}\left[y_c(t_k)y_s(t_{k-1}) - y_s(t_k)y_c(t_{k-1}) \right] \qquad (8.2\text{-}25)$$

$$= \text{constant} \times \sin\left[(\theta(t_k) - \theta(t_{k-1})\right]$$

$$\approx \text{constant} \times \sin\left[(\omega_i - \hat{\omega}_i)\,\Delta T\right]$$

with $t_k = k\,\Delta T$. For $(\omega_i - \hat{\omega}_i)\,\Delta T \ll 1$, the discriminator output error signal $D(\omega_i - \hat{\omega}_i)$ is proportional to the difference between the received signal and the local oscillator frequency.

Coming back to the continuous-time case we note that the ideal differentiators, which are physically unrealizable, may be replaced by realizable filters. As will be seen, it is possible to use a wide class of filters.

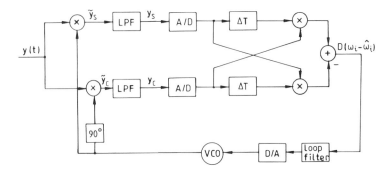

Figure 8.2-6. Discrete time equivalent of differentiator automatic frequency control.

If $y_c(t)$ is filtered by a filter having the transfer function $H(\omega)$ then the filtered signal results in $(n(t) = 0)$

$$y_{cf}(t) = AK_{m1}|H(\omega_i - \hat{\omega}_i)| \sin\left[(\omega_i - \hat{\omega}_i)t + \theta + \theta_H\right] \qquad (8.2\text{-}26a)$$

and, respectively

$$y_{sf}(t) = AK_{m1}|H(\omega_i - \hat{\omega}_i)| \cos\left[(\omega_i - \hat{\omega}_i)t + \theta + \theta_H\right] \qquad (8.2\text{-}26b)$$

where $\theta_H = \arg\{H(\omega_i - \hat{\omega}_i)\}$ and K_{m1} is the multiplier gain. Thus the error signal is given by

$$\begin{aligned} D(\omega_i - \hat{\omega}_i) &= A^2 K_{m1}^2 K_{m2}|H(\omega_i - \hat{\omega}_1)|\sin\theta_H \\ &= A^2 K_{m1}^2 K_{m2} \operatorname{Im}\{H(\omega_i - \hat{\omega}_i)\} \qquad (8.2\text{-}27) \end{aligned}$$

Since for real impulse responses the symmetry of $\operatorname{Im}\{H(\omega_i - \hat{\omega}_i)\}$ is always odd, *any* real filter produces the required odd characteristic of the error signal provided the bandwidth of $H(\omega)$ is considerably larger than the bandwidth of the AFC Circuit; i.e., the discriminator filter $H(\omega)$ must be allowed to settle at the new frequency before the frequency changes once again. This condition is called the "quasistatic assumption" [4]. Notice that the ideal differentiator is the only device which yields a linear error characteristic for all frequency errors since $\operatorname{Im}\{H(\omega_i - \hat{\omega}_i)\} = T_D(\omega_i - \hat{\omega}_i)$.

One of the multipliers of the balanced quadricorrelator may be removed. The resulting structure is known as the *unbalanced quadricorrelator* and is shown in Figure 8.2-7. Only one arm of the discriminator is differentiated. However, no ideal differentiator is needed since only a 90° phase-shift between the filtered quadrature signals is required ("relative differentiation"). This can be accomplished, for example, by the following filter pair

$$H_1(\omega) = \frac{sT_2}{1 + sT_1} \qquad (8.2\text{-}28a)$$

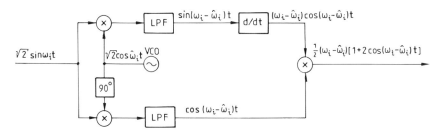

Figure 8.2-7. Unbalanced quadricorrelator.

$$H_2(\omega) = \frac{1}{1 + sT_1} \qquad (8.2\text{-}28b)$$

From Figure 8.2-7 we learn that saving a multiplier is possible at the expense of a ripple at double the frequency error.

If we have a phase or frequency modulated signal some modifications of the basic AFC structures may be necessary. We refer the reader to references [3] and [5].

Main Points of the Section

- An AFC circuit is used for frequency control in noncoherent transmission systems and as acquisition aid when coherent demodulation is employed.
- There are two basic types of frequency discriminators. The first is based on the maximum likelihood estimator of the signal frequency and the second calculates, in essence, the derivative with respect to time of the maximum likelihood estimate of the phase of the input signal.
- There are open-loop (ML estimator of signal frequency) and closed-loop tracking structures (DFTAFC, differentiator AFC).

8.3. PERFORMANCE IN THE PRESENCE OF ADDITIVE NOISE

This section analyzes the behavior of the AFC loop in the presence of additive noise. To limit the workload we will restrict the analysis to the frequently used AFC which has a balanced quadricorrelator as discriminator. The reader who is interested in more details about other AFC loops is referred to Natali [3, 5], who has studied the noise performance of some of the other AFC loops. We will first determine the noisy discriminator output and discuss some important aspects. Then we will close the loop and discuss the loop performance in terms of the variance of the frequency error.

The input to the frequency detector is assumed to be a pure sine wave plus gaussian noise (8.2-1). Without loss of generality we can set the unknown constant phase to $\theta_0 = 0$. If we further assume *bandpass* gaussian noise (see Chapter 3), the noise component can be expressed in terms of its low-pass equivalent

$$n(t) = \sqrt{2}\,n_c(t)\cos(\omega_i t) - \sqrt{2}\,n_s(t)\sin(\omega_i t) \qquad (8.3\text{-}1)$$

with

$$P_n = E[n^2(t)] = 2E[n_c^2(t)] = 2E[n_s^2(t)] \qquad (8.3\text{-}2)$$

Thereby we find (Figure 8.2-5)

$$y_s(t) = K_{m1}[A \cos(\omega_i - \hat{\omega}_i)t - n_c(t) \sin(\omega_i - \hat{\omega}_i)t - n_s(t) \cos(\omega_i - \hat{\omega}_i)t]$$

$$(8.3\text{-}3)$$

$$y_c(t) = K_{m1}[A \sin(\omega_i - \hat{\omega}_i)t + n_c(t) \cos(\omega_i - \hat{\omega}_i)t - n_s(t) \sin(\omega_i - \hat{\omega}_i)t]$$

Using the complex notation

$$\underline{y} = y_c + jy_s, \quad \underline{y}_f = y_{cf} + jy_{sf}, \quad \underline{n}(t) = n_c(t) + jn_s(t) \qquad (8.3\text{-}4)$$

the detector output becomes

$$D(\omega_i - \hat{\omega}_i) = e(t) = K_{m2} \operatorname{Im}\{\underline{y}(t)\underline{y}_f^*(t)\} = K_{m2} \operatorname{Im}\left\{\int_{-\infty}^{\infty} h(t-\tau)\underline{y}(t)\underline{y}^*(\tau)\, d\tau\right\}$$

$$(8.3\text{-}5)$$

where $h(t)$ is the impulse response of the filter with input y_c and y_s, respectively. (In Figure 8.2-5 the filter is an ideal differentiator.)

Using (8.3-3) in (8.3-5) yields the following result for the noisy discriminator output (after some lengthy algebraic manipulation)

$$e(t) = A^2 K_{m1}^2 K_{m2} \operatorname{Im}\left\{\int_{-\infty}^{\infty} h(t-\tau) \exp[-j(\omega_i - \hat{\omega}_i)(t-\tau)]\, d\tau\right\}$$

$$+ A K_{m1}^2 K_{m2} \operatorname{Im}\left\{j \int_{-\infty}^{\infty} h(t-\tau) \exp[-j(\omega_i - \hat{\omega}_i)(t-\tau)][\underline{n}(\tau) - \underline{n}^*(t)]\, d\tau\right\}$$

$$+ K_{m1}^2 K_{m2} \operatorname{Im}\left\{\int_{-\infty}^{\infty} h(t-\tau) \exp[-j(\omega_i - \hat{\omega}_i)(t-\tau)]\underline{n}^*(t)\underline{n}(\tau)\, d\tau\right\}$$

$$= e_0(t) + n_1(t) + n_2(t) \qquad (8.3\text{-}6)$$

$e_0(t)$ is the useful control signal which is the only term remaining in the noise-free case. This can easily be seen since the integral represents the Fourier transform of the impulse response $h(t)$ at the location $(\omega_i - \hat{\omega}_i)$ which is the well-known result of (8.2-27).

As a result of the nonlinear operations in the discriminator two additional noise terms arise: $n_1(t)$ comprises the product terms of the input signal and input noise ($s \times n$ term, signal × noise); $n_2(t)$ results from the product of the input noise with itself ($n \times n$ term, noise × noise).

The control signal is the dc-component of the discriminator output $e(t)$ or mathematically speaking the expected value. Taking the expected value of (8.3-6) now yields not only $e_0(t)$ but also a contribution from $n_2(t)$

$$E[n_2(t)] = K_{m1}^2 K_{m2} \operatorname{Im}\left\{\int_{-\infty}^{\infty} h(t-\tau) \exp[-j(\omega_i - \hat{\omega}_i)(t-\tau)]\right.$$

$$\left. \times E[\underline{n}^*(t)\underline{n}(\tau)]\, d\tau\right\} \neq 0 \qquad (8.3\text{-}7)$$

In (8.3-7) we make use of the autocorrelation function of the complex noise process $\underline{n}(t)$. The results on narrowband noise in Chapter 3 (3.1-6) lead to

$$R_{\underline{n}}(t - \tau) = E[\underline{n}^*(t)\underline{n}(\tau)] = 2R_{n_c}(t - \tau) + j2R_{n_c n_s}(t - \tau) \qquad (8.3\text{-}8)$$

When using this in (8.3-7)

$$E[n_2(t)] = K_{m1}^2 K_{m2} \text{ Im} \left\{ \int_{-\infty}^{\infty} h(u)R_{\underline{n}}(u) \exp\left[-j(\omega_i - \hat{\omega}_i)u\right] du \right\} \qquad (8.3\text{-}9)$$

we can recognize that (8.3-9) represents the imaginary part of the Fourier transform of $h(t)R_{\underline{n}}(t)$ at the location $(\omega_i - \hat{\omega}_i)$

$$E[n_2(t)] = K_{m1}^2 K_{m2} \text{ Im} \left\{ S_{\underline{n}}(\omega) * H(\omega) \big|_{\omega = \omega_i - \hat{\omega}_i} \right\} \qquad (8.3\text{-}10)$$

$S_{\underline{n}}(\omega)$ here represents the Fourier transform of $R_{\underline{n}}(\tau)$. As can easily be concluded from (8.3-8), $S_{\underline{n}}(\omega)$ is a real function.

We will now illustrate the result by the following example. Using ideal differentiators (see Figure 8.2-5) as discriminator filters

$$H(\omega) = T_D j\omega \qquad (8.3\text{-}11)$$

we find for (8.3-10)

$$E[n_2(t)] = K_{m1}^2 K_{m2} T_D \text{ Im} \left\{ j \int_{-\infty}^{+\infty} S_{\underline{n}}(u)(\omega_i - \hat{\omega}_i - u) du \right\}$$

$$= K_{m1}^2 K_{m2} T_D \left[(\omega_i - \hat{\omega}_i) \int_{-\infty}^{+\infty} S_{\underline{n}}(\omega) d\omega - \int_{-\infty}^{+\infty} \omega S_{\underline{n}}(\omega) d\omega \right] \qquad (8.3\text{-}12)$$

The first integral equals the power of the noise, P_n. Since the integral is multiplied by the frequency error signal $(\omega_i - \hat{\omega}_i)$ it looks as if noise increases the useful signal!

At first glance this is totally against intuition. Before solving this apparent paradox we first want to discuss the second integral which is independent of the error signal $(\omega_i - \hat{\omega}_i)$. It therefore generates a *bias* in the frequency detector output. This has important consequences. Suppose the frequency detector (FD) is used as an acquisition aid for a narrowband PLL (see Chapter 5). Due to the bias of the FD, the PLL frequency is shifted to an incorrect value and there exists the possibility that the PLL will never be brought into its pull-in range. (Remember that a bias caused by additive noise is *not* present in a PLL.)

The second integral can be interpreted as center of gravity of the noise spectrum at the input of the FD. From Chapter 3 we know that the Fourier transform of $R_{\underline{n}}(\tau)$ of (8.3-8) equals

$$S_n(\omega) = 2S_{n_c}(\omega) + j2S_{n_c n_s}(\omega) \tag{8.3-13}$$

Since

$$S_{n_c}(-\omega) = S_{n_c}(\omega) \tag{8.3-14a}$$

$$S_{n_c n_s}(-\omega) = -S_{n_c n_s}(\omega) \tag{8.3-14b}$$

the second integral in (8.3-12) vanishes only if $S_{n_c n_s} \equiv 0$ for all ω. And since (3.1-8)

$$S_n(\omega) = [S_{n_c}(\omega - \hat{\omega}_i) + S_{n_c}(\omega + \hat{\omega}_i)] - j[S_{n_c n_s}(\omega - \hat{\omega}_i) - S_{n_c n_s}(\omega + \hat{\omega}_i)] \tag{8.3-15}$$

(see Fig. 8.3-1) this requires a symmetrical power spectral density $S_n(\omega)$ with center frequency equal to the local reference frequency $\hat{\omega}_i$. Thus a bias will be generated if the center of gravity of the noise spectrum does not coincide with $\hat{\omega}_i$.

The possibility of a bias in frequency tracking loops was first recognized by Gardner [6]. He also showed how this bias can be avoided which we will explain here briefly. In many receivers the noise spectrum is shaped by passing wideband noise through bandpass filters in intermediate-frequency portions of the receiver.

There are two ways to close the frequency tracking loop, a short loop connection or a long loop. These options are illustrated in Figure 8.3-2. The bandpass filter (BPF) in the IF portion of the receiver has a fixed characteristic. If the frequency $\hat{\omega}_i$ is allowed to vary, as is necessary in a short loop, then the center of gravity of the noise spectrum cannot coincide with $\hat{\omega}_i$

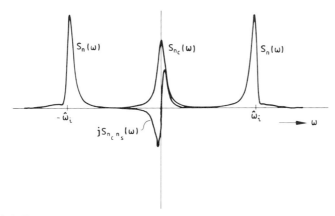

Figure 8.3-1. Example of the decomposition of the spectrum of $S_n(\omega)$ of the noise process $n(t)$ into two components.

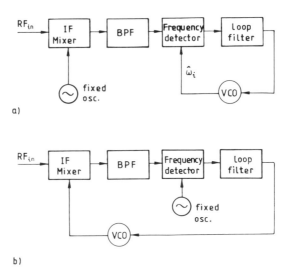

a)

b)

Figure 8.3-2. Frequency tracking loops. (*a*) Short loop; (*b*) long loop.

except by rare accident. Therefore, in general, a noise bias must be anticipated.

By contrast, in a long loop, both $\hat{\omega}_i$ and the filter center-frequency are fixed. Tracking is accomplished by controlling the frequency of an oscillator that precedes the reference source. If the filter and reference source are properly aligned and stable, then the center of gravity of the noise spectrum will always coincide with the reference frequency and no noise bias will be generated.

Superimposed on the useful signal is a random disturbance as can be seen in (8.3-6). The statistical properties of the disturbance determine the tracking properties of the closed frequency-locked loop. Our next step is to analyze the properties of this disturbance by means of the autocorrelation function and the power spectral density. To keep this to a minimum we will restrict our observation to the case where ideal differentiators (8.3-11) are used as discriminator filters. Using (8.3-11) in (8.3-6) we obtain

$$e(t) = K_f \left\{ (\omega_i - \hat{\omega}_i) \left[1 - 2\frac{n_s(t)}{A} + \left(\frac{n_c(t)}{A}\right)^2 + \left(\frac{n_s(t)}{A}\right)^2 \right] \right.$$

$$\left. + \frac{\dot{n}_c(t)}{A} + \frac{n_c(t)\dot{n}_s(t)}{A^2} - \frac{\dot{n}_c(t)n_s(t)}{A^2} \right\}$$

$$= e_0(t) + n'(t, \omega_i - \hat{\omega}_i) + m(t) \qquad (8.3\text{-}16)$$

with

$$K_f = K_{m1}^2 K_{m2} T_D A^2$$

We have grouped the noise terms in the last equation into a group that depends on the frequency error and a group that is independent of $(\omega_i - \hat{\omega}_i)$. Before we compute the power spectrum we subtract the expected value offset

$$E[e(t)] = K_f(\omega_i - \hat{\omega}_i)\left\{1 + E\left[\left(\frac{n_c(t)}{A}\right)^2 + \left(\frac{n_s(t)}{A}\right)^2\right]\right.$$
$$\left. + E\left[\frac{n_c(t)\dot{n}_s(t)}{A^2}\right] - E\left[\frac{\dot{n}_c(t)n_s(t)}{A^2}\right]\right\} \tag{8.3-17}$$

If the noise has a symmetrical power spectrum about the reference frequency $\hat{\omega}_i$ the cross-correlation $R_{n_c n_s}(\tau)$ is zero for all τ. Using the well-known result

$$E[n_c(t)\dot{n}_s(t)] = -\frac{d}{d\tau} R_{n_c n_s}(\tau)\big|_{\tau=0}$$

and

$$E[\dot{n}_c(t)n_s(t)] = \frac{d}{d\tau} R_{n_c n_s}(\tau)\big|_{\tau=0} \tag{8.3-18}$$

we can see that the process $m(t)$ has zero mean. For a nonsymmetrical spectrum the expected value of $m(t)$

$$E[m(t)] = -2 \frac{K_f}{A^2} \frac{d}{d\tau} R_{n_c n_s}(\tau) \tag{8.3-19}$$

equals a bias in the frequency estimate, as discussed previously. Returning to the case of a symmetric power spectrum we obtain

$$E[e(t)] = K_f(\omega_i - \hat{\omega}_i)(1 + \text{SNR}_i^{-1}) \tag{8.3-20}$$

where

$$\text{SNR}_i = \frac{P_s}{P_n} = \frac{A^2}{P_n}$$

equals the input signal-to-noise ratio. Equation (8.3-20) shows that the expected value increases with increasing noise, a paradox we pointed out earlier. In order to resolve this seeming paradox and to analyze the statistical properties of the frequency error we will have to calculate the autocorrelation function of the error signal $e(t)$.

Some portions of the noise in (8.3-16) depend on the frequency difference. When the statistical properties of the closed loop are calculated this dependency complicates the analysis since advanced mathematical procedures (Fokker–Planck techniques) are necessary. In the tracking mode the frequency error should be very small so that the part of the noise $n'(t, \omega_i - \hat{\omega}_i)$ which is proportional to the error signal may be neglected. (We will return to this point at a later stage.) Our aim now is to calculate the autocorrelation function of the remaining noise term denoted by $m(t)$.

The determination of the autocorrelation function of the last three terms in (8.3-16) requires knowledge of third- and fourth-order moments of gaussian random variables (see Appendix 8.3A)

$$R_m(\tau) = K_f^2 E \left\{ \left[\frac{\dot{n}_c(t+\tau)}{A} + \frac{n_c(t+\tau)\dot{n}_s(t+\tau)}{A^2} - \frac{\dot{n}_c(t+\tau)n_s(t+\tau)}{A^2} \right] \right.$$
$$\left. \times \left[\frac{\dot{n}_c(t)}{A} + \frac{n_c(t)\dot{n}_s(t)}{A^2} - \frac{\dot{n}_c(t)n_s(t)}{A^2} \right] \right\} \tag{8.3-21}$$

After some lengthy, but elementary algebraic manipulations, one finds the spectrum $S_m(\omega)$ assuming a rectangular spectrum $S_n(\omega)$ with center frequency $\hat{\omega}_i$ and bandwidth B_{IF} for $n(t)$ (see Figure 8.3-3)

$$S_m(\omega) = S_{m_1}(\omega) + S_{m_2}(\omega) \tag{8.3-22}$$

with

$$S_{m_1}(\omega) = \begin{cases} \dfrac{(K_f \omega)^2}{B_{IF}} \, \mathrm{SNR}_i^{-1}, & \left| \dfrac{\omega}{2\pi} \right| < \dfrac{B_{IF}}{2} \\ 0, & \text{otherwise} \end{cases} \tag{8.3-23a}$$

$$S_{m_2}(\omega) = \begin{cases} \dfrac{8\pi^2}{3} \dfrac{(K_f B_{IF})^2}{B_{IF}} \left(1 - \dfrac{|\omega/2\pi|}{B_{IF}} \right)^3 \mathrm{SNR}_i^{-2}, & \left| \dfrac{\omega}{2\pi} \right| < B_{IF} \\ 0, & \text{otherwise} \end{cases}$$
$$\tag{8.3-23b}$$

The first term $S_{m_1}(\omega)$ is due to the multiplication of the (signal × noise), it rises proportionally to ω^2. The second term is due to the (noise × noise) multiplication. It is maximum for $\omega = 0$ (see Figure 8.3-3).

With the help of (8.3-23) we want to show that, inspite of what (8.3-20) shows, increasing noise actually decreases the loop performance. To demonstrate this we simply have to show that the ratio of the expected value $E^2[e(t)]$ of the useful signal to the variance $\mathrm{Var}[m(t)]$ of the noise actually increases with increasing SNR_i. After some simple algebraic manipulations one finds

$$\frac{E[e^2(t)]}{\mathrm{Var}[m(t)]} = \frac{K_f^2(\omega_i - \hat{\omega}_i)^2 [1 + \mathrm{SNR}_i]^2}{\dfrac{(2\pi)^2}{3}(K_f B_{IF})^2 [1 + \mathrm{SNR}_i/4]} \tag{8.3-24}$$

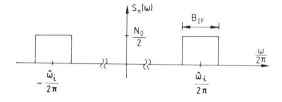

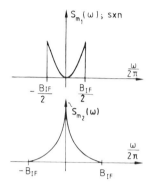

Figure 8.3-3. Noise spectra of input noise and discriminator output noise. (a) Spectrum of input noise; (b) spectrum of discriminator output noise.

After calculating the statistical properties of the error signal we will now close the loop in order to calculate the variance of noise-induced frequency fluctuations which is the value of particular interest. Using the same arguments as in Chapter 3.2 we can state: If the noise bandwidth B_{IF} is much larger than the bandwidth of the AFC loop B_L (which is of course generally the case) then the closed-loop statistics of $m(t)$ can be approximated by the statistics obtained under open-loop conditions.

The linear model of the AFC in the presence of additive noise is shown in Figure 8.3-4. An integrator is in general sufficient for the loop filter $H_f(s)$ since an initial frequency offset of the loop input will be completely eliminated then.

With $H_f(s) = K_I/s$ we obtain the following state equation describing the loop

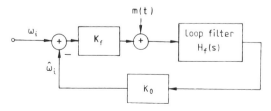

Figure 8.3-4. Linearized model of the frequency control loop in the presence of noise.

$$\frac{d(\omega_i - \hat{\omega}_i)}{dt} = -K_0 K_I K_f (\omega_i - \hat{\omega}_i) - K_0 K_I m(t) \qquad (8.3\text{-}25)$$

with the state variable $(\omega_i - \hat{\omega}_i)$.

The transfer function of the closed loop is thus $(m(t) = 0)$

$$H(s) = \frac{\hat{\omega}_i(s)}{\omega_i(s)} = \frac{1}{1 + sT_L} \qquad (8.3\text{-}26)$$

where $T_L = 1/(K_0 K_I K_f) = 1/(4B_L)$ and B_L is the loop bandwidth (cf. Chapter 2 for definition of B_L). $\hat{\omega}_i(s)$ and $\omega_i(s)$ denote the Laplace transforms of $\hat{\omega}_i$ and ω_i, respectively. The variance of the noise-induced frequency fluctuations of the VCO is given by (see Figure 8.3-4).

$$\sigma^2_{\Delta\omega} = \frac{1}{2\pi} \int_{-\infty}^{\infty} \frac{|H(\omega)|^2}{K_f^2} S_m(\omega)\, d\omega \qquad (8.3\text{-}27)$$

Integrating (8.3-27) with $(B_{IF}/B_L) > 15$ yields

$$\frac{\sigma^2_{\Delta f}}{B_L^2} \approx \underbrace{\frac{4}{\pi^2} \mathrm{SNR}_i^{-1}}_{s \times n} + \underbrace{\frac{1}{3} \mathrm{SNR}_i^{-2}\left(\frac{B_{IF}}{B_L}\right)}_{n \times n} \qquad (8.3\text{-}28)$$

The normalized variance is plotted in Figure 8.3-5 as a function of SNR_i. For a bandwidth ratio of $(B_{IF}/B_L) = 40$, the $(n \times n)$ dominates the $(s \times n)$ contribution up to high input signal-to-noise ratios. For a larger ratio (B_{IF}/B_L), the dominance extends even to higher SNR_is. We thus conclude

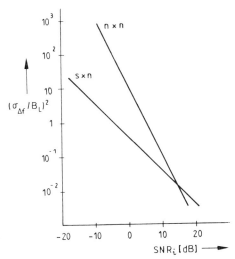

Figure 8.3-5. Numerical example of frequency variance $(B_{IF}/B_L = 40)$.

that the $(s \times n)$ contribution is negligible for all applications where B_{IF}/B_L is large.

This result could already be predicted from Figure 8.3-3. The $(n \times n)$ portion of the noise reaches its maximum at $\omega = 0$ and therefore even for smaller loop bandwidths it makes a considerable contribution, whereas the $(s \times n)$ portion vanishes at $\omega = 0$. Since the $(s \times n)$ component results through multiplication of signal with noise it accordingly becomes stronger with increasing SNR_i, in contrast to the $(n \times n)$ component.

When we divide both sides of (8.3-28) by $(B_{IF}/B_L)^2$

$$\frac{\sigma_{\Delta f}^2}{B_{IF}^2} \approx \frac{4}{\pi^2} \, \text{SNR}_i^{-1} \left\{ \frac{B_L}{B_{IF}} \right\}^2 + \frac{1}{3} \, \text{SNR}_i^{-2} \left(\frac{B_L}{B_{IF}} \right)$$

another important fact becomes more evident. $\sigma_{\Delta f}$ can be made arbitrarily small by choosing B_L sufficiently narrow (in practice there is always a lower limit on B_L because otherwise the loop might be too slow to follow signal fluctuations and initial acquisition might take too long). The ac-component $m(t)$ is effectively filtered out by the loop while the useful control signal, which is the dc-component of the frequency detector, is passed. Therefore this frequency locked loop can also operate at low SNR_i. When the useful control signal consists of the whole detector output, such as in a frequency modulated signal, the input SNR_i must of course be much higher.

Standard discriminators include limiters so as to gain full FM-noise advantage in presence of inevitable AM noise. But a limiter (or otherwise an ideal FM discriminator) causes a sharp threshold effect, which is not present in a quadricorrelator when the limiter is omitted. Performance is likely to deteriorate if limiters were included.

We now return to the problem of the frequency error dependent noise contributions in (8.3-16). Only as a result of an omission of this term are we allowed to use linear analysis to calculate the variance of the frequency error. If one takes the frequency error dependency into account, one is forced to use the theory of nonlinear stochastic differential equations. This advanced mathematical tool is discussed in Chapters 9 to 12. Calculations carried out by Kehren [7] show, however, that omission of the frequency dependent noise term in (8.3-16) leads to a negligible error for the SNR_is of practical interest. Therefore, we do not further elaborate on this topic.

Main Points of the Section

- The differentiator AFC tracks the center of gravity of the input spectrum. Thus a bias is produced if the center of gravity does not coincide with the reference frequency $\hat{\omega}_i$.
- A noise induced bias can be avoided with a "long loop" configuration.
- The error variance of the frequency estimate consists of a signal × noise $(s \times n)$ and a noise × noise $(n \times n)$ induced term. For most applications the $(n \times n)$ dominates the $(s \times n)$ contribution.

REFERENCES

1. M. Abramovitz and I. A. Stegun, *Handbook of Mathematical Functions*, Dover Publications, New York, 1965.

2. F. J. Harris, The Discrete Fourier Transform Applied to Time Domain Signal Processing, *IEEE Communications Magazine*, Vol. 20, No. 5, pp. 13–22 May 1982.

3. F. D. Natali, AFC Tracking Algorithms, *IEEE Transactions on Communications*, Vol. 32, No. 8, pp. 935–947, August 1984.

4. R. W. D. Booth, A Note on the Design of Baseband AFC Discriminators, Conference Record NTC 1980, Vol. 2, paper 24.2.

5. F. D. Natali, Noise Performance of a Cross-Product AFC with Decision Feedback for DPSK Signals, *IEEE Transactions on Communications*, Vol. 34, No. 3, pp. 303–307, March 1986.

6. F. M. Gardner, Properties of Frequency Difference Detectors, IEEE Transactions on Communications, Vol. 33, No. 2, pp. 131–138, February 1985.

7. H. Kehren, A Study of the Properties of Automatic Frequency Control Loops (in German), Internal Report, Department of Electrical Engineering, University of Technology, Aachen, February 1985.

8. H. L. Van Trees, *Detection, Estimation, and Modulation Theory*, Part I, Wiley, New York, 1968.

APPENDIX 8.2A
MAXIMUM LIKELIHOOD (ML) PARAMETER ESTIMATION

We start with the definition of the problem for the simplest case. Let there be a nonrandom unknown parameter c which may assume any value in an interval C. The parameter cannot be determined error-free by an observer since only a noisy measurement y is available. This noisy measurement (observation) is mathematically described by a probabilistic mapping $p(y \mid c)$ of the parameter c into the observation space Y. An *estimator* is then a deterministic rule $d(y)$ which assigns an *estimate* $\hat{c}$ to any measurement y (see Figure 8.2A-1). For the ML-estimator this rule is developed as follows.

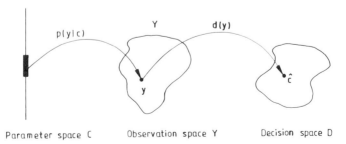

Figure 8.2A-1. Illustration of the mapping procedure.

The unknown parameter c is viewed as a trial parameter. Then $p(y \mid c)$ is called the *likelihood function**. The ML estimate $\hat{c}$ is that value of the trial parameter for which $p(y \mid c)$ assumes its greatest value

$$p(y \mid \hat{c}) > p(y \mid c), \quad \text{for all } c \in C \tag{8.2A-1}$$

In other words the ML estimate $\hat{c}$ is that value of the parameter which most likely has produced the observation y.

Notation. To avoid confusion we summarize the following notational conventions

c Trial parameter $c \in C$
$\hat{c}$ ML-estimate
c_0 Unknown, *true* value of the parameter

The estimation error $e = c_0 - \hat{c}$ is the difference between the (unknown) true value c_0 of the parameter c and the ML-estimate $\hat{c}$. To distinguish the true value of the parameter from the trial value the true value carries the index zero.

The ML principle is readily extended to the estimation of a set of k unknown parameters by replacing the parameter c by a k-tuple $c = \{c_1, \ldots, c_k\}$. Assume we want to estimate the parameter(s) c of $x(c)$ from a measurement y corrupted by additive gaussian noise n of variance σ^2

$$y = x(c) + n \tag{8.2A-2}$$

To find the ML estimate(s) $\hat{c}$ we have to maximize

$$p_y(y \mid c) = p_n(y - x(c) \mid c) = p_n(y - x(c))$$

$$= \frac{1}{\sqrt{(2\pi\sigma^2)}} \exp\left[-\frac{(y - x(c))^2}{2\sigma^2}\right] \tag{8.2A-3}$$

But maximizing (8.2A-3) is equivalent to maximizing the so-called log-likelihood function

$$\lambda(y \mid c) = -(y - x(c))^2 \tag{8.2A-4}$$

Now assume that in addition $x(c, \psi)$ is a function of the random variable ψ with probability density function $p(\psi)$

---

*Strictly speaking $p(y \mid c)$ is not a conditional probability since the parameter c is nonrandom. However, it is customary to write $p(y \mid c)$ as if it were a conditional probability density function.

$$y = x(c, \psi) + n \tag{8.2A-5}$$

Then, applying Bayes' rule yields

$$p_y(y \mid c) = \int_{-\infty}^{+\infty} p(y, \psi \mid c) \, d\psi = \int_{-\infty}^{+\infty} p(y \mid c, \psi) p(\psi) \, d\psi$$

From (8.2A-3) we find that the ML estimate(s) $\hat{c}$ are found by maximizing the likelihood function

$$\Lambda(y \mid c) = \int_{-\infty}^{+\infty} \exp\left[-\frac{(y - x(c, \psi))^2}{2\sigma^2} \right] p(\psi) \, d\psi \tag{8.2A-6}$$

Now we want to apply the ML principle to the estimation of the unknown parameter(s) c of a signal $x(t, c)$ in the interval $0 \le t \le T_E$ from a received signal $y(t)$ corrupted by *white* gaussian noise $n(t)$ with double-sided power spectral density $N_0/2$

$$y(t) = x(t, c) + n(t), \quad 0 \le t \le T_E \tag{8.2A-7}$$

Since our observation space is a time-continuous random waveform the probability $\Pr[y(t), t \in (0, T_E) \mid c]$ is not defined. The problem can be solved by the concept of "sufficient statistic" [8]. Basically, a sufficient statistic is a set of random variables (possibly a countably infinite set) that makes complete use of the information on $x(t, c)$ in $y(t)$. The ML estimate is then found by maximizing a function defined by this sufficient statistics. In the present case $\hat{c}$ is found by maximizing

$$\Lambda(y \mid c) = \exp\left\{ -\frac{1}{N_0} \int_0^{T_E} [y(t) - x(t, c)]^2 \, dt \right\} \tag{8.2A-8}$$

or equivalently by maximizing

$$\lambda(y \mid c) = -\int_0^{T_E} [y(t) - x(t, c)]^2 \, dt \tag{8.2A-9}$$

For more details the reader is referred to van Trees [8].

Finally, consider as in (8.2A-5) that $x(t, c, \psi)$ also depends on the random variable ψ. Then, it can be shown that the ML estimate(s) $\hat{c}$ are found by maximizing

$$\Lambda(y \mid c) = \int_{-\infty}^{+\infty} \exp\left\{ -\frac{1}{N_0} \int_0^{T_E} [y(t) - x(t, c, \psi)]^2 \, dt \right\} p(\psi) \, d\psi \tag{8.2A-10}$$

Note that the estimation problem is extended to the case of a m-tuple of random variables $\psi = \{\psi_1, \ldots, \psi_m\}$ in the same way as given with (8.2A-6).

APPENDIX 8.2B
OPTIMAL PHASE ESTIMATOR

Suppose we are given the noisy received signal

$$y(t) = \sqrt{2}A \sin(\omega_{i,0}t + \theta_0) + n(t) \qquad (8.2B\text{-}1)$$

in the time interval $0 \le t \le T_E$. In contrast to (8.2-1) let us suppose that the angular frequency $\omega_{i,0}$ is known. Here we are concerned with the problem of determining the best estimate $\hat{\theta}$ for the phase angle using the maximum likelihood principle. According to the derivation in Section 8.2.1 and the results of Appendix 8.2A we can find for the likelihood function $\Lambda(y \mid \theta)$ the same result as in (8.2-5)

$$\Lambda(y \mid \theta) = \exp\left[\frac{2\sqrt{2}A}{N_0} \int_0^{T_E} y(t) \sin(\omega_{i,0}t + \theta) \, dt \right] \qquad (8.2B\text{-}2)$$

The ML estimate $\hat{\theta}$ is the value of the trial parameter which maximizes the likelihood function (8.2B-2). Since the logarithm is a monotonically increasing function of its argument we can equivalently maximize the logarithm of (8.2B-2). The so-called log likelihood function is

$$\lambda(y \mid \theta) = \int_0^{T_E} y(t) \sin(\omega_{i,0}t + \theta) \, dt \qquad (8.2B\text{-}3)$$

Using elementary trigonometric formulas this function can be rewritten

$$\lambda(y \mid \theta) = \sin\theta \int_0^{T_E} y(t) \cos(\omega_{i,0}t) \, dt + \cos\theta \int_0^{T_E} y(t) \sin(\omega_{i,0}t) \, dt$$

$$= R \cos(\theta - \gamma) \qquad (8.2B\text{-}4)$$

where

$$R^2 = \left[\int_0^{T_E} y(t) \cos(\omega_{i,0}t) \, dt \right]^2 + \left[\int_0^{T_E} y(t) \sin(\omega_{i,0}t) \, dt \right]^2$$

and

$$\gamma = \cos^{-1}\left\{ \frac{1}{R} \int_0^{T_E} y(t) \sin(\omega_{i,0}t) \, dt \right\}$$

$$= \sin^{-1}\left\{ \frac{1}{R} \int_0^{T_E} y(t) \cos(\omega_{i,0}t) \, dt \right\} \qquad (8.2B\text{-}5)$$

It is obvious from (8.2B-4) that λ is maximum for $\theta = \gamma$. Thus, the ML estimate of the phase θ is

$$\hat{\theta} = \gamma = \tan^{-1} \left[\frac{\int_0^{T_E} y(t) \cos(\omega_{i,0} t)}{\int_0^{T_E} y(t) \sin(\omega_{i,0} t)} \right] \tag{8.2B-6}$$

Equation (8.2B-6) defines exactly the estimator we have given in Figure 8.2-4 as the optimal phase estimator. Note that γ actually is given uniquely in $[-\pi, \pi)$ by (8.2B-5). Therefore, $\tan^{-1}$ here is also defined in $[-\pi, \pi)$.

APPENDIX 8.3A
EVALUATION OF GAUSSIAN MOMENTS

The use of nonlinear functions involving gaussian random variables is widespread. For this reason a short list of frequently used expected values is tabulated in this appendix. The third- and fourth-order moments that are needed in (8.3-21) can be obtained by differentiation of the characteristic function. The results are included in the list below.

The quantity x, used in the list, is assumed to be a gaussian column vector of dimension N with statistical parameters defined by

$$m := E(x)$$

$$C := E(xx^T) - mm^T, \quad \text{covariance matrix}$$

$$\Lambda \triangleq C^{-1}$$

so that the probability density function is given by

$$p(x) = \frac{1}{[(2\pi)^N \det(C)]^{1/2}} \exp\left[-\frac{1}{2}(x-m)^T \Lambda (x-m)\right]$$

The following additional definitions are made

$$\xi = \text{an arbitrary (nonrandom) row vector}$$

$$v := \text{an arbitrary (nonrandom) column vector}$$

The list of gaussian moments is as follows:

1. $E(x_i x_j x_k) = C_{ij} m_k + C_{ik} m_j + C_{jk} m_i + m_i m_j m_k$

2. $E(x_i x_j x_k x_l) = C_{ij} C_{kl} + C_{ik} C_{jl} + C_{il} C_{jk} + C_{kj} m_k m_l$
   $\qquad + C_{ik} m_j m_l + C_{il} m_j m_k + C_{jk} m_i m_l$
   $\qquad + C_{jl} m_i m_k + C_{kl} m_i m_j + m_i m_j m_k m_l$

3. $E(xv^Txx^T) = (v^Tm)C + mv^TC + Cvm^T + mv^Tmm^T$

4. $E(e^{\xi x}) = \exp[\xi m + (\xi C\xi^T/2)]$

5. $E(xe^{\xi x}) = (m + C\xi^T)\exp[\xi m + (\xi C\xi^T/2)]$

6. $E(xx^Te^{\xi}x) = [C + (m + C\xi^T)(m + C\xi^T)^T]\exp[\xi m + (\xi C\xi^T/2)]$

7. $E(\sin(\xi x)) = \exp(-\xi C\xi^T/2)\sin(\xi m)$

8. $E(\cos(\xi x)) = \exp(-\xi C\xi^T/2)\cos(\xi m)$

9. $E(x\sin(\xi_x)) = [m\sin(\xi m) + C\xi^T\cos(\xi m)]\exp(-\xi C\xi^T/2)$

10. $E(x\cos(\xi x)) = [m\cos(\xi m) - C\xi^T\sin(\xi m)]\exp(-\xi C\xi^T/2)$

11. $E(xx^T\sin(\xi x)) = [(C - C\xi^T\xi C + mm^T)\sin(\xi m)$
$$+ (m\xi C + C\xi^Tm^T)\cos(\xi m)]\exp(-\xi C\xi^T/2)$$

12. $E(xx^T\cos(\xi x)) = [(C - C\xi^T\xi C + mm^T)\cos(\xi m)$
$$- (m\xi C + C\xi^Tm^T)\sin(\xi m)]\exp(-\xi C\xi^T/2)$$

PART 4

9

BRIEF REVIEW OF SOME
MATHEMATICAL FUNDAMENTALS

9.1. INTRODUCTION

In the remaining part of this volume the nonlinear theory of operation of synchronizers is developed. Basically, we need to characterize the error $\varphi(t)$ between a phase $\theta(t)$ of a received signal and a phase $\hat{\theta}(t)$ provided by an estimator[*]

$$\varphi(t) = \theta(t) - \hat{\theta}(t) \tag{9.1-1}$$

As the reader will discover, the performance of the estimator is assessed by decomposing the phase error into the sum of two components, a cyclic component $\phi(t)$ (measured modulo 2π) and a counting process $M(t)$ for cycle slips

$$\varphi(t) = \phi(t) + 2\pi M(t) \tag{9.1-2}$$

We will see that the phase error variance grows asymptotically with t

$$\sigma_\varphi^2 = \sigma_\phi^2 + (2\pi)^2 Dt \tag{9.1-3}$$

where D is a diffusion coefficient approximately equal to the average cycle slip rate $D \simeq N$.

Thus, the decomposition of (9.1-2) allows the performance of the estimator to be characterized in its different modes of operation. The phase error variance σ_ϕ^2 characterizes the tracking performance while the diffusion coefficient D provides the characterization of the threshold. While the

---

[*]*Estimator*: A device which generates a replica of the phase $\theta(t)$ based on a noisy measurement. The PLL is an example of such an estimator.

tracking performance can be assessed employing linear theory we must resort to nonlinear theory to determine the threshold parameter D.

In Chapter 9 we briefly review the mathematical tools needed for an engineering application of nonlinear theory. A working knowledge of random processes as presented, for example, by Papoulis [1] is presumed. For a detailed discussion of the fundamentals the reader is referred to the book by Risken [8] and the book by Gardiner [17].

Notational remark: We do not notationally distinguish between a random variable and its realization. It should be clear from the context. In distribution and density functions, the arguments identify the random variables under discussion.

9.2. STOCHASTIC DIFFERENTIAL EQUATIONS

Synchronization systems which are subject to random disturbances can be represented by a state variable description of the type

$$\frac{dx(t)}{dt} = g[x(t)] + F[x(t)]n(t) \qquad (9.2\text{-}1)$$

where x is the n-dimensional state vector and $g(x)$ is a nonlinear function of x, $F(x)$ is an $(n \times m)$ matrix function and $n(t)$ is a noise vector with dimension m. Q is the covariance matrix

$$Q(\tau) = E[n(t+\tau)n(t)^T] = \begin{bmatrix} R_{n_1}(\tau) & & & 0 \\ & R_{n_2}(\tau) & & \\ & & \ddots & \\ 0 & & & R_{n_n}(\tau) \end{bmatrix} \qquad (9.2\text{-}1a)$$

which has diagonal form since the components of $n(t)$ are assumed statistically independent.

For example, for a second-order PLL disturbed by additive noise the functions in (9.2-1) are given by

$$x = \begin{bmatrix} \phi \\ x_1 \end{bmatrix}, \quad g(x) = \begin{bmatrix} -2\zeta \sin \phi - x_1 + \gamma \\ \sin \phi \end{bmatrix}, \quad F(x) = \begin{bmatrix} -2\zeta \\ 1 \end{bmatrix}, \quad n(t) = n'(t)$$

$$(9.2\text{-}2)$$

$$E[n'(t+\tau)n'(t)] = R_{n'}(\tau) \qquad (9.2\text{-}2a)$$

The noise vector $n'(t)$ has a dimension of $m = 1$ and the noise intensity is independent of the state since F is a matrix with constant elements.

We now consider the differential equation (9.2-1) when $n(t)$ is a white noise process. The reason that we deal with physically nonrealizable white noise processes is that $x(t)$ then becomes a vector Markov process which is statistically completely specified by the *transition probability density function* $p(x, s \mid y, t)$ and the *initial probability density function* $p(x, o)$. For a white noise process two sample values $n(t + h)$ and $n(t)$ are statistically independent for any $h \neq 0$, which makes it impossible to define a derivative $\dot{x}(t)$ in the usual sense, since the value of $n(t + h)$ changes erratically as $h \rightarrow 0$.

There are two different approaches to give meaning to the random differential equation given by (9.2-1). First, we may assume $n(t)$ to be a physically realizable process; that is, the noise process has a very small but nonzero correlation time τ_n. In the limit as $\tau_n \rightarrow 0$ $n(t)$ approaches a white noise process. (We call this approach the "physical" approach.)

Mathematicians do not use the physical approach outlined above. They start with an integral formulation of (9.2-1) ("mathematical" approach)

$$x(t) - x(0) = \int_0^t g[x(t')] \, dt' + \int_0^t F[x(t')] \, d\beta(t') \qquad (9.2\text{-}3)$$

where we have used the fact that $n(t)$ is formally defined as the derivative of the brownian motion

$$\frac{d\beta(t)}{dt} = n(t) \quad \Rightarrow \quad d\beta(t) = n(t) \, dt \qquad (9.2\text{-}4)$$

with variance parameter σ^2.

Equation (9.2-3) is meaningful only insofar as the integrals on the right-hand side are defined in an appropriate way. Since $x(t)$ can be shown to be continuous in the mean square (m.s.) sense [4: Theorem 4.5] the integral of $g[x(t)]$ exists as a Riemann integral in the m.s. sense. Because of the erratic properties of the brownian motion, the second integral poses difficult problems. It has been defined by Itô [5] and is called the *Itô stochastic integral*. Another definition has been given by Stratonovich [6]. This integral is called a *Stratonovich integral*. Once the stochastic integral is defined the differential equation

$$dx(t) = g[x(t)] \, dt + F[x(t)] \, d\beta(t) \qquad (9.2\text{-}5)$$

can be considered as a formal notation for the integral equation (9.2-3) with

$$g(x) = \begin{bmatrix} g_1(x) \\ g_2(x) \\ \vdots \\ g_i(x) \\ \vdots \\ g_n(x) \end{bmatrix}, \quad F(x) = \begin{bmatrix} F_{11}(x) & F_{12}(x) & \cdots & F_{1m}(x) \\ F_{21}(x) & & & \\ \vdots & & & \vdots \\ \vdots & & & \vdots \\ F_{n1}(x) & F_{n2}(x) & \cdots & F_{nm}(x) \end{bmatrix}$$

$$x = \begin{bmatrix} x_1 \\ x_2 \\ \vdots \\ x_n \end{bmatrix}, \quad \beta(t) = \begin{bmatrix} \beta_1(t) \\ \beta_2(t) \\ \vdots \\ \beta_m(t) \end{bmatrix}, \quad E[d\beta_i(t)^2] = \sigma_i^2\, dt \qquad (9.2\text{-}6)$$

The question might arise which integral is "right." The answer of course is that both integrals are "right" since they are merely mathematically defined in a different way. More important is the fact (see later) that the Stratonovich integral must be used when modeling physical phenomena.

It can be shown that for the Itô equation

(I) $$dx(t) = g[x(t)]\, dt + F[x(t)]\, d\beta(t) \qquad (9.2\text{-}7)$$

there is an equivalent equation in the Stratonovich sense

(S) $$dx(t) = \left[g[x(t)] - \frac{1}{2}\, \Delta[x(t)] \right] dt + F[x(t)]\, d\beta(t) \qquad (9.2\text{-}8)$$

where the ith component of the vector $\Delta[x(t)]$ is given by

$$\Delta[x(t)]_i = \sum_{k=1}^{n} \sum_{j=1}^{m} F_{kj}(x) \frac{\partial F_{ij}(x)}{\partial x_k} \qquad (9.2\text{-}9)$$

9.3. FOKKER–PLANCK EQUATION

We have pointed out that the solution $x(t)$ of the stochastic differential equation (9.2-1) is a vector Markov process if $n(t)$ is a white gaussian vector process. The process is therefore statistically completely described by the transition probability density function $p(x, s \mid y, t)$ for $s > t$ and the initial distribution $p(x, 0)$.

For every Markov process there is an equation that describes the law of evolution of the transition probability density function in time.

9.3.1. Derivation of the Fokker–Planck Equation

There are different ways of deriving the Fokker–Planck equation from the stochastic differential equation. Here we will present an approach based on the Chapman–Kolmogorov equation. For simplicity the derivation is given for a first-order system. The generalization to a nth order system is conceptually straightforward.

We begin with the integral

$$I = \int_{-\infty}^{\infty} R(x) \frac{\partial p(x, t \mid y, t_0)}{dt}\, dx \qquad (9.3\text{-}1)$$

where $R(x)$ is a function upon whose derivative $R^{(n)}(x) = dR^n/dx^n$ we place certain conditions, to be stated below. We replace the partial derivative in (9.3-1) by a differential quotient

$$I_\Delta = \int_{-\infty}^{\infty} R(x) \left[\frac{p(x, t + \Delta t \mid y, t_0) - p(x, t \mid y, t_0)}{\Delta t} \right] dx$$

$$= \frac{1}{\Delta t} \int_{-\infty}^{\infty} R(x) p(x, t + \Delta t \mid y, t_0) \, dx - \frac{1}{\Delta t} \int_{-\infty}^{\infty} R(x) p(x, t \mid y, t_0) \, dx$$

$$(9.3\text{-}2)$$

Now according to the Chapman–Kolmogorov equation

$$p(x, t + \Delta t \mid y, t_0) = \int_{-\infty}^{\infty} p(x, t + \Delta t \mid z, t) p(z, t \mid y, t_0) \, dz \quad (9.3\text{-}3)$$

Using (9.3-3), the right-hand side of (9.3-2) becomes

$$I_\Delta = \frac{1}{\Delta t} \int_{-\infty}^{\infty} \int_{-\infty}^{\infty} R(x) p(x, t + \Delta t \mid z, t) p(z, t \mid y, t_0) \, dz \, dx$$

$$- \frac{1}{\Delta t} \int_{-\infty}^{\infty} R(x) p(x, t \mid y, t_0) \, dx \qquad (9.3\text{-}4)$$

After interchanging the order of integration in the first integral of (9.3-4) and replacing the variable of integration x in the second integral by z we obtain

$$I_\Delta = \frac{1}{\Delta t} \int_{-\infty}^{\infty} p(z, t \mid y, t_0) \left\{ \int_{-\infty}^{\infty} R(x) p(x, t + \Delta t \mid z, t) \, dx - R(z) \right\} dz$$

$$(9.3\text{-}5)$$

Expanding $R(x)$ into a Taylor series

$$R(x) = R(z) + \sum_{n=1}^{\infty} \frac{R^{(n)}(z)(x - z)^n}{n!} \qquad (9.3\text{-}6)$$

and assuming that summation and integration may be interchanged, we obtain the following expression

$$I_\Delta = \frac{1}{\Delta t} \int_{-\infty}^{\infty} p(z, t \mid y_0, t_0) \left\{ \int_{-\infty}^{\infty} R(z) p(x, t + \Delta t \mid z, t) \, dx \right.$$

$$\left. + \int_{-\infty}^{\infty} \sum_{n=1}^{\infty} \frac{R^{(n)}(z)}{n!} (x - z)^n p(x, t + \Delta t \mid z, t) \, dx - R(z) \right\} dz$$

$$= \frac{1}{\Delta t} \int_{-\infty}^{\infty} p(z, t \mid y, t_0) \left\{ \sum_{n=1}^{\infty} \frac{R^{(n)}(z)}{n!} \int_{-\infty}^{\infty} (x - z)^n p(x, t + \Delta t \mid z, t) \, dx \right\} dz$$

$$(9.3\text{-}7)$$

Now we denote the limit of the normalized nth conditional moment of the increment $(x - z)$ during the time Δt by

$$K_n(z) := \lim_{\Delta t \to 0} \frac{1}{\Delta t} E[(x(t + \Delta t) - z(t))^n \mid z(t)], \quad n \geq 1 \qquad (9.3\text{-}8)$$

which has the following integral representation

$$K_n(z) = \lim_{\Delta t \to 0} \frac{1}{\Delta t} \int_{-\infty}^{\infty} (x - z)^n p(x, t + \Delta t \mid z, t)\, dx, \quad n \geq 1 \qquad (9.3\text{-}9)$$

Then the substitution of (9.3-9) into (9.3-7) yields (for $\Delta t \to 0$)

$$
\begin{aligned}
I &= \int_{-\infty}^{\infty} p(z, t \mid y, t_0) \sum_{n=1}^{\infty} \frac{R^{(n)}(z)}{n!} K_n(z)\, dz \\
&= \sum_{n=1}^{\infty} \frac{1}{n!} \int_{-\infty}^{\infty} R^{(n)}(z) K_n(z) p(z, t \mid y, t_0)\, dz \qquad (9.3\text{-}10)
\end{aligned}
$$

Integrating the nth term in the last series n-fold by parts yields

$$
\begin{aligned}
I_n &= \int_{-\infty}^{\infty} R^{(n)}(z) K_n(z) p(z, t \mid y, t_0)\, dz \\
&= R^{(n-1)}(z) [K_n(z) p(z, t \mid y, t_0)] \Big|_{-\infty}^{\infty} \\
&\quad - R^{(n-2)}(z) \frac{\partial}{\partial z} [K_n(z) p(z, t \mid y, t_0)] \Big|_{-\infty}^{\infty} \\
&\qquad \vdots \\
&\quad + (-1)^{n-1} R(z) \frac{\partial^{n-1}}{\partial z^{n-1}} [K_n(z) p(z, t \mid y, t_0)] \Big|_{-\infty}^{\infty} \\
&\quad + \int_{-\infty}^{\infty} (-1)^n R(z) \frac{\partial^n}{\partial z^n} [K_n(z) p(z, t \mid y, t_0)]\, dz \qquad (9.3\text{-}11)
\end{aligned}
$$

Under the assumption that $R(z)$ and its derivatives decrease sufficiently fast to zero as $z \to \infty$, only the last term in the previous sum remains. Thus

$$I = \sum_{n=1}^{\infty} \frac{(-1)^n}{n!} \int_{-\infty}^{\infty} R(z) \frac{\partial^n}{\partial z^n} [K_n(z) p(z, t \mid y, t_0)]\, dz \qquad (9.3\text{-}12)$$

Finally subtracting (9.3-12) from (9.3-1) and changing the variable of integration in (9.3-12) from z to x we obtain

$$
\begin{aligned}
I - I &= 0 \\
&= \int_{-\infty}^{\infty} R(x) \left\{ \frac{\partial p(x, t \mid y, t_0)}{\partial t} - \sum_{n=1}^{\infty} \frac{(-1)^n}{n!} \frac{\partial^n}{\partial x^n} [K_n(x) p(x, t \mid y, t_0)] \right\} dx \\
&\qquad\qquad\qquad\qquad\qquad\qquad\qquad\qquad\qquad\qquad\qquad (9.3\text{-}13)
\end{aligned}
$$

Since $R(x)$ is an arbitrary function, except for the above condition on differentiability, in order for the integral to vanish the quantity within the braces must vanish and we arrive at the so-called *Kramers–Moyal* expansion

$$\frac{\partial p(x, t \mid y, t_0)}{\partial t} - \sum_{n=1}^{\infty} \frac{(-1)^n}{n!} \frac{\partial^n}{\partial x^n} [K_n(x)p(x, t \mid y, t_0)] = 0$$

(9.3-14)

Equation (9.3-14) describes the evolution of the transition probability density function $p(x, t \mid y, t_0)$ of $x(t)$ starting with the fixed value at $t = t_0$ of $x(t_0) = y$. Therefore, we can also view the solution of (9.3-14) as the probability density function $p(x, t)$ with the initial distribution $p(x, 0) = \delta(x - y)$ which permits the use of the simpler notation $p(x, t)$ in the last equation.

As will be seen in the next section, for an important class of processes $K_n(x) = 0$ for $n \geq 3$. Then (9.3-14) becomes the so-called *Fokker–Planck equation*

$$\frac{\partial p(x, t)}{\partial t} + \frac{\partial}{\partial x} [K_1(x)p(x, t)] - \frac{1}{2} \frac{\partial^2}{\partial x^2} [K_2(x)p(x, t)] = 0$$

$$p(x, 0) = \delta(x - y)$$

(9.3-15)

In this book we only deal with systems where the intensity coefficients $K_n(x)$ do not explicitly depend on time. This implies that the functions in the corresponding stochastic differential equations do not explicitly depend on time. A process with these properties is called *time-homogeneous*.

9.3.2. Intensity Coefficients

To obtain the Fokker–Planck equation we must compute the intensity coefficients $K_n(x)$ since they uniquely define the equation.

"Physical" approach: gaussian processes. Let us assume that $x(t)$ is the solution of the (first-order) differential equation

$$\frac{dx(t)}{dt} = g[x(t)] + F[x(t)]n(t)$$

(9.3-16)

where $n(t)$ is a physically realizable gaussian process which approaches white noise. We assume a normalized process here such that $R_n(\tau) \rightarrow \delta(\tau)$ in the limit. Furthermore, $g[x(t)]$ and $F[x(t)]$ are assumed to be time-invariant, i.e., they do not explicitly depend on time. We write the differential equation in the form of an integral equation

$$x(t) - x(t_0) = \int_{t_0}^{t} g[x(t')] \, dt' + \int_{t_0}^{t} F[x(t')]n(t') \, dt'$$

(9.3-17)

We expand g and F into Taylor series, retaining the first two terms only

$$g[x(t')] = g[x(t_0)] + g_x[x(t_0)][x(t') - x(t_0)]$$

$$F[x(t')] = F[x(t_0)] + F_x[x(t_0)][x(t') - x(t_0)]$$

(9.3-18)

with

$$g_x := \frac{\partial g[x(t)]}{\partial x}, \quad F_x := \frac{\partial F[x(t)]}{\partial x}$$

and insert into (9.3-17)

$$x(t) - x(t_0) = \int_{t_0}^{t} g[x(t_0)]\, dt' + \int_{t_0}^{t} g_x[x(t_0)][x(t') - x(t_0)]\, dt'$$

$$+ \int_{t_0}^{t} F[x(t_0)]n(t')\, dt' + \int_{t_0}^{t} F_x[x(t_0)]n(t')[x(t') - x(t_0)]\, dt'$$

(9.3-19)

We iterate the procedure for $x(t') - x(t_0)$ and obtain

$$x(t) - x(t_0) = \int_{t_0}^{t} g[x(t_0)]\, dt'$$

$$+ \int_{t_0}^{t} g_x[x(t_0)]\left\{ \int_{t_0}^{t'} g[x(t_0)]\, dt'' \right.$$

$$+ \int_{t_0}^{t'} g_x[x(t_0)][x(t'') - x(t_0)]\, dt''$$

$$+ \int_{t_0}^{t'} F[x(t_0)]n(t'')\, dt''$$

$$\left. + \int_{t_0}^{t'} F_x[x(t_0)]n(t'')[x(t'') - x(t_0)]\, dt'' \right\} dt'$$

$$+ \int_{t_0}^{t} F[x(t_0)]n(t')\, dt'$$

$$+ \int_{t_0}^{t} F_x[x(t_0)]n(t')\left\{ \int_{t_0}^{t'} g[x(t_0)]\, dt'' \right.$$

$$+ \int_{t_0}^{t'} g_x[x(t_0)][x(t'') - x(t_0)]\, dt''$$

$$+ \int_{t_0}^{t'} F[x(t_0)]n(t'')\, dt''$$

$$\left. + \int_{t_0}^{t'} F_x[x(t_0)]n(t'')[x(t'') - x(t_0)]\, dt'' \right\} dt'$$

(9.3-20)

Rearrangement of (9.3-20) yields

$$x(t) - x(t_0) = \int_{t_0}^{t} g[x(t_0)]\, dt'$$

$$+ \int_{t_0}^{t} g_x[x(t_0)] \int_{t_0}^{t'} g[x(t_0)]\, dt''\, dt'$$

$$+ \int_{t_0}^{t} g_x[x(t_0)] \int_{t_0}^{t'} F[x(t_0)]n(t'')\, dt''\, dt'$$

$$+ \text{ terms with } x(t'') - x(t_0)$$

$$+ \int_{t_0}^{t} F[x(t_0)]n(t')\, dt'$$

$$+ \int_{t_0}^{t} F_x[x(t_0)]n(t') \int_{t_0}^{t'} g[x(t_0)]\, dt''\, dt'$$

$$+ \int_{t_0}^{t} F_x[x(t_0)]n(t') \int_{t_0}^{t'} F[x(t_0)]n(t'')\, dt''\, dt'$$

$$+ \text{ terms with } x(t'') - x(t_0) \qquad (9.3\text{-}21)$$

By repeated application of this procedure we obtain integrals which only contain products of the known functions F and g and their derivatives at $x(t_0)$ and the noise $n(t)$. If we now take expected values of (9.3-21) we find for $t = t_0 + \Delta t$ (discarding the terms with $x(t'') - x(t_0)$ for the moment)

$$E[x(t_0 + \Delta t) - x(t_0) \mid x(t_0)]$$

$$= \int_{t_0}^{t_0 + \Delta t} g[x(t_0)]\, dt' + \int_{t_0}^{t_0 + \Delta t} \int_{t_0}^{t'} g_x[x(t_0)]g[x(t_0)]\, dt''\, dt'$$

$$+ \int_{t_0}^{t_0 + \Delta t} \int_{t_0}^{t'} F_x[x(t_0)]F[x(t_0)]E[n(t')n(t'') \mid x(t_0)]\, dt''\, dt' \qquad (9.3\text{-}22)$$

since $E[n(t')] = 0$.

If we take for the (unconditional) correlation function $R_n(t' - t'')$, a symmetric representation around the origin, we find in the limit that

$$E[n(t')n(t'') \mid x(t_0)] \to R_n(t' - t'') = \delta(t' - t'')$$

Thus

$$\int_{t_0}^{t_0 + \Delta t} \int_{t_0}^{t'} F_x[x(t_0)]F[x(t_0)]E[n(t')n(t'') \mid x(t_0)]\, dt'\, dt''$$

$$= F_x[x(t_0)]F[x(t_0)] \int_{t_0}^{t_0 + \Delta t} dt' \int_{t_0}^{t'} \frac{r}{2} \exp\left(-r \mid t' - t'' \mid\right) dt''$$

$$= F_x[x(t_0)]F[x(t_0)] \frac{\Delta t}{2} \quad \text{ as } r \to \infty \qquad (9.3\text{-}23)$$

To clarify the factor $1/2$ in (9.3-23) we first perform the integration over t'' from t_0 to t'. The integral then becomes

$$\frac{r}{2} \int_{t_0}^{t'} \exp\left(-r|t' - t''|\right) dt'' = \frac{r}{2} \int_{t'-t_0}^{0} \exp(-r|x|) \, dx$$

$$= \frac{1}{2} \quad \text{as } r \to \infty \tag{9.3-24}$$

which explains why we required a symmetric definition of the Dirac impulse. Inserting the result (9.3-23) into (9.3-22) finally yields in the limit $\Delta t \to 0$

$$\lim_{\Delta t \to 0} \frac{E[x(t_0 + \Delta t) - x(t_0) \mid x(t_0)]}{\Delta t} = g[x(t_0)] + \frac{1}{2} F_x[x(t_0)] F[x(t_0)] \tag{9.3-25}$$

This can be shown as follows: the k-fold integral not containing noise terms obtained by successive application of the iteration procedure is proportional to $(\Delta t)^k$, $k \geq 2$. Integrals containing noise terms either equal zero or give a contribution proportional to Δt^2.

By using the same arguments for the second moment $E[[x(t + \Delta t) - x(t_0)]^2 \mid x(t_0)]$ we find

$$E[[x(t_0 + \Delta t) - x(t_0)]^2 \mid x(t_0)] = \int_{t_0}^{t_0 + \Delta t} \int_{t_0}^{t_0 + \Delta t} F^2[x(t_0)] R_n(t' - t'') \, dt' \, dt''$$

$$= F^2[x(t_0)] \Delta t \tag{9.3-26}$$

All K_n vanish for $n \geq 3$. Since K_1 and K_2 do not explicitly depend on time, we omit t_0 in the notation and obtain as the final result

$$K_1(x) = g(x) + \frac{1}{2} F_x(x) F(x)$$

$$K_2(x) = F^2(x) \tag{9.3-27}$$

$$K_n(x) = 0 , \quad n \geq 3$$

Note: $n(t)$ is normalized such that $R_n(\tau) \to \delta(\tau)$.

"Physical" approach: non-gaussian processes. In the previous derivation for gaussian processes the factorization of higher order moments into first and second order moments was used to show that these moments do not contribute to the intensity coefficients as $\Delta t \to 0$. But factorization is a property of gaussian processes not shared by general random processes. The higher order moments which occur by repeated application of the iteration procedure might therefore be thought of to be important too in the non-gaussian case. Surprisingly, this turns out not to be the case as will be demonstrated now.

A glance at the derivation of the intensity coefficients shows that by repeated application of the iteration procedure the Kramers–Moyal expansion has an infinite number of terms. Also, additional terms in the first two intensity coefficients appear. All these terms are due to the higher order moments ($k \geq 3$) of the noise process. But an equation with an infinite number of terms can not be treated numerically and the question arises whether one can approximate the infinite expansion by truncating at a finite order. An important theorem due to Pawula [10] states that the Kramers–Moyal expansion either may stop at $n = 1$ or $n = 2$ or that it must contain an infinite number of terms. Truncation at $n = 2$ and neglecting terms due to higher order moments in $K_1(x)$ and $K_2(x)$ yields the so-called diffusion approximation (Fokker–Planck equation).

Example: If we perform a third iteration in equation (9.3-21) we obtain

$$E[x(t_0 + \Delta t) - x(t_0)|x(t_0)] = g(x)\,\Delta t + F(x)F_x(x)\Phi_2 + F(x)F_x^2(x)\Phi_3$$

$$E[(x(t_0 + \Delta t) - x(t_0))^2|x(t_0)] = 2F^2(x)\Phi_2 + 4F^2(x)F_x(x)\Phi_3$$

$$+ \text{ terms with } \Phi_4, \Phi_5, \Phi_6$$

$$E[(x(t_0 + \Delta t) - x(t_0))^3|x(t_0)] = 6F^3(x)\Phi_3 + \text{terms with } \Phi_4, \ldots, \Phi_9$$

with the correlation coefficient Φ_k defined by

$$\Phi_k := \int_{t_0}^{t_0 + \Delta t} \int_{t_0}^{t_1} \cdots \int_{t_0}^{t_{k-1}} E[n(t_1)n(t_2)\cdots n(t_k)]\, dt_1 \cdots dt_k$$

To demonstrate that the higher order moments, $k \geq 3$, do not contribute in the limit let us consider a noise process parameterized by a parameter $\varepsilon > 0$

$$n_\varepsilon(t) = \frac{1}{\varepsilon}\, y\!\left(\frac{t}{\varepsilon^2}\right)$$

where we assume $y(t)$ is a stationary, mean square continuous process with spectral density $S_y(\omega)$. The spectral density of $n_\varepsilon(t)$ is related to $S_y(\omega)$ by

$$S_{n_\varepsilon}(\omega) = S_y(\omega\varepsilon^2)$$

Note that the t/ε^2 time scaling spreads the bandwidth and the $1/\varepsilon$ amplitude scaling keeps the spectral density at the origin independent of ε, that is, $S_{n_\varepsilon}(0) = S_y(0)$ for all ε. Assuming that the limit for $\varepsilon \to 0$ exists, the bandwidth of $n_\varepsilon(t)$ goes to infinity while the spectral density converges to a constant (white noise). The second order correlation coefficient Φ_2 for $n_\varepsilon(t)$ is given by

$$\Phi_2 = \int_{t_0}^{t_0 + \Delta t} \int_{t_0}^{t_1} R_{n_\varepsilon}(t_2 - t_1)\, dt_1\, dt_2 = \frac{\Delta t}{2}\, S_{n_\varepsilon}(0);\ (\varepsilon \to 0)$$

Thus, by construction Φ_2 converges to a constant as $\varepsilon \to 0$.

The integrand $E[n_\varepsilon(t_1)n_\varepsilon(t_2)n_\varepsilon(t_3)]$ of the third correlation coefficient is essentially nonzero only for values of (t_2, t_3) in a rectangle of width proportional to τ_{n_ε} centered at t_1, thus yielding the following proportionality relation

$$\Phi_3 \sim \frac{1}{\varepsilon^3}\, \tau_{n_\varepsilon}^2$$

The factor $1/\varepsilon^3$ stems from the amplitude scaling of the threefold product. The correlation time τ_{n_ε} of the parameterized noise process is related to the correlation time τ_y of the original process $y(t)$ by

$$\tau_{n_\varepsilon} = \frac{1}{R_{n_\varepsilon}(0)} \int_0^\infty R_{n_\varepsilon}(\tau)\, d\tau = \frac{\varepsilon^2}{R_y(0)} \int_0^\infty R_y\left(\frac{\tau}{\varepsilon^2}\right) \frac{d\tau}{\varepsilon^2}$$

$$= \varepsilon^2 \tau_y$$

Using this equality yields for the third order correlation coefficient

$$\Phi_3 \sim \varepsilon \tau_y^2$$

One can similarly show that all higher order correlation coefficients, $k > 3$, are proportional to powers of ε. Hence the correction terms in $K_1(x)$ and $K_2(x)$ as well as the higher order intensity coefficients $K_k(x)$, $k \geq 3$, vanish as $\varepsilon \to 0$. We thus have shown that the diffusion approximation applies as $n_\varepsilon(t)$ approaches a white (not necessarily gaussian) process. The intensity coefficients are identical to those of the white gaussian process with spectral density $S_{n_\varepsilon}(0) = S_y(0)$ at the origin, which is independent of ε.

There is an alternative physical interpretation which gives valuable insight into the reasons why the diffusion approximation becomes independent of the amplitude distribution as $n_\varepsilon(t)$ approaches white noise. Let us consider a time interval Δt which is much larger than the correlation time τ_{n_ε} of the noise process $n_\varepsilon(t)$ but much smaller than the system time constant τ_x.

$$\tau_{n_\varepsilon} \ll \Delta t \ll \tau_x$$

Since Δt is much smaller than τ_x the process $x(t)$ is little affected when we replace $n_\varepsilon(t)$ by

$$z_\varepsilon(t) = \frac{1}{\Delta t} \int_{t-\Delta t}^{t} n_\varepsilon(\tau) \, d\tau$$

The process $z_\varepsilon(t)$ is obtained by low-pass filtering the actual noise input with a filter with rectangular impulse response

$$h(t) = \begin{cases} \dfrac{1}{\Delta t} & \text{for } 0 < t < \Delta t \\ 0 & \text{otherwise} \end{cases}$$

Independent of the amplitude distribution of $n_\varepsilon(t)$ the noise increment $z_\varepsilon(t)$ is approximately gaussian distributed. This is a consequence of the central limit theorem. Provided the conditions on τ_{n_ε}, τ_x and Δt hold we thus may replace the actual process $n_\varepsilon(t)$ by a gaussian process $z_\varepsilon(t)$ with the same spectral density at the origin. In summary: our derivation was based on the following assumption:

(a) $y(t)$ is a stationary, mean square continuous process. No assumption on the amplitude distribution was required.
(b) $n_\varepsilon(t) = 1/\varepsilon \; y(t/\varepsilon^2)$ converges to a white process for $\varepsilon \to 0$.

It was emphasized in the foregoing discussion that to apply the white noise approximation ($\varepsilon \to 0$) the ratio of the correlation time τ_{n_ε} of the noise to the system time constant τ_x must be sufficiently small. This is the case for virtually all synchronizer applications. For a rigorous derivation of the white noise limit we refer to the book by Gardiner [17, p. 210] and the paper by Kushner [18]. The difficult problems arising in obtaining an approximate Fokker–Planck equation where the noise is colored (the white noise approximation is not justified) is reviewed by van Kampen [19].

"*Mathematical*" approach. In contrast to the physical approach, where we performed the calculations with physically realizable noise and took the limit to white noise *after* taking expected values, let us perform the limit to white noise from the very beginning. The integral equation for $x(t)$ then reads

$$x(t) - x(t_0) = \int_{t_0}^{t} g[x(t')] \, dt' + \int_{t_0}^{t} F[x(t')] \, d\beta(t') \qquad (9.3\text{-}28)$$

with

$$E[d\beta^2(t)] = dt$$

where the second integral must be understood in either the sense of Itô or Stratonovich. If we employ the same iteration procedure as in the "physical" approach we have to substitute $n(t)\,dt$ by $d\beta(t)$ in all terms of (9.3-21). Taking (conditioned) expected values yields for $t = t_0 + \Delta t$

$$E[x(t_0 + \Delta t) - x(t_0) \mid x(t_0)]$$

$$= E\left[\int_{t_0}^{t_0 + \Delta t} g[x(t_0)]\,dt' \;\middle|\; x(t_0)\right]$$

$$+ E\left[\int_{t_0}^{t_0 + \Delta t} \int_{t_0}^{t'} g_x[x(t_0)]g[x(t_0)]\,dt''\,dt' \;\middle|\; x(t_0)\right]$$

$$+ E\left[\int_{t_0}^{t_0 + \Delta t} F_x[x(t_0)]\int_{t_0}^{t'} F[x(t_0)]\,d\beta(t'')\,d\beta(t') \;\middle|\; x(t_0)\right] \qquad (9.3\text{-}29)$$

(All integrals containing $d\beta(t)$ vanish since E [function of $x(t_0)\,d\beta(t)] = 0$). Using the following theorem for conditional expectations

$$E[f(x) \mid x = y] = f(y)$$

and the fact that the first two integrands are constants we find

$$E\left[\int_{t_0}^{t_0 + \Delta t} g[x(t_0)]\,dt' \;\middle|\; x(t_0)\right] = g[x(t_0)]\,\Delta t$$

$$(9.3\text{-}30)$$

$$E\left[\int_{t_0}^{t_0 + \Delta t} \int_{t_0}^{t'} g_x[x(t_0)]g[x(t_0)]\,dt''\,dt' \;\middle|\; x(t_0)\right] = g_x[x(t_0)]g[x(t_0)]\frac{\Delta t^2}{2}$$

While the first two integrals of (9.3-29) are ordinary Riemann integrals, the last integral must be understood in the sense of Itô or Stratonovich. As we will see, the result will be *different* for the two definitions!

Since $F_x[x(t_0)]$ and $F[x(t_0)]$ are constants we may take them out of the integral

$$\int_{t_0}^{t_0 + \Delta t} F_x[x(t_0)]\,d\beta(t')\int_{t_0}^{t'} F[x(t_0)]\,d\beta(t'')$$

$$(9.3\text{-}31)$$

$$= F_x[x(t_0)]F[x(t_0)]\int_{t_0}^{t_0 + \Delta t} d\beta(t')\int_{t_0}^{t'} d\beta(t'')$$

Since by definition

$$\int_{t_0}^{t'} d\beta(t'') = \beta(t') - \beta(t_0) \qquad (9.3\text{-}32)$$

we obtain for the integral of (9.3-31)

$$\int_{t_0}^{t_0+\Delta t} d\beta(t') \int_{t_0}^{t'} d\beta(t'') = \int_{t_0}^{t_0+\Delta t} d\beta(t')[\beta(t') - \beta(t_0)] \qquad (9.3\text{-}33)$$

The last integral can be evaluated for both definitions, see [4: p. 104, 117].

$$\int_{t_0}^{t_0+\Delta t} [\beta(t') - \beta(t_0)]\, d\beta(t') = \begin{cases} \dfrac{1}{2}\, [\beta(t_0 + \Delta t) - \beta(t_0)]^2 - \dfrac{1}{2}\, \Delta t & \text{(Itô)} \\[2mm] \dfrac{1}{2}\, [\beta(t_0 + \Delta t) - \beta(t_0)]^2 & \text{(Stratonovich)} \end{cases}$$

$$(9.3\text{-}34)$$

Finally taking expected values of the increment of the Brownian motion with unit variance parameter yields

$$E[\beta(t_0 + \Delta t) - \beta(t_0)]^2 = \Delta t \qquad (9.3\text{-}35)$$

we see that the expected value of the Itô integral vanishes while the expected value of the Stratonovich integral yields $\Delta t/2$. The resulting intensity coefficient $K_1(x)$ is thus different for the two types of integrals.

$$K_1(x) = \begin{cases} g(x) & \text{(Itô)} \\[2mm] g(x) + \dfrac{1}{2}\, F_x(x)F(x) & \text{(Stratonovich)} \end{cases} \qquad (9.3\text{-}36)$$

The Stratonovich integral agrees with the result for physically realizable noise. Therefore, we conclude that the Itô equation cannot be used as a model of a physical phenomenon where the physical noise is always "colored" or, stated otherwise, that the Stratonovich equation must be used as an idealized model of a physical phenomenon. From the previous discussions we know that for every Stratonovich equation (9.2-8) there exists a corresponding Itô equation (9.2-7) which can be used if one wishes to employ the Itô calculus. The term

$$\frac{1}{2}\, F_x(x)F(x) \qquad (9.3\text{-}37)$$

is called the *Itô* (or *Wong-Zakai*) *correction term*.

By the same kind of arguments that led to $K_1(x)$ one can show that the second intensity coefficient is the same for both types of integrals

$$K_2(x) = F^2(x) \qquad (9.3\text{-}38)$$

in agreement with the results obtained by the "physical" approach.

9.3.3. Physical Interpretation of the Fokker–Planck Equation

Let us consider a large number N of particles, i.e., a gas, concentrated in a small volume. When the particles are released at time zero, we observe that the size of the volume where the particles are found will gradually grow. The mechanism by which the particles are propagated through the volume is called *diffusion*.

In this physical picture the random differential equation (9.2-1) describes the motion of a single particle in the whole volume. The probability density function $p(x, t \mid x_0, 0)$ then either determines the probability of finding a single particle with starting point x_0 at time zero in a small volume around x or, alternatively, can be interpreted as the relative number of particles $\Delta N / N = p(x, t \mid x_0, 0) \, \Delta x$ found in that small volume around x at time t.

The Fokker–Planck equation represents the equation of the probability conservation law: The normalized number of particles remains the same for all times t, only the density of particles may change with time. Introducing the probability current $J(x, t)$

$$J(x, t) = \left[K_1(x) - \frac{1}{2} \frac{\partial}{\partial x} K_2(x) \right] p(x, t) \tag{9.3-39}$$

the Fokker–Planck equation reads

$$\frac{\partial p(x, t)}{\partial t} + \frac{\partial}{\partial x} J(x, t) = 0 \tag{9.3-40}$$

The probability current describes the amount of probability (particles) crossing the abscissa x in the positive direction per unit of time. Replacing the partial derivative in (9.3-39) by corresponding differential quotients yields

$$[p(x, t + \Delta t) - p(x, t)] \, \Delta x + [J(x + \Delta x, t) - J(x, t)] \, \Delta t = 0 \tag{9.3-41}$$

But

$$[p(x, t + \Delta t) - p(x, t)] \, \Delta x \tag{9.3-42}$$

equals the difference of particles in the volume Δx between t and $t + \Delta t$. Since no particle can be created or annihilated, this difference must be balanced by the difference of particles that entered the volume Δx at x minus the particles that left the volume at $x + \Delta x$ during the time interval Δt (Figure 9.3-1)

$$J(x, t) \, \Delta t \, , \qquad \begin{array}{l} \text{number of particles that entered} \\ \text{the volume } \Delta x \text{ at } x \text{ during } \Delta t \end{array}$$

$$J(x + \Delta x, t) \, \Delta t \, , \qquad \begin{array}{l} \text{number of particles that left } \Delta x \\ \text{at } x + \Delta x \text{ during } \Delta t \end{array}$$

Hence

$$[J(x, t) - J(x + \Delta x, t)] \, \Delta t$$

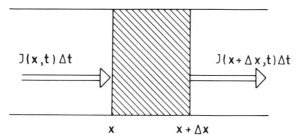

Figure 9.3-1. Physical interpretation: particle current.

equals the increase/decrease in the number of particles in the volume element Δx in the time interval Δt. Since the number of particles must be constant we must have

$$[p(x, t + \Delta t) - p(x, t)]\Delta x = [J(x, t) - J(x + \Delta x, t)]\,\Delta t$$

which is (9.3-41).

The first intensity coefficient multiplied by a small time interval Δt

$$K_1(x)\,\Delta t = E\left[\frac{x(t + \Delta t) - x(t)}{\Delta t}\,\middle|\,x(t)\right]\Delta t \qquad (9.3\text{-}43)$$

is roughly the average displacement of a particle during that interval. If $x(t)$ is the solution of a random differential equation, we have found for $K_1(x)$ (9.3-23)

$$K_1(x)\,\Delta t = \left[g(x) + \frac{1}{2}\,F_x(x)F(x)\right]\Delta t \qquad (9.3\text{-}44)$$

The average displacement comprises two terms. The first term is the average displacement of $x(t)$ due to the deterministic restoring force $g(x)$. The second term always appears when the noise intensity depends on the state $x(t)$ and is referred to as *noise induced drift* or *spurious drift* in the literature.

9.3.4. *N*-Dimensional Fokker–Planck Equation

By a conceptually straightforward generalization of the one-dimensional derivation given in Section 9.3.1 we can show that a vector Markov process satisfies the *n*-dimensional Chapman–Kolmogorov equation

$$p(x, t + \Delta t \mid y, 0) = \int_{-\infty}^{\infty} p(x, t + \Delta t \mid z, t)p(z, t \mid y, 0)\,dz \qquad (9.3\text{-}45)$$

where the integral is *n*-fold and $dz = dz_1\,dz_2\cdots dz_n$. From a generalization of the function $R(x)$ to n dimensions with appropriate boundary conditions we obtain the Fokker–Planck equation

$$\frac{\partial p(x, t)}{\partial t} = -\sum_{k=1}^{n} \frac{\partial}{\partial x_k} [K_k(x)p(x, t)]$$

$$+ \frac{1}{2} \sum_{k=1}^{n} \sum_{m=1}^{n} \frac{\partial^2}{\partial x_k \partial x_m} [K_{km}(x)p(x, t)] \qquad (9.3\text{-}46)$$

with the initial condition

$$p(x, 0) = \prod_{i=1}^{n} \delta(x_i - x_i(0)) \qquad (9.3\text{-}47)$$

For later purposes we introduce the Fokker–Planck operator (forward operator)

$$L := -\sum_{k=1}^{n} \frac{\partial}{\partial x_k} K_k(x) + \frac{1}{2} \sum_{k=1}^{n} \sum_{m=1}^{n} \frac{\partial^2}{\partial x_k \partial x_m} K_{km}(x) \qquad (9.3\text{-}48)$$

to obtain the compact notation for (9.3-46)

$$\frac{\partial p(x, t)}{\partial t} = Lp(x, t) \qquad (9.3\text{-}48a)$$

The intensity coefficient $K_k(x)$ is the expected value of the increment of the kth component of x given the vector x at time t

$$K_k(x) = \frac{1}{\Delta t} E\{x_k(t + \Delta t) - x_k(t) \mid x(t)\} \quad \text{as } \Delta t \to 0 \qquad (9.3\text{-}49)$$

Physically $K_k(x) \Delta t$ can be interpreted as the average drift of the particle in the kth direction when at time t it was observed at point x. The intensity coefficients $K_{km}(x)$

$$K_{km}(x) = \frac{1}{\Delta t} E\{[x_k(t + \Delta t) - x_k(t)][x_m(t + \Delta t) - x_m(t)] \mid x(t)\} \quad \text{as } \Delta t \to 0 \qquad (9.3\text{-}50)$$

are measures for the uncertainty about where a particle is Δt seconds after knowing its immediate position $x(t)$. With similar arguments as in the one-dimensional case one can show that the higher order coefficients

$$K_{km}^{(M, J)}(x) = \frac{1}{\Delta t} E\{[x_k(t + \Delta t) - x_k(t)]^M [x_m(t + \Delta t) - x_m(t)]^J \mid x(t)\}$$

$$\text{as } \Delta t \to 0 \qquad (9.3\text{-}51)$$

vanish for gaussian processes. (The superscript (M, J) is not written for $M, J \leq 1$.)

Notational Remark. For a number of reasons it has been customary to use different notations for the intensity coefficients of first- and higher order Fokker–Planck equations. For first-order equations the index i of $K_i(x)$ denotes the *ith conditional moment* (9.3-9).

In an *n*th order Fokker–Planck equation a single index of $K_i(x)$ denotes the *first* conditional moment of the *i*th component of the state vector (9.3-49). Double indices $K_{im}(x)$ denote the *second* moment of the *i*th and *m*th component of $x(t)$ (9.3-50). Higher order moments are denoted by exponents in parentheses (9.3-51).

Probability Current. In analogy to the one-dimensional case one defines the probability current $J(x, t)$

$$J^T(x, t) = [J_1(x, t), J_2(x, t), \ldots, J_n(x, t)] \tag{9.3-52}$$

The Fokker–Planck equation takes on the form of a continuity equation

$$\frac{\partial p(x, t)}{\partial t} + \nabla J(x, t) = 0 \tag{9.3-53}$$

with the Nabla operator

$$\nabla = \left[\frac{\partial}{\partial x_1}, \frac{\partial}{\partial x_2}, \ldots, \frac{\partial}{\partial x_n} \right]$$

We further dissect the probability current into a drift and diffusion component. The drift component

$$J_{i,s}(x, t) = K_i(x)p(x, t \mid y, 0) \tag{9.3-54}$$

represents the *i*th component of the total drift current $J_s(x, t)$ while

$$J_{i,D}(x, t) = -\frac{1}{2} \sum_{k=1}^{n} \frac{\partial}{\partial x_k} [K_{ik}(x)p(x, t \mid y, 0)] \tag{9.3-55}$$

represents the *i*th component due to diffusion.

9.3.5. Formal Derivation of the Intensity Coefficients for a Vector Process

In this section we want to summarize the methods of obtaining the intensity coefficients for the *n*-dimensional Fokker–Planck equation from the differential equation of the physical process. We start with the equation for the *n*-dimensional physical process which is repeated here for convenience

$$\frac{dx(t)}{dt} = g[x(t)] + F[x(t)]n(t) \tag{9.3-56}$$

with the covariance matrix $Q(\tau)$

$$Q(\tau) = E[n(t + \tau)n^T(t)] = \begin{bmatrix} R_{n_1}(\tau) & & 0 \\ & R_{n_2}(\tau) & \\ & & \cdots \\ 0 & & R_{n_n}(\tau) \end{bmatrix}$$

(9.3-57)

Since the components of the noise process are assumed to be independent the covariance matrix has a diagonal form.

Step 1: Every component of the physical noise process is approximated (we explicitly have to assume that this approximation is valid) by Brownian motion with unit variance parameter $\sigma^2 = 1$, multiplied by $\sqrt{(S_i(0))}$

$$n_i(t)dt = \sqrt{S_i(0)} \, d\beta(t)$$

(9.3-58)

Here $S_i(0)$ is the value of the spectral density of n_i at the origin

$$S_i(0) = \int_{-\infty}^{\infty} R_{n_i}(\tau) \, d\tau$$

(9.3-59)

As the noise process $n_i(t)$ becomes white we obtain

$$R_{n_i}(\tau) = S_i(0)R_\beta(\tau) = S_i(0)\delta(\tau)$$

(9.3-60)

where $\dot\beta := d\beta/dt$. For the whole vector we can write

$$n(t) \, dt = \begin{bmatrix} n_1(t) \\ \vdots \\ n_n(t) \end{bmatrix} = S^{1/2} \, d\beta(t)$$

(9.3-61)

where $S^{1/2}$ is a diagonal matrix with $\sqrt{(S_i(0))}$ at the diagonal entries

$$S^{1/2} = \begin{bmatrix} \sqrt{(S_1(0))} & & 0 \\ & \ddots & \\ 0 & & \sqrt{(S_n(0))} \end{bmatrix}$$

(9.3-62)

Step 2: Using the above approximation of $n(t)$ we obtain the stochastic differential equation

$$dx(t) = g[x(t)] \, dt + F[x(t)]S^{1/2} \, d\beta(t) \quad (S)$$

(9.3-63)

which must be understood in the sense of Stratonovich in order to properly model a physical phenomenon (see Section 9.3.2).

Step 3: Using a straightforward generalization of the methods outlined in Section 9.3.2 one finds that the ith intensity coefficient is related to the stochastic differential equation by

$$K_i(x) = g_i(x) + \frac{1}{2} \sum_{k=1}^{n} \sum_{j=1}^{m} [F(x)S^{1/2}]_{k,j} \frac{\partial}{\partial x_k} [F(x)S^{1/2}]_{i,j} \quad (9.3\text{-}64)$$

where $[\cdot]_{k,j}$ stands for the (k, j) element of the matrix $[FS^{1/2}]$. The intensity coefficients with double indices $K_{ij}(x)$ are the elements (i, j) of the following matrix

$$E[F(x)S^{1/2} \, d\beta(t) \, d\beta(t)^T S^{1/2} F(x)^T]$$
$$= F(x)S^{1/2} E[d\beta(t) \, d\beta(t)^T] S^{1/2} F(x)^T$$
$$= F(x)SF^T(x) \, dt \qquad (9.3\text{-}65)$$

Thus

$$K_{ij}(x) = [FSF^T]_{i,j} \qquad (9.3\text{-}66)$$

To arrive at the last result we have used

1. $E[d\beta(t) \, d\beta(t)^T] = I \, dt$ where I is the unity matrix.
2. $(S^{1/2})^T = S^{1/2} \rightarrow (S^{1/2})^T S^{1/2} = S$ where S is a diagonal matrix with entries $S_{i,i} = S_i(0)$.

Example: Second-order PLL   The state equation (3.4-3) is repeated here for convenience

$$\begin{bmatrix} \dfrac{d\phi(\tau)}{d\tau} \\[2ex] \dfrac{dx_2(\tau)}{d\tau} \end{bmatrix} = \begin{bmatrix} -2\zeta \sin \phi(\tau) - x_2(\tau) + \dfrac{\Delta\omega}{\omega_1} \\[2ex] \sin \phi(\tau) \end{bmatrix} + \begin{bmatrix} -2\zeta \\[1ex] 1 \end{bmatrix} n'(\tau)$$
$$(9.3\text{-}67)$$

We find for the intensity coefficients

$$K_1(x) = -2\zeta \sin \phi - x_2 + \frac{\Delta\omega}{\omega_1}$$
$$K_2(x) = \sin \phi \qquad (9.3\text{-}68)$$

The matrix S is a scalar in this case given by (3.4-13)

$$S_i(0) = \frac{1}{\rho\zeta} \frac{1}{1 + (1/4\zeta^2)} \qquad (9.3\text{-}69)$$

Thus

$$FSF^T = \begin{bmatrix} -2\zeta \\ 1 \end{bmatrix} \frac{1}{\rho\zeta[1 + (1/4\zeta^2)]} [-2\zeta \quad 1]$$

$$= \frac{1}{\rho\zeta[1 + (1/4\zeta^2)]} \begin{bmatrix} 4\zeta^2 & -2\zeta \\ -2\zeta & 1 \end{bmatrix} \tag{9.3-70}$$

from which we directly read off

$$K_{11}(x) = \frac{4\zeta^2}{\rho\zeta[1 + (1/4\zeta^2)]} \quad K_{22}(x) = \frac{1}{\rho\zeta[1 + (1/4\zeta^2)]}$$

$$K_{12}(x) = K_{21}(x) = -\frac{2\zeta}{\rho\zeta[1 + (1/4\zeta^2)]} \tag{9.3-71}$$

9.3.6. Initial and Boundary Conditions

The Fokker–Planck equation is a second-order partial differential equation with respect to each of n variables $[x_1, \ldots, x_n]$ and with respect to time t. Therefore, in order to determine its solution we must specify $2n$ independent *boundary conditions*. Also we need to specify the *initial distribution* $p(x, 0)$. For our purposes we specify the initial distribution by

$$p(x, 0) = \prod_{i=1}^{n} \delta(x_i - y_i) \tag{9.3-72}$$

with

$$x_i(0) = y_i$$

which says that the position of the particle at time $t = 0$ is at point $x(0) = y$.

Boundary conditions are more subtle to deal with. A general theory of permissible boundary conditions is known only for first-order equations. For our purposes we need three types of boundary conditions.

First, if a component x_k of the vector x extends from $-\infty$ to $+\infty$ the probability density function $p(x, t)$ must necessarily vanish sufficiently fast

$$p(x, t) \to 0 \quad \text{for } |x_k| \to \infty \tag{9.3-73}$$

in order to fulfill the normalization condition

$$\int_{-\infty}^{\infty} p(x, t) \, dx = 1$$

Besides the probability density function, the first partial derivative $(\partial p/\partial x_i)(x, t)$ must also converge to zero sufficiently fast as $|x_k| \to \infty$. Conditions which involve limits of $p(x, t)$ and $(\partial p/\partial x_i)(x, t)$ at $|x_k| \to \infty$ are called *natural boundary conditions*. Their exact form depends on the specific problem as we shall see later on.

The second type of boundary conditions can be interpreted as *periodicity constraints*. As an example consider the state variables of a PLL. If we are interested in the modulo $M2\pi$ reduced phase error only, then the transition probability density and also the probability current must be periodic with period $M2\pi$. This leads to *periodic boundary conditions* in ϕ.

$$p(\phi, x_2, x_3, \ldots, x_n, t) = p(\phi + M2\pi, x_2, x_3, \ldots, x_n, t)$$

$$\frac{\partial p}{\partial \phi}(\phi, x_2, x_3, \ldots, x_n, t) = \frac{\partial p}{\partial \phi}(\phi + M2\pi, x_2, x_3, \ldots, x_n, t)$$

(9.3-74)

(The reason for considering a period of $M2\pi$ instead of simply 2π will become clear later on.)

The last type of boundary conditions we are concerned with are of the *absorbing type*. Let us assume a particle starts inside a domain D with boundary ∂D of an n-dimensional space. A perfectly absorbing barrier is placed on the boundary ∂D of D so that the particle can never leave the boundary once it has touched the boundary for the first time. Mathematically this is expressed as follows

$$\Pr\left[x(t) = x(t_1) \mid x(0) \in D\right] = 1, \quad \text{for } t \geq t_1 \tag{9.3-75}$$

where t_1 is the time when the particle hits the boundary for the first time, i.e., $t_1 = \inf\{\tau \geq 0 \mid x(\tau) \in \partial D\}$.

To understand the form of absorbing boundary conditions of $p(x, t)$ we recall from earlier discussions that the variance of dx/dt of a Markov process is infinite. That means that the instantaneous velocity dx/dt of the particle changes its sign infinitely often in an arbitrarily short time interval. However, the displacement Δx remains finite since the particle moves in both directions. Assume now that the particle is in a thin layer very close to the boundary (see Figure 9.3-2). Since the particle "oscillates" around the position with infinite frequency it must have touched the boundary in the very recent past where it is observed. Therefore, there are practically no particles in the layer near the boundary and hence

$$\lim_{x \to \partial D} p(x, t \mid y, 0) = 0 \quad \text{for all } t \geq 0$$

$$x(0) = y \in D \tag{9.3-76}$$

It is interesting to note that the probability current is *not* zero on the boundary ∂D. Using the Gauss theorem of vector analysis we note that

$$\int_D \nabla J(x, t)\, dx = \int_{\partial D} e_n(x) J(x, t)\, dF = -\frac{\partial}{\partial t}\int_D p(x, t)\, dx \tag{9.3-77}$$

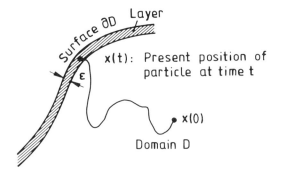

Figure 9.3-2. Absorbing boundary conditions.

where $e_n(x)$ is the unity vector normal to the surface ∂D and directed positively outwards (see Figure 9.3-3) and dF is an infinitesimal element of the surface ∂D. Equation (9.3-77) states that the rate of change of the particles which have not been absorbed, i.e., are inside D without ever having touched the surface ∂D before

$$-\frac{\partial}{\partial t} \int_D p(x, t)\, dx \qquad (9.3\text{-}78)$$

equals the integral flow of particles out of the volume D

$$\int_{\partial D} e_n(x) J(x)\, dF \qquad (9.3\text{-}79)$$

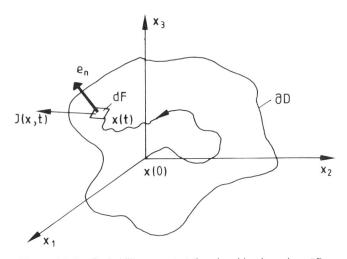

Figure 9.3-3. Probability current at the absorbing boundary ∂D.

9.3.7. Disturbance of Systems by Impulsive Noise

So far we have assumed that the noise is a stationary, mean square continuous process for which the diffusion approximation is valid as $\epsilon \to 0$. However, there are a number of important applications where the random perturbation can be modeled as impulsive noise. Such a process is *white* but not mean square continuous. As will be presently shown higher order intensity coefficients do *not* vanish in this case.

Following the procedure given in Section 3.2 with $n'(\tau, \phi)$ replaced by $n'(\tau, \phi) + n'_p(\tau)$ one formally arrives at the following differential equation

$$\frac{d\phi}{d\tau} = [-g(\phi) + \gamma] + n'(\tau, \phi) + n'_p(\tau) \tag{9.3-80}$$

where $n'_p(\tau)$ is the impulse noise given by

$$n'_p(\tau) = \frac{\sqrt{2}}{A} \sum_j x_j \delta(\tau - \tau_j) \cos(\omega_0 \tau + \hat{\theta}) \tag{9.3-81}$$

Time is normalized to $\tau = 4B_L t$ in the above equation. The train of impulses is generated by differentiating a Poisson process with normalized pulse density parameter $\lambda_p/4B_L$ [1: p. 239].

The random amplitudes x_j are assumed to be independent, identically distributed random variables. When an impulse occurs its resultant amplitude will vary with time due to the cosine factor. Assuming that the density of impulses $(\lambda_p/4B_L)$ is much less than $(\omega_0/4B_L)$ then the modulo 2π argument of the cosine factor is approximately uniformly distributed in $[-\pi, \pi]$. Therefore, we can write

$$n'_p(\tau) = \frac{\sqrt{2}}{A} \sum_j x_j \delta(\tau - \tau_j) \cos \Phi_j \tag{9.3-82}$$

where Φ_j is a uniformly distributed phase. The noise $n'(\tau, \phi)$ is assumed to be a gaussian process.

Kramers–Moyal Expansion. Since $\phi(t)$ is a Markov process the Kramers–Moyal expansion of (9.3-14) applies. The intensity coefficients $K_m(\phi)$ are given by (see (9.3-9))

$$K_m(\phi) = \lim_{\Delta\tau \to 0} \frac{1}{\Delta\tau} E\{[\phi(\tau + \Delta\tau) - \phi(\tau)]^m \mid \phi(\tau)\} \tag{9.3-83}$$

For $m = 1$ we obtain $K_1(\phi) = -g(\phi) + \gamma$, and for $m \geq 2$ we use the binomial theorem to write

$$K_m(\phi) = \lim_{\Delta\tau \to 0} \frac{1}{\Delta\tau} \, E\left\{ \left[\int_\tau^{\tau+\Delta\tau} n'(v, \phi)\, dv + \int_\tau^{\tau+\Delta\tau} n'_p(v)\, dv \right]^m \right\}$$

$$= \lim_{\Delta\tau \to 0} \frac{1}{\Delta\tau} \left\{ \sum_{i=0}^m \begin{bmatrix} m \\ i \end{bmatrix} E\left\{ \left[\int_\tau^{\tau+\Delta\tau} n'(v, \phi)\, dv \right]^i \right\} \right.$$

$$\left. \times\, E\left\{ \left[\frac{\sqrt{2}}{A} \sum_j x_j \int_\tau^{\tau+\Delta\tau} \delta(\tau' - \tau_j) \cos \Phi_j\, d\tau' \right]^{m-i} \right\} \right\} \qquad (9.3\text{-}84)$$

As $\Delta\tau$ approaches zero the increment of the Poisson process becomes either zero (no jump) with probability $1 - (\lambda_p/4B_L)\,\Delta\tau$ or one with probability $(\lambda_p/4B_L)\,\Delta\tau$. Thus

$$E\left\{ \left[\frac{\sqrt{2}}{A} \sum_j x_j \int_\tau^{\tau+\Delta\tau} \delta(\tau' - \tau_j) \cos \phi_j\, d\tau' \right]^{m-1} \right\}$$

$$= (\sqrt{2})^{m-i} M_{m-i} Q_{m-i} \, \Pr\,(\text{jump in } \Delta\tau) \qquad (9.3\text{-}85)$$

where

$$M_{m-i} = \frac{E(x_j^{m-i})}{A^{m-i}}, \quad Q_{m-i} = E[(\cos \phi_j)^{m-i}]$$

$$\Pr\,(\text{jump in } \Delta\tau) = (\lambda_p/4B_L)\Delta\tau \qquad (9.3\text{-}86)$$

The moment of $\cos \Phi_j$ can be expressed by the well-known gamma function

$$Q_k = \begin{cases} \dfrac{\Gamma(k+1)}{2^k \Gamma^2(k/2 + 1)}, & k \text{ even} \\[2ex] 0, & k \text{ odd} \end{cases}$$

Since $n'(\tau, \phi)$ is a gaussian process which approaches white noise we finally obtain

$$K_m(\phi) = \begin{cases} -g(\phi) + \gamma, & m = 1 \\[2ex] \dfrac{2}{\rho} + 2\left(\dfrac{\lambda_p}{4B_L} \right) M_2 Q_2, & m = 2 \\[2ex] (\sqrt{2})^m \left(\dfrac{\lambda_p}{4B_L} \right) M_m Q_m, & m = 2k, \quad k \geq 2 \\[2ex] 0, & m = 2k - 1, k \geq 2 \end{cases} \qquad (9.3\text{-}87)$$

with the signal-to-noise ratio in the loop ρ defined by $\dfrac{1}{\rho} := 2B_L S_{n'}(0,0)$.

The intensity coefficient $K_2(\phi)$ is comprised of the contribution of the gaussian noise and of the impulsive noise. The latter contribution is proportional to the normalized pulse density and the signal-to-noise ratio M_2 of the impulsive noise. The coefficients $K_m(\phi)$, $m \geq 4$, are due to the impulsive noise only.

Substitution of the intensity coefficients $K_m(\phi)$ into (9.3-14) leads to the Kramers–Moyal expansion for the probability density function of the phase error ϕ

$$\frac{\partial p}{\partial \tau}(\phi, \tau) = \frac{\partial}{\partial \phi}[(-g(\phi) + \gamma)p(\phi, \tau)] - \left[\frac{1}{\rho} + \left(\frac{\lambda_p}{4B_L}\right)M_2 Q_2\right]\frac{\partial^2}{\partial \phi^2} p(\phi, \tau)$$

$$+ \left(\frac{\lambda_p}{4B_L}\right)\sum_{m=2}^{\infty}\frac{(\sqrt{2})^{2m}}{(2m)!} M_{2m}Q_{2m}\frac{\partial^{2m}}{\partial \phi^{2m}} p(\phi, \tau) \qquad (9.3\text{-}88)$$

Truncating at $n = 2$ yields the diffusion approximation. To see under what conditions this approximation is valid we parameterize the impulsive noise in the following way

$$n'_{p,\varepsilon}(\tau) = \frac{\sqrt{2}}{A}\sum_j x_j \varepsilon \delta(\tau - \tau_j \varepsilon^2) \cos \phi_j$$

The normalized pulse rate and the moments M_m become

$$\frac{\lambda_{p,\varepsilon}}{4B_L} = \frac{\lambda_p}{4B_L\,\varepsilon^2}, \quad M_{m,\varepsilon} = M_m \varepsilon^m$$

It immediately follows that $K_2(\phi)$ is independent of ε. For $m = 2k$, $k \geq 2$, the intensity coefficients vanish as $\varepsilon \to 0$. The diffusion approximation is thus valid if the normalized pulse rate is much higher than one and if the mth moments, $m > 2$ of the normalized pulse amplitude become small, i.e.

$$\frac{\lambda_{p,\varepsilon}}{4B_L} \gg 1, \quad M_{m,\varepsilon} \ll 1$$

The result is interesting. Physically, it means that the perturbation by impulsive noise with high density but small amplitude approximates white gaussian noise. It is easy to verify that the power spectrum of $n'_{p,\varepsilon}(\tau)$ is white with a spectral density equal to the second term of $K_2(\phi)$ for all values of ε. For more general results on jump processes the reader is referred to the book by Gardiner [17].

REFERENCES

1. A. Papoulis, *Probability, Random Variables, and Stochastic Processes*, McGraw-Hill, New York, 2nd edn., 1984.

2. W. C. Lindsey, *Synchronization System in Communication and Control*, Prentice Hall, Englewood Cliffs, NJ, 1972.

3. A. J. Viterbi, *Principles of Coherent Communications*, McGraw-Hill, New York, 1966.

4. A. H. Jazwinski, *Stochastic Processes and Filtering Theory*, Academic Press, New York, London, 1970.

5. K. Itô, Stochastic Integral, *Proceedings of the Imperial Academy* (Tokyo), Vol. 20, pp. 519–524, 1944.

6. R. L. Stratonovich, A New Form of Representing Stochastic Integrals and Equations, *SIAM Journal on Control*, Vol. 4, pp. 362–371, 1966.

7. R. L. Stratonovich, *Topics in the Theory of Random Noise*, Gordon Breach, New York, 1963.

8. H. Risken, *The Fokker-Planck Equation*, Springer-Verlag, Berlin, 1984.

9. J. M. C. Clark, The Representation of Nonlinear Stochastic Systems with Applications to Filtering, Ph.D. Thesis, Electrical Engineering Department, Imperial College, London, England, 1966.

10. R. F. Pawula, *Physical Review*, Vol. 162, p. 186, 1967.

11. D. L. Snyder, *Random Point Processes*, John Wiley, New York, 1975.

12. J. L. Hibey, Cycle Slipping in an Optical Communication System Employing Subcarrier Angle Modulation, *IEEE Transactions on Information Theory*, Vol. 33, No. 2, pp. 203–209, March 1987.

13. R. M. Gagliardi, Synchronization Using Pulse Edge Tracking in Optical Pulse-Position Modulated Communication Systems, *IEEE Transactions on Communications*, Vol. 22, pp. 1693–1702, October 1972.

14. R. M. Gagliardi and S. Karp, *Optical Communications*, John Wiley, New York, 1976.

15. D. L. Snyder and I. B. Rhodes, Phase and Frequency Tracking Accuracy in Direct-Detection Optical Communications Systems, *IEEE Transactions on Communications*, Vol. 20, pp. 1139–1142, December 1972.

16. E. Wong, *Stochastic Processes in Information and Dynamical Systems*, McGraw-Hill, New York, 1971.

17. C. W. Gardiner, *Handbook of Stochastic Methods*, Springer-Verlag, Berlin, 1983.

18. H. J. Kushner, Diffusion Approximation to Output Processes of Nonlinear Systems with Wide-Band Inputs and Applications, *IEEE Transactions on Information Theory*, Vol. 26, pp. 715–725, November 1980.

19. N. G. van Kampen, "Langevin-Like Equation with Colored Noise", *Journal of Statistical Physics*, Vol. 54, Nos. 5/6, 1989.

10

RELAXATION TIMES, MEANTIME BETWEEN CYCLE SLIPS, TRANSITION RATES, AND EIGENVALUES OF FOKKER–PLANCK OPERATORS

10.1. INTRODUCTION

As a preparation for the detailed mathematical analysis to follow, we want to *qualitatively* relate the above parameters in this section. Let us start with an example. If we analyze the phase-locked loop by solving the Fokker–Planck equation we are faced with the problem in which interval do we want to study the phase error. A typical trajectory of the phase error is plotted in Figure 10.1-1. Due to the cycle slips the phase error trajectory extends over the entire axis with the passage of time. This is reflected in the build-up of the steady-state probability density function in Figure 10.1-2. Initially, the phase error is taken at its stable tracking point ϕ_0

$$p(\varphi, t) = \delta(\varphi - \phi_0) \qquad (10.1\text{-}1)$$

After a short time interval T_x the initial probability density function has diffused into the vicinity of ϕ_0. From then on the probability density function retains its shape over a long time and we speak of a "quasi-stationary" state. The time constant to reach this state is called the *relaxation time constant* T_x. It is typically in the order of magnitude of $1/B_L$. Eventually the loop slips a cycle therefore, the probability mass begins to build up in the adjacent 2π intervals (see Figure 10.1-2c). This build-up occurs much more slowly than the relaxation process mentioned before. One measure for this build-up is the *meantime between cycle slips*, $E(T_s)$, introduced in Chapter 6. With the passage of time the probability mass diffuses over the entire axis until, in the steady state

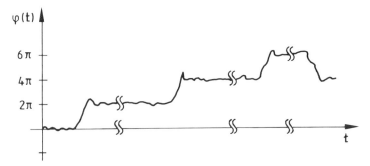

Figure 10.1-1. Unrestricted phase error trajectory $\varphi(t)$.

$$p(\varphi, t) \rightarrow 0 \quad \text{as } t \rightarrow \infty \tag{10.1-2}$$

Equation (10.1-2) says that it is equally probable to find the particle anywhere along the φ-axis. Thus, φ possesses an unbounded variance.

But since the steady-state behavior of the loop is of utmost interest we must look for a "modified" phase error process which retains the pertinent information of the original process yet yields a meaningful mathematical solution in the steady state, i.e. $t \rightarrow \infty$. There are various ways to define such a modified process.

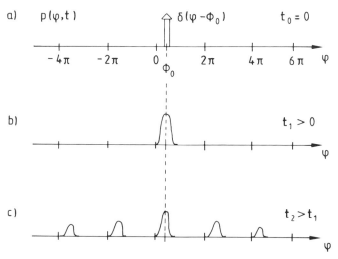

Figure 10.1-2. Evolution of the transition probability density $p(\varphi, t)$ with the passage of time. (a) Initially all probability is concentrated at $\varphi = \phi_0$. (b) *Diffusion of the phase error for $t > T_x$.* Nearly all probability mass is in the initial 2π interval. T_x, relaxation time constant. (c) Diffusion of the phase error for $t > E(T_s) \gg T_x$. Flow of probability mass to the kth 2π interval due to cycle slipping. $E(T_s)$, mean time between cycle slips.

10.2.   MODULO 2π PHASE ERROR PROCESS

A phase meter determines the phase error ϕ only within an interval $[-\pi, \pi]$; i.e., the unrestricted phase error is reduced modulo 2π (Figure 10.2-1a). Such a reduced process can be viewed as being defined on a circle S_1 which is more convenient than defining it on a $[-\pi, \pi]$ interval of the real axis R_1. On a circle the process becomes continuous as the jumps disappear. This is shown in Figure 10.2-1b for the example of a first-order phase-locked

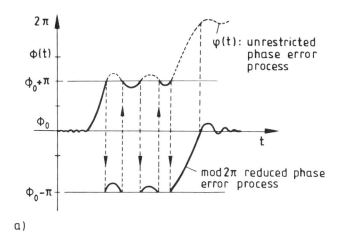

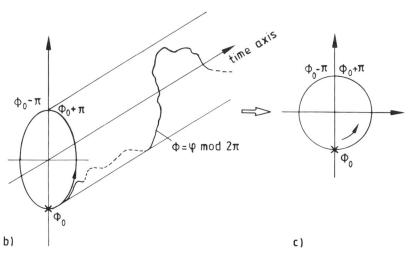

Figure 10.2-1. (a) Modulo 2π reduced phase error process $\phi(t)$. (b) Phase error trajectory on a cylinder obtained by joining the axis $\phi = \phi_0 + \pi$ and $\phi = \phi_0 - \pi$. (c) Projection of the phase error $\phi(t)$ on a circle. An error is counted positive if the circle is traversed counterclockwise.

loop. By construction, since all probability mass (or particles) moves on the surface of a cylinder, a steady-state probability density function with bounded variance exists. The probability density function $p(\phi, y, t)^*$ is periodic in ϕ with period 2π. From this it follows that $p(\phi, y, t)$ is the solution of the Fokker–Planck equation subject to periodic boundary conditions on ϕ

$$p(-\pi, y, t) = p(\pi, y, t)$$

$$\frac{\partial p}{\partial \phi}(-\pi, y, t) = \frac{\partial p}{\partial \phi}(\pi, y, t)$$

(10.2-1)

where y denotes the remaining state variables.

Several important properties can be inferred by viewing the phase error trajectory $\phi(t)$ as a movement on a cylinder. First, there exists only one stable equilibrium point on the cylinder. In analogy to physics we call this tracking point an *attractor* to express the fact that a particle is attracted by a force toward this point. Therefore, only the dynamics of the phase error within the domain of attraction of a *single* attractor is considered by the modulo 2π phase error process. Any evolution of the phase error trajectory always leads back to the same attractor. Transitions from one attractor to the neighboring attractor are not considered by this construction. Hence, from the modulo 2π phase error one obtains information on the relaxation times of a system but not on phenomena which involve transitions from one attractor to another, i.e., cycle slips.

10.3. RENEWAL PROCESS

In order to include cycle slips the modified phase error process must include at least two attractors so that particles can move from one attractor to a neighboring attractor. There are various ways to define a modified phase error process with the above property. Here we first discuss the so-called *renewal process*. The basic principle is explained for a first-order system (one-dimensional case) and later generalized to an nth order system.

We consider a particle which is released at the position ϕ_0 of the attractor (stable tracking point) at time zero. The particle randomly moves around the attractor for a while but eventually reaches one of the barriers located at $\pm 2\pi$ away from ϕ_0. By doing so the particle has made a cycle slip, either a positive one if it reaches $\phi_0 + 2\pi$ or a negative one in the other case. When the particle reaches either of these two barriers for the first time it is absorbed. In order to preserve the number of particles, a new particle is started simultaneously at $\phi = \phi_0$. This procedure is repeated every time a

---

*We write for the state vector $x^T = [\phi, y]$ where ϕ is the phase error and $y = [x_2, \ldots, x_n]$ are the remaining state variables.

particle reaches the barrier for the first time and yields the phase error trajectory $\phi(t)$ shown in Figure 10.3-1. Since each time a particle is absorbed a new particle is created the equivalent phase error process has been called a "*renewal process.*"

The renewal process retains *all* the information contained in the original process defined on the entire real axis, it is thus an *equivalent* phase error process. Recording the renewal epochs $t_1, t_2, \ldots, t_n$ and the magnitude of jumps $\eta_1, \eta_2, \ldots, \eta_n$ one can always reconstruct the original process from the renewal process. The random variable $T_k = t_k - t_{k-1}$ equals the lifetime of the kth particle and therefore, also the time between the $(k-1)$th and the kth cycle slips. The random jump variable η_k describes whether the slip was positive in which case $\eta_k = +2\pi$ or negative, $\eta_k = -2\pi$. Thus, the collection of pairs (T_k, η_k) represents the complete statistical description of the cycle slips. As the modulo 2π phase error is the union of two mutually exclusive events of the renewal process, namely

(A_1) $\phi \in [\phi, \phi + d\phi]$

  or   for $\phi \in [\phi_0, \phi_0 + \pi]$

  $\phi \in [\phi - 2\pi, \phi - 2\pi + d\phi]$

and

(A_2) $\phi \in [\phi, \phi + d\phi]$

  or   for $\phi \in [\phi_0 - \pi, \phi_0]$

  $[\phi + 2\pi, \phi + 2\pi + d\phi]$

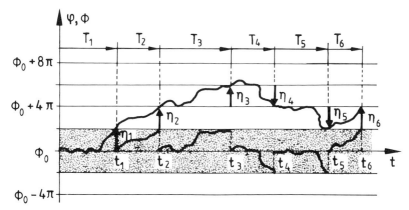

Figure 10.3-1. Sample trajectory of unrestricted process $\varphi(t)$ and of process $\phi(t)$ defined in 4π-interval $[\phi_0 - 2\pi, \phi_0 + 2\pi]$. Random variable T_k (lifetime of individual trajectory), renewal epochs t_k, and random variable η_k (magnitude of jumps at t_k) are also indicated.

the renewal process also carries the information on the relaxation phenomena which involve a single attractor only. In other words, the modulo 2π phase error information can always be obtained from the renewal process but not vice versa.

The extension to n-dimension is shown in Figure 10.3-2. When the kth particle is absorbed the $(k + 1)$th particle must start with the same values of the state variables $y = [x_2, \ldots, x_n]$ as its predecessor. Note that $x_2, \ldots, x_n$ assume random values at the renewal epochs. Therefore, the initial conditions for the kth particle are also random, which complicates the analysis.

The boundary conditions for the renewal process are of the absorbing type

$$p(\phi_0 - 2\pi, y, t) = p(\phi_0 + 2\pi, y, t) = 0 \qquad (10.3\text{-}2)$$

We repeat that the n-dimensional renewal process retains all the statistical information of the original process. For the sake of accuracy in terminology we emphasize that the n-dimensional process is *not* a renewal process in the strict mathematical sense. Mathematically, a renewal process is defined as a process which starts, in a statistical sense, anew at every absorption of a trajectory. This implies that the individual processes are statistically independent with the same initial distribution. But since the values of the state variables $[x_2, \ldots, x_n]$ at the time of absorption are the initial values of the following individual process, the repeated vector process is not a renewal process.

However, we believe that the repeated process of absorption and creation of particles is well described by the term "renewal process" and thus we have chosen to call the repeated vector process a *renewal process*. A "true" (in the mathematical sense) renewal process will be called a *renewal process in the strict sense*.

The renewal approach (in the strict sense) is not restricted to periodic nonlinearities, but equally well applies to systems with aperiodic nonlinearities such as delay-locked loops (DLLs) to be discussed in Section 11.3.

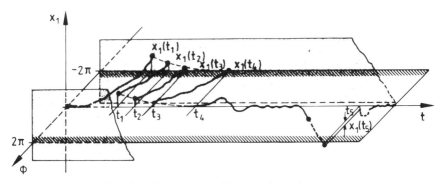

Figure 10.3-2. Sample trajectory of the two-dimensional renewal process.

10.4. BISTABLE AND MULTISTABLE CYCLIC MODELS

Instead of absorbing the particles at the boundaries $\phi = \pm 2\pi$ (as in the renewal process model) we measure the phase error modulo 4π. Thus, the trajectory of the particle is considered on a cylinder with circumference 4π. The key difference to the modulo 2π phase error process, as introduced previously, is that we consider two attractors on the cylinder. Thus, the model includes transitions from one attractor to the other (see Figure 10.4-1).

Multiple transitions as they occur in a burst of slips in higher order systems cannot be distinguished from a single (or no) transition in this model. For example, assume the PLL makes a burst of length two. Thus, the trajectory starts and ends at the same attractor bypassing the other attractor (i.e., the particle does not enter the domain of attraction of this attractor). Therefore, since starting and final attractor are identical, this event is not counted as a transition. In general, multiple transitions with an even number of transitions are ignored while odd numbers of transitions are counted as single transitions from one attractor to the next.

The bistable cyclic model can be extended to a model with M attractors in which the phase error is then considered modulo $M2\pi$. Such a model will be called *multistable cyclic model* (or *M-attractor cyclic model*). Clearly, using such a model, multiple transitions up to length $M - 1$ are distinguishable.

Since the phase error is measured modulo $M2\pi$, $M = 2, 3, 4 \ldots$ the transition probability density function and the probability current are periodic with period $M2\pi$. Thus, the boundary conditions are of the periodic type

$$p(-M\pi, y, t) = p(M\pi, y, t)$$

$$\frac{\partial p}{\partial \phi}(-M\pi, y, t) = \frac{\partial p}{\partial \phi}(M\pi, y, t)$$

(10.4-1)

Since all the probability mass is located on a cylinder for all time, a steady-state distribution with bounded variance exists. This steady-state

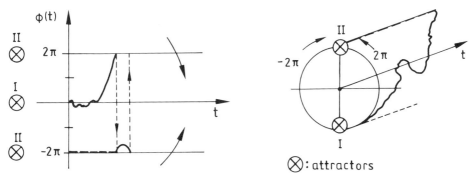

Figure 10.4-1. Construction of the modulo 4π phase error for a first-order system.

probability density function has a period of 2π (and *not* only $M2\pi$) and is identical to the modulo 2π probability density function apart from a constant of proportionality factor of $1/M$. This interesting fact is readily explained by considering a bistable system, $M = 2$ (see Figure 10.4-2).

Let us assume the particle is in the domain of attractor I at time zero. After having spent a long time around attractor I it makes a single transition to attractor II where it again spends a long time before it makes another (possibly multiple) transition which ends at either one of the two attractors. Clearly, as time passes, this process repeats infinitely often and it becomes equally probable to find the particle in either of the two domains of attractions

$$P_I = P_{II} = \frac{1}{2} \qquad (10.4\text{-}2)$$

Therefore, the attractors cannot be distinguished. This implies that the steady-state probability density function must be identical for the two attractors and, must equal the probability density function of the mod 2π reduced phase error. (apart from the factor $1/2$).

Notice that the reasoning given above is true irrespective of whether the particle has a preferred direction of motion, that is a tendency to make more positive or negative cycle slips. A preference for positive transitions, for example, means that the particle has the tendency to move counterclockwise on the cylinder but has no consequence with respect to the fact that both attractors are equally probable locations for the particle.

The extension of the above results to a multistable cyclic system is straightforward: the steady state probability density function is obtained by periodically repeating the modulo 2π probability density function over an $M2\pi$ interval. The scaling factor of $1/M$ guarantees that the integral over

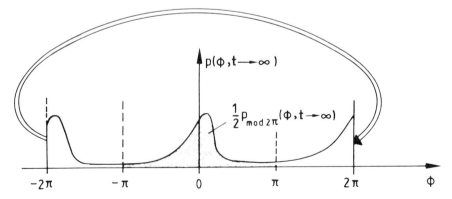

Figure 10.4-2. Steady state probability density function for a bistable system, $M = 2$. The modulo 2π probability density function multiplied by $1/2$ is shown "marked".

$M2\pi$ equals one, as is required by the normalization condition of the probability density function.

We want to close this discussion by emphasizing again that the bistable (multistable) cyclic model does not contain the full information on cycle slips since multiple transitions of up to length $(M - 1)$ only are distinguishable. It does, however, provide information on the *transition rates* which are the single most important performance measure when *both* nonlinearity and noise have to be taken into account.

10.5. "COARSE-GRAINED" MODEL

In properly operating communication systems, cycle slips are very rare events. We are, therefore, mostly interested in system behavior at moderate noise level where cycle slips are rare events and, as a consequence, where the stationary probability density function is concentrated within well separated domains R_I, each containing one attractor (stable operating point). Since transitions $R_i \to R_j$ are rare events two different time scales exist: the slow one for the transitions and the fast one for the motion inside the R_I.

Often we are interested primarily in the transition rate but not in details of the fast motion inside R_I. Thus, this splitting up of the time scale can be exploited to develop a simpler model which ignores the fine structure of the phase error process but retains the essential features of transitions between attractors. In this "coarse grained" model, the exact value of the phase error within R_I is not of interest. The system is thus approximated by an M-state model as shown in Figure 10.5-1, exemplary for $M = 4$. The phase error process $\phi(t)$ assumes only $M = 4$ different values. A value Φ_j is assumed if the system is in state $I = j$.

Since the attractors in the M-attractor model are indistinguishable it suffices to describe a transition event by the numbers of slipped attractors, $k = 1, 2, . M - 1$. With the convention that the states are numbered consecutively and that a positive label k denotes a transition from state I to state $(I - k)$ modulo M we have single, double and triple transitions as shown in Figure 10.5-1. (The transition rates are denoted by a_k, $k = 1, 2, \ldots M - 1$).

We should be aware that the M-attractor model does not contain the full statistical information on cycle slips. To see this fact we briefly digress to the unrestricted phase error process. In the picture of the unrestricted phase error $-\infty < \varphi < \infty$ we denote the transition rate by r_k where the sign of k indicates the direction of the jumps and k the number of slipped periods. The connection of r_k with the a_k is

$$a_k = \sum_{\nu = -\infty}^{\infty} r_{k + \nu M}, \quad 1 \le k < M$$

We observe the following phenomena

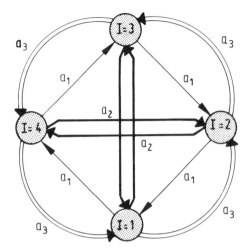

Figure 10.5-1. State transition diagram for a system with $M = 4$ attractors.

1. Cycle slips by M periods ($r_{\nu M}$) are fully ignored in the M-attractor cyclic model.
2. Cycle slip directions are indistinguishable. r_k and r_{k-M} contribute to the same a_k, for example, $M = 4$, r_1 and r_{-3} to a_1 or r_2 and r_{-2} to a_2.
3. Cycle slip length may also be misinterpreted, for example, r_1 and r_{-3} in a_1.

As will be shown in Chapter 12 the transition rates in a "coarse grained" multistable cyclic model with M-attractors are determined by the smallest $(M - 1)$ eigenvalues of the Fokker–Planck operator L. It is the great advantage of the cyclic model that these eigenvalues can be computed numerically.

Main Points of the Chapter

- Since the phase error $\phi(t)$ eventually exceeds all limits as time goes to infinity, various *modified phase error processes* are introduced which have a meaningful steady state probability distribution.
- The *renewal process* retains the full statistical information. No assumptions have to be made on the noise intensity.
- *The cyclic models*

  The modulo 2π process contains information of a phase meter only, but no useful information about cycle slips. The model applies to all noise intensities.

  The multistable $(M \geq 2)$ cyclic model contains the pertinent information on cycle slips. For moderate noise the transition rates a_k are determined by the first $(M - 1)$ smallest eigenvalues of the Fokker–Planck operator L.

11

RENEWAL PROCESS APPROACH

11.1. FIRST-ORDER SYSTEMS WITH PERIODIC PHASE DETECTOR CHARACTERISTIC

In this section we analyze the renewal process model for a first-order system. Even though first-order systems are very rarely used in applications, the theory of first-order systems is of great importance for various reasons.

From a mathematical point of view first-order systems are of interest because they are the only ones for which exact solutions to the Fokker–Planck equations exist. In many cases, the essential dynamic properties of a higher order system can be well approximated by a first-order system. Thus, the first-order system theory provides insight into the nonlinear behavior of higher order systems. The theory of higher order systems is a generalization of the fundamental concepts developed for first-order systems.

This section is made up as follows. In Section 11.1.1 the equivalent phase error model is defined and in Section 11.1.2 the probability law of the single process is derived. In the following section the basic idea of the renewal process approach is introduced. We derive basic recursion relations which relate the probability law of the renewal process to that of the defining single process. In Section 11.1.4 we use these relations to derive a modified Fokker–Planck equation of the renewal process. A particularly interesting and important role is played by the probability law of the process in the steady state. We deal with these aspects in Sections 11.1.5 and 11.1.6 where we relate stationary phase error distribution, meantime between cycle slips, and mean cycle slip rate.

Time-dependent statistics are discussed in the remaining sections. We first present mathematical techniques to solve the time-dependent Fokker–Planck equation. Based on these solutions we then obtain the distribution of the renewal epochs t_n and of the associated jumps and thus obtain the complete probability law of the renewal process.

11.1.1. Modeling the Phase Error as a Renewal Process

The basic idea to model the equivalent phase error process as a renewal process was introduced in Section 10.3. We now put these ideas into formal mathematics. The system to be analyzed obeys a stochastic differential equation of the form

$$d\varphi = g[\varphi(t)] \, dt + F[\varphi(t)] \, d\beta(t) \tag{11.1-1}$$

where $d\beta(t)$ is the increment of the Wiener process with unit variance parameter and $\varphi(t)$ is the phase error process. The use of the symbol $\varphi(t)$ for the unrestricted phase error process instead of ϕ, as is usual, will soon become clear. The functions g and F are assumed to have the following properties

(A$_1$) $g(\varphi + k2\pi) = g(\varphi)$

              $F(\varphi + k2\pi) = F(\varphi) \tag{11.1-2}$

(A$_2$) $g(\varphi), f(\varphi)$ do not explicitly depend on time

From the periodicity of $g(\varphi)$ and $F(\varphi)$, it follows that if $\varphi(t)$ is a solution of (11.1-1), then $\varphi(t) + k2\pi$ is also a solution for an arbitrary integer k. Property A$_2$ guarantees that $\varphi(t)$ is a time homogeneous Markov process. Let us assume that the process $\varphi(t)$ obeys the initial condition (ϕ_0 can be chosen arbitrarily)

$$\varphi(0) = \phi_0 \tag{11.1-3}$$

and that the process $\varphi(t)$ reaches one of the barriers $\varphi = \phi_0 \pm 2\pi$ for the first time at t_1 (see Figure 11.1-1), where

$$t_1 = \inf \{t \mid |\varphi(t) - \phi_0| \geq 2\pi\} \tag{11.1-4}$$

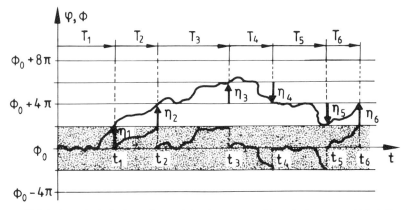

Figure 11.1-1. Sample trajectory of unrestricted process $\varphi(t)$ and of process $\phi(t)$ defined in 4π-interval $[\phi_0 - 2\pi, \phi_0 + 2\pi]$. Random variable T_k (lifetime of individual trajectory), renewal epochs t_k, and random variable η_k (magnitude of jumps at t_k) are also indicated.

We now define the equivalent phase error process $\phi(t)$, $\phi(t) \in (\phi_0 - 2\pi, \phi_0 + 2\pi)$ as follows

$$\phi(t) \begin{cases} \varphi(t) & \text{for } t \in [0, t_1) \\ \varphi(t) - \eta_1, & \text{for } t \in [t_1, t_2) \end{cases} \tag{11.1-5}$$

with

$$\eta_1 = \varphi(t_1) - \varphi(0)$$

$$t_2 = \inf \{t > t_1 \mid |(\varphi(t) - \eta_1) - \phi_0| \geq 2\pi\} \tag{11.1-6}$$

For an arbitrary $k \geq 1$, we similarly define

$$\phi(t) = \varphi(t) - \sum_{n=1}^{k-1} \eta_n, \quad \text{for } t \in [t_{k-1}, t_k) \tag{11.1-7}$$

with

$$\eta_k = \varphi(t_k) - \varphi(t_{k-1})$$

$$t_k = \inf \left\{ t > t_{k-1} \mid \left| \left(\varphi(t) - \sum_{n=1}^{k-1} \eta_n \right) - \phi_0 \right| \geq 2\pi \right\} \tag{11.1-8}$$

$$T_k = t_k - t_{k-1}$$

From the properties A_1 and A_2, it follows that $T_1, T_2, \ldots, T_n$ are independent random variables defined on $[0, \infty)$ with a common distribution $\psi(t)$. The variable T_n can be interpreted as the lifetime of an individual trajectory starting at $\phi = \phi_0$ and time t_{n-1}. Moreover, $\eta_1, \eta_2, \ldots, \eta_n$ are independent random variables assuming either of the two values $\pm 2\pi$. They correspond to a cycle slip of the tracking system. Note that the number of cycle slips to the right or left equals exactly the number of positive or negative values of η, respectively. Since $\varphi(t)$ is the unique solution of (11.1-1) and continuous with probability one, for every sample trajectory $\varphi(t)$ there exists exactly one trajectory $\phi(t)$ with probability one and vice versa. The process $\phi(t)$ shown in Figure 11.1-1 can be viewed as a *regenerative*, or *renewal*, process. As soon as the trajectory reaches one of the two barriers $\phi = \phi_0 \pm 2\pi$ for the first time it is absorbed and a new trajectory is started.

We often will find it useful to give a physical interpretation to results or to state a problem in physical terminology. For this reason we have listed the pertinent correspondences of the functions/quantities involved (Table 11.1-1).

TABLE 11.1-1 Correspondence of Physical and System Terminology

Quantity/Function	Physical Terminology	System Terminology
$\phi(t)$	Trajectory of a particle	Phase error, stochastic process
$p(\phi, t)\, d\phi$	Number of particles in $[\phi, \phi + d\phi]$ at time t	Probability of finding the phase error in $[\phi, \phi + d\phi]$ at time t
T_k	Lifetime of a particle	Time between the $(k-1)$th and kth cycle slip (random variable)
t_k	Epoch of absorption of a particle	Epoch of completion of the kth cycle slip
$J(\phi, t)$	Probability current	
$-J(\phi_0 - 2\pi, t) = \Gamma_-$ $J(\phi_0 + 2\pi, t) = \Gamma_+$	Number of particles per unit of time crossing the boundaries = absorption rate $\Gamma_{(\pm)}(t)$	
$\Psi(t)$	Point source intensity	
$E(T_s)$	Mean lifetime of a particle	Meantime between cycle slips
$N_{(\pm)} = \Gamma_{(\pm)}$	Average number of particles crossing the boundaries = average absorption rate	Average number of cycle slips per unit of time

11.1.2. Probability Laws of the Single Process

We denote the transition probability density function of the single process starting at $\phi = \phi_0$ and $t = 0$ by $q(\phi, t)$. (The reason for denoting a probability density function with $q(\phi, t)$ will soon become clear.) The function $q(\phi, t)$ satisfies the Fokker–Planck equation with

$$\frac{\partial q(\phi, t)}{\partial t} = Lq(\phi, t) \tag{11.1-9}$$

where

$$L = -\frac{\partial}{\partial \phi} K_1(\phi) + \frac{1}{2}\frac{\partial^2}{\partial \phi^2} K_2(\phi)$$

$$K_1(\phi) = g(\phi) + \frac{1}{2} F_\phi(\phi) F(\phi) \tag{11.1-10}$$

$$K_2(\phi) = F^2(\phi)$$

subject to absorbing boundary conditions

$$q(\phi_0 - 2\pi, t) = q(\phi_0 + 2\pi, t) = 0 \tag{11.1-11}$$

Because of the time homogeneity, the transition probability density function of the process starting at time t_k is simply $q(\phi, t - t_k)$. Since the particles are absorbed when they reach either of the two boundaries for the first time the integral over $q(\phi, t)$ at any time t equals the probability of finding a particle inside the interval $(\phi_0 - 2\pi, \phi_0 + 2\pi)$.

$$\Pr(T_1 > t) = \int_{\phi_0 - 2\pi}^{\phi_0 + 2\pi} q(\phi, t)\, d\phi \leq 1 \tag{11.1-12}$$

$\Pr(T_1 > t)$ is always a monotonically decreasing function of time t, at most equal to one. Hence, since $q(\phi, t)$ is not normalized to one, it is not a transition probability density function in the strict sense. However, the conservation law of probability is not violated. A particle that is not in the interval $(\phi_0 - 2\pi, \phi_0 + 2\pi)$ at time t must necessarily have been absorbed somewhere in the time interval $[0, t]$ with probability $1 - \Pr(T_1 > t)$. Hence, the probability mass $1 - \Pr(T_1 > t)$ is concentrated as point mass at the two points $\phi_0 \pm 2\pi$. This assertion can be verified by writing the Fokker–Planck equation in the "physical" form

$$\frac{\partial q(\phi, t)}{\partial t} + \frac{\partial j(\phi, t)}{\partial \phi} = 0 \tag{11.1-13}$$

where $j(\phi, t)$ is the probability current (9.3-39). Integrating equation (11.1-13) once over ϕ yields

$$\frac{\partial}{\partial t} \int_{\phi_0 - 2\pi}^{\phi_0 + 2\pi} q(\phi, t)\, d\phi + j(\phi_0 + 2\pi, t) - j(\phi_0 - 2\pi, t) = 0 \tag{11.1-14}$$

The integral equals the probability that the trajectory $\phi(t)$ has not yet reached the boundaries at time t

$$\Pr(T_1 > t) = \int_{\phi_0 - 2\pi}^{\phi_0 + 2\pi} q(\phi, t)\, d\phi \tag{11.1-15}$$

Thus, one minus this probability equals the probability that the trajectory has reached the boundaries at time t

$$\Pr(T_1 > t) = 1 - \Pr(T_1 \leq t)$$
$$= 1 - P_{T_1}(t) \tag{11.1-16}$$

where $P_{T_1}(t)$ is the probability distribution of the time to reach the boundaries for the first time

$$P_{T_1}(t) = 1 - \int_{\phi_0 - 2\pi}^{\phi_0 + 2\pi} q(\phi, t) \, d\phi \qquad (11.1\text{-}17)$$

Differentiating $P_{T_1}(t)$ with respect to t yields the probability density function of T_1

$$\psi(t) := \frac{dP_{T_1}(t)}{dt} = -\frac{\partial}{\partial t} \int_{\phi_0 - 2\pi}^{\phi_0 + 2\pi} q(\phi, t) \, d\phi \qquad (11.1\text{-}18)$$

Replacing the term with the integral in (11.1-14) by $\psi(t)$ finally yields

$$\psi(t) = j(\phi_0 + 2\pi, t) - j(\phi_0 - 2\pi, t) \qquad (11.1\text{-}19)$$

Equation (11.1-19) is the desired result. The probability that the particle is absorbed in $[t, t + dt]$ equals $\psi(t) \, dt$. Equation (11.1-13) tells us that $\psi(t) \, dt$ equals the flow of particles out of the interval $(\phi_0 - 2\pi, \phi_0 + 2\pi)$ in dt

$$\gamma_+(t) \, dt = j(\phi_0 + 2\pi, t) \, dt \quad \text{relative number of particles absorbed at the upper boundary in } [t, t + dt]$$

$$(11.1\text{-}19a)$$

$$\gamma_-(t) \, dt = -j(\phi_0 - 2\pi, t) \, dt \quad \text{relative number of particles absorbed at the lower boundary in } [t, t + dt]$$

Hence the loss of probability inside the interval $(\phi_0 - 2\pi, \phi_0 + 2\pi)$ during the time interval $[0, t]$ is compensated by the build-up of point masses at the boundaries

$$\int_0^t \psi(t') \, dt' = -\int_0^t j(\phi_0 - 2\pi, t') \, dt' + \int_0^t j(\phi_0 + 2\pi, t') \, dt' \qquad (11.1\text{-}19b)$$

The functions $\gamma_+(t)$ and $\gamma_-(t)$ introduced above are called the positive/negative *absorption rate* of the single process, respectively.

Note that since the probability current is a vectorial quantity which is counted positive in the positive ϕ direction, the negative sign of the current $j(\phi_0 - 2\pi, t)$ is necessary to obtain a vector that points outward. We notice that (11.1-14) is the one-dimensional representation of the Gauss theorem of vector analysis cited earlier (9.3-77).

The expected time to reach one of the two boundaries for the first time is

$$E(T_1) = \int_0^\infty -t \frac{\partial}{\partial t} [1 - P_{T_1}(t)] \, dt = -t [1 - P_{T_1}(t)] \Big|_0^\infty + \int_0^\infty [1 - P_{T_1}(t)] \, dt \qquad (11.1\text{-}20)$$

If the probability distribution $1 - P_{T_1}(t)$ approaches zero faster than $1/t$, the first term on the right-hand side of (11.1-20) is zero. This must be the case if the integral in the second term is to exist. Inserting $1 - P_{T_1}(t)$ from (11.1-17) in the last equation yields

$$E(T_1) = \int_0^\infty dt \int_{\phi_0 - 2\pi}^{\phi_0 + 2\pi} q(\phi, t)\, d\phi \qquad (11.1\text{-}21)$$

If we define

$$Q(\phi) = \int_0^\infty q(\phi, t)\, dt \qquad (11.1\text{-}21a)$$

and interchange the order of integration we can write $E(T_1)$ as

$$E(T_1) = \int_{\phi_0 - 2\pi}^{\phi_0 + 2\pi} Q(\phi)\, d\phi \qquad (11.1\text{-}22)$$

Let us next derive a differential equation whose solution is $Q(\phi)$. Integrating $Q(\phi)$ over ϕ then leads to the average time to reach the boundaries, $E(T_1)$. (In our "physical" analogy we will also often call $E(T_1)$ the mean lifetime of a particle.)

Integrating both sides of (11.1-9) with respect to t over the infinite interval we obtain

$$q(\phi, \infty) - q(\phi, 0) = -\frac{d}{d\phi} K_1(\phi)Q(\phi) + \frac{1}{2}\frac{d^2}{d\phi^2} K_2(\phi)Q(\phi)$$

$$= LQ(\phi) \qquad (11.1\text{-}23)$$

Since sooner or later all particles are absorbed, we have $q(\phi, \infty) = 0$. Since $\phi(0) = \phi_0$ initially

$$q(\phi, 0) = \delta(\phi - \phi_0)$$

Therefore, $Q(\phi)$ is the solution of the linear, ordinary differential equation

$$\delta(\phi - \phi_0) = +\frac{d}{d\phi} K_1(\phi)Q(\phi) - \frac{1}{2}\frac{d^2}{d\phi^2} K_2(\phi)Q(\phi) \qquad (11.1\text{-}24)$$

with the boundary conditions

$$Q(\phi_0 - 2\pi) = \int_0^\infty q(\phi_0 - 2\pi, t)\, dt = 0$$

$$\qquad (11.1\text{-}25)$$

$$Q(\phi_0 + 2\pi) = \int_0^\infty q(\phi_0 + 2\pi, t)\, dt = 0$$

which follow from (11.1-11).

Let us summarize: the transition probability density function $q(\phi, t)$ of the single process which stops when it reaches the boundaries for the first time is obtained by solving a Fokker–Planck equation with absorbing boundary conditions. All probability laws can be inferred from knowledge of $q(\phi, t)$. For example, the probability distribution of the first time to reach the boundaries is one minus the integral of $q(\phi, t)$ over ϕ.

11.1.3. Basic Recurrence Relations for the Renewal Process

Without explicitly stating it all the time, we assume that all the processes start at ϕ_0 and are absorbed at $\phi_0 \pm 2\pi$ when they reach the boundaries for the first time. Let us define the following events for the nth process

A_n: the trajectory of the nth process is in $[\phi, \phi + d\phi]$ at time t

$$(11.1\text{-}26)$$

B_n: the nth process was stopped in $[t, t + dt]$

These two events can be used to form both conditioned and combined events

$A_n B_1$:   the nth process is in $[\phi, \phi + d\phi]$ at time t
   and the first process was stopped in $[t', t' + dt']$, $(t' < t)$

$$(11.1\text{-}27)$$

$A_n \mid B_1$:   the nth process is in $[\phi, \phi + d\phi]$ at time t under the *condition* that the first process was stopped in $[t', t' + dt']$

We assign the following probabilities to the single events

$$\Pr(A_n) = w_n(\phi, t)\, d\phi$$
$$\Pr(B_n) = \psi_n(t)\, dt$$

$$(11.1\text{-}28)$$

For $n = 1$ we recover the probability density functions of the single process

$$\Pr(A_1) = w_1(\phi, t)\, d\phi := q(\phi, t)\, d\phi$$
$$\Pr(B_1) = \psi(t)\, dt$$

$$(11.1\text{-}29)$$

The individual processes are time-homogeneous and statistically independent (Markov property). Therefore, the conditional probability $\Pr(A_n \mid B_1)$ equals the probability that the $(n-1)$th process is in $(\phi, \phi + d\phi)$ if the first process started at t'. Formally

$$\Pr(A_n \mid B_1) = w_{n-1}(\phi, t - t')\, d\phi \qquad (11.1\text{-}30)$$

Using Bayes rule we find for the combined event

$$\Pr(A_n B_1) = P(A_n \mid B_1) P(B_1)$$

$$\begin{cases} w_{n-1}(\phi, t - t') \psi(t') \, d\phi \, dt' , & 0 \le t' < t \\ 0 , & \text{otherwise} \end{cases} \tag{11.1-31}$$

By integration over all possible values of t' we obtain the *fundamental recursion* between w_{n-1} and w_n (Figure 11.1-2)

$$\Pr(A_n) = w_n(\phi, t) \, d\phi$$

$$= d\phi \int_0^t w_{n-1}(\phi, t - t') \psi(t') \, dt' \tag{11.1-32}$$

The second line in (11.1-32) is a convolutional integral which becomes a product if formulated in the Laplace domain with

$$w_n(\phi, s) = \int_0^\infty e^{-st} w_n(\phi, t) \, dt \tag{11.1-33}$$

and similarly for all other functions the recursion reads

$$w_n(\phi, s) = w_{n-1}(\phi, s) \psi(s) \tag{11.1-34}$$

Replacing $w_{n-1}(\phi, s)$ by $w_{n-2}(\phi, s) \psi(s)$ and so on until $n - 1 = 1$, we obtain

$$w_n(\phi, s) = w_1(\phi, s) \psi^{n-1}(s)$$

$$= q(\phi, s) \psi^{n-1}(s) \tag{11.1-35}$$

Now, the probability that any trajectory, $n = 1, 2, \ldots$, is in a given interval $d\phi$ at time t is, by construction, the probability of finding the trajectory of the renewal process in this interval.

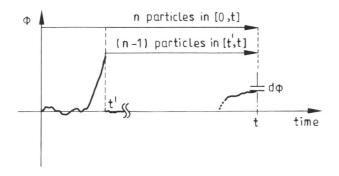

Figure 11.1-2. Fundamental recursion between $w_n(\phi, t)$ and $w_{n-1}(\phi, t)$.

$$p(\phi, s)\, d\phi = \Pr\left[\bigcup_{n=1}^{\infty} A_n\right]$$

$$= \sum_{n=1}^{\infty} w_n(\phi, s)\, d\phi$$

$$= q(\phi, s) \sum_{n=1}^{\infty} \psi^{n-1}(s)\, d\phi \qquad (11.1\text{-}36)$$

The infinite series on the right-hand side is a geometric series for which a closed form exists

$$\sum_{n=1}^{\infty} \psi^{n-1}(s) = \frac{1}{1 - \psi(s)} \qquad (11.1\text{-}37)$$

Finally, using this result we find the *fundamental relation* between the *single* and *repeated* process in the stationary as well as in the transient mode

$$p(\phi, s) = \frac{q(\phi, s)}{1 - \psi(s)} \qquad (11.1\text{-}38)$$

By a very similar reasoning one obtains a recursive relation between $\psi_{n-1}(t)$ and $\psi_n(t)$. Since

$$\Pr(B_n \mid B_1) = \psi_{n-1}(t - t')\, dt \qquad (11.1\text{-}39)$$

the probability of the combined event equals

$$\Pr(B_n B_1) = P(B_n \mid B_1) P(B_1)$$

From this we obtain

$$Pr(B_n) = \psi_n(t)\, dt$$

$$= dt \int_0^t \psi_{n-1}(t - t') \psi(t')\, dt' \qquad (11.1\text{-}40)$$

or in Laplace space

$$\psi_n(s) = \psi_{n-1}(s)\psi(s)$$
$$= \psi^n(s) \qquad (11.1\text{-}41)$$

The event that *any* trajectory $n = 1, 2, \ldots$ is absorbed within $[t, t + dt]$ has the probability

$$\Psi(t)\,dt := \Pr\left[\bigcup_{n=1}^{\infty} B_n\right]$$

$$= \sum_{n=1}^{\infty} \psi_n(t)\,dt \qquad (11.1\text{-}42)$$

Using (11.1-41) we obtain in Laplace space

$$\Psi(s) = \frac{\psi(s)}{1 - \psi(s)} \qquad (11.1\text{-}43)$$

In summary: we have shown how the probability density function of the phase error and the probability density function of the absorption of the renewal process are related to the corresponding functions of the generating single process.

11.1.4. Modified Fokker–Planck Equation of the Renewal Process

We now develop the partial differential equation for the transition probability density function of the renewal process. Taking the Laplace transform of the Fokker–Planck equation of the single process (11.1-9) we obtain

$$sq(\phi, s) - Lq(\phi, s) - q(\phi, t = 0) = 0 \qquad (11.1\text{-}44)$$

with initial condition

$$q(\phi, t = 0) = \delta(\phi - \phi_0)$$

and boundary conditions

$$q(\phi_0 - 2\pi, t) = q(\phi_0 + 2\pi, t) = 0$$

Replacing $q(\phi, s)$ with $p(\phi, s)$ by using the result of (11.1-38) gives

$$sp(\phi, s) - Lp(\phi, s) - \frac{\delta(\phi - \phi_0)}{1 - \psi(s)} = 0 \qquad (11.1\text{-}45)$$

The boundary conditions for $p(\phi, s)$ are the same as for $q(\phi, s)$

$$p(\phi_0 - 2\pi, s) = p(\phi_0 + 2\pi, s) = 0$$

Using the identity

$$\frac{\delta(\phi - \phi_0)}{1 - \psi(s)} = \delta(\phi - \phi_0) + \delta(\phi - \phi_0)\frac{\psi(s)}{1 - \psi(s)} \qquad (11.1\text{-}46)$$

and the function $\Psi(s)$ defined previously, (11.1-45) takes on the form

$$sp(\phi, s) - Lp(\phi, s) - \delta(\phi - \phi_0) = \delta(\phi - \phi_0) \frac{\psi(s)}{1 - \psi(s)}$$

$$= \delta(\phi - \phi_0)\Psi(s) \qquad (11.1\text{-}47)$$

It is interesting to observe that $p(\phi, s)$ satisfies the same Fokker–Planck equation as $q(\phi, s)$ modified only by a *source term*

$$\delta(\phi - \phi_0)\Psi(s) \qquad (11.1\text{-}48)$$

We claim that this source term arises from the fact that every particle absorbed at the boundaries is immediately replaced by a particle created by the point source. To prove this assertion we consider the particles of the renewal process absorbed at the boundaries. The probability current associated with the modified Fokker–Planck equation (11.1-47) is, in analogy with (9.3-45), given by

$$J(\phi, t) = K_1(\phi)p(\phi, t) - \frac{1}{2} \frac{\partial}{\partial \phi} K_2(\phi)p(\phi, t) \qquad (11.1\text{-}49)$$

and the absorption rate analogous to $\gamma_\pm(t)$ is

$$\Gamma_-(t) + \Gamma_+(t) = -J(\phi_0 - 2\pi, t) + J(\phi_0 + 2\pi, t) \qquad (11.1\text{-}50)$$

Since $p(\phi_0 \pm 2\pi, t) = 0$, the total absorption rate is given by

$$\Gamma_-(t) + \Gamma_+(t) = +\frac{1}{2} \frac{\partial}{\partial \phi} K_2(\phi)p(\phi, t)\Big|_{\phi_0 - 2\pi} - \frac{1}{2} \frac{\partial}{\partial \phi} K_2(\phi)p(\phi, t)\Big|_{\phi_0 + 2\pi}$$

$$(11.1\text{-}51)$$

Going from the time domain to the Laplace domain and replacing $p(\phi, s)$ with the right-hand side of (11.1-38) we obtain

$$\Gamma_-(s) + \Gamma_+(s) = \frac{1}{1 - \psi(s)} \left[\frac{1}{2} \frac{\partial}{\partial \phi} K_2(\phi)q(\phi, s)\Big|_{\phi_0 - 2\pi} \right.$$

$$\left. - \frac{1}{2} \frac{\partial}{\partial \phi} K_2(\phi)q(\phi, s)\Big|_{\phi_0 + 2\pi} \right]$$

But the expression in the square brackets is the absorption rate of the single process, see (11.1-19)

$$\gamma_-(s) + \gamma_+(s) = -j(\phi_0 - 2\pi, s) + j(\phi_0 + 2\pi, s)$$

$$= \frac{1}{2} \frac{\partial}{\partial \phi} K_2(\phi)q(\phi, s)\Big|_{\phi_0 - 2\pi} - \frac{1}{2} \frac{\partial}{\partial \phi} K_2(\phi)q(\phi, s)\Big|_{\phi_0 + 2\pi}$$

$$= \psi(s) \qquad (11.1\text{-}52)$$

Therefore

$$\Gamma_-(s) + \Gamma_+(s) = \frac{\psi(s)}{1 - \psi(s)} = \Psi(s) \qquad (11.1\text{-}53)$$

Going back to the modified Fokker–Planck equation (11.1-47) we see that the point source on the right-hand side has intensity $\Psi(t)$. Therefore, the absorption rate $\Gamma_-(t) + \Gamma_+(t)$ at the boundaries equals the intensity $\Psi(t)$ of the point source at any time t

$$\Gamma_-(t) + \Gamma_+(t) = \Psi(t) \qquad (11.1\text{-}54)$$

$$\underset{\substack{\text{absorption} \\ \text{rate at the} \\ \text{boundaries}}}{} \qquad \underset{\substack{\text{source} \\ \text{intensity}}}{}$$

In fact, we have just proven that $p(\phi, t)$ is normalized to one since the loss of particles at $\phi_0 \pm 2\pi$ is exactly compensated by the source at $\phi = \phi_0$ at any time t.

11.1.5. Equations for the Steady State

To obtain steady-state functions the final value theorem of Laplace transform is of great help. The theorem states that for any function $f(t)$ in the limit $t \to \infty$

$$\lim_{t \to \infty} f(t) = \lim_{s \to 0} sF(s) \qquad (11.1\text{-}55)$$

where $F(s)$ is the Laplace transform of $f(t)$, provided that the limit exists. Thus, we can obtain steady-state values directly in the Laplace domain without first computing the inverse Laplace transform, and subsequently evaluate the result for $t \to \infty$. We multiply both sides of (11.1-47) by s

$$s^2 p(\phi, s) - L\,sp(\phi, s) - s\delta(\phi - \phi_0) = \delta(\phi - \phi_0)\frac{s\psi(s)}{1 - \psi(s)} \qquad (11.1\text{-}56)$$

Since $\lim_{s \to \infty} sp(\phi, s)$ equals, by definition, the stationary probability density function, $s^2 p(\phi, s)$ vanishes as $s \to 0$. The same is true for $s\delta(\phi - \phi_0) = 0$ as $s \to 0$ and we obtain for the left-hand side of (11.1-56)

$$s^2 p(\phi, s) - L\,sp(\phi, s) - s\delta(\phi - \phi_0) \to -Lp(\phi), \quad \text{as } s \to 0 \qquad (11.1\text{-}57)$$

The interesting term is the source intensity on the right-hand side of the modified Fokker–Planck equation. To evaluate it for $s \to 0$ we recall the general definition of an expected value

$$E[m(x)] = \int m(x)p(x)\,dx \qquad (11.1\text{-}58)$$

and the definition of the Laplace transform of $\psi(s)$

$$\psi(s) = \int_0^\infty e^{-st}\psi(t)\,dt \qquad (11.1\text{-}59)$$

Comparing the two expressions we see that $\psi(s)$ can be written as the expected value of $\exp(-sT_1)$

$$\psi(s) = E[\exp(-sT_1)] \qquad (11.1\text{-}60)$$

where T_1 is the time of absorption of the first process (random variable). By expanding the exponential function into a Taylor series

$$\exp(-sT_1) = 1 - sT_1 + \left(\frac{sT_1}{2}\right)^2 + \cdots + \frac{(-sT_1)^n}{n!} + \cdots \qquad (11.1\text{-}61)$$

the limit can easily be computed

$$\begin{aligned}
\frac{s\psi(s)}{1-\psi(s)} &= \frac{sE[1 - sT_1 + [(sT_1)^2/2] + \cdots]}{1 - E[1 - sT_1 + [(sT_1)^2/2] + \cdots]} \\
&= \frac{E[1 - sT_1 + [(sT_1)^2/2] + \cdots]}{E[T_1 - (sT_1^2/2) + \cdots]} \\
&= \frac{1}{E(T_1)} \quad \text{as } s \to 0 \qquad (11.1\text{-}62)
\end{aligned}$$

By definition $E(T_1)$ is the meantime for a trajectory to reach the boundaries for the first time, or, speaking in terms of phase-locked loop theory, the meantime between two cycle slips. (In the mathematical literature $E(T_1)$ is called *mean first exit time*.)

In our physical analogy we can state the result as follows: the intensity of the point source $\Psi(t \to \infty)$ at $\phi = \phi_0$ is inversely proportional to the mean lifetime of a particle and equals the mean number of particles absorbed at the boundaries. Of course, this is just the special case for $t \to \infty$ of the general result that in an arbitrary interval $[t_1, t_2]$ the number of particles absorbed at the boundaries is always exactly compensated by the particles emitted by the source (see (11.1-54)).

Combining the findings of (11.1-57) and (11.1-62) and replacing $E(T_1)$ with $E(T_s)$ the modified Fokker–Planck equation in the steady state reads

$$Lp(\phi) = -\frac{\delta(\phi - \phi_0)}{E(T_s)} \qquad (11.1\text{-}63)$$

where $E(T_s)$ is the meantime between cycle slips. A word about replacing $E(T_1)$ with $E(T_s)$ is necessary. Here T_1 denotes the random time when the trajectory number one reaches the boundaries for the first time. However, since the probability laws of the individual processes are identical (only the starting time being different), the number of the process is insignificant and we are allowed to replace $T_1, T_2, \ldots, T_k$ with T_s below.

Using (11.1-60) the fundamental relation between the probability density function of the single process and the renewal process can be written in the form

$$p(\phi, s) = \frac{q(\phi, s)}{1 - E[\exp(-sT_s)]} \tag{11.1-64}$$

Multiplying by s and expanding the denominator into a Taylor series the limit $s \to 0$ becomes

$$sp(\phi, s) = \frac{sq(\phi, s)}{1 - E[1 - sT_s + [(sT_s)^2/2] + \cdots]}$$

$$= \frac{q(\phi, s)}{E(T_s)}$$

$$= \frac{q(\phi, s = 0)}{E(T_s)} \quad \text{as } s \to 0 \tag{11.1-65}$$

Note that $q(\phi, s = 0)$ is not the stationary distribution of $q(\phi, t \to \infty)$ (which would be identical to zero) but

$$q(\phi, s = 0) = \lim_{s \to 0} \int_0^\infty e^{-st} q(\phi, t) \, dt$$

$$= \int_0^\infty q(\phi, t) \, dt \tag{11.1-66}$$

The integral of $q(\phi, t)$ was encountered previously and given the symbol $Q(\phi)$ (see (11.1-21a)). Using $Q(\phi)$ the stationary probability density function $p(\phi)$ can be written as

$$p(\phi) = \frac{Q(\phi)}{E(T_s)} \tag{11.1-67}$$

Equation (11.1-67) verifies that $p(\phi)$ is normalized to one. Indeed, $Q(\phi)$ and $p(\phi)$ are solutions of the same type of modified Fokker–Planck equation with the same boundary conditions (compare (11.1-24) and (11.1-63)). The only difference is in the weight of the Dirac impulse which are 1 and $1/E(T_s)$, respectively. But since (see (11.1-24))

$$E(T_s) = \int_{\phi_0 - 2\pi}^{\phi_0 + 2\pi} Q(\phi) \, d\phi$$

it follows from (11.1-67) that $p(\phi)$ is normalized to one, as it should be for a proper PDF. Again, we remark that in Section 11.1.3 we have proved the more general result that $p(\phi, t)$ is normalized for all times t.

11.1.6. Stationary Phase Error Distribution, Meantime between Cycle Slips and Mean Cycle Slip Rate

Stationary Probability Density Function p(φ). If we integrate both sides of (11.1-63) with respect to ϕ we arrive at

$$K_1(\phi)p(\phi) - \frac{1}{2}\frac{d}{d\phi}K_2(\phi)p(\phi) = \frac{1}{E(T_s)}[u(\phi - \phi_0) - D_0] \tag{11.1-68}$$

where $u(\phi - \phi_0)$ is the unit step function and D_0 is a constant of integration. The general solution of the linear, inhomogeneous first-order differential equation (11.1-68) is the sum of homogeneous and particular solution

$$p(\phi) = \frac{\exp[-U(\phi)]}{K_2(\phi)}\left\{D_1 + \frac{2}{E(T_s)}\int_{\phi_0-2\pi}^{\phi}[D_0 - u(x - \phi_0)]\exp[U(x)]\,dx\right\} \tag{11.1-69}$$

where $U(x)$ is a potential function given by

$$U(\phi) = -2\int^{\phi}\frac{K_1(x)}{K_2(x)}\,dx \tag{11.1-70}$$

Like any potential function, $U(\phi)$ is only defined up to an arbitrary constant. For a constant intensity coefficient $K_2(\phi)$ the potential function has the physical interpretation as the integral of the nonlinear restoring force $-2K_1(\phi)/K_2$.

By making use of the boundary conditions

$$p(\phi_0 - 2\pi) = p(\phi_0 + 2\pi) = 0$$

the constants are easily found to be

$$D_1 = 0$$

$$D_0 = \frac{\displaystyle\int_{\phi_0}^{\phi_0+2\pi}\exp[U(\phi)]\,d\phi}{\displaystyle\int_{\phi_0-2\pi}^{\phi_0+2\pi}\exp[U(\phi)]\,d\phi} \tag{11.1-71}$$

Equation (11.1-69) then becomes

$$p(\phi) = \frac{2}{E(T_s)} \frac{\exp[-U(\phi)]}{K_2(\phi)} \int_{\phi_0 - 2\pi}^{\phi} [D_0 - u(x - \phi_0)] \exp[U(x)]\, dx$$

$$(11.1\text{-}72)$$

For numerical purposes it is necessary to use the normalized intensity coefficients of Appendix 11.1A. Since the resulting expressions become quite cumbersome we advise the reader at first reading to skip the following details up to the next example.

Using the normalized intensity coefficients of Appendix 11.1A the solution for $p(\phi)$ can be written as

$$p(\phi) = \frac{\rho}{4B_L E(T_s)} \exp[-U(\phi)] \int_{\phi_0 - 2\pi}^{\phi} \frac{D_0 - u(x - \phi_0)}{S_{n'}(0, \phi)/S_{n'}(0,0)} \exp[U(x)]\, dx$$

$$(11.1\text{-}73)$$

with

$$U(x) = -\rho \int^{x} \frac{-g(y) + \gamma + \dfrac{1}{2\rho} \dfrac{d}{dy}[S_{n'}(0, y)/S_{n'}(0,0)]}{S_{n'}(0, y)/S_{n'}(0,0)}\, dy$$

$$= \rho \int^{x} \frac{g(y) - \gamma}{S_{n'}(0, y)/S_{n'}(0,0)}\, dy - \frac{1}{2} \int^{x} \frac{\dfrac{d}{dy}[S_{n'}(0, y)/S_{n'}(0,0)]}{S_{n'}(0, y)/S_{n'}(0,0)}\, dy$$

$$= \rho \int^{x} \frac{g(y) - \gamma}{S_{n'}(0, y)/S_{n'}(0,0)}\, dy - \frac{1}{2} \ln[S_{n'}(0, x)/S_{n'}(0,0)]$$

$$= U_{eq}(x) - \frac{1}{2} \ln[S_{n'}(0, x)/S_{n'}(0,0)] \qquad (11.1\text{-}74)$$

where

$$U_{eq}(x) := \rho \int^{x} \frac{g(y) - \gamma}{S_{n'}(0, y)/S_{n'}(0,0)}\, dy$$

Using (11.1-74), $p(\phi)$ can be transformed into

$$p(\phi) = \frac{\rho}{4B_L E(T_s)[S_{n'}(0, \phi)/S_{n'}(0,0)]^{1/2}}$$

$$\times \exp[-U_{eq}(\phi)] \int_{\phi_0 - 2\pi}^{\phi} \frac{D_0 - u(x - \phi_0)}{[S_{n'}(0, x)/S_{n'}(0,0)]^{1/2}} \exp[U_{eq}(x)]\, dx$$

$$(11.1\text{-}75)$$

When $S_{n'}(0, \phi)$ is constant for all ϕ then $p(\phi)$ reduces to the simpler expression

$$p(\phi) = \frac{\rho}{4B_L E(T_s)} \exp\left[-U(\phi)\right] \int_{\phi_0 - 2\pi}^{\phi} \left[D_0 - u(x - \phi_0)\right] \exp\left[U(x)\right] dx$$

$$(11.1\text{-}76)$$

As the integrand in (11.1-74) is 2π-periodic we make the following expansion

$$\frac{g(y) - \gamma}{S_{n'}(0, y)/S_{n'}(0, 0)} = g_{eq}(y) - \gamma_{eq} \tag{11.1-77}$$

where $-\gamma_{eq}$ is the constant term of the above expansion and $g_{eq}(y)$ is a periodic function which yields zero when integrated over an interval of length 2π. The constant D_0 can be written in the form

$$D_0 = \frac{\displaystyle\int_{\phi_0}^{\phi_0 + 2\pi} \exp\left[U(\phi)\right] d\phi}{\displaystyle\int_{\phi_0 - 2\pi}^{\phi_0} \exp\left[U(\phi)\right] d\phi + \int_{\phi_0}^{\phi_0 + 2\pi} \exp\left[U(\phi)\right] d\phi}$$

$$= \frac{1}{\displaystyle 1 + \int_{\phi_0 - 2\pi}^{\phi_0} \exp\left[U(\phi)\right] d\phi \Big/ \int_{\phi_0}^{\phi_0 + 2\pi} \exp\left[U(\phi)\right] d\phi}$$

$$(11.1\text{-}78)$$

But

$$\int_{\phi_0 - 2\pi}^{\phi_0} \exp\left[U(\phi)\right] d\phi = \int_{\phi_0}^{\phi_0 + 2\pi} \exp\left[U(\phi + 2\pi)\right] d\phi$$

$$= \exp\left(2\pi\rho\gamma_{eq}\right) \int_{\phi_0}^{\phi_0 + 2\pi} \exp\left[U(\phi)\right] d\phi$$

$$(11.1\text{-}79)$$

which follows from (11.1-74) and (11.1-77), and the periodicity of $S_{n'}(0, \phi)$ and $g_{eq}(\phi)$. Therefore, D_0 can be written in the form

$$D_0 = \frac{1}{1 + \exp\left(2\pi\rho\gamma_{eq}\right)} \tag{11.1\text{-}80}$$

When $S_{n'}(0, \phi)/S_{n'}(0, 0)$ for all ϕ, then $g_{eq}(\phi)$ and γ_{eq} reduce to $g(\phi)$ and γ, respectively.

Example: Sinusoidal phase detector characteristic. With $g(\phi) = \sin \phi$, $\phi_0 = 0$, $\gamma = 0$, and $S_{n'}(0, \phi) = $ constant, the stationary probability density function is given by (Figure 11.1-3)

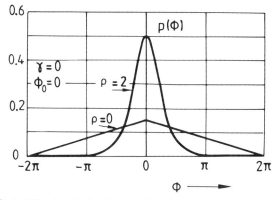

Figure 11.1-3. Probability density function of the phase-locked loop with sinusoidal phase detector characteristic without detuning.

$$p(\phi) = \frac{\rho}{4B_L E(T_s)} \exp(\rho \cos \phi) \left[\pi I_0(\rho) - \frac{1}{2} \int_0^{|\phi|} \exp(-\rho \cos x) \, dx \right]$$
$$(11.1\text{-}81)$$

where $I_0(\rho)$ is the Bessel function. In the limiting case $\rho \to 0$, the probability density function is triangular and not uniform.

Meantime Between Cycle Slips, $E(T_s)$ and Average Cycle Slip Rate. A graphic physical interpretation can be given for (11.1-68), where the right-hand side equals the probability current $J(\phi)$

$$J(\phi) = \frac{1}{E(T_s)} \left[u(\phi - \phi_0) - D_0 \right] \qquad (11.1\text{-}82)$$

The probability current is constant both in the (open) left and in the (open) right 2π interval and jumps at the location of the point source and at the boundaries (sinks) (see Figure 11.1-4). With the exception of the location of point source and the two sinks, the current must be constant in the steady state in order to fulfil the Fokker–Planck equation. Since $\partial p(\phi, t)/\partial t = 0$ we must necessarily have $\partial J(\phi, t)/\partial \phi = 0$.

The average number (per unit of time) of particles flowing to the right is given by $(1 - D_0)/E(T_s)$ whereas the average number of particles to the left is given by $D_0/E(T_s)$ (see Figure 11.1-4). Because creation and absorption of particles is always in balance we have the equality

$$\Gamma_- + \Gamma_+ = -J(\phi_0 - 2\pi) + J(\phi_0 + 2\pi) = \Psi \qquad (11.1\text{-}83)$$

total absorption rate	source intensity

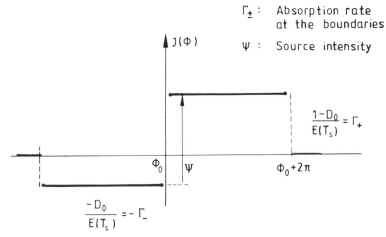

Figure 11.1-4. Probability current in the steady state.

Since the event "absorption of a particle at the left/right boundary" is the physical analog of a cycle slip to the left/right, the average number of cycle slips per unit of time is

$$N_+ = \Gamma_+ = (1 - D_0)/E(T_s)$$

$$N_- = \Gamma_- = D_0/E(T_s) \tag{11.1-84}$$

From (11.1-83) and (11.1-84) we make the important observation that the average slip rate is always equal to the inverse of the meantime between cycle slips

$$N = N_+ + N_- = \frac{1}{E(T_s)} \tag{11.1-85}$$

Using (11.1-80) the ratio N_+/N_- is given by

$$\frac{N_+}{N_-} = \exp\left(2\pi\rho\gamma_{eq}\right) \tag{11.1-85a}$$

This result shows the extreme dependence of the cycle slip direction on loop detuning γ_{eq}.

We next derive an expression for $E(T_s)$. Integration of both sides of (11.1-75) and subsequently solving for $E(T_s)$ yields

$$E(T_s) = \frac{\rho}{4B_L} \int_{\phi_0-2\pi}^{\phi_0+2\pi} dx \int_{\phi_0-2\pi}^{x} \frac{\exp\left[U_{eq}(y) - U_{eq}(x)\right]}{\left[S_{n'}(0,x)S_{n'}(0,y)/S_{n'}^2(0,0)\right]^{1/2}} [D_0$$
$$- u(y - \phi_0)] \, dy \tag{11.1-86}$$

The double integral must be evaluated numerically. An asymptotic expression which is accurate for all signal-to-noise ratios ρ of practical interest and which provides good insight into the dependence of N_+, N_- on the various loop parameters has been developed by Moeneclaey [11] and Chie [12]. The basic idea of this approximation is that the major contribution to $E(T_s)$ in (11.1-86) comes from the integration over a small region centered about the point $(x = \phi_s, y = \phi_u)$ where ϕ_s and ϕ_u indicate the arguments for which $U_{eq}(\phi)$ reaches the minimum and maximum value, respectively (Figure 11.1-5). Note that ϕ_s and ϕ_u correspond to the stable and unstable equilibrium point, respectively. Within this small region the potential function is well approximated by the first two terms of its Taylor series expansion. We obtain the following approximation

$$E(T_s) = \frac{\pi}{2B_L} \, |g_\phi(\phi_s)g_\phi(\phi_u)|^{-1/2} \, \frac{\exp\left[U_{eq}(\phi_u) - U_{eq}(\phi_s)\right]}{1 + \exp\left(-2\pi\rho\gamma_{eq}\right)}$$

(11.1-87)

where $g_\phi(x)$ is the derivative of the nonlinearity $g(\phi)$ with respect to ϕ evaluated at x. The above expression can be transformed into

$$E(T_s) = \frac{\pi}{2B_L} \, |g_\phi(\phi_s)g_\phi(\phi_u)|^{-1/2} \, \frac{\exp\left(\rho A_0\right)}{1 + \exp\left(-2\pi\rho|\gamma_{eq}|\right)}$$

(11.1-88)

where A_0 is a positive quantity given by

$$A_0 = \begin{cases} \displaystyle\int_{\phi_s}^{\phi_u} \left[g_{eq}(x) - \gamma_{eq}\right] dx = \int_{\phi_s}^{\phi_u} \frac{g(x) - \gamma}{S_{n'}(0, x)/S_{n'}(0, 0)} \, dx, & \gamma_{eq} \geq 0 \\[4mm] \displaystyle\int_{\phi_u - 2\pi}^{\phi_s} \left[\gamma_{eq} - g_{eq}(x)\right] dx = \int_{\phi_u - 2\pi}^{\phi_s} \frac{\gamma - g(x)}{S_{n'}(0, x)/S_{n'}(0, 0)} \, dx, & \gamma_{eq} < 0 \end{cases}$$

(11.1-89)

For $\gamma_{eq} \geq 0$, A_0 indicates the shaded area shown in Figure 11.1-6. A similar expression holds for negative γ_{eq}. Several important conclusions can be drawn from (11.1-87) and (11.1-88)

1. For all first-order systems with periodic intensity coefficients the meantime between slips depends exponentially on the potential difference between the stable and unstable locking point.

2. The potential difference is proportional to the signal-to-noise ratio in the loop, ρ. The proportionality factor A_0 is a *global* function of the phase detector characteristic. Notice the important difference to the tracking performance parameter $\sigma_\phi^2 = 1/\rho$ which, as a function of the

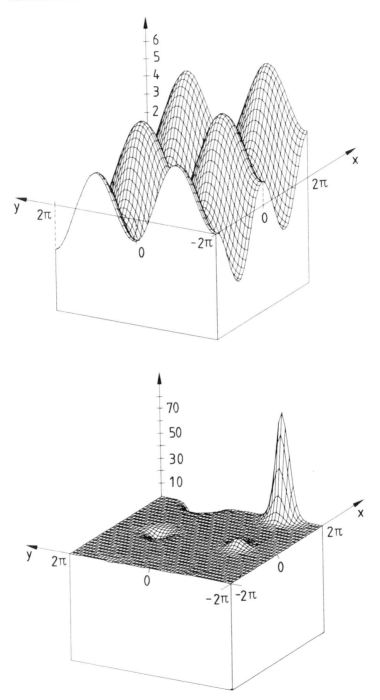

Figure 11.1-5. (*a*) Potential difference $U_{eq}(y) - U_{eq}(x)$ for a sinusoidal phase detector characteristic with slight detuning. (*b*) Corresponding integrand of integral formula for $E(T_s)$.

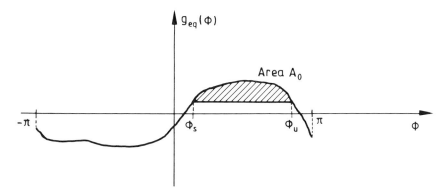

Figure 11.1-6. Construction of the area A_0.

derivative of the phase detector characteristic, depends only on the local property of the phase detector. Thus, minimization of σ_ϕ^2 does not guarantee maximization of $E(T_s)$.

3. Frequency detuning reduces A_0 and thus greatly affects $E(T_s)$.

We finally touch upon an important point. Since the limit of integration in (11.1-86) depends on ϕ_0 one might expect that $E(T_s)$ depends on ϕ_0. From the Markov property of the renewal process, it follows, however, that $E(T_s)$ must be independent of the arbitrary choice of ϕ_0. A formal proof has been given by Chie [12]. We want to mention that formula (11.1-88) has been tabulated for various $g(\phi)$ in Table 6.3-2.

An asymptotic expression for $E(T_s)$ for small ρ can be obtained as follows. For $\rho \to 0$ the phase error detector characteristic approaches a triangular form (see Figure 11.1-3).

$$p(\phi) = \frac{\rho}{4B_L E(T_s)} \frac{1}{2} [2\pi - |\phi|], \quad \phi_0 = 0$$

$$S_n \cdot (0, \phi) = \text{constant}$$

$$(11.1\text{-}90)$$

Integration over ϕ and solving for $E(T_s)$ yields

$$E(T_s) = \frac{\rho \pi^2}{2B_L} \qquad (11.1\text{-}91)$$

which says that for small ρ the meantime between cycle slips becomes inversely proportional to the loop bandwidth.

Modulo 2π Reduced Phase Error. We now develop the interconnection between the phase error of the renewal process and the modulo 2π reduced phase error discussed in Section 10.2. The probability density function $p_r(\phi)$ of the modulo 2π reduced phase error measures the phase within an interval of $[-\pi, \pi]$ (the range of a phase meter). Since the events

A_1: the phase error is within $[\phi, \phi + d\phi]$ or
$[\phi - 2\pi, \phi - 2\pi + d\phi]$ for $\phi \in [0, \pi]$

A_2: the phase error is within $[\phi, \phi + d\phi]$ or
$[\phi + 2\pi, \phi + 2\pi + d\phi]$ for $\phi \in [-\pi, 0]$ (11.1-92)

are mutually exclusive we obtain the probability density function $p_r(\phi)$ by simply adding the probabilities of the respective $d\phi$ intervals

$$p_r(\phi) = p(\phi) + \begin{cases} p(\phi - 2\pi), & 0 \leq \phi \leq \pi \\ p(\phi + 2\pi), & -\pi \leq \phi \leq 0 \end{cases} \qquad (11.1\text{-}93)$$

(By $p(\phi)$ we denote the probability density function of the renewal process.)

This summation is equivalent to folding down intervals in the range beyond $[-\pi, \pi]$ as illustrated in Figure 11.1-7. By inspection of Figure

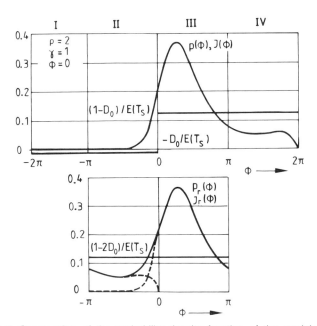

Figure 11.1-7. Construction of the probability density function of the modulo 2π reduced phase error from the PDF of the renewal process.

11.1-7 we see without any calculation that the probability current $J(\phi)$ is now constant and equals the difference

$$J(\phi) = (1 - 2D_0)/E(T_s) \tag{11.1-94}$$

The values of $p_r(\phi)$ at $\phi = \pm\pi$ are by definition

$$p_r(-\pi) = p(-\pi) + p(\pi)$$

$$p_r(\pi) = p(\pi) + p(-\pi) \tag{11.1-94a}$$

Thus

$$p_r(-\pi) = p_r(\pi)$$

$$\frac{dp_r(-\pi)}{d\phi} = \frac{dp_r(\pi)}{d\phi} \tag{11.1-95}$$

Since the area under the integral remains the same, $p_r(\phi)$ is normalized to one as it should be

$$\int_{-\pi}^{\pi} p_r(\phi)\, d\phi = \int_{-\pi}^{0} [p(\phi) + p(\phi + 2\pi)]\, d\phi + \int_{0}^{\pi} [p(\phi) + p(\phi - 2\pi)]\, d\phi$$

$$= \int_{-2\pi}^{2\pi} p(\phi)\, d\phi \tag{11.1-96}$$

Note that the modulo 2π reduced probability density functions contain only incomplete information about the average cycle slip rate, namely the difference $N_+ - N_-$

$$J(\phi) = (1 - 2D_0)/E(T_s) = N_+ - N_- \tag{11.1-97}$$

which vanishes for $\gamma = 0$.

It is easy to verify that $p_r(\phi)$ obtained from the renewal process probability density function $p(\phi)$ is identical to the solution $p_{ex}(\phi)$ obtained directly by solving the homogeneous Fokker–Planck equation

$$\frac{d}{d\phi} K_1(\phi) p_{ex}(\phi) - \frac{1}{2} \frac{d^2}{d\phi^2} K_2(\phi) p_{ex}(\phi) = 0$$

subject to periodic boundary conditions. We mention that $p_{ex}(\phi)$ is called the "periodic extension solution" in the literature.

11.1.7. Time-Dependent Solution of the Fokker–Planck Equation of the Single Process

To solve the time-dependent Fokker–Planck equation (11.1-9) subject to absorbing boundary conditions, we apply the method of separation of variables, i.e., we look for a solution of the form

$$q(\phi, t) = r(t)v(\phi) \tag{11.1-98}$$

where $r(t)$ and $v(\phi)$ are functions dependent only on t and $\phi \in [\phi_0 - 2\pi, \phi_0 + 2\pi]$, respectively. Using the Ansatz (11.1-98) in the Fokker–Planck equation yields

$$\dot{r}(t)v(\phi) = r(t)Lv(\phi) \tag{11.1-99}$$

Letting $(-\lambda)$ be the separation constant

$$\frac{\dot{r}(t)}{r(t)} = -\lambda \tag{11.1-100}$$

we obtain

$$-\lambda r(t)v(\phi) = r(t)Lv(\phi)$$

$$\Rightarrow -\lambda v(\phi) = Lv(\phi) \tag{11.1-101}$$

$$r(t) = \exp(-\lambda t)$$

In general, (11.1-101) has nontrivial solutions which fulfill the boundary conditions only for certain values of the separation constant λ. The values of the parameters are called the *eigenvalues* of L and the corresponding solutions of (11.1-101) are called the *eigenfunctions* of L since the operator L maps on eigenfunction on the same function.

Using the definition of L, (11.1-101) reads

$$[K_2(\phi)v(\phi)]'' - 2[K_1(\phi)v(\phi)]' + 2\lambda v(\phi) = 0 \tag{11.1-102}$$

where $'$ denotes the differentiation with respect to ϕ. Thus, to obtain $v(\phi)$ we have to solve an eigenvalue problem (posed above) subject to the boundary conditions

$$v(\phi_0 - 2\pi) = v(\phi_0 + 2\pi) = 0 \tag{11.1-103}$$

By means of the transformation

$$v(\phi) = a(\phi)u(\phi) \tag{11.1-104}$$

with

$$a(\phi) = \exp\left[-\int_{\phi_0}^{\phi} \frac{K_1(x)}{K_2(x)}\, dx\right] \tag{11.1-105}$$

this equation can be written in the form

$$u''(\phi) + I(\phi)u(\phi) = 0 \tag{11.1-106}$$

with the boundary conditions (assuming $K_2(\phi) > 0$ for all ϕ)

$$u(\phi_0 - 2\pi) = u(\phi_0 + 2\pi) = 0$$

The function $I(\phi)$ is the so-called *invariant*

$$I(\phi) = Z(\phi) + \lambda R(\phi)$$

$$= -\left[\frac{K_1(\phi)}{K_2(\phi)}\right]' - \left[\frac{K_1(\phi)}{K_2(\phi)}\right]^2 + \lambda\,\frac{2}{K_2(\phi)} \tag{11.1-107}$$

Equation (11.1-106) represents a boundary value problem of the Sturm–Liouville type in the self-adjoint form. This type of problem has been extensively studied in the field of mathematics. The following theorem is of fundamental importance and is frequently used below.

Theorem 1 Under the condition that $R(\phi)$ and $Z(\phi)$ are continuous functions in $[\phi_0 - 2\pi, \phi_0 + 2\pi]$ and that $R(\phi) > 0$, a denumerable sequence of eigenvalues (discrete spectrum) exists

$$\lambda_0 < \lambda_1 < \lambda_2 \cdots < \lambda_k < \cdots \to \infty \tag{11.1-108}$$

with the associated eigenfunctions $u_0(\phi), u_1(\phi), \ldots, u_k(\phi), \ldots$ which are nontrivial solutions of

$$u''(\phi) + [Z(\phi) + \lambda R(\phi)]u(\phi) = 0$$

subject to the boundary conditions

$$u_k(\phi_0 - 2\pi) = u_k(\phi_0 + 2\pi) = 0 . \tag{11.1-109}$$

The eigenfunctions $u_k(\phi)$ are uniquely determined up to a normalization constant and have exactly k zeros inside the interval $[\phi_0 - 2\pi, \phi_0 + 2\pi]$. Furthermore, the eigenfunctions form a complete orthonormal set of functions; i.e.,

$$\int_{\phi_0 - 2\pi}^{\phi_0 + 2\pi} R(\phi)u_k(\phi)u_m(\phi)\, d\phi = \begin{cases} 0, & \text{for } m \neq k \\ 1, & \text{for } m = k \end{cases} \tag{11.1-110}$$

The proof of this theorem is given by Gröbner and Lesky [7].

Theorem 1 tells us that a denumerable set of constants $\{\lambda_0, \lambda_1, \ldots, \lambda_k, \ldots\}$ exist for which a corresponding solution of (11.1-106) exists. The general solution of the Fokker–Planck equation is then the superposition of the individual solutions $u_k(\phi)$, $k \geq 0$. Since $K_2(\phi)$ is strictly positive and since $K_1(\phi)$ is bounded, the inverse of the transformation function $a(\phi)$ always exists (see (11.1-105)). Therefore, the mapping of $u_k(\phi)$ and $v_k(\phi)$ is one-to-one. For the same reason the set of eigenvalues of the original problem is identical to that of the transformed problem (11.1-106).

With the aid of the definition of $v_n(\phi)$ and of $r(t) = c \exp(-\lambda t)$ we obtain for the time-dependent solution of the Fokker–Planck equation

$$q(\phi, t) = \sum_{n=0}^{\infty} c_n \exp(-\lambda_n t) a(\phi) u_n(\phi) \qquad (11.1\text{-}111)$$

We can interpret this solution of $q(\phi, t)$ as a series expansion of $q(\phi, t)$ with respect to the complete set of orthonormal functions $\{u_n(\phi)\}$ for every time t. However, the solution is formal since we have to assume that piecewise differentiation of the infinite series is permissible. Inserting (11.1-111) into the Fokker–Planck equation yields

$$\frac{\partial}{\partial t} \left[\sum_{n=0}^{\infty} c_n \exp(-\lambda_n t) a(\phi) u_n(\phi) \right] = L \left[\sum_{n=0}^{\infty} c_n \exp(-\lambda_n t) a(\phi) u_n(\phi) \right] \qquad (11.1\text{-}112)$$

Assuming that summation and differentiation may be interchanged we obtain

$$\sum_{n=0}^{\infty} \{(-\lambda_n) c_n \exp(-\lambda_n t) a(\phi) u_n(\phi) - c_n \exp(-\lambda_n t) L[a(\phi) u_n(\phi)]\} = 0 \qquad (11.1\text{-}113)$$

$$\Leftrightarrow \sum_{n=0}^{\infty} [c_n \exp(-\lambda_n t)] \{\lambda_n a(\phi) u_n(\phi) + L[a(\phi) u_n(\phi)]\} = 0$$

The coefficients c_n are determined by the initial conditions $q(\phi, 0) = \delta(\phi - \phi_0)$. Since the Dirac impulse is not an ordinary function, the question arises whether the series

$$\sum_{n=0}^{\infty} c_n a(\phi) u_n(\phi) \qquad (11.1\text{-}114)$$

is a valid representation of this distribution. The answer is affirmative and a proof is given by Lindsey and Meyr [14]. Multiplying $q(\phi, 0) = \delta(\phi - \phi_0)$ by $u_k(\phi) R(\phi) a^{-1}(\phi)$ and integrating over the interval $[\phi_0 - 2\pi, \phi_0 + 2\pi]$ yields, using the orthonormality of the set $\{u_n(\phi)\}$

$$\int_{\phi_0-2\pi}^{\phi_0+2\pi} \delta(\phi - \phi_0)u_k(\phi)R(\phi)a^{-1}(\phi)\,d\phi = \sum_{n=0}^{\infty} c_n \int_{\phi_0-2\pi}^{\phi_0+2\pi} R(\phi)u_k(\phi)u_n(\phi)\,d\phi$$

(11.1-115)

$$u_k(\phi_0)R(\phi_0)a^{-1}(\phi_0) = c_k$$

(11.1-116)

Using this result and $a(\phi_0) = 1$, $q(\phi, t)$ can be written in the form

$$q(\phi, t) = \sum_{n=0}^{\infty} \exp(-\lambda_n t)u_n(\phi_0)R(\phi_0)a(\phi)u_n(\phi)$$

$$= \sum_{n=0}^{\infty} \exp(-\lambda_n t)R(\phi_0)v_n(\phi_0)v_n(\phi)$$

(11.1-117)

It has been shown by Lindsey and Meyr [14] that the eigenvalues have the following properties

1. $$\lambda_n > 0 \text{ for all } n$$

(11.1-118)

2. $$\lambda_n = 0(n^2) \text{ as } n \to \infty$$

11.1.8. Distribution of Renewal Epochs t_n and the Associated Jumps η_n

The probability law of the renewal process is completely specified by the knowledge of $q(\phi, t)$, the probability laws of the renewal epochs

$$t_n = \sum_{k=1}^{n} T_k$$

(11.1-119)

and the jumps η_n occurring at the epochs t_n. Having found $q(\phi, t)$ one must now determine the probability law of the random variables t_n and η_n.

The renewal epoch t_n shown in Figure 11.1-1 is the sum of the lifetimes of the preceding n trajectories. The t_n are the sum of independent random variables T_k with a common distribution $\psi(t)$. The probability density function of the nth trajectory to stop in $[t, t + dt]$ was found to be (see (11.1-41))

$$\psi_n(s) = \psi^n(s)$$

(11.1-120)

The nth power of $\psi(s)$ corresponds to the n-fold convolution of $\psi(t)$ in the time domain. The probability that the nth trajectory is absorbed within $[0, t]$ is found by integrating $\psi_n(t)$ over $[0, t]$.

From a system point of view it is more interesting to know the probability of how many cycle slips occur in a given time interval. But the event that the

nth trajectory is stopped within $[0, t]$ is equivalent to the event that at least n-renewal epochs (or cycle slips) fall within $[0, t]$. For this reason

$$\Pr(Z \geq n, t) = \int_0^t \psi_n(t') \, dt' \qquad (11.1\text{-}121)$$

where Z is the random number of renewal epochs. Again, this probability is embedded in the knowledge of $q(\phi, t)$. Using Laplace transform allows (11.1-121) to be written in the form

$$\Pr(Z \geq n, s) = \frac{1}{s} \psi^n(s) \qquad (11.1\text{-}122)$$

From (11.1-122) the corresponding probability density function for the discrete random variable n is easily found to be

$$p_Z(n, s) := \Pr(Z \geq n, s) - \Pr(Z \geq n - 1, s)$$
$$= \frac{1}{s} \psi^{n-1}(s)[\psi(s) - 1] \qquad (11.1\text{-}123)$$

In order to obtain $\psi(s)$ from $q(\phi, t)$ the Laplace transform of the series expansion solution of $q(\phi, t)$ must be computed. We find from (11.1-117)

$$q(\phi, s) = \sum_{n=0}^{\infty} R(\phi_0) \frac{v_n(\phi_0) v_n(\phi)}{s + \lambda_n} \qquad (11.1\text{-}124)$$

Through the use of the differentiation law, the transform $\psi(s)$ is subsequently found to be (see 11.1-18)

$$\psi(s) = 1 - s \sum_{n=0}^{\infty} R(\phi_0) d_n \frac{1}{s + \lambda_n} \qquad (11.1\text{-}125)$$

with

$$d_n = v_n(\phi_0) \int_{\phi_0 - 2\pi}^{\phi_0 + 2\pi} v_n(\phi) \, d\phi \qquad (11.1\text{-}126)$$

In arriving at (11.1-125) we have made use of the fact that the series is uniformly convergent, thus allowing an interchange of summation and integration.

Of particular interest is the probability of finding at least one renewal epoch (cycle slip) within $[0, t]$. Inserting (11.1-125) into (11.1-122) yields

$$\Pr(Z \geq 1, s) = \frac{1}{s} \left[1 - s \sum_{n=0}^{\infty} R(\phi_0) d_n \frac{1}{s + \lambda_n} \right] \qquad (11.1\text{-}127)$$

This probability can easily be transformed back into the time domain

$$\Pr\left(Z \geq 1, t\right) = 1 - R(\phi_0) \sum_{n=0}^{\infty} d_n \exp\left(-\lambda_n t\right) \qquad (11.1\text{-}128)$$

Assuming $\lambda_0 \ll \lambda_1, \ldots, \lambda_n, \ldots$ (an assumption we will verify later), we need to retain only the first term in the series

$$\Pr\left(Z \geq 1, t\right) \approx 1 - R(\phi_0) d_0 \exp\left(-\lambda_0 t\right) \qquad (11.1\text{-}129)$$

Under the assumption just made, we can also write

$$q(\phi, t) \approx R(\phi_0) v_0(\phi_0) v_0(\phi) \exp\left(-\lambda_0 t\right) \qquad (11.1\text{-}130)$$

Note that this last approximation is meaningful because the eigenfunction $v_0(\phi)$ is strictly positive in the entire interval $[\phi_0 - 2\pi, \phi_0 + 2\pi]$.

   Thus, for $1/\lambda_1 \ll t < 1/\lambda_0$ a "*quasi-stationary*" regime exists characterized by a slow exponential decay of the restricted probability density function $q(\phi, t)$. However, the shape of the probability density remains unchanged. Nearly all probability mass is concentrated near ϕ_0 during this interval, hence

$$\int_{\phi_0 - 2\pi}^{\phi_0 + 2\pi} q(\phi, t)\, d\phi \approx R(\phi_0) v_0(\phi_0) \int_{\phi_0 - 2\pi}^{\phi_0 + 2\pi} v_0(\phi)\, d\phi \approx 1$$
$$(11.1\text{-}131)$$

From the last equation and the definition of d_0 (11.1-126) it follows that the coefficient $R(\phi_0) d_0$ of (11.1-129) must be approximately 1, thus leading to

$$\Pr\left(Z \geq 1, t\right) \approx 1 - \exp\left(-\lambda_0 t\right) \qquad (11.1\text{-}132)$$

Notice that we would have arrived at the same result if the cycle slipping events had formed a Poisson process with

$$\psi(t) = \lambda_0 \exp\left(-\lambda_0 t\right) \qquad (11.1\text{-}133)$$

where $1/\lambda_0$ is the *mean number of cycle slips per unit of time*

$$\lambda_0 = 1/E(T_s) \qquad (11.1\text{-}133a)$$

This confirms the experimental results of Chapter 6 indicating the Poisson nature of the cycle slip events.

   The discovery of the existence of a "quasi-stationary" regime is a key result of the theory. The mean lifetime of a particle, or equivalently, the meantime between cycle slips is related to the first eigenvalue λ_0. The second eigenvalue is (as will be seen) in the order of magnitude of the loop bandwidth, $\lambda_1 \sim B_L$. It thus describes the motion around the stable tracking point. In other words, we have a separation into two scales. A "fast" time scale for motions around the stable tracking point and a "slow" scale for transitions from the stable tracking point to the boundaries.

The η_n form a set of independent, identically distributed random variables assuming the values of 2π or -2π. Thus, the process M

$$M = \frac{1}{2\pi} \sum_{k=1}^{n} \eta_k$$

is a random walk process described by the probability that $\eta_k/2\pi$ assumes either $+1$ or -1, respectively.

Let us use p^- to denote the probability that the trajectory starting at time $t = 0$ arrived at the point $\phi = \phi_0 - 2\pi$ at the first renewal epoch t_1. Similarly, let p^+ be the probability that the trajectory reached the upper bound $\phi = \phi_0 + 2\pi$. From the definition of η_n and its properties, the desired distribution immediately follows

$$\Pr[\eta_n = 2\pi] = p^+$$

$$\Pr[\eta_n = -2\pi] = p^-$$

(11.1-134)

To obtain p^+ and p^-, we use a technique outlined by Gichman and Skorochod [6]. Let $\phi(t)$ be a solution of the stochastic differential equation (11.1-1) for $t \in [0, t_1]$ and let $y(\phi)$ be a function with the following properties:

1. $y(\phi)$ does not explicitly depend on t.
2. $y(\phi)$ has a continuous second derivative.

Then the Itô differential rule implies (see [2, p.112]*

$$dy[\phi(t)] = \left\{ \frac{1}{2} y_{\phi\phi}[\phi(t)]K_2[\phi(t)] + y_\phi[\phi(t)]K_1[\phi(t)] \right\} dt$$

$$+ y_\phi[\phi(t)]\sqrt{K_2[\phi(t)]} \, d\beta(t)$$

(11.1-135)

Integrating (11.1-135) from $t = 0$ to the first renewal epoch yields

$$y[\phi(t_1)] - y(\phi_0) = \int_0^{t_1} \left\{ \frac{1}{2} y_{\phi\phi}[\phi(t)]K_2[\phi(t)] + y_\phi[\phi(t)]K_1[\phi(t)] \right\} dt$$

$$+ \int_0^{t_1} \sqrt{K_2[\phi(t)]} \, y_\phi[\phi(t)] \, d\beta(t)$$

(11.1-136)

---

*Notice that $K_1(x)$, $K_2(x)$ are the intensity coefficients of the Fokker–Planck equation and $y_\phi = \partial y/\partial \phi$ and $y_{\phi\phi} = \partial^2 y/\partial \phi^2$.

According to our assumption, for all $t < t_1$, we have $\phi(t) \in (\phi_0 - 2\pi, \phi_0 + 2\pi)$. We now choose the function $y(\phi)$ such that it is a solution of the differential equation

$$\frac{1}{2} K_2(\phi) \frac{d^2}{d\phi^2} y(\phi) + K_1(\phi) \frac{d}{d\phi} y(\phi) = 0 \qquad (11.1\text{-}137)$$

where $\phi \in [\phi_0 - 2\pi, \phi_0 + 2\pi]$. The notation $d/d\phi$ emphasizes that y is viewed as a function of ϕ only, and thus (11.1-137) is an ordinary differential equation. We further demand that $y(\phi)$ does not equal a constant. Then, taking expected values on both sides of (11.1-136), we obtain

$$E\{y[\phi(t_1)]\} = y(\phi_0)$$

By definition, $\phi(t_1)$ assumes either of the two values $\phi = \phi_0 \pm 2\pi$; hence,

$$E\{y[\phi(t_1)]\} = y(\phi_0 - 2\pi)p^- + y(\phi_0 + 2\pi)p^+$$
$$= y(\phi_0) \qquad (11.1\text{-}138)$$

where $y(\phi)$ is the general solution of the linear second-order differential equation (11.1-137) with two free constants of integration. We choose these constants such that the solution $\tilde{y}$ satisfies the boundary conditions

$$\tilde{y}(\phi_0 - 2\pi) = 1$$
$$\tilde{y}(\phi_0 + 2\pi) = 0 \qquad (11.1\text{-}139)$$

Then the probabilities p^-, p^+ are given by

$$p^- = \tilde{y}(\phi_0)$$
$$p^+ = 1 - p^- \qquad (11.1\text{-}140)$$

It is easy to verify that $\tilde{y}(\phi)$ can be manipulated to assume the form

$$\tilde{y}(\phi) = \frac{\displaystyle\int_{\phi}^{\phi_0 + 2\pi} \exp\left[-2 \int_{\phi_0 - 2\pi}^{x} \frac{K_1(x')}{K_2(x')}\, dx'\right] dx}{\displaystyle\int_{\phi_0 - 2\pi}^{\phi_0 + 2\pi} \exp\left[-2 \int_{\phi_0 - 2\pi}^{x} \frac{K_1(x')}{K_2(x')}\, dx'\right] dx} \qquad (11.1\text{-}141)$$

Because of the periodicity of $K_1(x)$ and of $K_2(x)$ it can easily be shown that $\tilde{y}(\phi_0)$ is independent of ϕ_0. Therefore, p^-, p^+ are also independent of the arbitrary starting point ϕ_0 of the trajectory. Also, p^-, p^+ do not depend on the random time t_1 when the trajectory reaches the boundaries.

Furthermore, substitution of $\phi = \phi_0$ in (11.1-141) yields $\tilde{y}(\phi_0) = D_0$ (see (11.1-78)), hence $p^- = D_0$ and $p^+ = 1 - D_0$. From (11.1-84) then follows

$$p^+ = \frac{N_+}{N_+ + N_-}, \quad p^- = \frac{N_-}{N_+ + N_-} \tag{11.1-142}$$

where N_+, N_- are the average cycle slip rate (positive/negative). By (11.1-80) we obtain

$$\frac{p_+}{p_-} = \frac{N_+}{N_-} = \exp\left(2\pi\rho\gamma_{eq}\right) \tag{11.1-143}$$

This last result shows the extreme dependence of the cycle slipping direction on a frequency offset γ.

The number of positive (negative) slips among a total number of slips is binomially distributed

$$\Pr\left(Z_+ = k \mid Z = n\right) = \binom{n}{k}(p^+)^k(1 - p^+)^{n-k}$$

By the law of large numbers we find asymptotically that the ratio of the positive to the negative cycle slips approaches p^+/p^- as $n \to \infty$.

11.1.9. Time-Dependent Probability Density Function of the Renewal Process

The probability density function $p(\phi, s)$ is related to the $q(\phi, s)$ and $\psi(s)$ (11.1-38)

$$p(\phi, s) = \frac{q(\phi, s)}{1 - \psi(s)}$$

Taking the Laplace transform of the series expansion of $q(\phi, t)$ yields

$$q(\phi, s) = \sum_{n=0}^{\infty} R(\phi_0) \frac{v_n(\phi_0)v_n(\phi)}{s + \lambda_n} \tag{11.1-144}$$

Using the differentiation law, the Laplace transform $\psi(s)$ is found from $q(\phi, s)$ to be

$$\psi(s) = 1 - s \sum_{n=0}^{\infty} R(\phi_0)d_n \frac{1}{s + \lambda_n} \tag{11.1-145}$$

with d_n given by (11.1-126).

Substituting the right-hand side of (11.1-144) and (11.1-145), respectively for $q(\phi, s)$ and $\psi(s)$ leads to

$$p(\phi, s) = \frac{1}{s} \frac{\displaystyle\sum_{n=0}^{\infty} \frac{v_n(\phi_0)v_n(\phi)}{s + \lambda_n}}{\displaystyle\sum_{n=0}^{\infty} d_n \frac{1}{s + \lambda_n}} \tag{11.1-146}$$

The stationary probability density function can be easily obtained using the final value theorem

$$p(\phi) = \frac{\displaystyle\sum_{n=0}^{\infty} \frac{v_n(\phi_0)v_n(\phi)}{\lambda_n}}{\displaystyle\sum_{n=0}^{\infty} \frac{d_n}{\lambda_n}} \tag{11.1-147}$$

Modulo 2π Reduced Phase Error. The probability density function of the modulo 2π reduced phase error is related to $p(\phi, t)$ by (see 11.1-93)

$$p_r(\phi, t) = p(\phi, t) + \begin{cases} p(\phi - 2\pi, t), & 0 \le \phi \le \pi \\ p(\phi + 2\pi, t), & -\pi \le \phi \le 0 \end{cases} \tag{11.1-148}$$

It can be shown [14] that $p_r(\phi, t)$ satisfies the homogeneous Fokker–Planck equation

$$\frac{\partial p_r(\phi, t)}{\partial t} - Lp_r(\phi, t) = 0 \tag{11.1-149}$$

subject to periodic boundary conditions

$$p_r(-\pi, t) = p_r(\pi, t)$$
$$\frac{\partial p_r(-\pi, t)}{\partial \phi} = \frac{\partial p_r(\pi, t)}{\partial \phi} \tag{11.1-150}$$

and initial condition $p_r(\phi, 0) = \delta(\phi)$.

11.1.10. Numerical Example: First-Order Phase-Locked Loop

For a first-order phase-locked loop with sinusoidal phase detector characteristic, the intensity coefficients read (see Appendix 11.1A)

$$K_1(\phi) = -\sin\phi + \gamma, \quad S_{n'}(0, \phi) = \text{constant}$$
$$K_2(\phi) = \frac{2}{\rho} \tag{11.1-151}$$

For these coefficients the invariant $I(\phi)$ (11.1-107) becomes

$$I(\phi) = \frac{1}{2} \rho \cos \phi - \frac{1}{4} \rho^2 [-\sin \phi + \gamma]^2 + \left(\frac{\lambda}{4B_L}\right)\rho \quad (11.1\text{-}152)$$

The first five eigenfunctions obtained by numerical integration are plotted in Figure 11.1-8 for $\gamma = 0$. As stated in Theorem 1 (11.1-108 ff) for every eigenvalue λ_n, exactly one eigenfunction exists with n zeroes inside the 4π interval. Note that the first eigenvalue $(\lambda_0/4B_L)$ is orders of magnitude smaller than the other eigenvalues, which clearly demonstrates the existence of a "quasi-stationary" regime. The average cycle slip rate as a function of loop detuning is plotted in Figure 11.1-9.

Main Points of the Section

- The absolute phase error is described by a stochastic differential equation

$$d\varphi = g[\varphi(t)]\, dt + F[\varphi(t)]\, d\beta(t) \quad (11.1\text{-}1)$$

  where the noise intensity may depend on the phase error.
- *Model.* The key idea is to define a *renewal process* $\phi(t)$ which is a statistically equivalent process that retains all the information, yet yields a meaningful solution in the steady state.
- The probability density function $q(\phi, t)$ of the single process is the solution of the Fokker–Planck equation

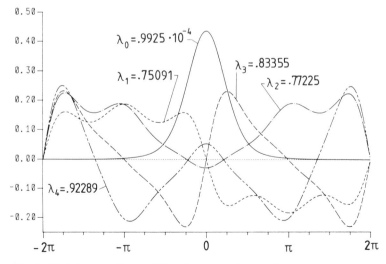

Figure 11.1-8. First eigenfunctions $u_k(\phi)$ and associated normalized eigenvalues λ_k for $\gamma = 0$, $\phi_0 = 0$, $\rho = 4$.

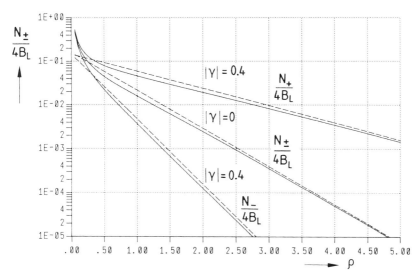

Figure 11.1-9. Average cycle slip rate versus signal-to-noise ratio in the loop with γ as parameter (loop detuning). Dashed curves correspond to the approximation (11.1-88).

$$\frac{\partial q(\phi, t)}{\partial t} = Lq(\phi, t)$$

subject to absorbing boundary conditions

$$q(\phi_0 - 2\pi, t) = q(\phi_0 + 2\pi, t) = 0$$

and initial distribution

$$q(\phi, 0) = \delta(\phi - \phi_0)$$

- The probability law of the renewal process is derived from the probability law of the generating single process via *recursive equations*. The fundamental equation relating renewal and single process is (in Laplace space)

$$p(\phi, s) = \frac{q(\phi, s)}{1 - \psi(s)} \tag{11.1-38}$$

where $p(\phi, s)$ is the probability density function of the renewal process and $q(\phi, s)$ is the probability density function of the single process. $\psi(t)$ is the probability density function attributed to the event that the trajectory is absorbed at time t. It is related to $q(\phi, t)$ by

$$\psi(t) = -\frac{\partial}{\partial t} \int_{\phi_0 - 2\pi}^{\phi_0 + 2\pi} q(\phi, t) \, d\phi \tag{11.1-18}$$

- The Fokker–Planck equation of the renewal process contains a *source term* which exactly compensates for the particles absorbed at the boundaries

$$sp(\phi, s) - Lp(\phi, s) - \delta(\phi - \phi_0) = \delta(\phi - \phi_0)\Psi(s)$$

with

$$\Psi(s) \quad = \quad \Gamma_+(t) + \Gamma_-(t)$$

| source | absorption rate |
| intensity | at the boundaries |

The probability density function $p(\phi, t)$ obeys absorbing boundary conditions

$$p(\phi_0 - 2\pi, t) = p(\phi_0 + 2\pi, t) = 0$$

The initial distribution is

$$p(\phi, 0) = \delta(\phi - \phi_0)$$

- *Steady-State Statistics.* The Fokker–Planck equation of the renewal process reads

$$Lp(\phi) = -\frac{\delta(\phi - \phi_0)}{E(T_s)} \tag{11.1-63}$$

The intensity of the source is constant and inversly proportional to the meantime between cycle slips, $E(T_s)$. The average absorption rate $\Gamma_\pm$ then equals the average rate of cycle slips to the right/left, respectively.

$$\frac{1}{E(T_s)} = \Gamma_+ + \Gamma_-$$

$$\frac{1}{E(T_s)} = N = N_+ + N_-$$

where N is the total average cycle slip rate and $N_\pm$ the positive/negative average cycle slip rate. $E(T_s)$, $N_\pm$ can be shown to be independent of the arbitrary location of the source ϕ_0.

- *Time-Dependent Statistics.* The Fokker–Planck equation is solved by the separation of variables technique

$$q(\phi, t) = r(t)v(\phi)$$

$q(\phi, t)$ can be represented by a series expansion

$$q(\phi, t) = \sum_{n=0}^{\infty} R(\phi_0)v_n(\phi_0)v_n(\phi) \exp(-\lambda_n t) \qquad (11.1\text{-}117)$$

The set of functions $\{v_n(\phi)\}$ is complete and orthonormal. It is obtained by solving a Sturm–Liouville boundary value problem.

- A key result is the discovery of the existence of a *quasi-stationary regime*. For a reasonably large signal-to-noise ratio in the loop the first eigenvalue λ_0 is orders of magnitude smaller than the other eigenvalues $\lambda_1, \ldots, \lambda_k, \ldots$. Thus, after a short time, only one term of the infinite series remains

$$q(\phi, t) \approx R(\phi_0)v_0(\phi_0)v_0(\phi) \exp(-\lambda_0 t)$$

Since the eigenfunction $v_0(\phi)$ is strictly positive in the entire interval $(\phi_0 - 2\pi, \phi_0 + 2\pi)$ one can interpret the above expression as the probability density function of the particles which have not yet reached one of the boundaries. The probability mass decays exponentially but the shape of the probability density function remains.

- The mean lifetime of a particle or, equivalently, the meantime between cycle slips is related to the first eigenvalue λ_0 as

$$E(T_s) = 1/\lambda_0 \qquad (11.1\text{-}133a)$$

The second eigenvalue is in the order of magnitude of the loop bandwidth

$$\lambda_1 \sim B_L$$

It thus describes the fast relaxation phenomena around the stable tracking point. We thus have a separation into two time scales: a "fast" time scale for motions around the stable tracking point; and a "slow" scale for transitions from the stable tracking point to the boundaries. All the remaining eigenvalues $\lambda_2, \ldots, \lambda_k, \ldots$ describe rather unsignificant details of how the quasi-stationary regime is reached from the initial distribution $\delta(\phi - \phi_0)$. The eigenvalues increase asymptotically as n^2

$$\lambda_n = 0(n^2) \quad \text{as } n \to \infty$$

- The ratio of positive to negative average cycle slip rate is a function of ρ and the normalized loop detuning γ_{eq}

$$\frac{N_+}{N_-} = \exp(2\pi\rho\gamma_{eq}) \qquad (11.1\text{-}143)$$

- The modulo 2π reduced phase error distribution (as seen by a phase meter) contains only partial information on the cycle slip phenomena, namely the difference between positive and negative average slip rate, $N_+ - N_-$. It is obtained from the probability density function of the renewal process by "folding down" the probability density function in the range beyond $[-\pi, \pi]$, see (11.1-92) and following equations.

11.2. HIGHER ORDER SYSTEMS WITH PERIODIC PHASE DETECTOR CHARACTERISTIC

For higher order systems there is an interaction between the phase error and the other state variables which causes effects such as bursts of slips, i.e., events where the tracking system tends to slip several cycles within a short period of time before relocking (if at all). In this section we generalize the theory for first-order systems to take these phenomena into account. While it is possible to develop a general theory, unfortunately, no general method of solution for these equations is known at present.

11.2.1. Modeling of the Phase Error as a Vector Renewal Process*

The modeling of the phase error as a renewal process is a generalization of the one-dimensional case which takes into account the interaction between the phase error and the other state variables. The unrestricted phase error $\varphi(t)$ together with the other state variables $[x_2, x_3, \ldots, x_n]$ obey a stochastic differential equation of the form

$$dx(t) = g[x(t)] \, dt + F[x(t)] \, d\beta(t) \qquad (11.2\text{-}1)$$

where x is the n-dimensional state vector. We write x in the form $x^T = [\varphi, y]$ where the first component is always the phase error while the remaining state variables are denoted by y.

The solution of (11.2-1) is a vector Markov process completely described by the joint probability density function $p(x, t) = p(\varphi, y, t)$. The stochastic differential equation (11.2-1) is periodic in φ and homogeneous in time. Whenever the trajectory reaches one of the hyperplanes $\varphi = \pm 2\pi$ (this event *defines* the positive or negative cycle slips), it is stopped (absorbed) and immediately restarted at $\varphi = 0$ (Figure 11.2-1). This purely fictitious restarting is permissible because of the periodicity of (11.2-1) with respect to φ. The state variables y are not changed thereby, which means that the random value at the jumping epochs t_i are the starting values for the next part of the process. By the Markov property these random initial values are sufficient to determine the further law of probability of the state vector. Due

---

*Definition is given in Section 10.3.

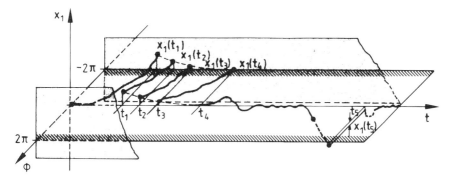

Figure 11.2-1. Sample trajectory of the renewal process of a second-order system with state variables (ϕ, x_1).

to time homogeneity the individual parts of the process between two cycle slips are statistically identical, apart from their initial conditions. The values of the state variables y taken at the renewal epochs are statistically dependent. It is this dependence which we must adequately take into account in the theory.

11.2.2. Probability Laws of the Single Process

The probability density function $q(\phi, y, t)$ attributed to the event that a state vector starting from $\phi(0) = 0$, $y(0) = y_0$ arrives at $[\phi(t), y(t)]$ without ever having touched the hyperplanes at $\phi = \pm 2\pi$ in $[0, t]$ is the solution of the Fokker–Planck equation

$$\frac{\partial q(x, t)}{\partial t} = Lq(x, t) \tag{11.2-2}$$

with the operator

$$L = -\sum_{k=1}^{n} \frac{\partial}{\partial x_k} K_k(x) + \frac{1}{2} \sum_{k=1}^{n} \sum_{m=1}^{n} \frac{\partial^2}{\partial x_k \partial x_m} K_{km}(x) \tag{11.2-3}$$

(where x_1 stands for ϕ and $\partial/\partial x_1$ stands for $\partial/\partial \phi$) subject to the initial conditions

$$q(x, 0) = \delta(\phi) \prod_{i=1}^{n-1} \delta[y_i - y_i(0)] \tag{11.2-4}$$

and the absorbing boundary conditions (on ϕ)

$$q(\pm 2\pi, y, t) = 0 \tag{11.2-5}$$

For the remaining state variables natural boundary conditions apply (Section

9.3.6). Since, by assumption, the process is time homogeneous the symbol t in $q(\phi, y, t)$ may stand for the time elapsed since an arbitrary starting epoch. The integral of q with respect to the state variables $[\phi, y]$ decays with time and equals the probability that the trajectory $x(t)$ has not been absorbed in the time interval $[0, t]$.

As pointed out before, the distribution of the random vector y at the time of absorption is of particular interest since it determines the starting condition of the next single process. However, the stopping time is itself a random variable. We therfore introduce the joint probability density function $\gamma(y, t \mid y_0)$ of the combined events $A_1 A_2 A_3$

A_1: absorption occurs at time t

A_2: absorption occurs at $\phi = +2\pi$ or $\phi = -2\pi$, respectively

(11.2-6)

A_3: $y(t) = y$

with initial conditions $\phi(0) = 0$, $y(0) = y_0$.

The quantities $\gamma_\pm(y, t \mid y_0) \, dy \, dt$ equal the amount of probability reaching the hyperplanes $\phi = \pm 2\pi$ within the differential area $dy = dx_2 \, dx_3 \ldots dx_n$ during the time interval $[t, t + dt]$, hence

$$\gamma_\pm(y, t \mid y_0) = \pm j_1(\pm 2\pi, y, t \mid y_0) \tag{11.2-7}$$

The probability current j_1 is the first component in ϕ direction of the probability current vector $j^T = [j_1 \quad j_2 \quad \cdots \quad j_n]$ (see Figure 11.2-2)

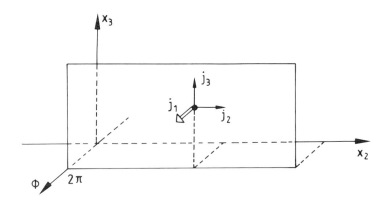

$$j_k = j_k(2\pi, y, t \mid y_0), \qquad k = 1, 2, 3$$

Figure 11.2-2. Components j_1, j_2, j_3 of the three-dimensional probability current vector.

$$j_1(\pm 2\pi, y, t \mid y_0)$$

$$= K_1(\pm 2\pi, y) q(\pm 2\pi, y, t \mid y_0)$$

$$-\frac{1}{2} \sum_{k=1}^{n} K_{1k}(\pm 2\pi, y) \frac{\partial}{\partial x_k} q(\phi, y, t \mid y_0)\big|_{\phi = \pm 2\pi} \qquad (11.2\text{-}8)$$

Because of the absorbing boundary conditions on q, all but one term in (11.2-8) vanishes

$$j_1(\pm 2\pi, y, t \mid y_0) = -\frac{1}{2} K_{11}(\pm 2\pi, y) \frac{\partial}{\partial \phi} q(\phi, y, t \mid y_0)\big|_{\phi = \pm 2\pi} \qquad (11.2\text{-}9)$$

Thus the $\gamma_\pm$ are expressed in terms of q. Their sum

$$\gamma_+(y, t \mid y_0) + \gamma_-(y, t \mid y_0) =: \psi(y, t \mid y_0) \qquad (11.2\text{-}10)$$

equals the probability that the trajectory is absorbed in $[t, t + dt]$ at $y(t) = y$. Integrating ψ over all possible values of the state variables y yields

$$\int_{-\infty}^{\infty} \psi(y, t \mid y_0) \, dy = \psi(t \mid y_0) \qquad (11.2\text{-}11)$$

where we maintain that $\psi(t \mid y_0) \, dt$ is the probability that the trajectory has been absorbed in $[t, t + dt]$. We verify this conjecture by integrating the Fokker–Planck equation first over $dy = dx_2 dx_3 \ldots dx_n$

$$-\frac{\partial}{\partial t} \int_{-\infty}^{\infty} q(\phi, y, t \mid y_0) \, dy = \frac{\partial}{\partial \phi} \int_{-\infty}^{\infty} j_1(\phi, y, t \mid y_0) \, dy$$

$$+ \sum_{k=2}^{n} \int_{-\infty}^{\infty} \left[\frac{\partial}{\partial x_k} j_k(\phi, y, t \mid y_0) \right] dy \qquad (11.2\text{-}12)$$

The terms in the sum are all zero because the current vanishes at $x_k = \pm\infty$

$$\int_{-\infty}^{\infty} \frac{\partial}{\partial x_k} j_k(\phi, y, t \mid y_0) \, dy =$$

$$\int_{-\infty}^{\infty} [j_k(\phi, y, t \mid y_0)\big|_{x_k = \infty} - j_k(\phi, y, t \mid y_0)\big|_{x_k = -\infty}] \, dx_1 \ldots dx_{k-1} \, dx_{k+1} \ldots dx_n$$

$$= 0 \qquad (11.2\text{-}13)$$

Integrating the remaining terms (11.2-12) over ϕ yields

$$-\frac{\partial}{\partial t} \int_{-2\pi}^{2\pi} \int_{-\infty}^{\infty} q(\phi, y, t \mid y_0) \, d\phi \, dy = \int_{-\infty}^{\infty} [j_1(2\pi, y, t \mid y_0)$$

$$- j_1(-2\pi, y, t \mid y_0)] \, dy$$

$$= \psi(t \mid y_0) \qquad (11.2\text{-}14)$$

where the right-hand side equals $\psi(t \mid y_0)$ defined by (11.2-11). Clearly, the left-hand side of (11.2-14) is the rate of decrease of the probability for no absorption to occur in $[0, t]$. It is therefore, the probability density function for the first slip to occur at time t.

The reader will have noticed that $\psi(y, t \mid y_0)$ and $\gamma_{\pm}(y, t \mid y_0)$ are the generalization to n-dimensions of the quantities defined in the one-dimensional case by (11.1-18) and (11.1-19a), respectively.

11.2.3. Basic Recurrence Relations of the Vector Renewal Process

We define the following events which are generalizations of the events defined for the one-dimensional case

A_n: the nth trajectory is in the volume $d\phi \, dy$ at point $[\phi, y]$ at time t. The starting position of the first trajectory was at $y(0) = y_0$ (11.2-15)

B_n: the nth trajectory is stopped at $\phi = +2\pi$ or -2π, $y(t) = y$ in the interval $[t, t + dt]$. The starting position of the first trajectory $(n = 1)$ is $y(0) = y_0$ (11.2-16)

These events can be used to define both conditioned and combined events, e.g.

$B_{n+1} \mid B_1$: the $(n + 1)$th trajectory is stopped at $\phi = +2\pi$ or -2π, $y(t) = y$ in the interval $[t, t + dt]$. The starting position of the first trajectory $(n = 1)$ is $y(0) = y_0$.
Condition: the first trajectory $(n = 1)$ is stopped at $\phi = +2\pi$ or -2π, $y(\tau) = z$ in the interval $[\tau, \tau + d\tau]$. The starting position is $y(0) = y_0$.

We now assign probabilities to the events B_n

$$Pr(B_n) = \psi_n(y, t \mid y_0) \, dy \, dt \qquad (11.2-18)$$

The first $(n = 1)$ of these probabilities equals (by definition) the function $\psi_1(y, t \mid y_0) := \psi(y, t \mid y_0)$ of the single process. The other functions $\psi_n(y, t \mid y_0)$ are defined by a fundamental recursion which we now develop using

1. The time homogeneity of the vector process.
2. The Markov property of $x^T = [\phi, y]$.

Because of Property 1 the probability that the nth trajectory is stopped in a given time interval depends on the difference between absorbing and starting epoch only, not on absolute time. Suppose now we know that the first $(n = 1)$ trajectory is absorbed at time $\tau > 0$ at $y(\tau) = z$ (see Figure 11.2-3). Then, by the Markov property the probability that the $(n + 1)$th trajectory is absorbed in $[t, t + dt]$ depends on $y(\tau) = z$ but not $y(0) = y_0$. From this and the time homogeneity, it follows that the conditional probability $\Pr(B_{n+1} \mid B_1)$ is equal to the probability that the nth trajectory is absorbed in $[t, t + dt]$ when the first trajectory was started at $y(\tau) = z$

$$\Pr(B_{n+1} \mid B_1) = \psi_n(y, t - \tau \mid z) \, dy \, dt \qquad (11.2\text{-}19)$$

with

$$\Pr(B_1) = \psi(z, \tau \mid y_0) \, dz \, d\tau$$

and using Bayes rule for the combined events we find

$$\Pr(B_{n+1} B_1) = \psi_n(y, t - \tau \mid z) \psi(z, \tau \mid y_0) \, dy \, dz \, d\tau \, dt \qquad (11.2\text{-}20)$$

By integrating over all possible values of τ and z we obtain the fundamental recursion

$$\psi_{n+1}(y, t \mid y_0) = \int_0^t d\tau \int_{-\infty}^\infty \psi_n(y, t - \tau \mid z) \psi(z, \tau \mid y_0) \, dz \qquad (11.2\text{-}21)$$

Instead of conditioning on the first process we can equivalently condition on the last process. Then

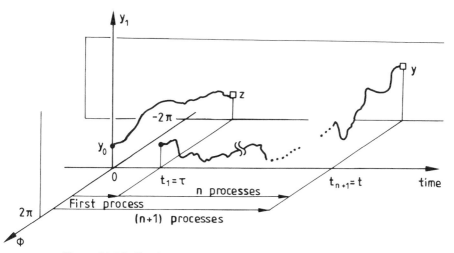

Figure 11.2-3. Fundamental recursion of the vector Markov process.

$$\Pr(B_{n+1} \mid B_n) = \psi(y, t - \tau \mid z) \, dy \, dt \tag{11.2-22}$$

and

$$\Pr(B_n) = \psi_n(z, \tau \mid y_0) \, dz \, d\tau$$

Hence, we find the equivalent form of the recursion

$$\psi_{n+1}(y, t \mid y_0) = \int_0^t d\tau \int_{-\infty}^{\infty} \psi(y, t - \tau \mid z) \psi_n(z, \tau \mid y_0) \, dz \tag{11.2-23}$$

We now assign probabilities to the event of finding the nth trajectory in volume $d\phi \, dy$ at point $x^T = [\phi \quad y]$ at time t

$$\Pr(A_n) = w_n(\phi, y, t \mid y_0) \, d\phi \, dy \tag{11.2-24}$$

The first of these probabilities, w_1, equals the restricted probability density function of the single process

$$w_1(\phi, y, t \mid y_0) \, d\phi \, dy := q(\phi, y, t \mid y_0) \, d\phi \, dy \tag{11.2-25}$$

while the others are related to q and ψ_n by the recursion

$$w_{n+1}(\phi, y, t \mid y_0) = \int_0^t d\tau \int_{-\infty}^{\infty} \psi_n(z, \tau \mid y_0) q(\phi, y, t - \tau \mid z) \, dz \tag{11.2-26}$$

The derivation of the last recursion follows the same path as that of ψ_n and is therefore omitted.

Next we consider the probability density function of $y(t_n)$ at the jumping epochs t_n without regard to when the jumps occur. Clearly

$$\Pr[y(t_1) = y \mid y(0) = y_0] = dy \int_0^{\infty} \psi(y, t \mid y_0) \, dt$$

$$=: \hat{\psi}(y \mid y_0) \, dy \tag{11.2-27}$$

The $y(n) := y(t_n)$ form a Markov sequence (Figure 11.2-4) where the random variables $y(n)$ are of the continuous type. The sequence is time homogeneous with a transition probability given by the integral of (11.2-27)

$$p(y, n + 1 \mid z, n) = \hat{\psi}(y \mid z) \tag{11.2-28}$$

where $(n + 1)$ and n stand for discrete time epochs. We denote the station-

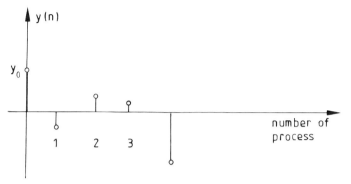

Figure 11.2-4. Markov chain defined by the state variables $y = [x_2 \quad \cdots \quad x_n]$ at the time of absorption of a particle. Transition probability density function is $\hat{\psi}(y \mid z)$.

ary distribution of the Markov sequence $y(n)$ by $V(y)$ (if it exists), in order to clearly distinguish it from the stationary distribution of the state vector $x^T = [\phi \quad y]$. Then, $V(y)$ is the solution of the linear, homogeneous integral equation

$$V(y) = \int_{-\infty}^{\infty} \hat{\psi}(y \mid z) V(z) \, dz \qquad (11.2\text{-}29)$$

The solution of (11.2-29) is determined only up to an arbitrary constant which is given by the normalization condition

$$\int_{-\infty}^{\infty} V(y) \, dy = 1 \qquad (11.2\text{-}30)$$

We pause here for a moment to stress that the functions $w_n(\phi, y, t \mid y_0)$, $\psi_n(y, t \mid y_0)$ and $V(y)$ together with the corresponding recursion relations form the basis on which the theory is derived.

For example, the probability density function of the state vector $x(t)$ without regard to the number of the process—or equivalently the number of cycle slips that have occurred in $[0, t]$—is found by adding w_n over all values of n, i.e.

$$p(\phi, y, t \mid y_0) = \sum_{n=1}^{\infty} w_n(\phi, y, t \mid y_0) \qquad (11.2\text{-}31)$$

Consider the function

$$\Psi(y, t \mid y_0) = \sum_{n=1}^{\infty} \psi_n(y, t \mid y_0) \qquad (11.2\text{-}32)$$

which according to (11.2-21) and (11.2-23) is determined by either of the linear integral equations

$$\Psi(y, t \mid y_0) = \psi(y, t \mid y_0) + \int_0^t d\tau \int_{-\infty}^{\infty} \Psi(y, t - \tau \mid z)\psi(z, \tau \mid y_0)\, dz \tag{11.2-33}$$

$$\Psi(y, t \mid y_0) = \psi(y, t \mid y_0) + \int_0^t d\tau \int_{-\infty}^{\infty} \Psi(z, \tau \mid y_0)\psi(y, t - \tau \mid z)\, dz \tag{11.2-34}$$

Then, summing both sides of (11.2-26) yields the probability density function of the state vector

$$p(\phi, y, \mid y_0) = q(\phi, y, t \mid y_0) + \int_0^t d\tau \int_{-\infty}^{\infty} \Psi(z, \tau \mid y_0)q(\phi, y, t - \tau \mid z)\, dz \tag{11.2-35}$$

Notice that $\Psi(y, t \mid t_0)$ is the probability density function that at time t *any* cycle slip occurs, and that $y(t) = y$. Physically speaking, the integral of $\Psi(y, t \mid y_0)$ over y is the absorption rate of the renewal process.

We urge the reader to compare the n-dimensional results, (11.2-32) and (11.2-34), with the much simpler results for the one-dimensional case, (11.1-43) and (11.1-38), respectively.

11.2.4. Modified Fokker–Planck Equation of the Renewal Process

We now develop the partial differential equation for the transition probability density function of the repeated process. Taking the Laplace transform of (11.2-35) reads

$$p(\phi, y, s \mid y_0) = q(\phi, y, s \mid y_0) + \int_{-\infty}^{\infty} \Psi(z, s \mid y_0)q(\phi, y, s \mid z)\, dz \tag{11.2-36}$$

Applying the operator $(s - L)$ on both sides of the last equation yields

$$(s - L)p(\phi, y, s \mid y_0) = (s - L)q(\phi, y, s \mid y_0)$$
$$+ (s - L)\int_{-\infty}^{\infty} \Psi(z, s \mid y_0)q(\phi, y, s \mid z)\, dz \tag{11.2-37}$$

The first term on the right-hand side

$$(s - L)q(\phi, y, s \mid y_0) = \delta(\phi)\prod_{i=1}^{n-1} \delta[y_i - y_i(0)] \tag{11.2-38}$$

as is easily verified by taking the Laplace transform of (11.2-2) with initial conditions given by (11.2-4). To be concise we write (11.2-38) in the form

$$(s - L)q(\phi, y, s \mid y_0) = \delta(\phi)\delta(y - y_0) \qquad (11.2\text{-}38a)$$

In evaluating the second term on the right-hand side of (11.2-37) we remember that L operates on the variables ϕ and y. Assuming that integration and the operator L commute, we therefore obtain

$$(s - L) \int_{-\infty}^{\infty} \Psi(z, s \mid y_0)q(\phi, y, s \mid z) \, dz$$

$$= \int_{-\infty}^{\infty} \Psi(z, s \mid y_0)(s - L)q(\phi, y, s \mid z) \, dz$$

which because of (11.2-38a) simplifies to

$$\int_{-\infty}^{\infty} \Psi(z, s \mid y_0)\delta(\phi)\delta(y - z) \, dz = \delta(\phi)\Psi(y, s \mid y_0) \qquad (11.2\text{-}39)$$

Combining the results for the two terms on the right-hand side of (11.2-37) we finally obtain the modified Fokker–Planck equation of the repeated process

$$(s - L)p(\phi, y, s \mid y_0) - \delta(\phi)(y - y_0) = \delta(\phi)\Psi(y, s \mid y_0)$$
$$(11.2\text{-}40)$$

As $q(\pm 2\pi, y, t \mid y_0) = 0$ and according to (11.2-36) the transition probability density function of the repeated process also obeys absorbing boundary conditions

$$p(-2\pi, y, t \mid y_0) = p(+2\pi, y, t \mid y_0) = 0$$

As in the one-dimensional case (11.1-47) we observe that $p(\phi, y, s \mid y_0)$ obeys the same Fokker–Planck equation as the transition probability density function of the single process, $q(\phi, y, s \mid y_0)$, modified only by a *point source* term at $\phi = 0$ arising from the phase error being reset to zero.

Physically speaking, the point source exactly compensates the loss of particles due to absorption at the boundaries. The emission of a new particle by the source occurs simultaneously with the absorption of a particle. Thus, at every time t the number of particles inside the hyperplanes at $\phi = \pm 2\pi$ is the same. In other words, the function $p(\phi, y, t \mid y_0)$ is normalized to one for every time t.

We can prove this formally as follows. We first introduce the probability current associated with the modified Fokker–Planck equation, in analogy to (11.2-8), by

$$J_1(\phi, y, t \mid y_0) = K_1(\phi, y)p(\phi, y, t \mid y_0)$$

$$-\frac{1}{2} \sum_{k=1}^{n} \frac{\partial}{\partial x_k} K_{1k}(\phi, y)p(\phi, y, t \mid y_0)$$

$$x_1 := \phi \tag{11.2-41}$$

and the loss function (absorption rate) analogous to (11.2-7)

$$\Gamma_\pm(y, t \mid y_0) := \pm J_1(\pm 2\pi, y, t \mid y_0)$$

$$= \mp \frac{1}{2} \frac{\partial}{\partial \phi} K_{11}(\phi, y)p(\phi, y, t \mid y_0)\big|_{\phi = \pm 2\pi}$$

$$= \mp \frac{1}{2} K_{11}(\pm 2\pi, y) \frac{\partial}{\partial \phi} p(\phi, y, t \mid y_0)\big|_{\phi = \pm 2\pi} \tag{11.2-42}$$

We now have to prove that

$$\Gamma_-(y, t \mid y_0) + \Gamma_+(y, t \mid y_0) = \Psi(y, t \mid y_0) \tag{11.2-43}$$

absorption rate at the
boundaries   source intensity

Indeed, replacing $p(\phi, y, t \mid y_0)$ with the right-hand side of (11.2-35) with the aid of (11.2-9) and (11.2-10), we obtain after some algebraic manipulations

$$\Gamma_-(y, t \mid y_0) + \Gamma_+(y, t \mid y_0) = \psi(y, t \mid y_0)$$

$$+ \int_0^t d\tau \int_{-\infty}^{\infty} \Psi(z, \tau \mid y_0)\psi(y, t - \tau \mid z)\, dz \tag{11.2-44}$$

But due to (11.2-34), the right-hand side of this equation equals Ψ. In fact, we have just shown that $p(\phi, y, t \mid y_0)$ is normalized to one for any time t, as maintained.

Modulo 2π Phase Error Process. If one is interested only in the phase error reduced modulo 2π the corresponding probability density function can easily be obtained from $p(\phi, y, t \mid y_0)$ as

$$p_r(\phi, y, t \mid y_0) = p(\phi, y, t \mid y_0) + \begin{cases} p(\phi - 2\pi, y, t \mid y_0), & 0 \le \phi \le \pi \\ p(\phi + 2\pi, y, t \mid y_0), & -\pi \le \phi \le 0 \end{cases} \tag{11.2-45}$$

It immediately follows that $p_r(\phi, y, t \mid y_0)$ satisfies the periodic boundary conditions

$$p_r(-\pi, y, t \mid y_0) = p_r(\pi, y, t \mid y_0)$$

$$\frac{\partial p_r}{\partial \phi}(-\pi, y, t \mid y_0) = \frac{\partial p_r}{\partial \phi}(\pi, y, t \mid y_0) \tag{11.2-46}$$

As in the one-dimensional case, it can be proved that $p_r(\phi, y, t \mid y_0)$ is identical to the "periodic extension solution" $p_{ex}(\phi, y, t)$ obtained by solving the homogeneous Fokker–Planck equation.

$$\frac{\partial p(\phi, y, t)}{\partial t} = Lp(\phi, y, t) \tag{11.2-47}$$

subject to periodic boundary conditions and initial conditions $\delta(\phi)\delta(y - y_0)$. Since $p_{ex}(\phi, y, t \mid y_0)$ can be computed without knowledge of $\Psi(y, s \mid y_0)$, it contains only partial information about the cycle slips.

11.2.5. Equations for the Steady State

We begin the section with the comment that we have to assume the existence of the stationary solution. No conditions are known today which guarantee the existence of a stationary solution.

In order to determine the stationary probability density function $p(\phi, y \mid y_0)$, we make use of the final value theorem of Laplace transform which gives the asymptotic value of a function provided this limit exists. In Laplace space, (11.2-34) reads

$$\Psi(y, s \mid y_0) = \psi(y, s \mid y_0) + \int_{-\infty}^{\infty} \psi(y, s \mid z)\Psi(z, s \mid y_0) \, dz \tag{11.2-48}$$

In the limit $s \to 0$ we obtain

$$\lim_{s \to 0} \begin{cases} s\Psi(y, s \mid y_0) = \Psi(y \mid y_0) \\ s\psi(y, s \mid y_0) = 0 \\ s\psi(y, s \mid z)\Psi(z, s \mid y_0) = \hat{\psi}(y \mid z)\Psi(z \mid y_0) \end{cases} \tag{11.2-49}$$

where $\hat{\psi}(z \mid y)$ is the integral of $\psi(z, t \mid y)$ over t (see (11.2-27)). Hence, multiplying both sides of (11.2-48) by s and letting $s \to 0$ yields

$$\Psi(y \mid y_0) = \int_{-\infty}^{\infty} \hat{\psi}(y \mid z)\Psi(z \mid y_0) \, dz \tag{11.2-50}$$

Comparing (11.2-50) with (11.2-29) we realize that the source function $\Psi(y \mid y_0)$ is determined by the same integral equation of the Fredholm type as the stationary probability density function $V(y)$. If $V(y)$ is the unique (determined up to a constant) solution of this integral equation, then it follows that

$$\Psi(z \mid y_0) = V(z)h(y_0) \tag{11.2-51}$$

where $h(y_0)$ is a function to be determined. To do so we integrate (11.2-36) over ϕ and y to obtain

$$\frac{1}{s} = \int_{-2\pi}^{2\pi} d\phi \int_{-\infty}^{\infty} q(\phi, y, s \mid y_0) \, dy$$
$$+ \int_{-\infty}^{\infty} dz \Psi(z, s \mid y_0) \int_{-2\pi}^{2\pi} d\phi \int_{-\infty}^{\infty} q(\phi, y, s \mid z) \, dy \tag{11.2-52}$$

After multiplying by s and letting $s \to 0$ we obtain with the above result

$$1 = \int_{-\infty}^{\infty} \Psi(z \mid y_0) \, dz \int_{-2\pi}^{2\pi} d\phi \int_{-\infty}^{\infty} q(\phi, y, s = 0 \mid z) \, dy$$
$$= h(y_0) \int_{-\infty}^{\infty} V(z) \, dz \int_{-2\pi}^{2\pi} d\phi \int_{-\infty}^{\infty} q(\phi, y, s = 0 \mid z) \, dy \tag{11.2-53}$$

This shows that $h(y_0)$ does not depend on y_0 and equals the reciprocal of the meantime between two cycle slips, $E(T_s)$. The last statement can be proved as follows. By definition

$$q(\phi, y, s = 0 \mid z) = \int_0^{\infty} q(\phi, y, t \mid z) \, dt \tag{11.2-54}$$

Integrating the last equation over ϕ and y equals the mean lifetime of a single process starting at $y(0) = z$

$$\int_{-2\pi}^{2\pi} d\phi \int_{-\infty}^{\infty} q(\phi, y, s = 0 \mid z) \, dy = E(T_s \mid z) \tag{11.2-55}$$

Using this result in (11.2-53) yields

$$1 = h(y_0) \int_{-\infty}^{\infty} V(z) E(T_s \mid z) \, dz$$
$$= h(y_0) E(T_s) \Rightarrow E(T_s) = 1/h(y_0) \tag{11.2-56}$$

Thus, the source function becomes independent of y_0 so that $\Psi(z \mid y_0)$ (11.2-51) can be written as

$$\Psi(z) = \frac{V(z)}{E(T_s)} \tag{11.2-57}$$

Finally, the probability density function of the state vector in the steady state is obtained from (11.2-35) using the final value theorem

$$sp(\phi, y, s \mid y_0) = sq(\phi, y, s \mid y_0) + \int_{-\infty}^{\infty} s\Psi(z, s \mid y_0) q(\phi, y, s \mid z) \, dz$$

$$(11.2\text{-}58)$$

In the limit $s \to 0$ we obtain with

$$sp(\phi, y, s \mid y_0) \to p(\phi, y)$$
$$sq(\phi, y, s \mid y_0) \to 0 \qquad\qquad \text{as } s \to 0 \quad (11.2\text{-}59)$$
$$s\Psi(z, s \mid y_0) \to \Psi(z \mid y_0) = V(z)/E(T_s)$$

$$q(\phi, y, s \mid z) \to q(\phi, y, 0 \mid z) = \int_{0}^{\infty} q(\phi, y, t \mid z) \, dt$$

the following result

$$p(\phi, y) = \frac{1}{E(T_s)} \int_{0}^{\infty} dt \int_{-\infty}^{\infty} q(\phi, y, t \mid z) V(z) \, dz \qquad (11.2\text{-}60)$$

This is the natural extension of the corresponding result for the first-order system (11.1-67)

$$p(\phi) = \frac{1}{E(T_s)} \int_{0}^{\infty} q(\phi, t) \, dt$$

Note that the stationary probability density function is independent of the initial value y_0, which is plausible. Equivalently, the stationary probability density function may be obtained from (11.2-40) as $s \to 0$

$$Lp(\phi, y) = -\frac{\delta(\phi)V(y)}{E(T_s)} \qquad (11.2\text{-}61)$$

11.2.6. Meantime between Cycle Slips

Since the absorption rate of trajectories equals the source intensity at any time we obtain from (11.2-43) as $t \to \infty$

$$\Gamma_-(y) + \Gamma_+(y) = \Psi(y) \qquad (11.2\text{-}62)$$

The source function $\Psi(y)$ is found to be independent of the initial value (11.2-57). Hence, the absorption rate $\Gamma_-(y) + \Gamma_+(y)$ is also independent of y_0. Replacing $\Psi(y)$ by $V(y)/E(T_s)$ and integrating over y yields

$$\Gamma_- + \Gamma_+ = \frac{1}{E(T_s)} \qquad (11.2\text{-}63)$$

The total mean number of cycle slips always equals the reciprocal of the meantime between cycle slips. The mean number of positive (negative) cycle slips is given by N_+ and N_-, respectively. There is an interesting interpretation of this result. Writing the Fokker–Planck equation of the repeated process in terms of the probability current (11.2-61) reads

$$\frac{\partial}{\partial \phi} J_1(\phi, y) + \sum_{k=2}^{n} \frac{\partial}{\partial x_k} J_k(\phi, y) = \frac{\delta(\phi) V(y)}{E(T_s)} \qquad (11.2\text{-}64)$$

Let us first integrate the Fokker–Planck equation over the state variables $y = [x_2, \ldots, x_n]$

$$\int_{-\infty}^{\infty} dx_2 \cdots dx_n \frac{\partial}{\partial x_k} J_k(\phi, y)$$

$$= \int_{-\infty}^{\infty} dx_2 dx_3 \cdots dx_{k-1} dx_{k+1} \cdots dx_n \int_{-\infty}^{\infty} dx_k \left[\frac{\partial}{\partial x_k} J_k(\phi, y) \right]$$

$$= \int_{-\infty}^{\infty} dx_2 dx_3 \cdots dx_{k-1} dx_{k+1} \cdots dx_n \left[J_k(\phi, y)\big|_{x_k = +\infty} - J_k(\phi, y)\big|_{x_k = -\infty} \right] = 0$$

$$(11.2\text{-}65)$$

Since the current must vanish at infinity (natural boundary conditions), the contribution of all terms in the sum is zero. The integral over the first component of the current remains

$$J(\phi) := \int_{-\infty}^{\infty} dx_2 \cdots dx_n J_1(\phi, y) \qquad (11.2\text{-}66)$$

$J(\phi)$ is the total probability current across the hyperplane $\phi = $ constant. Using the total current $J(\phi)$ just introduced we see that integration over the state variables $x_2, \ldots, x_n$ leads to a one-dimensional Fokker–Planck equation in ϕ

$$\frac{\partial}{\partial \phi} J(\phi) = \frac{\delta(\phi)}{E(T_s)} \qquad (11.2\text{-}67)$$

Unfortunately, the result is of little help in computing the stationary probability density function of the phase error, $p(\phi)$ and $E(T_s)$, the quantities of greatest interest in application. The reason is that computation of $J(\phi)$ requires knowledge of the n-dimensional probability density function $p(\phi, y) = p(\phi) p(y \mid \phi)$

$$J(\phi) = p(\phi) \int_{-\infty}^{\infty} dx_2 \cdots dx_n \left[K_1(\phi, y) - \frac{1}{2} \sum_{k=1}^{n} \frac{\partial}{\partial x_k} K_{1k}(\phi, y) \right] p(y \mid \phi)$$

$$(11.2\text{-}68)$$

which is unknown.

11.2.7. Approximative Use of Renewal Theory in the Strict Sense*

Since so far no analytical or numerical solution methods are known for the equations of the renewal process, approximations are of interest. The equations for the renewal process can be greatly simplified if the cycle slip events are approximately independent. The individual processes are statistically independent for a first-order system only, modeling the repeated process as a renewal process in the strict sense therefore tacitly assumes that the dynamics of the nth-order system can be approximated by those of a first-order system. A renewal process in the strict sense requires that the initial distribution of any process is identical and independent of the past. This implies that $\psi(y, t \mid y_0)$ must be replaced by

$$\psi(y, t \mid y_0)\, dt = p_E(y)\psi(t \mid y_0) \quad \text{renewal process in the strict sense}$$

(11.2-69)

where $\psi(t \mid y_0)$ given by (see 11.2-14)

$$\psi(t \mid y_0) = -\frac{\partial}{\partial t} \int_{-2\pi}^{2\pi} d\phi \int_{-\infty}^{\infty} q(\phi, y, t \mid y_0)\, dy$$

(11.2-70)

equals the probability that the first particle has been absorbed at time t and $p_E(y)$ is the initial PDF of the state variables $y(n) = y(t_n)$ of the following trajectory.

Averaging over the starting value y_0 with the probability density function $p_E(y_0)$ (which is identical for all trajectories) "removes" the condition on y_0

$$\psi(y, t) = p_E(y)\psi(t)$$

(11.2-71)

with

$$\psi(t) = \int_{-\infty}^{\infty} \psi(t \mid y_0)p_E(y_0)\, dy_0$$

The source density function $\psi(y, z)$ of the renewal process immediately follows from (11.2-33) (in Laplace space)

$$\Psi(y, s) = p_E(y)\,\frac{\psi(s)}{1 - \psi(s)}$$

(11.2-72)

The probability density function then is obtained from either (11.2-40) or (11.2-35) employing $\psi(y, s)$ given by (11.2-71).

For example, averaging (11.2-35) over y_0 yields in the Laplace domain

$$p(\phi, y, s) = q(\phi, y, s) + \int_{-\infty}^{\infty} \Psi(z, s)q(\phi, y, s \mid z)\, dz$$

(11.2-73)

with

*See definition of renewal process in the strict sense in Chapter 10.3.

$$p(\phi, y, s) = \int_{-\infty}^{\infty} p(\phi, y, s \mid y_0) p_E(y_0) \, dy_0$$

$$q(\phi, y, s) = \int_{-\infty}^{\infty} q(\phi, y, s \mid y_0) p_E(y_0) \, dy_0$$

Replacing $\Psi(z, s)$ by the right-hand side of (11.2-72) we arrive at the following result

$$p(\phi, y, s) = \frac{q(\phi, y, s)}{1 - \psi(s)} \tag{11.2-74}$$

A particularly interesting fact is observed by applying the operator $(s - L)$ on (11.2-73). After some simple manipulations we obtain in the steady state

$$Lp(\phi, y) = - \frac{\delta(\phi) p_E(y)}{E(T_s)} \tag{11.2-74}$$

It is interesting that formally the same equation is obtained for the renewal process in the strict sense as for the general case where statistical dependence exists between the individual processes of the repeated process (cf. (11.2-61)).

Let us now address the problem of under what conditions the dynamics of a higher order system can be approximated by those of a first-order system. Experimental evidence given in Chapter 6 revealed that burstings in second-order PLLs become very unlikely under the following two conditions:

1. For overdamped loops for all signal-to-noise ratios of interest.
2. For underdamped loops at very high signal-to-noise ratios.

Let us first consider the case of an overdamped loop ($\zeta > 1$). The poles of the loop are real; with increasing damping ratio we have a fast pole at (-2ζ) which determines the response of the phase error (and thus the cycle slip events) and a slow pole at $-1/2\zeta$ which determines the response of the variable x_1. Hence, replacing the slow variable by its expected value is feasible as supported by the experimental data given in Figure 6.32. To achieve a given meantime between cycle slips $E(T_s B_L)$ the difference in signal-to-noise ratio ρ between a first-order loop and a second-order loop (to be approximated) is only 1.1 dB for $\zeta = 1.0$ and 0.7 dB for a damping ratio of $\zeta = 1.5$. This confirms that the difference is indeed small and that the approximation improves with increasing damping ratio.

Let us now consider the second condition. From the same Figure 6.32 we infer that for a sufficiently large signal-to-noise ratio ρ for the frequently used damping ratio of $\zeta = 0.707$ the difference is about 2.0 dB. But for an undercritically damped loop we can no longer justify an approximation of a first-order model based on separation into a fast and slow variable. No mathematically founded order reduction technique is known for this case.

11.2.8. Stability, Persistence, and Steady-State Distribution

The issues involved in the stability of a nonlinear system perturbed by noise are difficult ones and not resolved in general. Here we consider only systems described by a stochastic differential equation of the form

$$dx(t) = g[x(t)]\, dt + F[x(t)]\, d\beta(t)$$

where $g[x(t)]$ and $F[x(t)]$ are periodic functions of the variable $x_1 = \phi$ since it is this periodicity which is responsible for the unique dynamic system behavior.

We start by briefly surveying some concepts from the field of deterministic stability theory [9]. Because of the periodicity in the state variable ϕ, the system may have an infinite number of *critical points* defined by $g(x_0) = 0$. The critical point is an asymptotically stable equilibrium point if all solutions $x(t)$ starting inside a domain D always converge to x_0, i.e., $x(t_0) \in D$ implies that $\|x(t) - x_0\| \to 0$ as $t \to \infty$. The point x_0 is called an *attractor* and D is the domain of attraction (Figure 11.2-5).

The above definition of a stable equilibrium point requires that the solution $x(t)$ starts inside D. We cannot, therefore, speak of the stability of a nonlinear system but may only speak of the stability of a solution $x(t)$. If $x(t)$ starts outside D no prediction is made on the dynamic behavior of the system. The state $x(t)$ could either converge to another attractor, perform a stable oscillation or increase above all limits.

To include random perturbations of a dynamic system one can try to modify the convergence concepts. For instance, if a deterministic equation solution were to reach a given limit point asymptotically, we would ask whether all the sample solutions (with the exception of a set of samples having total probability of zero) of the corresponding stochastic equation converge to a limit point. Instead, we might ask whether the probability that all trajectories that begin sufficiently close to the deterministic stable equilibrium point x_0 remain for all times in the neighborhood of x_0 with the exception of a set of trajectories of arbitrarly small probability.

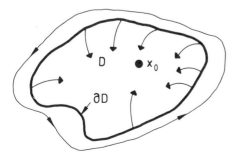

Figure 11.2-5. Domain of attraction D and critical point x_0.

But since the smallest random perturbation causes the phase error to increase above all limits with probability one, the concept of stability no longer applies and must be replaced by a measure which takes into account the spread of diffusion of probability mass out of a given domain. For instance, it can be replaced by the expected time elapsed before leaving the domain of attraction, or, for synchronizers by the meantime between cycle slips. We shall refer to this time as the *persistence of the system*, a term first used by Ludwig [20]. Typically, the persistence of the system must be many orders of magnitude larger than the relaxation time constants for proper operation of the system.

As a final topic we want to address the question of the existence of a stationary distribution, which we explicitly assumed to exist in Section 11.2. This topic is closely related to the previously addressed topics of stability and persistence. It is clear, for example, that no stationary probability density function exists if the deterministic system has no stable equilibrium point. In general, however, many important problems remain unsolved. Assume a trajectory starts inside the domain of attraction of a stable equilibrium point with an initial value of $y(0) = z$. The trajectory will cross with probability one the plane $\phi = \pm 2\pi$ at a random value of $y(t_1) = y$. The steady-state distribution $V(y)$, if it exists, was found to be the solution of the linear integral equation (11.2-29)

$$V(y) = \int_{-\infty}^{\infty} \hat{\psi}(y \mid z) V(z) \, dz$$

At present, no sufficient conditions are known on the kernel $\hat{\psi}(y \mid z)$ to guarantee a stationary solution.

Main Points of the Section

- *Model.* The trajectory is absorbed when the phase error reaches $\phi = \pm 2\pi$ for the first time. The random value of the other state variables at the time absorption are the initial value for the following trajectory.

- *Single Process.* The probability density functions that describe the absorption of the trajectory are generalizations of the functions defined for first-order systems which take into account starting and end position of the trajectories: $\gamma_{(\pm)}(y, t \mid y_0) \, dy \, dt$ equals the probability that the trajectory is absorbed at time t at point $y(t_1) = y$ and $\phi = (\pm)2\pi$. This probability depends on the initial starting value $y(0) = y_0$. The probability

$$\psi(y, t \mid y_0) \, dt = [\gamma_+(y, t \mid y_0) + \gamma_-(y, t \mid y_0)] \, dt \quad (11.2\text{-}10)$$

is the y-dependent probability of absorption.

- *Repeated Process*
  - *Fundamental Functions.* There are basic functions which relate the repeated and the single process. From these three functions all other quantities can be derived

1. The probability *law of absorption* of the $(n+1)$th trajectory is related to that of the defining law of the single process by a fundamental recursion

$$\psi_{n+1}(y, t \mid y_0) = \int_0^t d\tau \int_{-\infty}^{\infty} \psi_n(y, t - \tau \mid z)\psi(z, \tau \mid y_0) \, dz$$
$$(11.2\text{-}21)$$

where $\psi_{n+1}(y, t \mid y_0) \, dy \, dt$ is the probability that the $(n+1)$th trajectory is absorbed in volume dy at point y at time t and $\psi_1(y, t \mid y_0) := \psi(y, t \mid y_0)$.

2. The probability of finding the $(n+1)$th trajectory in volume $d\phi \, dy$ at point (ϕ, y) at time t

$$w_{n+1}(\phi, y, t \mid y_0) \, d\phi \, dy$$

is related to the probability density function of the single process and to ψ_n by

$$w_{n+1}(\phi, y, t \mid y_0) = \int_0^t d\tau \int_{-\infty}^{\infty} \psi_n(z, \tau \mid y_0)q(\phi, y, t - \tau \mid z) \, dz$$
$$(11.2\text{-}26)$$

3. The random values of the state variables y at the time of absorption t_n form a Markov sequence with transition probability density function

$$\hat{\psi}(y \mid y_0) := \int_0^{\infty} \psi(y, t \mid y_0) \, dt \qquad (11.2\text{-}27)$$

If the stationary distribution of y (denoted by $V(y)$) exists, it is the solution of the integral equation

$$V(y) = \int_{-\infty}^{\infty} \hat{\psi}(y \mid z)V(z) \, dz \qquad (11.2\text{-}29)$$

The derivation of the above equations is based on the two main properties of the repeated process

(*a*) Time homogeneity.
(*b*) Markov property of $x(t)$.

- *Derived Functions.* If one is not interested in the actual number n of the process one finds for the repeated process

$$p(\phi, y, t \mid y_0) = \sum_{n=1}^{\infty} w_n(\phi, y, t \mid y_0)$$

$$\Psi(y, t \mid y_0) = \sum_{n=1}^{\infty} \psi_n(y, t \mid y_0)$$

$p(\phi, y, t \mid y_0)\, d\phi\, dy$ is the probability of finding any $(n = 1, 2, \ldots)$ trajectory in a given volume. $\Psi(y, t \mid y_0)$ is the source intensity which equals the absorption rate at the boundaries at any time t

$$\Psi(y, t \mid y_0) = \Gamma_-(y, t \mid y_0) + \Gamma_-(y, t \mid y_0)$$

$p(\phi, y, t)$ obeys a Fokker–Planck equation

$$(s - L)p(\phi, y, s \mid y_0) - \delta(\phi)\delta(y - y_0) = \delta(\phi)\Psi(y, s \mid y_0) \tag{11.2-40}$$

which is modified by a source term on the right-hand side. If a stationary probability density function exists it is independent of y_0 and the solution of the modified Fokker–Planck equation

$$Lp(\phi, y) = -\frac{\delta(\phi)V(y)}{E(T_s)} \tag{11.2-61}$$

11.3. SYSTEMS WITH APERIODIC PHASE DETECTOR CHARACTERISTIC

11.3.1. Mathematical Model of the Delay-Locked Loop

Estimating and tracking the delay between two versions of the same signal is encountered in various fields such as radar, sonar, radio astronomy, and ranging. In many cases the following signal model applies

$$y_1(t) = A_1 s(t) + n_1(t)$$
$$y_2(t) = A_2 s(t - D) + n_2(t) \tag{11.3-1}$$

In this model $y_1(t)$ and $y_2(t)$ are received signals and $n_1(t)$, $n_2(t)$ are two (stationary) noise signals statistically independent from each other and the signal $s(t)$. The useful signal is $s(t)$ and its delayed replica $s(t - D)$, respectively, where D denotes a slowly varying time delay.

The useful signal $s(t)$ can either be a known signal or a random signal. The case where $s(t)$ is a random signal is encountered in passive sonar [23] or in noncontact speed measurement [24] to mention two examples. In ranging the signal $s(t)$ is a pseudo-noise (PN) signal generated by a shift register with appropriate state feedback. The transmitted signal is reflected at the target and arrives at the receiver with delay D. By comparing the received signal $y_1(t)$ with a locally generated replica of the PN sequence with known delay $\hat{D}$, $y_2(t) = s(t - \hat{D})$, one obtains an estimate for the unknown delay D which is related to the range d by $D = 2d/c$, where c is the velocity of light.

The unknown delay D can, for example, be obtained by delaying the leading signal $y_1(t)$ electronically by $\hat{D}$ and by cross-correlating the two signals $y_1(t - \hat{D})$ and $y_2(t)$.

Let us first consider the case where $s(t)$ is a random signal. Then the correlation function for independent noise signals $n_1(t)$, $n_2(t)$ equals

$$E[y_1(t - \hat{D})y_2(t)] = E[A_1 s(t - \hat{D})A_2 s(t - D)]$$

$$= A_1 A_2 R_s(D - \hat{D}) \tag{11.3-2}$$

The correlation function has its maximum for $D = \hat{D}$ and a maximum seeking correlator could be used to estimate D. Since the location of the maximum is of interest only, in many cases an alternative to a maximum seeking device is an error tracking device. To obtain an error signal the two signals $y_1(t - \hat{D})$ and $y_2(t)$ are passed through a linear, physically realizable filter pair $H_1(\omega)$, $H_2(\omega)$ (see Figure 11.3-1). This filter pair optimally prefilters (for example, in the maximum likelihood sense) the incoming signals such that the cross-correlation function of the signals $x_1(t)$ and $x_2(t)$ is an odd function of the error $\phi = D - \hat{D}$

$$R_{x_1,x_2}(D - \hat{D}) = -R_{x_1,x_2}[-(D - \hat{D})] \tag{11.3-3}$$

The zero value of $R_{x_1,x_2}(D - \hat{D})$ can be tracked using feedback techniques to adjust the variable delay line. In analogy to the PLL, the feedback system is called *delay-locked loop* (DLL).

The same principles also apply for tracking a pseudo-noise signal. The essential of a DLL for tracking a PN signal is a shift register that generates two time displaced versions of a PN sequence. Each of the reference signals $A_2 s[t - \hat{D}(T_c/2)]$ and $A_2 s(t - \hat{D} + (T_c/2)]$ is multiplied by the received signal $A_1 s(t - D) + n_1(t)$. The difference between the multiplier outputs are obtained by a subtraction device that together with the multipliers form part of the cross-correlator delay detector circuit. Using the properties of the PN sequences, it can easily be shown that the expected value of the cross-correlator output

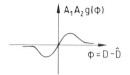

Detector characteristic

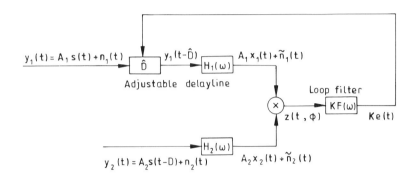

$x_1(t); \tilde{n}_1(t):$ Signals $s(t), n_1(t)$ filtered with $H_1(\omega)$

$x_2(t); \tilde{n}_2(t):$ Signals $s(t), n_2(t)$ filtered with $H_2(\omega)$

multiplier output

$z(t, \phi) = A_1 A_2 g(\phi) + n_n(t) + n_i(t, \phi)$

Figure 11.3-1. Block diagram of the delay-locked loop to track the delay between random signals.

$$E[A_1 s(t-D) + n_1(t)]\left[A_2 s\left(t - \hat{D} - \frac{T_c}{2}\right) - A_2 s\left(t - \hat{D} + \frac{T_c}{2}\right)\right]$$

$$= A_1 A_2 \left[R_s\left(D - \hat{D} + \frac{T_c}{2}\right) - R_s\left(D - \hat{D} - \frac{T_c}{2}\right)\right] = A_1 A_2 g(\phi) \qquad (11.3\text{-}4)$$

vanishes for $D = \hat{D}$ and is an odd function of the tracking error $\phi = D - \hat{D}$ (see Figure 11.3-2). Thus, for $\phi \neq 0$ an error signal is generated to adjust the clock frequency of the VCO such that (on the average) $D = \hat{D}$.

We next derive a mathematically equivalent model of the DLL. The main idea leading to this model is based on the decomposition of the multiplier output $z(t, \phi)$ in two parts

$$z(t, \phi) = E[z(t, \phi)] + \{z(t, \phi) - E[z(t, \phi)]\} \qquad (11.3\text{-}5)$$

The expected value $E[z(t, \phi)]$ is the useful signal to control the variable delay line

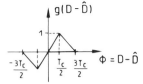

Detector characteristic

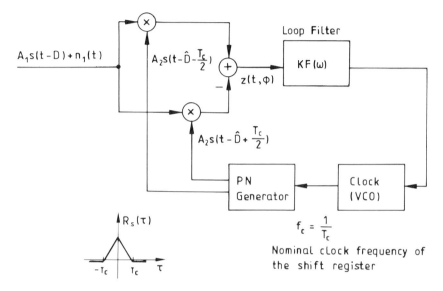

$R_s(\tau)$: Autocorrelation function of s(t)

Figure 11.3-2. Block diagram of the delay-locked loop to track the delay of a pseudo noise signal.

$$E[z(t, \phi)] = A_1 A_2 R_{x_1 x_2}(\phi) := A_1 A_2 g(\phi) \qquad (11.3\text{-}6)$$

while the mean free part

$$n(t, \phi) := z(t, \phi) - E[z(t, \phi)]$$
$$= n_i(t, \phi) + n_n(t) \qquad (11.3\text{-}7)$$

can be considered as noise. Note that $n(t, \phi)$ never vanishes—even if the two additive noise signals $n_1(t)$ and $n_2(t)$ are identical to zero—since it contains a term $n_i(t, \phi)$ which is solely dependent on the product $A_1 A_2 x_1(t) x_2(t)$. This component of $n(t, \phi)$ is called *intrinsic* or *self-noise*. The component $n_n(t)$ describes the contribution of the additive noise process at the multiplier output. This component is independent of the tracking error while the intensity of the intrinsic noise depends strongly on ϕ. We will give a numerical example later on.

We assume that the control law of the delay line is given by

$$\frac{d\hat{D}(t)}{dt} = Ke(t) \tag{11.3-8}$$

where K is a constant and the control voltage $e(t)$ equals the output of the loop filter with impulse response $f(t)$

$$Ke(t) = Kf(t) * z(t, \phi) \tag{11.3-9}$$

If we write $z(t, \phi)$ as the sum of useful signal plus noise then after replacing $\hat{D}$ with $D\text{-}\phi$, (11.3-8) assumes the form

$$\frac{d\phi}{dt} = -Kf(t) * [A_1 A_2 g(\phi) + n(t, \phi)] + \frac{dD}{dt} \tag{11.3-10}$$

Expanding the loop transfer function in partial fractions, (11.3-10) can always be written in the form of a first-order vector random differential equation (see state variable descriptions of loops in Chapter 2).

The mathematically equivalent model of the loop is shown in Figure 11.3-3. The model has much similarity with the baseband model of the PLL with two important differences:

1. The detector characteristic $g(\phi)$ is an aperiodic function of ϕ. (For a PN sequence $g(\phi)$ actually is periodic with a period much larger than the tracking range. Thus, it may well be approximated by an aperiod detector characteristic.)
2. The power spectrum of $n(t, \phi)$ is also an aperiodic function of ϕ.

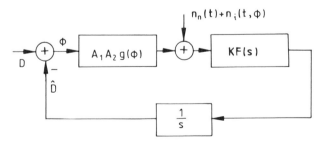

Figure 11.3-3. Equivalent model of the delay-locked loop.

11.3.2. Fokker–Planck Equation

Formally replacing $n(t, \phi)$ by a white gaussian noise process one arrives at a vector stochastic differential equation. Since $n(t, \phi)$ is the product of two Gaussian processes it is non-gaussian. Computer simulations of the actual system agree to an excellent degree with results predicted by Fokker–Planck equation, that is, $K_n(x) = 0$, $n \geq 3$, thus confirming the validity of the diffusion approximation. In order to properly model a physical process, the equation must be understood in the sense of Stratonovich. For the sake of simplicity we begin with a discussion of a first-order system. The generalization to nth order systems is straightforward.

We obtain ($a = dD/dt$, constant, $A = A_1 A_2$ and $F(\omega) = 1$)

$$d\phi = [-KAg(\phi) + a] \, dt + K\sqrt{S_n(0, \phi)} \, d\beta(t) \qquad (11.3\text{-}11)$$

The intensity coefficients of the Fokker–Planck equation are given by

$$K_1(\phi) = -KAg(\phi) + a + \frac{1}{4} K^2 \frac{d}{d\phi} S_n(0, \phi) \qquad (11.3\text{-}11a)$$

$$K_2(\phi) = K^2 S_n(0, \phi)$$

An interesting question is whether the Fokker–Planck equation has a meaningful stationary solution. A necessary condition for a stationary solution is that the $\partial p/\partial t$ vanishes. Assuming $\partial p/\partial t = 0$, integration of the Fokker–Planck equation over ϕ yields the stationary probability current j

$$j = K_1(\phi)p(\phi) - \frac{1}{2} \frac{d}{d\phi} K_2(\phi)p(\phi) \qquad (11.3\text{-}12)$$

But since j must be constant for a stationary solution and since the delay error extends over the entire line, the current must be zero. Thus, $p(\phi)$ is the solution of the homogeneous differential equation (11.3-12) given by

$$p(\phi) = \frac{C}{K_2(\phi)} \exp\left[2 \int^{\phi} \frac{K_1(x)}{K_2(x)} \, dx\right] \qquad (11.3\text{-}13)$$

The constant C is the normalization constant defined by

$$\frac{1}{C} = \int_{-\infty}^{\infty} \frac{1}{K_2(\phi)} \exp\left[2 \int^{\phi} \frac{K_1(x)}{K_2(x)} \, dx\right] d\phi \qquad (11.3\text{-}14)$$

Assuming the intensity coefficient $K_2(\phi)$ is strictly positive with a finite maximum, the integral of (11.3-14) exists only if the exponential function is asymptotically bounded by

$$\exp\left[2\int^{\phi}\frac{K_1(x)}{K_2(x)}\,dx\right] < \frac{B}{|\phi|^p} \quad \text{as } |\phi| \to \infty \tag{11.3-15}$$

where B is a constant and $p > 1$. Taking the natural logarithm on both sides the last inequality can equivalently be written in the form

$$2\int^{\phi}\frac{K_1(x)}{K_2(x)}\,dx < \ln\frac{B}{|\phi|^p} \quad \text{as } |\phi| \to \infty \tag{11.3-16}$$

We see, that asymptotically the right-hand side of the last inequality approaches $(-\infty)$. Hence the left-hand side must also become negative and approach $(-\infty)$. But this is impossible for any realizable tracking system as will be seen shortly. Inserting the expressions for $K_1(x)$ and $K_2(x)$ the integral of (11.3-16) becomes

$$\int^{\phi}\frac{K_1(x)}{K_2(x)}\,dx = \frac{-KA}{K^2}\int^{\phi}\frac{g(x)}{S_n(0,x)}\,dx + \frac{a}{K^2}\int^{\phi}\frac{1}{S_n(0,x)}\,dx$$

$$+ \frac{1}{4}\int^{\phi}\frac{d/dx\,S_n(0,x)}{S_n(0,x)}\,dx$$

$$= \frac{-KA}{K^2}\int^{\phi}\frac{g(x)}{S_n(0,x)}\,dx$$

$$+ \frac{a}{K^2}\int^{\phi}\frac{1}{S_n(0,x)}\,dx + \frac{1}{4}\ln S_n(0,\phi), \quad \phi \to \pm\infty \tag{11.3-17}$$

We first notice that $a \neq 0$ violates the inequality either for $\phi \to \infty$ or $\phi \to -\infty$, depending on the sign of a. Hence, this case is ruled out. Since $S_n(0,x)$ is a limited function for any ϕ, only the first integral could possibly tend toward $(-\infty)$. However, this is not the case. By definition, $g(\phi)$ equals (up to a constant) the cross-correlation of the signals $x_1(t)$ and $x_2(t)$. Since $x_1(t)$ and $x_2(t)$ are the output signals of physically realizable filters $H_1(\omega)$ and $H_2(\omega)$, respectively, the cross-correlation $R_{x_1x_2}(\phi)$ is integrable

$$\int_{-\infty}^{\infty}R_{x_1x_2}(\phi)\,d\phi < \infty \tag{11.3-18}$$

and therefore

$$\int_{-\infty}^{\infty}\frac{g(x)}{S_n(0,x)}\,dx < \infty$$

Hence, all terms of the integral remain bounded $(a = 0)$ which violates the inequality. While the proof has been given for a first-order system it is evident that the same reasoning applies for systems of arbitrary order.

This is an important result. It tells us that *no* physically realizable filter pair $H_1(\omega)$, $H_2(\omega)$ exists to generate a detector characteristic $g(\phi)$, such that a meaningful stationary probability density function $p(\phi)$ would exist. In analogy to the PLL we conclude that the delay error of $\phi(t)$ eventually exceeds every limit with the passage of time, that is, the loop falls *out-of-lock*.

There is an appealing physical interpretation of this result. As previously mentioned the error $\phi(t)$ may be thought of as the trajectory of a particle undergoing a random motion. The motion of the particle is governed by a force $2K_1(x)/K_2(x)$ which is the negative gradient of the potential $U_0(\phi)$

$$U_0(\phi) = -2 \int^{\phi} \frac{K_1(x)}{K_2(x)} \, dx \qquad (11.3\text{-}19)$$

Since the potential well has a finite depth, the restoring force tends too rapidly toward zero as $|\phi| \to \infty$ and the particle reaches any position with probability one (Figure 11.3-4), that is, the loop falls out-of-lock.

11.3.3. Modeling the Operation of the Delay-Locked Loop as a Renewal Process

As we have just demonstrated, the tracking error $\phi(t)$ exceeds every bound with probability one. Therefore, no meaningful stationary probability density function exists. In analogy to the PLL we must define a modified error process that models the actual behavior of the system thus leading to a meaningful result for the steady state, i.e. $t \to \infty$. This can be done as follows: at the beginning of each measurement the DLL has to be brought within the "in-lock" region. This can be done, for example, by sweeping the electronic delay line over the possible range of D. If $d\hat{D}/dt$ is suitably chosen, the DLL locks when $D = \hat{D}$. The "in-lock" condition is then

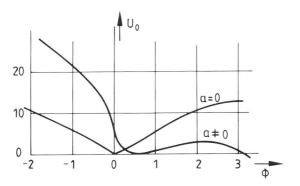

Figure 11.3-4. Typical potential function for zero loop detuning ($a = 0$) and for loop detuning, $a \neq 0$.

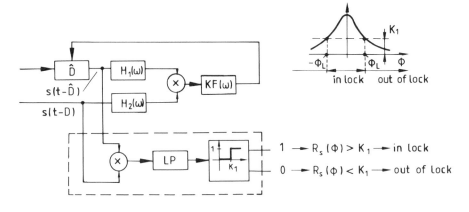

LP : Low Pass Filter

Figure 11.3-5. Delay-locked loop and associated lock detector.

supervised by a lock detector (Figure 11.3-5) which basically estimates the even correlation function $R_s(\phi)$, $\phi = D - \hat{D}$, which is maximum for $\phi = 0$. As soon as $R_s(\phi)$ falls below a certain threshold the lock detector signals "out of lock." Notice, that the lock detector indicates "out-of-lock" at the points $\pm\phi_L$ (corresponding to $R_s(\pm\phi_L) = K_1$) only on average and that for every ϕ, a probability exists that the circuit indicates out-of-lock. Thus, the behavior can be modeled as follows.

At time $t = 0$ we start a random process near the lock point (ϕ_0, y_0). At time t_1 the lock detector signals "out-of-lock" for the first time and the trajectory is stopped. Since the mean lifetime of a trajectory for a reasonable loop design is always much greater than the re-acquisition time, we assume that a new trajectory is started near (ϕ_0, y_0) immediately after the absorption of its predecessor. A typical trajectory of this repeated process is shown in Figure 11.3-6 for first-order loop.

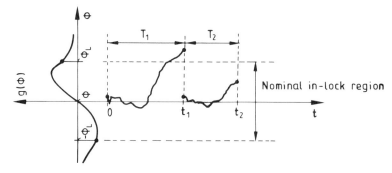

Figure 11.3-6. Sample trajectory of the renewal process for a first-order delay-locked loop.

We assume that the initial distributions of the individual processes are identical

$$p(\phi, y, t_i) = p_e(\phi, y), \quad i = 1, 2, \ldots \tag{11.3-20}$$

The regenerative process is therefore a renewal Markov process in the strict sense for which we developed a theory in Section 11.2.7. The probability density function $p(\phi, y, t)$ of the renewal process is related to that of the single process $r(\phi, y, t)$ by (11.2-74)

$$p(\phi, y, s) = \frac{q(\phi, y, s)}{1 - \psi(s)} \tag{11.3-21}$$

What remains to be done is the derivation of a Fokker–Planck equation for the probability density function $q(\phi, y, t)$ of the single process. The essential difference from all previously studied problems is the fact that the trajectory is no longer absorbed at the fixed positions $\phi = \pm \phi_L$ but may be absorbed at any point with a certain probability. Physically, we may therefore speak of absorption of particles at a *distributed sink*.

11.3.4. Fokker–Planck Equation for a Process with Distributed Sinks*

We assume that the lock detector probability depends only on the present position of the trajectory, but not on the past $x(\tau), \tau < t$

$$\Pr\left\{\begin{matrix}\text{lock-detector signals} \\ \text{out-of-lock in } [t, t + \Delta t]\end{matrix} \middle| x(t) = x\right\} = c\alpha(x) \, \Delta t \tag{11.3-22}$$

This assumption is only approximately valid for a realizable lock detector since it neglects the memory of the low-pass filtered lock detector signal (see Figure 11.3-5). The function $\alpha(x)$ is the lock detector characteristic which ideally should be zero inside the lock-in region and infinite outside. The combined tracker lock detector system is mathematically modeled as follows. We consider particles in the n-dimensional space X whose transition probability density $p(x, t \mid y, 0)$ satisfies a Fokker–Planck equation

$$\frac{\partial p(x, t \mid y, 0)}{\partial t} - Lp(x, t \mid y, 0) = 0 \tag{11.3-23}$$

These particles can "collide" (= lock detector signals "out-of-lock") with a certain probability. We demand that such a collision does not alter the paths of the particles but leaves a mark on them. In other words the probability

*This section is based on the paper of Darling and Siegert [22].

density function $p(x, t \mid y, 0)$ to find a particle in volume dx at point x is the same whether the particle has previously been marked or not.

We further demand that the probability to mark a particle at a position x in a time interval $[t, t + s]$, $s > 0$ is:

1. Independent of whatever happened to the particle before.
2. Depends only on the length of the interval but not on absolute time (time homogeneity). (11.3-24)
3. The probability for more than one mark in $\Delta t \rightarrow 0$ is zero.

We denote

$$P_1(\Delta t) := \Pr\left\{ \begin{array}{l} \text{marking of a particle in } [t, t + \Delta t] \\ \text{(lock detector signals "out-of-lock")} \end{array} \middle| \begin{array}{l} \text{particle in position} \\ x(t) = x \end{array} \right\}$$

$$= c\alpha(x)\Delta t \quad \text{as } \Delta t \rightarrow 0 \tag{11.3-25}$$

Under these assumptions the probability of finding the particle unmarked on a given path (sample trajectory) with starting point $x(0) = y$ and final position $x(t) = x$ equals

$$P_0(t) = \exp\left[-c \int_0^t \alpha[x(t')] \, dt' \right] \tag{11.3-26}$$

To derive this result we first make use of the property that the marking of a particle in $[t, t + s]$, $s > 0$ is statistically independent of what happened before. Thus, the probability of finding an unmarked particle in $[0, t + s]$ is

$$P_0(t + s) = P_0(t) P_0(s) \tag{11.3-27}$$

Notice that (11.3-27) holds for any s. Furthermore, the probability of finding an unmarked particle depends only on the length of the time interval but not on time t, see property (2) of (11.3-24). Since the probability of more than one mark in $\Delta t \rightarrow 0$ is negligible we have

$$P_0(\Delta t) = 1 - P_1(\Delta t)$$

$$= 1 - c\alpha[x(t)] \Delta t \tag{11.3-28}$$

Inserting (11.3-28) into (11.3-27) yields, with s replaced by Δt

$$P_0(t + \Delta t) = P_0(t)[1 - c\alpha[x(t)] \Delta t] \tag{11.3-29}$$

From which it follows

$$\frac{dP_0(t)}{dt} = -c\alpha[x(t)] P_0(t) \quad \text{as } \Delta t \rightarrow 0 \tag{11.3-30}$$

Notice that (11.3-30) holds for a *known* sample trajectory $x(t')$, $t' \leq t$. Hence, $\alpha[x(t')]$ is also known for all times $t' \leq t$. The solution of (11.3-30) with initial condition $P_0(0) = 1$ is easily found to agree with (11.3-26).

If we average over all possible sample paths which start at point $x(0) = y$ and end at $x(t) = x$ then

$$E\left\{ \exp\left[-c \int_0^t \alpha[x(t')]\, dt' \right] \, \Big| \, x(t) = x, x(0) = y \right\} \qquad (11.3\text{-}31)$$

equals the conditional probability that a particle that starts at point $x(0) = y$ at time zero reaches the point $x(t) = x$ at time $t > 0$ unmarked. Notice that the average is conditioned on the initial and present position of the particle. Multiplying (11.3-31) with the probability $p(x, t \mid y, 0)$ that $x(t)$ is in the volume element dx at x equals the probability of finding an unmarked particle in dx at point x at time t (starting point $x(0) = y$)

$$q(x, t \mid y, 0) := E\left[\exp\left[-c \int_0^t \alpha[x(t')]\, dt' \right] \, \Big| \, x(t) = x, x(0) = y \right] p(x, t \mid y, 0) \qquad (11.3\text{-}32)$$

We next derive an integral equation for $r(x, t \mid y, 0)$. The event that a particle is marked for the first time at point x' in the time interval $[t', t' + dt']$ is the union of two independent events

A_1: the particle starting at $x(0) = y$ has reached x' at time t'
and has never been marked before
$\Rightarrow \Pr(A_1) = q(x', t' \mid y, 0)\, dx'$ \qquad (11.3-33)

A_2: the particle is marked at point x' in $[t', t' + dt']$
$\Rightarrow \Pr(A_2) = c\alpha(x')\, dt'$

Thus

$$\Pr(A_1 \cap A_2) = q(x', t' \mid y, 0) c\alpha(x')\, dx'\, dt' \qquad (11.3\text{-}34)$$

If a particle has one or more marks at time t then there must be a point x' at time t' where it is marked for the first time. The probability for a particle to move unmarked from $x(0) = y$ to $x(t') = x'$ at time t', obtaining a mark and then to travel on to $x(t) = x$ suffering an irrelevant number of collisions is

$$\Pr\left\{ \begin{array}{l} \text{particle in position } x \\ \text{at time } t \end{array} ; \quad \text{first mark at } x' \text{ at } [t', t' + dt'] \right\}$$
$$\qquad (11.3\text{-}35)$$
$$= p(x, t \mid x', t') c\alpha(x') q(x', t' \mid y, 0)\, dx'\, dx\, dt'$$

Integrating this over the whole space X' gives the probability of finding a particle at point x at time t which was marked in the time interval $[t', t' + dt']$.

Now, since a particle can be marked anywhere in $t' \in [0, t]$ the probability that a particle has been marked anywhere from time $t = 0$ to the present time equals the integral over t'

$$\Pr \left\{ \begin{array}{ll} \text{particle in position } x \, ; & \text{first mark of particle anywhere in } X' \\ \text{at time } t & \text{space at } t' \in [0, t] \\ & = \text{at least one mark on the particle} \\ & \text{at time } t \end{array} \right\}$$

$$= dx \int_0^t dt' \int_{X'} p(x, t \mid x', t') c\alpha(x') q(x', t' \mid y, 0) \, dx' \qquad (11.3\text{-}36)$$

Finally, subtracting this probability (11.3-36) from the probability of finding the particle in dx at point x at time t (marked or unmarked) yields the probability of the unmarked particle being at point x. Therefore,

$$q(x, t \mid y, 0) = p(x, t \mid y, 0)$$

$$- c \int_0^t dt' \int_{X'} p(x, t \mid x', t') \alpha(x') q(x', t' \mid y, 0) \, dx' \qquad (11.3\text{-}37)$$

To find a corresponding differential equation we formally apply the operator $(-L + (\partial/\partial t))$ on both sides of (11.3-37)

$$\left(-L + \frac{\partial}{\partial t} \right) q(x, t \mid y, 0)$$

$$= \left(-L + \frac{\partial}{\partial t} \right) p(x, t \mid y, 0)$$

$$- \frac{\partial}{\partial t} c \int_0^t dt' \int_{X'} p(x, t \mid x', t') \alpha(x') q(x', t' \mid y, 0) \, dx'$$

$$+ c \int_0^t dt' \int_{X'} L p(x, t \mid x', t') \alpha(x') q(x', t' \mid y, 0) \, dx' \qquad (11.3\text{-}38)$$

(The operator L acts on x.) The first term on the right-hand side is zero according to (11.3-23). The second term can be rewritten as follows

$$\frac{\partial}{\partial t} c \int_0^t dt' \int_{X'} (\cdot) dx' = c \int_{X'} p(x, t \mid x', t) \alpha(x') q(x', t \mid y, 0) \, dx'$$

$$+ c \int_0^t dt' \int_{X'} \frac{\partial}{\partial t} p(x, t \mid x', t') \alpha(x') q(x', t' \mid y, 0) \, dx'$$

$$(11.3\text{-}39)$$

Substituting (11.3-39) into (11.3-38), combining terms and applying (11.3-23) we find

$$\left(-L + \frac{\partial}{\partial t}\right) q(x, t \mid y, 0) = -c \int_{X'} p(x, t \mid x', t)\alpha(x')q(x', t \mid y, 0)\, dx'$$

$$(11.3\text{-}40)$$

Note that $p(x, t \mid x', t)$ has the same time t on the left and right sides of the vertical bar. But by definition

$$p(x, t \mid x', t) = \delta(x - x') \tag{11.3-41}$$

where $\delta(x - x')$ stands for the product of Dirac impulses for the n components of the vector x.

Inserting (11.3-41) on the right-hand side of (11.3-40) yields

$$c \int_{X'} p(x, t \mid x', t)\alpha(x')q(x', t \mid y, 0)\, dx' = c\alpha(x)q(x, t \mid y, 0)$$

$$(11.3\text{-}42)$$

and we finally arrive at the partial differential equation

$$\frac{\partial q(x, t \mid y, 0)}{\partial t} - Lq(x, t \mid y, 0) + c\alpha(x)q(x, t \mid y, 0) \tag{11.3-43}$$

The modified Fokker–Planck equation (11.3-43) has a very appealing physical interpretation. The number of particles in the volume dx at point x changes in dt by

$$\frac{\partial}{\partial t} q(x, t \mid y, 0) = Lq(x, t \mid y, 0) - c\alpha(x)q(x, t \mid y, 0)$$

The first term on the right-hand side is the change through diffusion and drift, the second is the probability of being marked. Thus $q(x, t \mid y, 0)$ describes the probability law of the unmarked particles.

11.3.5. Numerical Example: First-Order Delay-Locked Loop

Model. (Figure 11.3-7). For a DLL used for noncontact speed measurement the following intensity coefficients were found (for details of the derivation see [24])

$$K_1(x) = -g(x) + \gamma + \frac{1}{2}\frac{d}{dx}\left[\frac{1}{\rho_s} N_i(x) + \frac{1}{\rho_n}\right]$$

$$(11.3\text{-}44)$$

$$K_2(x) = 2\left[\frac{1}{\rho_s} N_i(x) + \frac{1}{\rho_n}\right]$$

with:

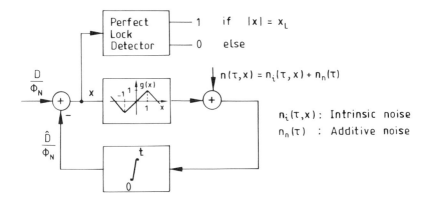

$$\frac{D}{\Phi_N}, \frac{\hat{D}}{\Phi_N} \quad : \text{Normalized delays}$$

$$x = \frac{D-\hat{D}}{\Phi_N} \quad : \text{Normalized error}$$

$$\tau = 4B_L t : \text{Normalized time}$$

$$B_L \qquad : \text{Loop bandwidth of the linearized loop}$$

Figure 11.3-7. Equivalent model of the delay-locked loop and the lock detector for normalized variables and parameters.

$x = \phi/\phi_N$ normalized delay error
ϕ_N normalization constant (s)
$g(x)$ triangular detector characteristic
γ normalized loop detuning
$1/\rho_s$ normalized loop bandwidth, $1/\rho_s = 2B_L\phi_N$
$N_i(x)$ normalized power density spectrum of the intrinsic noise

The intensity coefficient $K_2(x)$ is comprised of two terms

The term $1/\rho_n$ equals the signal-to-noise ratio of the additive noise process and the useful signal. This term is independent of ϕ.

The term $1/\rho_s \, N_i(x)$ is due to the intrinsic noise $n_i(t, \phi)$. The intrinsic noise contribution strongly depends on the error x as shown in Figure 11.3-8. The function $N_i(x)$ equals the normalized power density spectrum $S_{n_i}(0, x)$ at the origin. The factor $1/\rho_s = 2B_L\phi_N$ equals the normalized loop bandwidth.

Of particular interest is the value of the intensity coefficient $K_2(x)$ in the vicinity of the nominal tracking point, $x = 0$. By proper choice of the prefilters $H_1(\omega)$ and $H_2(\omega)$ of Figure 11.3-5 one can always achieve that $N_i(x)$ vanishes at $x = 0$. This is a very desirable property since it is this part which is not filtered out by the loop filter $F(\omega)$.

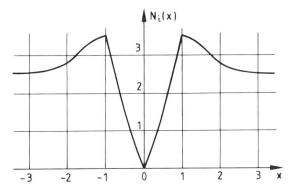

Figure 11.3-8. Normalized power density spectrum $N_i(x)$ of the intrinsic noise at the origin ($\omega = 0$) (from H. Meyr, Delay Lock Tracking of Stochastic Signals, *IEEE Transactions on Communications*, Vol. 24, No. 3, pp. 331–339, © IEEE 1976.)

The last term of the first intensity coefficient $K_1(x)$ is the noise-induced drift due to the x-dependence of $n_i(\tau, x)$. It is numerically insignificant compared to the restoring force $g(x)$.

We assume a perfect lock detector (Figure 11.3-9). For $|x| < x_L$ the probability of a false alarm equals zero. As soon as $x = \pm x_L$ for the first time the lock detector signals out-of-lock and the new trajectory is restarted at $x = \gamma$.

Modified Fokker–Planck Equation. For the data given, the steady-state probability density function of the renewal process obeys a modified Fokker–Planck equation

$$Lp(x) = -\frac{\delta(x - \gamma)}{4B_L E(T_s)} \tag{11.3-45}$$

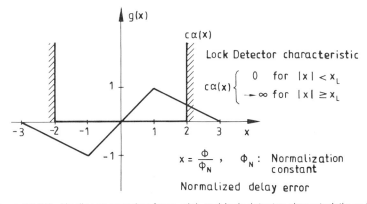

Figure 11.3-9. Nonlinear restoring force $g(x)$ and lock detector characteristic $c\alpha(x)$.

where $4B_L E(T_s)$ is the normalized meantime between falling out-of-lock events with L given by $K_1(x)$ and $K_2(x)$ in (11.3-44). The initial distribution is $\delta(x - \gamma)$, where γ denotes the value of x at the stable equilibrium point. The equation must be solved for absorbing boundary conditions at $x = \pm x_L$ (perfect lock detector assumption).

Results. Without belaboring the details one finds [24]

$$p(x) = \frac{1}{4B_L E(T_s)} \frac{1}{[(1/\rho_s)N_i(x) + (1/\rho_n)]^{1/2}} \exp\left[-\phi_N U_0(x)\right]$$

$$\times \int_{-x_L}^{x} [D_0 - u(x' - x_0)] \frac{1}{[(1/\rho_s)N_i(x) + (1/\rho_N)]^{1/2}} \exp\left[\phi_N U_0(x')\right] dx'$$

$$(11.3\text{-}46)$$

with the potential function

$$U_0(x) = -\frac{1}{\phi_N} \int^{x} \frac{1}{(1/\rho_s)N_i(x') + (1/\rho_n)} [-g(x') + \gamma] dx'$$

$$(11.3\text{-}47)$$

D_0 is a constant of integration determined by the boundary conditions

$$D_0 = \frac{\displaystyle\int_{x_0}^{x_L} \frac{1}{[(1/\rho_s)Ni(x) + (1/\rho_n)]^{1/2}} \exp\left[\phi_N U_0(x)\right] dx}{\displaystyle\int_{-x_L}^{x_L} \frac{1}{[(1/\rho_s)Ni(x) + (1/\rho_n)]^{1/2}} \exp\left[\phi_N U_0(x)\right] dx}$$

$$(11.3\text{-}48)$$

The stationary probability density function $p(x)$ is shown in Figure 11.3-10. Due to the fact that $N_i(x)$ vanishes for $x = 0$ we observe a sharp peak for the normalized detuning $\gamma = 0$. For the same reason the variance σ_x^2 is strongly dependent on γ, if the intrinsic noise is dominant. The normalized meantime between an out-of-lock event $4B_L E(T_S)$ is shown in Figure 11.3-11.

In order to verify the results of the theory and the validity of the assumptions made, a computer simulation program of the *actual* loop configuration (not the equivalent model) was written. The results for two runs with $\gamma = 0$ and $\gamma = 0.5$ are shown in Figure 11.3-12. The simulation confirms the strong dependence of $p(x)$ on $N_i(x)$. Also, the validity of the assumptions leading to the approximate Fokker–Planck equations are confirmed. It is worthwhile to note that the effect of a state dependent noise term can only be understood in the context of nonlinear theory.

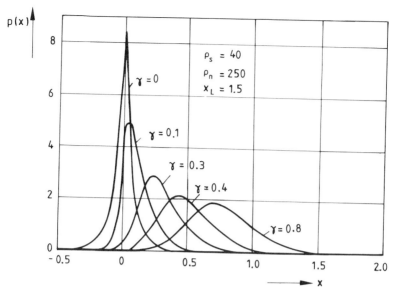

Figure 11.3-10. Steady-state probability density function $p(x)$ (from H. Meyr, Delay Lock Tracking of Stochastic Signals, *IEEE Transactions on Communications*, Vol. 24, No. 3, pp. 331–339, © IEEE 1976.)

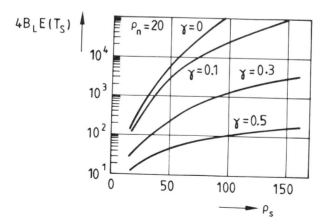

Figure 11.3-11. $E(T_s)$ in function of ρ_s for several values of γ (from H. Meyr, Delay Lock Tracking of Stochastic Signals, *IEEE Transactions on Communications*, Vol. 24, No. 3, pp. 331–339, © IEEE 1976.)

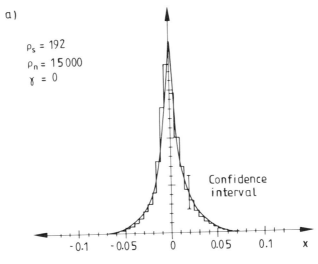

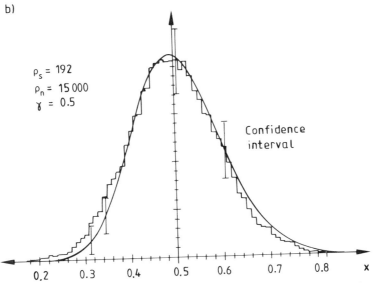

Figure 11.3-12. Computer simulated and theoretical probability density function for delay-locked loop with triangular nonlinearity (from H. Meyr, Delay Lock Tracking of Stochastic Signals, *IEEE Transactions on Communications*, Vol. 24, No. 3, pp. 331–339, © IEEE 1976.)

REFERENCES

1. W. C. Lindsey, *Synchronization System in Communication and Control*, Prentice Hall, Englewood Cliffs, NJ, 1972.
2. A. H. Jazwinski, *Stochastic Processes and Filtering Theory*, Academic Press, New York and London, 1970.

3. K. Itô, Stochastic Integral, *Proceedings of the Imperial Academy*, Tokyo, Vol. 20, pp. 519–524, 1944.

4. R. L. Stratonovich, A New Form of Representing Stochastic Integrals and Equations, *SIAM Journal on Control*, Vol. 4, pp. 362–371, 1966.

5. A. J. Viterbi, *Principles of Coherent Communications*, McGraw-Hill, New York, 1966.

6. I. I. Gichmann and A. W. Skorochod, *Stochastische Differential-gleichungen*, Akademie-Verlag, Berlin, 1971.

7. W. Gröbner and P. Lesky, *Mathematische Methoden der Physik*, Vol. II., Mannheim: BI, Hochschultaschenbücher Nr. 89/90, 90a, 1968.

8. R. Courant and D. Hilbert, *Methoden der mathematischen Physik*, Springer-Verlag, Berlin, 1931/1937.

9. P. S. Maybeck, *Stochastic Models, Estimation, and Control*, Vols. 1–3, Academic Press, New York, 1979.

10. E. Wong, *Stochastic Processes in Information and Dynamical Systems*, McGraw Hill, New York, 1971

11. M. Moeneclaey, The Influence of Phase-Dependent Loop Noise on the Cycle Slipping of Symbol Synchronizers, *IEEE Transactions on Communications*, Vol. 33, No. 12, pp. 1234–1239, December 1985.

12. C. M. Chie, New Results on Mean Time-to-First-Slip for a First-Order Loop, *IEEE Transactions on Communications*, Vol. 33, No. 9, pp. 897–903, September 1985.

13. H. Meyr, Nonlinear Analysis of Correlative Tracking Systems Using Renewal Process Theory, *IEEE Transactions on Communications*, Vol. 23, pp. 192–204, February 1975.

14. W. C. Lindsey and H. Meyr, Complete Statistical Description of the Phase-Error Process Generated by Correlative Tracking Systems, *IEEE Transactions on Information* Theory, Vol. 23, No. 2, pp. 194–202, March 1977.

15. D. Ryter and H. Meyr, Theory of Phase Tracking Systems of Arbitrary Order: Statistics of Cycle Slips and Probability Distribution of the State Vector, *IEEE Transactions on Information Theory*, Vol. 24, pp. 1–7, January 1978.

16. W. C. Lindsey, Nonlinear Analysis of Generalized Tracking Systems, *Proceedings of the IEEE*, Vol. 57, pp. 1705–1722, October 1969.

17. J. R. LaFrieda and W. C. Lindsey, Transient Analysis of Phase-Locked Tracking Systems in the Presence of Noise, *IEEE Transactions on Information Theory*, Vol. 19, pp. 155–164, March 1973.

18. F. J. Charles and W. C. Lindsey, Some Analytical and Experimental Phase-Locked Loop Results for Low Signal-to-Noise Ratios, *Proceedings of the IEEE*, Vol. 54, pp. 1152–1166, September 1966.

19. W. C. Lindsey and M. K. Simon, Detection of Digital FSK and PSK Using a First-Order Phase-Locked Loop, *IEEE Transactions on Communications*, Vol. 25, pp. 200–214, February 1977.

20. D. Ludwig, Persistence of Dynamical Systems under Random Perturbations, *SIAM Review*, Vol. 17, No. 4, pp. 605–640, October 1975.

21. A. Friedman, The Asymptotic Behavior of the First Real Eigenvalue of a Second Order Elliptic Operator with a Small Parameter in the Highest Deriva-

tives, *Indiana University Mathematics Journal*, Vol. 22, No. 10, pp. 1005–1015, 1973.

22. D. A. Darling and A. J. F. Siegert, A Systematic Approach to a Class of Problems in the Theory of Noise and Other Random Phenomena, Part I and Part II, Examples, *IRE Transactions on Information Theory*, pp. 32–43, March 1957; Part III, Examples, *IRE Transactions on Information Theory*, pp. 4–14, March 1958.

23. C. H. Knapp and G. C. Carter, The Generalized Correlation Method for Estimation of Time Delay, *IEEE Transactions on Acoustics, Speech, and Signal Processing*, Vol. 24, August 1976.

24. H. Meyr, Delay Lock Tracking of Stochastic Signals, *IEEE Transactions on Communications*, Vol. 24, No. 3, pp. 331–339, March 1976.

25. G. Spies and H. Meyr, The Structure and Performance of Estimators for Real-Time Estimation of Randomly Varying Time Delay, *IEEE Transactions on Acoustics, Speech, and Signal Processing*, Vol. 32, No. 1, pp. 81–94, February 1984.

26. J. J. Spilker, Jr., Delay-Lock Tracking of Binary Signals, *IEEE Transactions on Space Electronics and Telemetry*, Vol. 9, pp. 1–8, March 1963.

27. J. J. Spilker, Jr. and D. T. Magill, The Delay-Lock Discriminator—An Optimum Tracking Device, *Proceedings of the IRE*, Vol. 49, pp. 1403–1416, September 1961.

28. M. K. Simon, et al., Spread-Spectrum Communications, Rockville, MD: *Computer Science*, 1984.

29. W. J. Gill, A comparison of binary delay-lock loop implementations, *IEEE Transactions Aerospace and Electrical Systems*, Vol. AES-2, pp. 415–426, July 1966.

30. M. K. Simon, Non-coherent pseudonoise code tracking performance of spread spectrum receivers, *IEEE Transactions on Communications*, Vol. COM-25, pp. 327–345, March 1977.

31. R. Sampaio-Neto and R. A. Scholtz, Precorrelation Filter Design for Spread-Spectrum Code Tracking in Interference, *IEEE Journal on Selected Areas in Communications*, Vol. SAC-3, No. 5, September 1985.

32. A. Polydoros and C. L. Weber, Analysis and Optimization of Correlative Code-Tracking Loops in Spread Spectrum Systems, *IEEE Transactions on Communications, COM*-33, *pp*. 30–43, *January* 1985.

APPENDIX 11.1A
NORMALIZED STOCHASTIC DIFFERENTIAL EQUATION OF FIRST-ORDER SYSTEMS WITH STATE-DEPENDENT NOISE INTENSITY

The differential equation of a first-order system is (see Section 3.2.2)

$$\frac{d\phi(t)}{dt} = -K_0 K_D [g[\phi(t)] + n'(t, \phi)] + \frac{d\Theta}{dt} \qquad (11.1\text{A-}1)$$

where $g(\phi)$ is assumed to have a unity gain at the origin, $g'(0) = 1$, and $n'(t, \phi)$ is a physically realizable process. (Any $g'(0) \neq 1$ is easily absorbed in K_D and $n'(t, \phi)$). We first bring (11.1A-1) into the standard form of (9.2-1). We define

$$n'(t, \phi) = \sqrt{S_{n'}(0, \phi)}\, w'(t) \qquad (11.1A\text{-}2)$$

where $S_{n'}(\omega, \phi)$ is the power spectrum of $n'(t, \phi)$ (dependent on state ϕ) and $w'(t)$ is the normalized noise process with $S_{w'}(\omega) = 1$ for all frequencies of interest. Normalization of time as

$$\tau = K_0 K_D t = 4B_L t$$

yields

$$\frac{d\phi(\tau)}{d\tau} = -g[\phi(\tau)] - \sqrt{S_{n'}(0, \phi)}\, w'(\tau/4B_L) + \gamma \qquad (11.1A\text{-}3)$$

where $\gamma = \Delta\omega/4B_L$ is normalized detuning. Since the power spectrum of $w'(\tau/4B_L)$ in normalized frequency $\Omega = \omega/4B_L$ is (see Table 3.4-1)

$$\text{Spectrum of } w'(\tau/4B_L) = 4B_L S_{w'}(4B_L\Omega) \qquad (11.1A\text{-}4)$$

in normalized frequency Ω we can replace $w'(\tau/4B_L)$ in (11.1A-3) by a process $\sqrt{4B_L}\, m(\tau)$ with $S_m(\Omega) = 1$ (for all frequencies of interest)

$$\frac{d\phi(\tau)}{dt} = -g[\phi(\tau)] + \sqrt{4B_L S_{n'}(0, \phi)}\, m(\tau) + \gamma \qquad (11.1A\text{-}5)$$

we derive the signal-to-noise ratio in the loop as

$$\rho := \frac{1}{2}\, 4B_L S_{n'}(0, 0) \qquad (11.1A\text{-}6)$$

Solving the last equation for $4B_L$ and replacing $4B_L$ in (11.1A-5) with the result, we obtain the normalized differential equation in the desired form

$$\frac{d\phi(\tau)}{d\tau} = -g[\phi(\tau)] + \gamma + \left[\frac{2}{\rho}\, \frac{S_{n'}(0, \phi)}{S_{n'}(0, 0)}\right]^{1/2} m(\tau) \qquad (11.1A\text{-}7)$$

In a final step we replace $m(\tau)\, d\tau^*$ with a white noise process $d\beta(\tau)$ with unit variance parameter $E[d\beta^2(\tau)] = d\tau$. This yields the following equation in the sense of Stratonovich

---

*We have to assume that this replacement is permissible

$$d\phi(\tau) = [-g[\phi(\tau)] + \gamma] \, d\tau + \left[\frac{2}{\rho} \frac{S_{n'}(0, \phi)}{S_{n'}(0, 0)} \right]^{1/2} d\beta(\tau) \quad (11.1A\text{-}8)$$

The corresponding equation in the Itô sense is

$$d\phi(\tau) = \left\{ -g[\phi(\tau)] + \gamma + \frac{1}{2\rho} \frac{d}{d\phi} \left[\frac{S_{n'}(0, \phi)}{S_{n'}(0, 0)} \right] \right\} d\tau + \left[\frac{2}{\rho} \frac{S_{n'}(0, \phi)}{S_{n'}(0, 0)} \right] d\beta(\tau)$$

$$(11.1A\text{-}9)$$

The intensity coefficients of the Fokker–Planck equation (in normalized time) can directly be read off from the Itô equation

$$K_1(\phi) = -g(\phi) + \gamma + \frac{1}{2\rho} \frac{d}{d\phi} \left[\frac{S_{n'}(0, \phi)}{S_{n'}(0, 0)} \right]$$

$$(11.1A\text{-}10)$$

$$K_2(\phi) = \frac{2}{\rho} \left[\frac{S_{n'}(0, \phi)}{S_{n'}(0, 0)} \right]$$

If the noise is independent of ϕ the result simplifies to

$$K_1(\phi) = -g(\phi) + \gamma$$

$$(11.1A\text{-}11)$$

$$K_2(\phi) = \frac{2}{\rho}$$

since $S_{n'}(0, \phi) = S_{n'}(0, 0)$.

12

THE MATRIX EIGENVALUE APPROACH

As we have learned in Section 11.2, it is possible to extend the renewal process approach to higher order systems. We thus have at our disposal a complete statistical characterization of the dynamics of a higher order system. However, the extension of the renewal approach to higher order systems leads to equations for which no analytical solution is known nor have numerical methods of solutions been put forward until now. The *eigenvalue approach* is conceptually fundamentally different (see Section 10.4). It provides information on the transition rates which are the most important single performance measure, when both nonlinearity and noise have to be taken into account. The great advantage of the eigenvalue approach is that the transition rates can be found numerically by standard methods, namely by evaluating some eigenvalues of a matrix*. In this approach the noise level must be limited ("moderate") in such a way that cycle slips occur on a slower time scale than for example the relaxation to a stable state. Clearly, this condition holds for all synchronization systems of practical interest. The method is presented in a rather terse and occasionally intuitive way, which should be sufficient for using it in straightforward applications. The interested reader will find more detailed arguments in reference [12].

Well-known *weak noise methods*, which are based on different ideas, are briefly outlined in Appendix 12. They may be useful at very high signal-to-noise ratios, where the present approach requires large matrices in the numerical computation of the eigenvalues.

Notational remarks: For the sake of clarity, *vectors* and *matrices* are written in bold-type letters in this section. The kth element of a vector **c** is denoted by c_k. The (k, n)th element of a matrix **A** is denoted by $A_{k\,n}$.

455

12.1. EIGENFUNCTIONS OF THE OPERATOR L

Here we consider Fokker–Planck operators L

$$L = -\sum_{i=1}^{n} \frac{\partial}{\partial x_i} K_i(x) + \frac{1}{2} \sum_{i=1}^{n} \sum_{j=1}^{n} \frac{\partial^2}{\partial x_i \, \partial x_j} K_{ij}(x) \qquad (12.1\text{-}1)$$

with intensity coefficients $K_i(x)$, $K_{ij}(x)$. The coefficients are supposed to be time-independent and to admit a unique stationary probability density $p_0(x) > 0$, satisfying $Lp_0(x) = 0$. We now focus on those functions $p_n(x)$ which are reproduced by L up to a (complex) scaling constant $-\lambda_n$

$$Lp_n(x) = -\lambda_n p_n(x) \qquad (12.1\text{-}2)$$

The $p_n(x)$ are called eigenfunctions and the λ_n eigenvalues. Clearly, (12.1-2) is a partial differential equation of second order; since it is linear and homogeneous, the $p_n(x)$ are only determined up to a constant factor. The boundary conditions are those of the probability density function. They leave this factor undetermined, but they specify the eigenvalues (together with L), usually as a countable set $\{\lambda_0, \lambda_1, \lambda_2 \ldots\}$, where $\lambda_0 = 0$ belongs to $p_0(x)$. The evaluation of some of the λ_n will be the main numerical task of the present approach.

The $p_n(x)$ are linearly independent and usually also complete. This means that any probability density $p(x, t)$ can be represented as a linear combination

$$p(x, t) = \sum_n c_n(t) p_n(x) \qquad (12.1\text{-}3)$$

which is unique when the $p_n(x)$ are normalized somehow. Inserting (12.1-3) into $\partial p(x, t)/\partial t = Lp(x, t)$ gives

$$\sum_n \dot{c}_n(t) p_n(x) = \sum_n c_n(t)(-\lambda_n) p_n(x)$$

and by the independence of the $p_n(x)$ this holds for every n separately. As a consequence

$$\dot{c}_n(t) = -\lambda_n c_n(t) \quad \text{or} \quad c_n(t) = c_n(0) \exp(-\lambda_n t) \qquad (12.1\text{-}4)$$

hence

$$p(x, t) = \sum_n c_n(0) \exp(-\lambda_n t) p_n(x) \qquad (12.1\text{-}5)$$

The eigenvalues thus indicate the temporal decay of every contribution $p_n(x)$. Note that the assumed existence of a unique stationary distribution entails both

$$\text{Re } \lambda_n > 0 \quad (n \neq 0) \quad \text{and} \quad c_0(0) = 1 \qquad (12.1\text{-}6)$$

(the minus sign in (12.1-2) was introduced in order to obtain positive Re λ_n).

The integral $\int p_n(x)\, dx = 1$ for $n = 0$ (probability of the sure event), and for $n \neq 0$ (hence $\lambda_n \neq 0$) it is obtained from (12.1-2). The integral over Lp_n reduces to boundary terms that vanish [12] so that

$$\int p_n(x)\, dx = \delta_{n0} \qquad (12.1\text{-}7)$$

12.2. MODERATE NOISE AND COARSE-GRAINING TO A MARKOVIAN JUMP PROCESS

We now suppose that the system has M stable states. A stable state is a subset of x-space (point, cycle, etc.) which by the noiseless motion $\dot{x}_i = K_i(x)$ is approached and never left again. Geometrically it is an attractor of the drift field $\mathbf{K}(x)$, i.e., a point, cycle etc. with several ingoing field lines (infinitely many in multidimensional x-spaces) and no outgoing ones.

Noise of a "moderate" level causes the system to fluctuate mainly in the vicinity of the stable states, which large excursions—including transitions from one stable state to another—are rare events. This immediately characterizes $p_0(x)$ as being concentrated on to regions R_I ($I = 1, \ldots, M$), each containing a stable state. These R_I need not be small, but they must be well separated from each other. Clearly, the stationary occupation probabilities

$$p_{0I} := \int_{R_I} p_0(x)\, dx \qquad (12.2\text{-}1)$$

add up to one with a negligible error

$$\sum_{I=1}^{M} p_{0I} \approx 1$$

For any actual motion one can distinguish three different time scales. The fast scale is mainly associated with the deterministic motion, such as the approach to an R_I and possible systematic rotations (i.e., oscillations of particular variables) therein. The medium one holds for the erratic (noisy) contribution, which causes the trajectory to lose its memory while it stays within *one* R_I (relaxation). The slow one is that of the transitions between the R_I, which are of course also induced by the noise. Note that the memory loss within each of the R_I implies that the transitions are markovian.

To understand this more quantitatively, we translate the above picture into the time evolution of the probability density $p(x, t)$. On the fast scale an initial deltafunction is shifted to an R_I and continues to move towards the corresponding attractor, possibly on a spiraling curve. The medium scale applies for the widening of the probability density, until it covers the R_I initially chosen. The ultimate spreading to the remaining R_Is (until each of them is occupied with the probabilities (12.2-1)) occurs on the slow scale.

Comparison with (12.1-5) shows that these time scales are associated with large, medium, and small eigenvalues, respectively. The eigenfunctions differ in the following way: while those of the fast scale may be appreciable everywhere in x-space, those of the medium and slow scales are concentrated onto the regions R_I, where the latter are proportional to $p_0(x)$, with weight factors depending on both n and I (see Figure 12.2-1). Clearly, only the slow scale is relevant for the transition rates; the present statement about the pertinent eigenfunctions are substantiated by Ryter [12].

To derive the formula for the transition rates we integrate (12.1-5) over each of the R_I

$$p_I(t) = \sum_n c_n(0) \exp(-\lambda_n t) p_{nI}$$

with

$$p_I(t) := \int_{R_I} p(x, t)\, dx, \quad p_{nI} := \int_{R_I} p_n(x)\, dx \qquad (12.2\text{-}2)$$

The summation can be confined to the contributions with small eigenvalues. To determine their number, we consider the vectors $\mathbf{p}_n$, each consisting of M elements $\mathbf{p}_n^T = [p_{n1} \cdots p_{nM}]$. Clearly, there exist at most M linearly independent vectors. Since the pertinent eigenfunctions $p_n(x)$ are all proportional to $p_0(x)$ over the R_I, they are unambiguously specified by the $\mathbf{p}_n$. Hence, by the completeness of the eigenfunctions there are exactly M

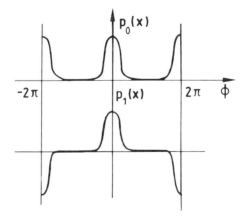

Figure 12.2-1. Qualitative sketch of the eigenfunction $p_0(x)$ and $p_1(x)$ for $M = 2$.

vectors $\mathbf{p}_n$. With an appropriate choice of the labels n, the sum runs thus from 0 to $M - 1$.

To proceed, we adopt a vector and matrix notation. With the column vector $\mathbf{c}(t)$ consisting of the elements $c_n(0) \exp(-\lambda_n t)$, $0 \le n \le M - 1$, and with the diagonal matrix Λ containing the eigenvalues λ_n on the main diagonal, we have

$$\dot{\mathbf{c}}(t) = -\Lambda \mathbf{c}(t)$$

Furthermore, with the matrix $\mathbf{P}$ consisting of the column vectors $\mathbf{p}_n$ ($0 \le n \le M - 1$), the above relation for the $p_I(t)$—now written as a column vector $\mathbf{p}(t)$—assumes the form

$$\mathbf{p}(t) = \mathbf{P}\mathbf{c}(t)$$

It immediately follows that

$$\dot{\mathbf{p}}(t) = -\mathbf{P}\Lambda \mathbf{c}(t)$$

and since the columns of $\mathbf{P}$ are linearly independent, $\mathbf{P}^{-1}$ exists, so that $\mathbf{c}(t) = \mathbf{P}^{-1}\mathbf{p}(t)$, giving

$$\dot{\mathbf{p}}(t) = -\mathbf{P}\Lambda \mathbf{P}^{-1}\mathbf{p}(t) \tag{12.2-3}$$

This is a key result. First, it states that the rate of change of the occupation probabilities $p_I(t)$ is determined by the $p_I(t)$ themselves (in fact, by a linear combination thereof), which implies that the occupation of the R_I is a markovian (jump) process. This is due to coarse-graining of two kinds: a *temporal* one, which focuses on the slow phenomena, ignoring the medium and fast effects; and a *spatial* one, which from the actual trajectories, only retains the information in which of the R_I they stay. Second, the transition rates are now expressed by eigenvalues and eigenfunctions of L. The elements of the matrix

$$\mathbf{a} := -\mathbf{P}\Lambda \mathbf{P}^{-1} \tag{12.2-4}$$

are the transition rates, namely a_{IJ}, from R_J to R_I ($J \ne I$). By a general property of markovian jump processes with continuous time, the sums over each column of $\mathbf{a}$ must vanish, which determines the diagonal elements as

$$a_{JJ} = -\sum_{I \ne J} a_{IJ} \quad \text{(each } J\text{)} \tag{12.2-5}$$

A more formal verification of (12.2-5) can be obtained from the coarse-grained version of (12.1-7)

$$\sum_{I=1}^{M} P_{nI} = \delta_{n0} \tag{12.2-6}$$

The column sums of $\mathbf{P}$ are thus zero, except for $n = 0$. By $\lambda_0 = 0$ all column sums of $\mathbf{P\Lambda}$ vanish, and this property persists after multiplying from the right by $\mathbf{P}^{-1}$ (or by any other matrix).

As a consequence of (12.2-5), the negative trace of $\mathbf{a}$ equals the sum of all transition rates, and since the trace of a product remains unchanged under a cyclic permutation of the factors, it follows from (12.2-4) that

$$-\mathrm{tr}\,\mathbf{a} = \mathrm{tr}\,\mathbf{\Lambda} = \sum_{n=1}^{M} \lambda_n = \sum_{I \neq J} a_{IJ} \tag{12.2-7}$$

where the last sum runs over all pairs of unequal indices.

It is important to note that $\mathbf{p}_n$ and $-\lambda_n$ are eigenvectors and eigenvalues of $\mathbf{a}$

$$\mathbf{a p}_n = -\lambda_n \mathbf{p}_n \tag{12.2-8}$$

which is readily seen on multiplying (12.2-4) from the right by $\mathbf{P}$ and by subsequently comparing column by column with the matrix equality

$$\mathbf{aP} = [\mathbf{ap}_0 | \mathbf{ap}_1 | \cdots | \mathbf{ap}_{M-1}] = -[\lambda_0 \mathbf{p}_0 | \lambda_1 \mathbf{p}_1 | \cdots | \lambda_{M-1} \mathbf{p}_{M-1}]$$

The matrix $\mathbf{a}$ is thus determined by its eigenvalues and eigenvectors. The unspecified scales of the $\mathbf{p}_n$ (except for $\mathbf{p}_0$) are irrelevant for this: a rescaling amounts to multiplying $\mathbf{P}$ from the right by a diagonal matrix $\mathbf{T}$ with arbitrary nonzero diagonal elements, and in

$$\mathbf{PT\Lambda T}^{-1}\mathbf{P}^{-1} = \mathbf{P\Lambda P}^{-1}$$

$\mathbf{T}$ cancels, since $\mathbf{T}$ and $\mathbf{\Lambda}$ commute. We further mention that a permutation of the indices n does not change $\mathbf{a}$ either.

The bistable case ($M = 2$) is particularly simple. By (12.2-7) it follows that

$$\lambda_0 = 0, \quad \lambda_1 = a_{12} + a_{21}$$

and in the symmetric case ($a_{12} = a_{21}$) this already gives

$$a_{12} = \lambda_1 / 2 \tag{12.2-9}$$

The asymmetric case can be treated by taking the condition for stationarity (equal probability flows in both directions)

$$p_{01}a_{21} = p_{02}a_{12}$$

as a second equation for the rates, which leads to

$$a_{12} = \lambda_1 p_{01}, \quad a_{21} = \lambda_1 p_{02} \tag{12.2-10}$$

For illustrating purposes we may also make direct use of (12.2-4), observing (12.2-5) and the fact that by (12.2-6) $\mathbf{p}_1$ must be proportional to the vector $[1 \ \ -1]^T$ (note that rescaling is immaterial!)

$$\begin{pmatrix} -a_{21} & a_{12} \\ a_{21} & -a_{12} \end{pmatrix} = -\begin{pmatrix} p_{01} & 1 \\ p_{02} & -1 \end{pmatrix}\begin{pmatrix} 0 & 0 \\ 0 & \lambda_1 \end{pmatrix}\begin{pmatrix} p_{01} & 1 \\ p_{02} & -1 \end{pmatrix}^{-1}$$

Straightforward evaluation of the right-hand side restates (12.2-10).

12.3. *M*-ATTRACTOR CYCLIC MODELS

We now consider systems that are periodic in one of the variables x_i (for phase trackers this is the phase error ϕ) and assume that there is one stable state in each period. If ϕ is taken from $-\infty$ to $+\infty$, this results in an infinite array of stable states and of regions R_l. There is a fundamental difference, whether ϕ is the only system variable (the one-dimensional case) or not. In the first case transitions can only go to the neighboring R_l, while in several dimensions bypassing is possible, so that the second nearest or even a more distant R_l can be reached in a single transition ("multiple jumps" or "burst of cycle slips"). Clearly, the transition rates can now be labeled by a single index k, which indicates the jump direction by its sign and the number of slipped periods by its absolute value. In the picture $-\infty < \varphi < +\infty$ we denote them by r_k.

The previous theory can be applied by reducing ϕ modulo M (≥ 2) periods. The effect of this step is particularly evident in two dimensions: there it amounts to considering the motion on the surface of a cylinder with a circumference of M periods of ϕ, while the cylinder axis spans the second variable. (Note the difference to Figure 10.2-1, where the cylinder axis spanned the time.)

Since the reduced picture (*M*-state model) identifies stable states separated by M periods, the transition rates (denoted by a_k in this case) can be labeled by $1 \leq k < M$. The connection with the r_k is

$$a_k = \sum_{\nu=-\infty}^{\infty} r_{k+\nu M}, \quad 1 \leq k < M \tag{12.3-1}$$

Clearly, M should be chosen such that only one or a few terms actually contribute to this sum. We shall return to this question below.

We now discuss the evaluation of the a_k. For the sake of simplicity, we start with the four attractor model $(M = 4)$ shown in Figure 10.5-1a. Assuming that a positive k denotes a transition to a lower state index and that only transitions by one, two or three steps occur in each direction (r_1, r_2, r_3, r_{-1}, r_{-2}, r_{-3}), we have the following relations between the r_k, the a_k and the matrix **a**

$$a_1 = r_1 + r_{-3}$$
$$a_2 = r_2 + r_{-2}$$
$$a_3 = r_3 + r_{-1}$$

In particular, the matrix **a** assumes the form

$$\mathbf{a} = \begin{bmatrix} a_0 & a_1 & a_2 & a_3 \\ a_3 & a_0 & a_1 & a_2 \\ a_2 & a_3 & a_0 & a_1 \\ a_1 & a_2 & a_3 & a_0 \end{bmatrix}, \quad M = 4$$

in which the rows (and columns) follow from each other by cyclic permutation. The quantity a_0 (defined herewith) equals

$$a_0 = -(a_1 + a_2 + a_3)$$

in view of (12.2-5).

The generalization to arbitrary M (≥ 2) is straightforward. With the convention that the states are numbered consecutively and that a positive k denotes a transition from state I to state $(I - K) \bmod M$, the matrix **a** is

$$\mathbf{a} = \begin{bmatrix} a_0 & a_1 & a_2 & a_3 & \cdots & a_{M-1} \\ a_{M-1} & a_0 & a_1 & a_2 & \cdots & a_{M-2} \\ a_{M-2} & a_{M-1} & a_0 & a_1 & \cdots & a_{M-3} \\ \vdots & \vdots & \vdots & \vdots & & \vdots \\ a_1 & a_2 & a_3 & a_4 & \cdots & a_0 \end{bmatrix}, \quad (M\text{-state model}) \tag{12.3-2}$$

with

$$a_0 = -\sum_{k=1}^{M-1} a_k \tag{12.3-3}$$

Due to the cyclic structure of the matrix the eigenvectors $\mathbf{p}_n$ only involve the root of unity

$$\kappa = \exp(j2\pi/M), \quad \kappa^M = 1 \tag{12.3-4}$$

$$\mathbf{p}_n^T \sim [1 \quad \kappa^n \quad \kappa^{2n} \quad \cdots \quad \kappa^{(M-1)n}] \tag{12.3-5}$$

while the a_k determine the eigenvalues

$$-\lambda_n = \sum_{k=0}^{M-1} \kappa^{nk} a_k \qquad (12.3\text{-}6)$$

The verification of these statements is straightforward. To obtain the a_k, it is sufficient to know the eigenvalues (and M). A simple formula results on inverting (12.3-6). By

$$\sum_{k=0}^{M-1} \kappa^{nk} = 0, \quad 1 \le n \le M-1$$

it is easily seen that

$$a_k = -\frac{1}{M} \sum_{n=1}^{M-1} \kappa^{-kn} \lambda_n \qquad (12.3\text{-}7)$$

When applying this result, it is important to observe the appropriate numbering of the λ_n. This numbering has in fact been specified by (12.3-5) and (12.3-6). For example, it follows from (12.3-6) that

$$\lambda_{M-n} = \lambda_n^* \qquad (12.3\text{-}8)$$

where λ_n^* denotes the complex conjugate of λ_n. This relation is already sufficient in the cases $M = 3$, 4 (up to a global redefinition of the jump direction), but for $M \ge 5$ the numerical eigenfunctions have to be compared with (12.3-5) to identify n. This becomes very simple with the partial matrices introduced in Section 12.8.

It is interesting to consider the result of (12.3-7) for $k = 0$

$$a_0 = -\frac{1}{M} \sum_{n=1}^{M-1} \lambda_n \qquad (12.3\text{-}9)$$

and to combine it with (12.3-3)

$$\sum_{k=1}^{M-1} a_k = \frac{1}{M} \sum_{n=1}^{M-1} \lambda_n \qquad (12.3\text{-}10)$$

The left-hand side gives the mean occurrence of transitions of any kind, and this "frequency of undesired events" is just the mean of the small eigenvalues of the operator L (note that here their numbering is immaterial and that only the real parts contribute in view of (12.3-8)).

We now try to reconstruct the cycle slip rates r_k of the *unreduced* model, i.e., to invert (12.3-1). Generally, this can only be done in an approximate way, but the remaining error can be minimized by exploiting some a priori knowledge, and by use of an appropriate M. Criteria for choosing M will be obtained in this context.

Three phenomena are to be observed:

1. Cycle slips over M periods (or a multiple of it) are fully ignored in the reduced picture.

2. Cycle slip directions may be indistinguishable: r_k and $r_{-(M-k)}$ contribute to the same a_k (e.g., $M = 4$, r_1 and r_{-3} or r_2 and r_{-2}). With $M = 2$ no direction can be identified.

3. Cycle slips lengths may also be misinterpreted, see the above example $M = 4$, r_1, and r_{-3}.

When the maximum cycle-slip lengths in both directions are known (k_{min} and k_{max}, so that $r_k \neq 0$ only for $-k_{min} \leq k \leq k_{max}$), the full reconstruction is possible with $M > |k_{min}| + k_{max}$. For tuned systems ($r_{-k} = r_k$) this slightly reduces to $M \geq 2k_{max}$ (equality allowed now).

In many applications only the occurrence of cycle slips, but not their direction or length, is of interest (see (12.3-10)). Then only point 1 is relevant, leading to $M > \max(|k_{min}|, k_{max})$, in the tuned case of $M > k_{max}$.

In practice, however, the maximum cycle-slip lengths are only known a priori for first-order (one-dimensional) systems or for qualitatively similar ones (see Section 11.2.7) where $-k_{min} = k_{max} = 1$.

For general models it may be appropriate to start with $M = 4$, which is the smallest M that indicates multiple jumps (by a_2). If a_2 is not negligible compared with a_1 and a_3, one should increase M to an extent that depends on the desired information: if all r_k are of interest, until two sets of nonzero a_k appear, one with small k, giving $a_k = r_k$, and one with large indices, giving $a_{M-k} = r_{-k}$; if only the sum of the r_k matters, until the value of (12.3-10) stops increasing (which is the case, when M exceeds the greatest jump length).

In this context the following statement can be useful: if M is increased by a factor i, then the new eigenvalues λ'_{in} equal the old eigenvalues λ_n. To show this, we first combine (12.3-1) and (12.3-6)

$$-\lambda_n = \sum_{k=0}^{M-1} \kappa^{nk} \sum_{\nu=-\infty}^{\infty} r_{k+\nu M}$$

With $M' = iM$ and $\kappa' = \kappa^{1/i}$ (see (12.3-4)), we have

$$-\lambda'_{in} = \sum_{k=0}^{M'-1} (\kappa')^{ink} \sum_{\nu=-\infty}^{\infty} r_{k+\nu M'}$$

$$= \sum_{k=0}^{iM-1} \kappa^{nk} \sum_{\nu=-\infty}^{\infty} r_{k+\nu iM}$$

When k runs from 0 to $iM - 1$, the factor κ^n assumes i times the same values, and this exactly compensates for the fact that in the sum over ν the r_k are sampled i times less often. Therefore

$$\lambda'_{in} = \lambda_n \tag{12.3-11}$$

as claimed. As examples, we mention that for any even M the eigenvalues $\lambda_{M/2}$ equal the nonzero eigenvalue with $M = 2$; for $M = 6$ one obtains λ_3 from $M = 2$, and λ_2, λ_4 from $M = 3$, so that only λ_1 ($\lambda_5 = \lambda_1^*$) is specific information.

Example: First-Order Systems   In a first-order system, which is the simplest possible, only two rates are nonzero

$$a_1 \quad \text{and} \quad a_{M-1}$$

For $M = 2$ the only transition rate a_1 ($a_1 = a_{M-1}$) equals the sum of the positive, r_1, and negative, r_{-1}, cycle slip rate (see (12.3-1))

$$a_1 = r_1 + r_{-1}$$

It is related to the eigenvalue λ_1 by (12.3-7) which reads in this case

$$a_1 = \frac{\lambda_1}{2}, \quad \kappa = -1$$

The eigenvectors are $\mathbf{p}_0^T = [1 \quad 1]$ and $\mathbf{p}_1^T = [1 \quad -1]$.

If we wish to obtain the positive and negative cycle slip rates individually, the number of attractors must be at least $M \geq 3$. If, for example we choose $M = 4$, then

$$a_1 = r_1, \quad a_2 = 0, \quad a_3 = r_{-1}$$

(The reader should be well aware that the relation of the transition rates a_i to the cycle slip rate r_1 and r_{-1} depends on the number of attractors M). The eigenvalues λ_n are related to the transition rates by (12.3-6)

$$\lambda_n = a_1(1 - \kappa^n) + a_{M-1}(1 - \kappa^{-n}), \quad M \geq 3, \quad n = 1, 2, \ldots, M - 1$$

While the transition rates are always real the eigenvalues λ_n for $M \geq 3$ can be complex. For $M = 4$ we obtain (verify) $\lambda_1 = \lambda_3^*$ and $\lambda_2/2 = \mathrm{Re}\,(\lambda_1)$. The eigenvectors are given by

$$\mathbf{p}_0^T = [1 \quad 1 \quad 1 \quad 1], \qquad \mathbf{p}_1^T = [1 \quad j \quad -1 \quad -j]$$

$$\mathbf{p}_2^T = [1 \quad -1 \quad 1 \quad -1], \quad \mathbf{p}_3^T = [1 \quad -j \quad -1 \quad j]$$

The simple expression for the (small) eigenvalues lends itself to the discussion of large M: for $a_1 = a_{M-1}$, the λ_n are easily seen to become dense in the whole interval $[0, 2a_1]$ as $M \to \infty$. This clearly shows that with the unreduced phase error $(-\infty < \phi < \infty)$ the eigenvalue approach would fail, hence the reduction to an M-attractor model is crucial.

12.4. NUMERICAL COMPUTATION OF THE EIGENVALUES

The eigenvalues and eigenfunctions of the Fokker–Planck operator can be expressed in closed form in exceptional cases only. They must, therefore, be computed by numerical methods. For this purpose we introduce a linearly independent and complete set of functions $\{\Psi_n(x)\}$ (notice that $\Psi_n(x)$ is a function of the N variables $(\phi, x_2, \ldots, x_N)$) by which every probability density function $p(x, t)$ can be represented as an infinite series (with coefficients depending on t). We also require that every image $L\Psi_n(x)$ is expressible by a linear combination of the $\Psi_n(x)$ (for reasons which will become clear soon)

$$L\Psi_n(x) = \sum_m A_{m\,n} \Psi_m(x) \qquad (12.4\text{-}1)$$

Every member of the function set $\{\Psi_n(x)\}$ must obey the boundary conditions imposed on the probability density function $p(x, t)$. Since in an M-attractor cyclic model we are interested in *periodic* solutions in the state variable ϕ, this immediately implies that every $\Psi_n(x)$ must be periodic in ϕ. By the periodicity of $\{\Psi_n(x)\}$ and by (12.4-1) it is evident that the image $L\Psi_n(x)$ also fulfils the periodic boundary conditions. For the remaining state variables going to infinity we impose natural boundary conditions.

   Equation (12.4-1) is the key to obtain a numerically tractable formulation for the computation of the eigenvalues of (12.1-2). Representing the kth eigenfunction $p_k(x)$ by the series

$$p_k(x) = \sum_n c_{n\,k} \Psi_n(x)$$

and inserting the right-hand side into the eigenvalue equation (12.1-2) yields

$$L\left[\sum_n c_{n\,k} \Psi_n(x)\right] = -\lambda_k \sum_n c_{n\,k} \Psi_n(x)$$

Assuming that summation and L operator commute, and using the relation of (12.4-1) the left-hand side can be written in the form

$$L\left[\sum_n c_{n\,k} \Psi_n(x)\right] = \sum_n c_{n\,k} \sum_m A_{m\,n} \Psi_m(x)$$

$$= \sum_m \Psi_m(x) \sum_n A_{m\,n} c_{n\,k}$$

and we obtain (changing the dummy summation variable n to m on the right-hand side)

$$\sum_m \Psi_m(x) \sum_n A_{m\,n} c_{n\,k} = -\lambda_k \sum_m c_{m\,k} \Psi_m(x)$$

Since the set $\{\Psi_n(x)\}$ was assumed to be linearly independent, the above equality holds for every m individually

$$\sum_n A_{m\,n} c_{n\,k} = -\lambda_k c_{m\,k} \qquad (12.4\text{-}2)$$

Clearly, the λ_k are the eigenvalues of the matrix **A** with elements $A_{m\,n}$ and the vectors $\mathbf{c}_k^T = [\cdots \quad c_{-r\,k} \quad \cdots \quad c_{0\,k} \quad \cdots \quad c_{r\,k} \quad \cdots]$ are eigenvectors of **A**.

Thus, the problem of finding the eigenvalues of the Fokker–Planck operator L is equivalent to evaluating the eigenvalues of an (infinite) matrix **A** defined by the mapping $L\Psi_n(x)$. This task requires the choice of a suitable set $\{\Psi_n(x)\}$ and the numerical computation of the matrix eigenvalues.

The infinite matrix **A** has to be truncated. Large orders are usually required at a weak noise. In such cases one can often calculate the transition rates at a higher noise level and then extrapolate by means of an asymptotic scaling law [15].

As was mentioned above, the periodicity of $\Psi_n(x)$ and of its image $L\Psi_n(x)$ is the key property that allows a series expansion by a suitable function set $\{\Psi_n(x)\}$. For absorbing boundary conditions and a general operator L it seems unfeasible to find a function set such that (i) (12.4-1) is fulfilled *and* (ii) zero boundary conditions are fulfilled simultaneously for $\Psi_n(x)$ and for the image $L\Psi_n(x)$ for any n. We show this by an example.

Example: *First-Order Phase-Locked Loop with Absorbing Boundary Conditions at* $\phi = \pm\pi$ We take $\Psi_n(x) = \sin\left[(n/2)(\phi + \pi)\right]$. Clearly, any 4π-periodic function that fulfills zero boundary conditions at $\phi = \pm\pi$ can be represented by the set $\{\sin\left[(n/2)(\phi + \pi)\right]\}$. For zero loop detuning ($\gamma = 0$) the image $L\sin\left[(n/2)(\phi + \pi)\right]$ can also be represented by $\{\sin\left[(n/2)(\phi + \pi)\right]\}$ (verify!). However, for $\gamma \neq 0$ this is no longer true since the image contains $\cos\left[(n/2)(\phi + \pi)\right]$ and, thus, the series "Ansatz" is meaningless.

An obvious idea to comply with the requirement of (12.4-1) is to use the function set $\{\sin\left[(n/2)(\phi + \pi)\right], \cos\left[(n/2)(\phi + \pi)\right]\}$. However, while we then indeed fulfill (12.4-1), the series expansion of the eigenfunctions $p_k(\phi)$ contains functions which do not fulfill the absorbing boundary conditions at $\phi = \pm\pi$. Hence, then this series "Ansatz" is applicable for periodic boundary conditions only.

12.5. THE MATRIX *A* FOR FIRST-ORDER SYSTEMS (*M* = 2 ATTRACTORS)

For a first-order system, the matrix-eigenvalue approach offers no additional insight over the renewal process approach. The justification to discuss first-order systems within the matrix-eigenvalue context is based on the fact that the fundamental properties can be most clearly presented for first-order systems, and that these findings can be generalized to an Nth order system.

We will proceed as follows. We first develop the matrix $\mathbf{A}$ for the familiar first-order PLL with sinusoidal phase detector characteristics. Later on we generalize the results to include arbitrary phase detector characteristics $g(\phi)$ and phase error dependent noise intensity coefficient $K_2(\phi)$.

The Fokker–Planck equation for first-order PLLs with sinusoidal phase detector characteristics is given by (11.1A-11)

$$L = - \frac{\partial}{\partial \phi} [-\sin \phi + \gamma] + \frac{1}{\rho} \frac{\partial^2}{\partial \phi^2} \tag{12.5-1}$$

We consider a bistable system, $M = 2$, for which we need a complete and orthogonal set of functions $\{z_n(\phi)\}^*$ of period 4π. A possible candidate is the set of trigonometric functions

$$\left\{ \sin \left(n \frac{\phi}{2} \right), \cos \left(n \frac{\phi}{2} \right) \right\}, \quad n \geq 0 \tag{12.5-2}$$

Usually, it is simpler to work with the equivalent set of complex exponentials

$$\{z_n(\phi)\} = \left\{ \exp \left(jn \frac{\phi}{2} \right) \right\}, \quad n \in Z \tag{12.5-3}$$

We map $z_n(\phi)$ by L

$$Lz_n(\phi) = - \frac{\partial}{\partial \phi} [-\sin (\phi) + \gamma] z_n(\phi) + \frac{1}{\rho} \frac{\partial^2}{\partial \phi^2} z_n(\phi) \tag{12.5-4}$$

Expressing $\sin \phi$ by means of the Euler relation

$$\sin \phi = \frac{\exp (j\phi) - \exp (-j\phi)}{2j} \tag{12.5-5}$$

and $z_n(\phi)$ by $\exp (jn\phi/2)$ yields

$$Lz_n(\phi) = - \frac{\partial}{\partial \phi} \left[- \frac{\exp (j\phi) - \exp (-j\phi)}{2j} + \gamma \right] \exp \left(jn \frac{\phi}{2} \right)$$

$$+ \frac{1}{\rho} \frac{\partial^2}{\partial \phi^2} \exp \left(jn \frac{\phi}{2} \right) \tag{12.5-6}$$

*For reasons which will become clear when treating higher order systems we use $\{z_n(\phi)\}$ for the function set of first-order systems.

But the right-hand side of (12.5-6) can be expressed as a linear combination of functions of the set $\{z_k(\phi)\}$. Using

$$\exp(j\phi)\exp\left(jn\,\frac{\phi}{2}\right) = \exp\left[j(2+n)\,\frac{\phi}{2}\right] = z_{n+2}(\phi) \qquad (12.5\text{-}7)$$

and

$$\frac{\partial}{\partial\phi}\,z_n(\phi) = j\,\frac{n}{2}\,z_n(\phi)$$

we obtain after some simple algebraic manipulations

$$Lz_n(\phi) = \frac{1}{2}\,\frac{n+2}{2}\,z_{n+2}(\phi) - \frac{1}{2}\,\frac{n-2}{2}\,z_{n-2}(\phi) - j\gamma\,\frac{n}{2}\,z_n(\phi)$$
$$- \frac{1}{\rho}\left(\frac{n}{2}\right)^2 z_n(\phi) \qquad (12.5\text{-}8)$$

for $n = 0,\ \pm 1,\ \pm 2,\ \ldots$.

According to the definition of (12.4-1) we find the coefficients of (12.5-8) in one column of the matrix **A**. Thus, the (complex) elements in the *n*th column/row of **A** read

k	A_{kn}
$n-2$	$-\dfrac{1}{2}\,\dfrac{n-2}{2}$
n	$-\dfrac{1}{\rho}\left(\dfrac{n}{2}\right)^2 - \dfrac{j\gamma n}{2}$
$n+2$	$\dfrac{1}{2}\,\dfrac{n+2}{2}$
else	0

*n*th column

k	A_{nk}
$n-2$	$+\dfrac{1}{2}\,\dfrac{n}{2}$
n	$-\dfrac{1}{\rho}\left(\dfrac{n}{2}\right)^2 - \dfrac{j\gamma n}{2}$
$n+2$	$-\dfrac{1}{2}\,\dfrac{n}{2}$
else	0

*n*th row

$$(12.5\text{-}9)$$

The matrix **A** has a tridiagonal structure as shown in Figure 12.5-1. However, the matrix is not symmetric, i.e., $A_{nk} \neq A_{kn}^{*}$. Notice that the row A_{0i} contains all zeroes. This implies that $\lambda = 0$ is an eigenvalue of the matrix.

The eigenvalues λ_n are found by evaluating the (complex) system of linear equations

$$-\lambda\mathbf{c} = \mathbf{Ac} \qquad (12.5\text{-}10)$$

with

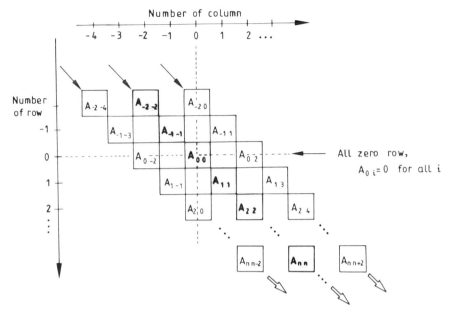

Figure 12.5-1. Tridiagonal structure of the matrix **A**. The index numbering of the rows/columns increases in direction of the arrows. All entries of the matrix outside the three diagonals are zero.

$$\mathbf{c}^T = [\cdots \quad c_{-r} \quad \cdots \quad c_0 \quad \cdots \quad c_r \quad \cdots]$$

or written in components

$$-\lambda c_n = A_{n\,n-2}\, c_{n-2} + A_{n\,n}\, c_n + A_{n\,n+2}\, c_{n+2}, \quad n = 0, \pm 1, \pm 2, \ldots$$

$$(12.5\text{-}10a)$$

An interesting property of the system dynamics is discovered by transforming the complex eigenvalue equation in two real, coupled equations. Since the operator L is real we obtain the image $Lz_n^*(\phi)$ as

$$Lz_n^*(\phi) = [Lz_n(\phi)]^* = \left[\sum_m A_{m\,n} z_m(\phi)\right]^*$$

$$= \sum_m A_{m\,n}^* z_m^*(\phi) \qquad (12.5\text{-}11)$$

Using the Euler relation for the trigonometric functions and the linearity of L we immediately arrive at

$$L \cos\left(\frac{n}{2}\,\phi\right) = L\left[\frac{z_n(\phi) + z_n^*(\phi)}{2}\right] = \frac{1}{2}\left[L\,z_n(\phi) + Lz_n^*(\phi)\right]$$

$$= \mathrm{Re}\left[\sum_n A_{m\,n} z_m(\phi)\right]$$

$$= \sum_m \mathrm{Re}\,A_{m\,n} \cos\left(m\,\frac{\phi}{2}\right) - \mathrm{Im}\,A_{m\,n} \sin\left(m\,\frac{\phi}{2}\right)$$

$$(12.5\text{-}12)$$

and

$$L \sin\left(\frac{n}{2}\,\phi\right) = \mathrm{Im}\left[\sum_m A_{m\,n} z_m(\phi)\right]$$

$$= \sum_m \mathrm{Re}\,A_{m\,n} \sin\left(\frac{m}{2}\,\phi\right) + \mathrm{Im}\,A_{m\,n} \cos\left(\frac{m}{2}\,\phi\right) \quad (12.5\text{-}12a)$$

Collecting $\cos[(n\phi/2)]$ and $\sin[(n\phi/2)]$ into a tuple

$$\mathbf{z}_n = \begin{bmatrix} \cos\left(n\,\dfrac{\phi}{2}\right) \\[2mm] \sin\left(n\,\dfrac{\phi}{2}\right) \end{bmatrix} \qquad (12.5\text{-}13)$$

and writing the results of (12.5-12) in matrix notation yields for $n = 0, 1, 2 \dots$

$$L\mathbf{z}_n = \sum_m \mathbf{A}_{m\,n}\,\mathbf{z}_m = \mathbf{A}_{n-2\,n}\,\mathbf{z}_{n-2} + \mathbf{A}_{n\,n}\,\mathbf{z}_n + \mathbf{A}_{n+2\,n}\,\mathbf{z}_{n+2} \qquad (12.5\text{-}14)$$

where $\mathbf{A}_{m\,n}$ is a (2×2) submatrix defined by

$$\mathbf{A}_{m\,n} = \begin{bmatrix} \mathrm{Re}\,A_{m\,n} & -\mathrm{Im}\,A_{m\,n} \\[1mm] \mathrm{Im}\,A_{m\,n} & \mathrm{Re}\,A_{m\,n} \end{bmatrix} \qquad (12.5\text{-}14a)$$

Thus for the nth row of $\mathbf{A}$ we obtain from (12.5-9)

$$\mathbf{A}_{n\,n} = \begin{bmatrix} -\dfrac{1}{\rho}\left(\dfrac{n}{2}\right)^2 & \gamma\,\dfrac{n}{2} \\[3mm] -\gamma\,\dfrac{n}{2} & -\dfrac{1}{\rho}\left(\dfrac{n}{2}\right)^2 \end{bmatrix}, \quad \mathbf{A}_{n\,n+2} = \begin{bmatrix} (\overset{+}{-})\dfrac{1}{2}\,\dfrac{n}{2} & 0 \\[3mm] 0 & (\overset{-}{+})\dfrac{1}{2}\,\dfrac{n}{2} \end{bmatrix}$$

$$(12.5\text{-}14b)$$

The eigenvalues λ_n are found by evaluating the (real) system of linear equations

$$-\lambda \mathbf{c}_n = \mathbf{A}_{n\ n-2}\,\mathbf{c}_{n-2} + \mathbf{A}_{n\ n}\,\mathbf{c}_n + \mathbf{A}_{n\ n+2}\,\mathbf{c}_{n+2}, \quad n = 0, 1, 2 \ldots$$

$$(12.5\text{-}15)$$

where $\mathbf{c}_n$ is a two-dimensional vector $\mathbf{c}_n^T = [\,b_n \quad d_n\,]$.

The indices in the complex matrix shown in Figure 12.5-1 run from negative to positive numbers. Using the real function set $\{\sin(n\frac{\phi}{2}), \cos(n\frac{\phi}{2})\}$ we have positive indices only. Since

$$\mathbf{z}_{-m} = \begin{bmatrix} \cos\left(m\,\dfrac{\phi}{2}\right) \\[2mm] -\sin\left(m\,\dfrac{\phi}{2}\right) \end{bmatrix} = \mathbf{T}\mathbf{z}_m \quad \text{with} \quad \mathbf{T} = \begin{bmatrix} 1 & 0 \\ 0 & -1 \end{bmatrix}$$

we can always replace the matrices with negative indices in the recursion of (12.5-15) by using the following relation

$$\mathbf{A}_{-m\ n}\,\mathbf{z}_{-m} = \mathbf{A}_{-m\ n}\begin{bmatrix} 1 & 0 \\ 0 & -1 \end{bmatrix}\mathbf{z}_m \qquad (12.5\text{-}16)$$

For a bistable model $(M = 2)$ with sinusoidal phase detector characteristic only the map of $\mathbf{z}_0$ and $\mathbf{z}_1$ lead to negative indices

$$L\mathbf{z}_0 = \mathbf{A}_{-2\ 0}\,\mathbf{z}_{-2} + \mathbf{A}_{0\ 0}\,\mathbf{z}_0 + \mathbf{A}_{2\ 0}\,\mathbf{z}_2$$

$$= \mathbf{A}_{0\ 0}\,\mathbf{z}_0 + [\mathbf{A}_{-2\ 0}\,\mathbf{T} + \mathbf{A}_{2\ 0}]\mathbf{z}_2 \qquad (12.5\text{-}16a)$$

and

$$L\mathbf{z}_1 = [\mathbf{A}_{-1\ 1}\,\mathbf{T} + \mathbf{A}_{1\ 1}]\mathbf{z}_1 + \mathbf{A}_{3\ 1}\,\mathbf{z}_3 \qquad (12.5\text{-}16b)$$

Thus the submatrices $\mathbf{A}_{2\ 0}$ and $\mathbf{A}_{1\ 1}$ must be replaced by the respective sums in the square brackets (see Figure 12.5-2). Clearly, the general case with more than two elements with negative indices can be treated accordingly.

Notice that the recursion relations of (12.5-14) for zero loop detuning, $\gamma = 0$, become decoupled and identical. To compute the eigenvalues one can therefore work with a matrix with $(N \times N)$ instead of $4(N \times N)$ elements which is advantageous from a numerical point of view.

We next show that a general phase detector characteristic* $g(\phi)$ also leads to a tridiagonal matrix structure where the elements are (generally complex) submatrices instead of complex scalars.

We approximate the phase detector characteristic $g(\phi)$ by a finite Fourier series

---

*The same approach is applicable if the intensity coefficient $K_2(\phi)$ is a function of ϕ.

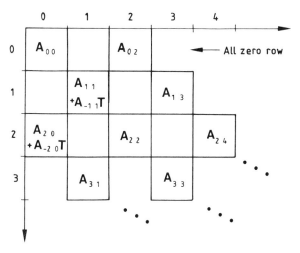

Figure 12.5-2. Real valued matrix **A**.

$$g(\phi) = \sum_{r=1}^{2R} g_r \sin{(r\phi)} \qquad (12.5\text{-}17)$$

For simplicity we assume $R = 1$; the generalization to $R > 1$ is trivial. Proceeding exactly as before we find for the matrix elements $A_{n\,k}$ in the *n*th row

k	$A_{n\,k}$
$n - 4$	$-\dfrac{1}{2}\dfrac{n}{2} g_2$
$n - 2$	$-\dfrac{1}{2}\dfrac{n}{2} g_1$
n	$-\dfrac{1}{\rho}\left(\dfrac{n}{2}\right)^2 - j\dfrac{n}{2}\gamma$
$n + 2$	$\dfrac{1}{2}\dfrac{n}{2} g_1$
$n + 4$	$\dfrac{1}{2}\dfrac{n}{2} g_2$

*n*th row

Clearly, the matrix has not the tridiagonal form (although it has bandstructure). Let us now show that this bandmatrix can also be written in tridiagonal form. We write down the corresponding recursions for c_n and c_{n+2}

$$-\lambda c_n = -\frac{1}{2}\frac{n}{2}g_2 c_{n-4} - \frac{1}{2}\frac{n}{2}g_1 c_{n-2} - \left[j\frac{n}{2}\gamma + \frac{1}{\rho}\left(\frac{n}{2}\right)^2\right]c_n$$

$$+ \frac{1}{2}\frac{n}{2}g_1 c_{n+2} + \frac{1}{2}\frac{n}{2}g_2 c_{n+4} \tag{12.5-19}$$

and

$$-\lambda c_{n+2} = -\frac{1}{2}\frac{n+2}{2}g_2 c_{n-2} - \frac{1}{2}\frac{n+2}{2}g_1 c_n$$

$$- \left[j\frac{n+2}{2}\gamma + \frac{1}{\rho}\left(\frac{n+2}{2}\right)^2\right]c_{n+2}$$

$$+ \frac{1}{2}\frac{n+2}{2}g_1 c_{n+4} + \frac{1}{2}\frac{n+2}{2}g_2 c_{n+6}$$

If we now introduce the (complex) column vectors

$$\mathbf{c}_n = \begin{bmatrix} c_n \\ c_{n+2} \end{bmatrix}, \quad \mathbf{c}_{n-2} = \begin{bmatrix} c_{n-4} \\ c_{n-2} \end{bmatrix}, \quad \mathbf{c}_{n+2} = \begin{bmatrix} c_{n+4} \\ c_{n+6} \end{bmatrix} \tag{12.5-20}$$

we obtain the vector recurrence equation

$$-\lambda \mathbf{c}_n = \mathbf{A}_{n\,n-2}\,\mathbf{c}_{n-2} + \mathbf{A}_{n\,n}\,\mathbf{c}_n + \mathbf{A}_{n\,n+2}\,\mathbf{c}_{n+2}, \quad n = 0, \pm 1, \pm 2 \tag{12.5-21}$$

where the matrices $\mathbf{A}_{n\,k}$ are given by

$$\mathbf{A}_{n\,n-2} = \begin{bmatrix} A_{n\,n-4} & A_{n\,n-2} \\ 0 & A_{n+2\,n-2} \end{bmatrix}, \quad \mathbf{A}_{n\,n} = \begin{bmatrix} A_{n\,n} & A_{n\,n+2} \\ A_{n+2\,n} & A_{n+2\,n+2} \end{bmatrix}$$

$$\mathbf{A}_{n\,n+2} = \begin{bmatrix} A_{n\,n+4} & 0 \\ A_{n+2\,n+4} & A_{n+2\,n+6} \end{bmatrix} \tag{12.5-22}$$

Thus, the matrix $\mathbf{A}$ has again the tridiagonal structure when the scalar elements $A_{n\,k}$ are replaced by the (2×2) submatrices $\mathbf{A}_{n\,k}$ (see Figure 12.5-1).

The generalization to arbitrary $2R$ is obvious. Introducing the vector

$$\mathbf{c}_n = \begin{bmatrix} c_n \\ c_{n+2} \\ \vdots \\ c_{n+2R} \end{bmatrix} \tag{12.5-23}$$

one again arrives at a vector recurrence relation.

12.6. DECOMPOSITION OF THE MATRIX *A* AND GEOMETRIC INTERPRETATIONS (*M* = 2 ATTRACTORS)

In the recurrence equation for c_n (scalar or vector) only c_n and c_{n+2}, c_{n-2} appear. Therefore two independent recursions exist: one recursion for *even* indices and one for *odd* indices. The existence of two decoupled recursions also nicely manifests itself in the structure of the matrix **A** which can be written as the sum of the two matrices. A first ("even") matrix $\mathbf{A}^{(0)}$ is obtained by setting all rows/columns with odd number equal to zero. The second ("odd") matrix $\mathbf{A}^{(1)}$ is obtained by setting the even rows/columns equal to zero (see Figure 12.6-1).

Since the odd numbered rows/columns of the even matrix are all zero they can be omitted for the numerical computation of the eigenvalues. Thus, a reduction by a factor four in the number of elements of the matrix is obtained. Setting $n = 2k$ the original recursion for c_n reads

$$-\lambda c_{2k} = A_{2k\ 2k-2}\ c_{2k-2} + A_{2k\ 2k}\ c_{2k} + A_{2k\ 2k+2}\ c_{2k+2} \qquad (12.6\text{-}1)$$

which is equivalent to the one-step recursion

$$-\lambda c_k^{(0)} = B_{k\ k-1}^{(0)}\ c_{k-1}^{(0)} + B_{k\ k}^{(0)}\ c_k^{(0)} + B_{k\ k+1}^{(0)}\ c_{k+1}^{(0)} \qquad (12.6\text{-}2)$$

with

$$c_k^{(0)} := c_{2k}$$

$$B_{k\ l}^{(0)} := A_{2k\ 2l}$$

Proceeding in exactly the same way for odd indices, $n = 2k + 1$, yields the one-step recursion

$$-\lambda c_k^{(1)} = B_{k\ k-1}^{(1)}\ c_{k-1}^{(1)} + B_{k\ k}^{(1)}\ c_k^{(1)} + B_{k\ k+1}^{(1)}\ c_{k+1}^{(1)} \qquad (12.6\text{-}3)$$

with

$$c_k^{(1)} := c_{2k+1}$$

$$\qquad (12.6\text{-}4)$$

$$B_{k\ l}^{(1)} := A_{2k+1\ 2l+1}$$

The existence of two decoupled (independent) recursions implies that the set of eigenvalues of the matrix **A** is the sum of two sets of eigenvalues determined by the matrices $\mathbf{A}^{(0)}$ and $\mathbf{A}^{(1)}$, respectively*. It is not difficult to show that every eigenvalue of either of the two submatrices also is an eigenvalue of **A**.

---

*For computational reasons the eigenvalues of $\mathbf{A}^{(0)}$ and $\mathbf{A}^{(1)}$ should of course be calculated with the help of the corresponding smaller matrices $\mathbf{B}^{(0)}$ and $\mathbf{B}^{(1)}$.

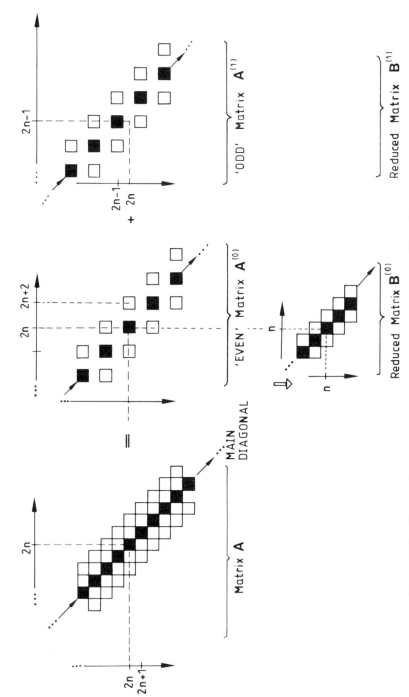

Figure 12.6-1. The matrix **A** is the sum of two nonoverlapping matrices $\mathbf{A}^{(0)} + \mathbf{A}^{(1)}$, omitting the all zero rows and columns in the matrices $\mathbf{A}^{(0)}$, $\mathbf{A}^{(1)}$ leads to the reduced matrices $\mathbf{B}^{(0)}$ and $\mathbf{B}^{(1)}$.

Without loss of generality we consider the matrix $\mathbf{A}^{(0)}$. Let us denote by $\mathbf{e}^{(0)}$ an eigenvector of $\mathbf{A}^{(0)}$ and by $\lambda^{(0)}$ the corresponding eigenvalue, i.e.

$$-\lambda^{(0)}\mathbf{e}^{(0)} = \mathbf{A}^{(0)}\mathbf{e}^{(0)} \tag{12.6-5}$$

Since the matrix $\mathbf{A}^{(0)}$ has nonzero elements at the even-numbered columns/rows only the odd component e_{2k+1} of the eigenvector may be chosen arbitrarily. We define $e_{2k+1} \equiv 0$

$$\mathbf{e}^{(0)} := \begin{bmatrix} \vdots \\ e_{2k-2} \\ 0 \\ e_{2k} \\ 0 \\ e_{2k+2} \\ \vdots \end{bmatrix} \tag{12.6-6}$$

We now map the eigenvector $\mathbf{e}^{(0)}$ by the matrix $\mathbf{A}$. Since $\mathbf{A} = \mathbf{A}^{(0)} + \mathbf{A}^{(1)}$ we find

$$\begin{aligned} \mathbf{A}\mathbf{e}^{(0)} &= [\mathbf{A}^{(0)} + \mathbf{A}^{(1)}]\mathbf{e}^{(0)} \\ &= \mathbf{A}^{(0)}\mathbf{e}^{(0)} + \mathbf{A}^{(1)}\mathbf{e}^{(0)} \end{aligned} \tag{12.6-7}$$

For the first term on the right-hand side of (12.6-7) we obtain $-\lambda^{(0)}\mathbf{e}^{(0)}$ while from (12.6-6) for the second term follows

$$\mathbf{A}^{(1)}\mathbf{e}^{(0)} = 0$$

Therefore

$$\mathbf{A}\mathbf{e}^{(0)} = \mathbf{A}^{(0)}\mathbf{e}^{(0)} = -\lambda^{(0)}\mathbf{e}^{(0)} \tag{12.6-8}$$

which proves that $\mathbf{e}^{(0)}$ is also an eigenvector of the matrix $\mathbf{A}$.

The abstract result can be given an illustrative geometrical interpretation. Any orthogonal and complete set of functions $\{z_n(\phi)\}$ can be viewed as a coordinate system (or base) which spans the infinite dimensional Hilbert space. Every single $z_n(\phi)$ is a base vector and every function is a vector in this base. The coefficients of the series expansion of an arbitrary function $p_m(\phi)$

$$p_m(\phi) = \sum_n c_{n\,m} z_n(\phi) \tag{12.6-9}$$

are the components of vector $p_m(\phi)$ with respect to the given base.

In our case, the space $\mathscr{S}$ comprises two spaces $\mathscr{S}^{(0)}$ and $\mathscr{S}^{(1)}$

$$\mathscr{S} = \mathscr{S}^{(0)} \oplus \mathscr{S}^{(1)} \tag{12.6-10}$$

spanned by the base

$$\beta^{(0)} = \{z_{2n}\} \qquad \text{base of subspace } \mathscr{S}^{(0)}$$

and

$$\beta^{(1)} = \{z_{2n+1}\} \qquad \text{base of subspace } \mathscr{S}^{(1)}$$

respectively. The two subspaces are orthogonal as immediately follows from the orthogonality of the base functions

$$\langle z_{2n}(\phi), z_{2m+1}(\phi) \rangle = \int_0^{4\pi} z_{2n}(\phi) z_{2m+1}(\phi) \, d\phi$$

$$= \int_0^{4\pi} \exp\left[j \frac{\phi}{2} (2n + 2m + 1) \right] d\phi$$

$$= 0 \quad \text{for all } n, m \tag{12.6-11}$$

The most interesting aspect about the two orthogonal subspaces lies in the fact that each of them represents one of the two fundamental aspects of the system dynamics. The vectors in $\mathscr{S}^{(0)}$ describe the motion within a single attractor for which we coined the term *relaxation dynamics*. The vectors in $\mathscr{S}^{(1)}$ describe the motion between adjacent attractors, i.e., the *transition dynamics* (Figure 12.6-2).

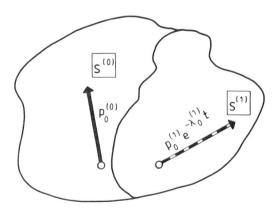

Relaxation Dynamics   Transition Dynamics

Figure 12.6-2. Decomposition of the space $\mathscr{S}$ into two orthogonal subspaces $\mathscr{S}^{(0)}$ and $\mathscr{S}^{(1)}$. In both subspaces the eigenvector corresponding to the smallest eigenvalues $\{\lambda_0, \lambda_1\}$ are indicated.

Why is this so? As we know, the system dynamics are described by the motions along the eigenfunctions (natural modes) (see (12.1-5))

$$p(\phi, t) = \sum c_n(0) p_n(\phi) \exp(-\lambda_n t) \qquad (12.6\text{-}12)$$

As shown before, each eigenvalue is determined either by $\mathbf{A}^{(0)}$ or $\mathbf{A}^{(1)}$. Hence, the corresponding eigenfunctions (vectors) belong to exactly one of the two subspaces $\mathscr{S}^{(0)}$ (if defined by $\mathbf{A}^{(0)}$) or $\mathscr{S}^{(1)}$, otherwise. It remains to be seen why $\mathscr{S}^{(0)}$ may be attributed to the relaxation dynamics and $\mathscr{S}^{(1)}$ to the transition dynamics.

The answer to this question is found by inspection of the base $\beta^{(0)}$ and the corresponding matrix $\mathbf{A}^{(0)}$. The base $\beta^{(0)}$ is exactly the one we would obtain for the one-attractor ($M = 1$) model

$$\beta^{(0)} = \{z_{2n}\} = \{\exp(j\phi n)\}$$

the eigenfunctions have a period of 2π (*not* only 4π) since the fundamental harmonics $z_2(\phi)$ of $\beta^{(0)}$ has a period of 2π. Therefore, the eigenfunctions/eigenvalues pertain to the relaxation dynamics only. As we know, the stationary probability density function $p_0^{(0)}(\phi)$ with eigenvalue $\lambda_0^{(0)} = 0$ is an eigenfunction of the relaxation dynamics space. Indeed, the matrix $\mathbf{A}^{(0)}$ has an all zero row for $2n = 0$ and hence $\lambda_0^{(0)} = 0$ is an eigenvalue. (Notice that $p_0^{(0)}(\phi)$ is identical to the PDF of the mod 2π reduced phase error, apart from a factor $1/2$ which takes the 4π interval into account.)

The eigenvectors belonging to $\mathscr{S}^{(1)}$ are spanned by the base

$$\beta^{(1)} = \left\{ \exp\left[j\, \frac{\phi}{2}\, (2n + 1) \right] \right\}$$

The base functions with index $n = -1$ and $n = 0$ have the maximum period of 4π which is required to represent 4π periodic eigenfunctions in the bistable model. There is no eigenfunction which has a period of 2π. The smallest eigenvalue $\lambda_0^{(1)} = 2a_1$ is related to the most important parameter of the transition dynamics, the transition rate a_1.

Let us finally relate the results for the continuous model to those of the coarse-grained model.

In the coarse-grained model only one vector in each subspace is observed, namely the one corresponding to the smallest eigenvalue, $\lambda_0^{(0)} = 0$, $\lambda_0^{(1)} = 2a_1$, respectively. Furthermore, the exact location of the vector in $\mathscr{S}^{(0)}$ (or $\mathscr{S}^{(1)}$) is ignored. One can therefore say that the "coarse-grained" model ignores all the fine structure (temporal and spatial) of the system dynamics while retaining the most important dynamic parameters represented by the smallest eigenvalue in each subspace.

Generalization for M > 2 attractors. The generalization to more than $M = 2$ attractors is straightforward. The map $Lz_n(\phi)$ reads

$$Lz_n(\phi) = \frac{1}{2} \frac{n+M}{M} z_{n+M}(\phi) - \frac{1}{2} \frac{n-M}{M} z_{n-M}(\phi)$$

$$- j\gamma \frac{n}{M} z_n(\phi) - \frac{1}{\rho} \left(\frac{n}{M} \right)^2 z_n(\phi)$$

with

$$z_n(\phi) = \exp \left(\frac{jn\phi}{M} \right)$$

Thus, the matrix $\mathbf{A}$ retains its tridiagonal structure. Since there are M independent recursions the matrix $\mathbf{A}$ is the sum of M submatrices $\mathbf{A}^{(m)}$, $m = 0, 1, \ldots, M - 1$. The vector space $\mathscr{S}$ accordingly comprises M orthogonal subspaces. The smallest eigenvalue of each subspace is related to the transition rate a_k as discussed in the example at the end of Section 12.3.

12.7. THE MATRIX *A* FOR THE *N*TH ORDER SYSTEM

We first consider a two dimensional Fokker–Planck equation and generalize to N variables later on. We assume that the orthogonal function set for the two variables ϕ and y allows a factorization into two complete and orthogonal sets $\{z_n(\phi)\}$ and $\{h_m(y)\}$, respectively, and develop the eigenfunction into a series with respect to these functions

$$p_k(\phi, y) = \sum_n \sum_m c_{n\,m} z_n(\phi) h_m(y) \tag{12.7-1}$$

We require that $z_n(\phi)$ and $h_m(y)$ as well as $L[z_n(\phi)h_m(y)]$ fulfill the imposed boundary conditions

1. $z_n(-M\pi) = z_n(M\pi)$,   periodic boundary conditions for the M-attractor cyclic model   (12.7-2)
2. natural boundary conditions for $h_m(y)$

Truncating the series (12.7-1) at $m = Q$ we may write

$$[p_k(\phi, y)]_{\text{truncated}} = \sum_n \sum_{m=0}^{Q} c_{n\,m} z_n(\phi) h_m(y)$$

$$= \sum_n \mathbf{c}_n^T z_n(\phi) \mathbf{h}(y) \tag{12.7-3}$$

with the vectors $\mathbf{c}_n$, $\mathbf{h}(y)$ defined as

$$\mathbf{c}_n = \begin{bmatrix} c_{n,0} \\ c_{n,1} \\ \vdots \\ c_{n,Q} \end{bmatrix}, \quad \mathbf{h}(y) = \begin{bmatrix} h_0(y) \\ h_1(y) \\ \vdots \\ h_Q(y) \end{bmatrix} \tag{12.7-4}$$

From now on the path to develop the matrix **A** parallels that for the first-order system discussed earlier.

The matrix **A** is found by mapping $z_n(\phi)\mathbf{h}(y)$ by the operator L. If we assume that the function sets $\{z_n(\phi)\}$ and $\{h_m(y)\}$ are chosen such that the image $L[z_n(\phi)\mathbf{h}(y)]$ can be expressed as a linear combination of $\{z_m(\phi)\mathbf{h}_+(y)\}$, then we obtain

$$L[z_n(\phi)\mathbf{h}(y)] = \sum_m \mathbf{A}_{m\,n}\, z_m(\phi)\mathbf{h}_+(y)$$

In general, the vector $\mathbf{h}_+(y)$ has dimension $(Q + m)$, $m \geq 1$ with the first $(Q + 1)$ components identical to those of $\mathbf{h}(y)$. The elements $\mathbf{A}_{m\,n}$ of the matrix **A** are themselves submatrices of order $(Q + 1) \times (Q + m)$.

We assume now that the dimension of $\mathbf{h}(y)$ is large enough such that the contribution of the components h_{Q+m}, $m \geq 2$ of the vector $\mathbf{h}_+(y)$ can be neglected. Then $\mathbf{A}_{m\,n}$ is a quadratic matrix and we obtain

$$L[z_n(\phi)\mathbf{h}(y)]_{\text{truncated}} = \sum_m \mathbf{A}_{m\,n}\, z_m(\phi)\mathbf{h}(y) \tag{12.7-5}$$

The eigenvalues and eigenvectors are obtained by solving the equation

$$\mathbf{Ac} = -\lambda\mathbf{c} \tag{12.7-6}$$

Each element of the vector **c** is itself a $(Q + 1)$-dimensional column vector (see (12.7-4)). We can formally verify (12.7-6) by some simple manipulations. Since $p_k(\phi, y)$ is an eigenfunction we have

$$Lp_k(\phi, y) = -\lambda_k p_k(\phi, y) \tag{12.7-7}$$

Expressing $p_k(\phi, y)$ by the (truncated) series of (12.7-3) we obtain for (12.7-7)

$$L\left[\sum_n \mathbf{c}_n^T z_n(\phi)\mathbf{h}(y)\right] = -\lambda_k \sum_n z_n(\phi)\mathbf{h}^T(y)\mathbf{c}_n \tag{12.7-8}$$

Again we assume that the L-operator and summation commute. Then, applying L to the series expansion of $p_k(\phi, y)$, but using the definition of $\mathbf{A}_{m,n}$ (truncated) given in (12.7-5), yields for the right-hand side of the last equation

$$L\left[\sum_n \mathbf{c}_n^T z_n(\phi)\mathbf{h}(y)\right] = \sum_n \mathbf{c}_n^T L[z_n(\phi)\mathbf{h}(y)]$$

$$= \sum_n \mathbf{c}_n^T\left[\sum_m \mathbf{A}_{m\,n}\, z_m(\phi)\mathbf{h}(y)\right]$$

$$= \left[\sum_m z_m(\phi)\mathbf{h}^T(y)\left[\sum_n \mathbf{A}_{m\,n}^T\, \mathbf{c}_n\right]\right]^T \qquad (12.7\text{-}9)$$

Comparing coefficients of the two series of (12.7-8) and (12.7-9) yields the eigenvalue equation (12.7-6) in component notation (corresponding to (12.4-2))

$$\sum_n \mathbf{A}_{m\,n}^T\, \mathbf{c}_n = -\lambda_k \mathbf{c}_m \qquad (12.7\text{-}10)$$

Structure of the matrix A. We want to show that under the following two conditions

1. M-attractor cyclic model
2. L operator 2π periodic in ϕ $\qquad\qquad\qquad\qquad$ (12.7-11)

the matrix **A** can be cast into a tridiagonal structure. As we will see later on, this property is responsible for the fact that the space can be decomposed into M-orthogonal subspaces, where each subspace describes a specific aspect of the systems dynamics.

We start by writing the mapping of $z_n(\phi)\mathbf{h}(y)$ by L in component notation

$$L[z_n(\phi)\mathbf{h}(y)] = L\begin{bmatrix} z_n(\phi)h_0(y) \\ z_n(\phi)h_1(y) \\ \vdots \\ z_n(\phi)h_l(y) \\ \vdots \\ z_n(\phi)h_Q(y) \end{bmatrix} \qquad (12.7\text{-}12)$$

We claim that under the conditions, (12.7-11) the (truncated) mapping of (12.7-12), can be written in the form

$$L[z_n(\phi)\mathbf{h}(y)] = \mathbf{A}_{n-M\,n}\, z_{n-M}(\phi)\mathbf{h}(y) + \mathbf{A}_{n\,n}\, z_n(\phi)\mathbf{h}(y)$$
$$+ \mathbf{A}_{n+M\,n}\, z_{n+M}(\phi)\mathbf{h}(y) \qquad (12.7\text{-}13)$$

where the matrices **A** have order $(Q+1)\times(Q+1)$ and $\mathbf{h}(y)$ is a $(Q+1)$ vector.

Proof. Since the Fokker–Planck operator equals the sum of operators of the type

$$\frac{\partial}{\partial x_k} K_k(\phi, y), \quad 1 \le k, m \le 2$$

$$\frac{\partial^2}{\partial x_k \partial x_m} K_{km}(\phi, y) \tag{12.7-14}$$

it is sufficient to discuss a single operator $\partial^2/\partial x_m \partial x_k K_{km}(\phi, y)$. The map of the vector $z_n(\phi)\mathbf{h}(y)$ is obtained by mapping every component of the vector by L (see (12.7-12)). Thus, we have to evaluate

$$\frac{\partial^2}{\partial x_k \partial x_m} [K_{km}(\phi, y)z_n(\phi)h_l(y)] \tag{12.7-15}$$

for all n and $0 \le l \le Q$.

Since the intensity coefficients are periodic functions of ϕ we can develop $K_{km}(\phi, y)$ into a Fourier series (truncated at R)

$$K_{km}(\phi, y) = \sum_{r=-R}^{R} e_r(y) \exp(jr\phi) \tag{12.7-16}$$

Replacing $K_{km}(\phi, y)$ by the right-hand side of the last equation and using $z_n(\phi) = \exp(jn\phi/M)$ yields for (12.7-15)

$$\frac{\partial^2}{\partial x_k \partial x_m} [K_{km}(\phi, y)z_n(\phi)h_l(y)]$$

$$= \frac{\partial^2}{\partial x_k \partial x_m} \left[\sum_{r=-R}^{R} e_r(y) \exp(jr\phi)z_n(\phi)h_l(y) \right]$$

$$= \sum_{r=-R}^{R} \frac{\partial^2}{\partial x_k \partial x_m} \left[e_r(y) \exp(jr\phi) \exp\left(j\frac{n\phi}{M}\right)h_l(y) \right]$$

$$= \sum_{r=-R}^{R} \frac{\partial^2}{\partial x_k \partial x_m} \left[e_r(y) \exp\left[j\frac{(rM+n)\phi}{M}\right]h_l(y) \right]$$

$$= \sum_{r=-R}^{R} \frac{\partial^2}{\partial x_k \partial x_m} [e_r(y)z_{n+rM}(\phi)h_l(y)] \tag{12.7-17}$$

By assumption $L[z_n(\phi)\mathbf{h}(y)]$ is expressible as a linear combination of $\{z_n(\phi)\mathbf{h}(y)\}$ (see (12.7-5)). Thus, this must also hold for the image of the (partial) operator of L in (12.7-17). Since differentiation with respect to $x_1 = \phi$ does not change the index of $z_{n+rM}(\phi)$ we obtain

$$\sum_{r=-R}^{R} \frac{\partial^2}{\partial x_k \, \partial x_m} \left[e_r(y) z_{n+rM}(\phi) h_l(y) \right] = \sum_{r=-R}^{R} \mathbf{f}_r^T z_{n+rM}(\phi) \mathbf{h}(y)$$

$$(12.7\text{-}18)$$

where $\mathbf{f}_r^T$ contains the weights of the $h_l(y)$, $l = 0, \ldots, Q$. Now assume for the moment $R = 1$. Then the sum in (12.7-17) contains only three terms

$$\sum_{r=-1}^{1} \mathbf{f}_r^T z_{n+rM}(\phi) \mathbf{h}(y) = \mathbf{f}_{-1}^T z_{n-M}(\phi) \mathbf{h}(y) + \mathbf{f}_0^T z_n(\phi) \mathbf{h}(y) + \mathbf{f}_1^T z_{n+M}(\phi) \mathbf{h}(y)$$

$$(12.7\text{-}19)$$

where $\mathbf{f}_r^T$ is the lth row of the matrix $\mathbf{A}_{n+rM \, n}$, $r = -1, 0, 1$. From this immediately follows our assertion for the special case $R = 1$. The general case of $R > 1$ can always be reduced to the case of $R = 1$ by the method developed for a first-order system (see (12.5-17) and succeeding equations).

In conclusion, the vector recurrence relation can be cast into

$$-\lambda \mathbf{c}_n = \mathbf{A}_{n \, n-M}^T \, \mathbf{c}_{n-M} + \mathbf{A}_{n \, n}^T \, \mathbf{c}_n + \mathbf{A}_{n \, n+M}^T \, \mathbf{c}_{n+M} \qquad (12.7\text{-}20)$$

Example: Second-order PLL with sinusoidal phase detector characteristic. M-attractor cyclic model
We choose the following complete set of base functions

1. $\{z_n(\phi)\} = \left\{ \exp\left(jn \, \frac{\phi}{M} \right) \right\}$

2. $\{h_m(y, \alpha)\} = \left\{ \dfrac{1}{\left[(2\alpha)^m m! \sqrt{\dfrac{\pi}{\alpha}} \right]^{1/2}} \exp\left(-\alpha \, \frac{y^2}{2} \right) H_m(y, \alpha) \right\}$

where α represents a freely choosable parameter. The Hermite polynomials H_m of the functions $h_m(y, \alpha)$ are defined by

$$H_m(y, \alpha) = (-1)^m \exp(\alpha y^2) \frac{d^m}{dy^m} \exp(-\alpha y^2)$$

Note that the first element $h_0(y, \alpha)$ is proportional to a gaussian probability density function

$$h_0(y, \alpha) = \left(\frac{\alpha}{\pi} \right)^{1/4} \exp\left(-\alpha \, \frac{y^2}{2} \right)$$

The generalization to an Nth order Fokker–Planck equation is obvious, at least formally. One develops the eigenfunctions into an N-fold series truncated at $Q_2, \ldots, Q_N$ for the x_i, $i \geq 2$ variables

$$p_k(\phi, x_2, \ldots, x_N) = \sum_n \left\{ \sum_{r=0}^{Q_2} \cdots \sum_{i=0}^{Q_N} c_{n,r,\ldots,i} [h_r^{(2)}(x_2) \cdots h_i^{(N)}(x_N)] \right\} z_n(\phi) \tag{12.7-21}$$

For each variable x_i the set of functions $\{h_k^{(i)}(x_i)\}$ must be complete and orthogonal and must fulfill the boundary conditions imposed on the variable x_i. Each (vector) element $\mathbf{c}_n$ has dimension

$$\dim \mathbf{c}_n = \prod_{i=2}^{N} (1 + Q_i) \tag{12.7-22}$$

which leads to the size of each submatrix $\mathbf{A}_{n\,k}$ of

$$\left[\prod_{i=2}^{N} (1 + Q_i) \right] \times \left[\prod_{i=2}^{N} (1 + Q_i) \right]$$

Notice that the ordering of the elements in $\mathbf{c}_n$ is arbitrary. While from a purely formal point of view the ordering is immaterial it plays a decisive role in the numerical calculations. Much research in the numerical eigenvalue evaluation of Nth order Fokker–Planck equations remains to be done.

Obviously it is of interest to exploit order-reducing methods to approximate higher order systems by lower order ones. A useful approach for the M-attractor cyclic model seems to be the so-called "conditional expectation method" discussed by Lindsey [1: p. 507ff].

12.8. DECOMPOSITION OF THE MATRIX A FOR AN M-ATTRACTOR MODEL OF AN NTH ORDER SYSTEM

As shown in the previous section, the matrix $\mathbf{A}$ has tridiagonal form with two secondary diagonals $\pm M$ apart from the main diagonal. Therefore, the matrix $\mathbf{A}$ defines M independent recursions. The first recursion comprises all indices $k = rM$. The second recursion comprises all indices $k = rM + 1$ and so on up to the Mth recursion with $k = rM + (M - 1)$. Each of the M-recursions is defined by one of the M-submatrices $\mathbf{A}^{(i)}$

$$\mathbf{A} = \mathbf{A}^{(0)} + \cdots + \mathbf{A}^{(M-1)} \tag{12.8-1}$$

By a straightforward extension of the proof given for $M = 2$, it is shown that

1. Every eigenvalue $\lambda_k^{(i)}$ defined by $\mathbf{A}^{(i)}$, $i = 0, \ldots, M - 1$ is an eigenvalue of the sum matrix $\mathbf{A}$.

2. Every eigenvector $\mathbf{c}^{(i)}$ belongs to the *kernel* of the matrices $\mathbf{A}^{(m)}$, $m \neq i$, i.e.

$$\mathbf{A}^{(m)}\mathbf{c}^{(i)} = 0 , \quad m \neq i \tag{12.8-2}$$

As in the bistable case one concludes that every eigenfunction belongs to exactly one of the orthogonal subspaces spanned by the base vectors

$$\beta^{(i)} = \{z_{rM+i}(\phi)\mathbf{h}(y)\} , \quad i = 0, \ldots, M-1 \tag{12.8-3}$$

The orthogonality of the subspaces follows from the orthogonality of the base vectors.

Example: For dim $\mathbf{x} = 2$, $(\mathbf{x}^T = [\phi \quad y])$ we obtain

$$\langle z_{rM+m}(\phi)\mathbf{h}(y), z_{qM+i}(\phi)\mathbf{h}(y)\rangle$$

$$:= \int_0^{M2\pi} d\phi \int_{-\infty}^{\infty} z_{rM+m}(\phi) z_{qM+i}^*(\phi)\mathbf{h}(y)\mathbf{h}^T(y) \, dy$$

$$= \int_0^{M2\pi} d\phi \; z_{rM+m}(\phi) z_{qM+i}^*(\phi) \int_{-\infty}^{\infty} dy \begin{bmatrix} h_0(y) \\ \vdots \\ h_Q(y) \end{bmatrix} [h_0(y) \quad \cdots \quad h_Q(y)]$$

$$= \begin{cases} M\mathbf{I} , & \text{if } r = q \text{ and } m = i \\ 0 , & \text{otherwise} \end{cases} , \quad \mathbf{I} = \text{unity matrix} \tag{12.8-4}$$

by the orthogonality of the functions $z_n(\phi)$ and $h_l(y)$.

Geometrical Interpretation. The whole space $\mathscr{S}$ is the direct sum of the M orthogonal subspaces. Space $\mathscr{S}^{(0)}$ describes the relaxation dynamics, irrespective of the number of attractors. Formally, this follows from the fact that the base $\beta^{(0)}$ is identical for every M

$$\beta^{(0)} = \left\{\exp\left(j\frac{\phi}{M}k\right)\mathbf{h}(y)\right\} = \{\exp(j\phi r)\mathbf{h}(y)\} , \quad k = rM$$

The remaining $(M-1)$ subspaces describe the dynamics of transitions between the attractors. By increasing the number of attractors M one obtains an increasingly detailed picture of the transition dynamics. The smallest eigenvalue $\lambda_0^{(i)}$ of every subspace is related to the transition rates a_k by (12.3-7) (justification to be given soon).

However, the transition rates can be computed only when we can nonambiguously assign the eigenvalues $\lambda_0^{(i)}$ to those of the coarse-grained model (which have a different numbering!). To establish this correspondence we exploit the fact that the first M eigenfunctions $p_m(x) := p_m(\phi, y)$ of the Fokker–Planck operator L are (approximately) proportional to the steady-state probability density function $p_0(x)$ in any region R_I with a weight factor depending on i (number of subspace) and the location I (see beginning of Section 12.2 for a refresher).

Let us now pick out the subspace $\mathcal{S}^{(i)}$ with base $\{\beta^{(i)}\}$ defined by (12.8-3). Without loss of generality we assign the first attractor region R_1 to the fundamental 2π-intervals centered at the origin. Since the eigenfunction $p_0^{(i)}(x)$ (corresponding to $\lambda_0^{(i)}$) is (approximately) proportional to $p_0(x)$ we can write

$$p_0^{(i)}(x) = D_1^{(i)} p_0(x), \quad x \in R_1$$

where $D_1^{(i)}$ is a constant of proportionality depending on the location I and the number i of the subspace $\mathcal{S}^{(i)}$. For an arbitrary region R_I the factor $D_I^{(i)}$ is readily related to the reference constant $D_1^{(i)}$ by replacing the variable ϕ in the base $\beta^{(i)}$ by

$$\phi = \phi - (I-1)2\pi + (I-1)2\pi$$
$$= \xi + (I-1)2\pi, \quad |\xi| < \pi, I = 1, \ldots, M$$

Inserting for ϕ, the right-hand side of the previous equation, using (12.8-3) yields

$$\beta^{(i)} = \left\{ \exp\left[\frac{j\phi}{M} (rM + i) \right] \mathbf{h}(y) \right\}$$
$$= \left\{ \exp\left[\frac{j2\pi}{M} i(I-1) \right] \exp\left[\frac{j\xi}{M} (rM + i) \right] \right\}$$
$$= \kappa^{i(I-1)} \left\{ \exp\left[\frac{j\xi}{M} (rM + i) \right] \mathbf{h}(y) \right\}$$

from which it immediately follows that the constant $D_I^{(i)}$ is related to $D_1^{(i)}$ by

$$D_I^{(i)} = \kappa^{i(I-1)} D_1^{(i)}, \quad I = 1, \ldots, M$$

Collecting the relative weights in a vector

$$\begin{bmatrix} 1 & \kappa^i & \kappa^{2i} & \cdots & \kappa^{i(M-1)} \end{bmatrix}$$

we recognize the eigenvector $\mathbf{p}_i^T$ (12.3-5) of the coarse-grained model. Thus, we have established the desired correspondence. We know that there are exactly M linearly independent eigenvectors $\mathbf{p}_i$ in the coarse-grained model and, from the above, that every $\mathbf{p}_i$ is uniquely related to the subspace $\mathcal{S}^{(i)}$ of the "full" model. Therefore, exactly *one* eigenvalue (the smallest one) in every subspace $\mathcal{S}^{(i)}$ corresponds to one eigenvalue in the coarse-grained model. The correspondence is

$$\lambda_0^{(i)} \qquad \text{corresponds to} \qquad \lambda_i, \quad i = 0, \ldots, M-1$$

smallest eigenvector   coarse-grained model
in subspace $\mathcal{S}^{(i)}$ with
base $\{\beta^{(i)}\}$, $i = 0, \ldots, M-1$

12.9. NUMERICAL EXAMPLES

Example 1: First-Order PLL with Sinusoidal Phase Detector Characteristic ($M = 2$ attractors)

Model. The Fokker–Planck operator reads

$$L = -\frac{\partial}{\partial \phi} \left[-\sin \phi + \gamma \right] + \frac{1}{2} \frac{\partial^2}{\partial \phi^2} \frac{2}{\rho}$$

The (2×2) matrices $\mathbf{A}_{n\,k}$ in the recursion are given by (12.5-14b).

Results. Table 12.9-1 illustrates that even for very low signal-to-noise ratios of no practical significance the smallest eigenvalue is well separated from the next eigenvalue. Thus, the main assumption of the matrix eigenvalue approach is well justified.

In Table 12.9-2 we have listed the first few eigenvalues of the transition and relaxation dynamics. For the sake of a comparison in the third column we have listed the eigenvalues for the renewal process which are obtained as the solution of a boundary value problem with entirely different boundary conditions, namely

TABLE 12.9-1 Smallest Eigenvalue $\lambda_0^{(1)}$ and the Ratio of $\lambda_1^{(1)}/\lambda_0^{(1)}$ of the Transition Dynamics as a Function of ρ and Loop Detuning

γ	ρ	2 dB	4 dB	6 dB
0	$\lambda_0^{(1)}$	0.235×10^{-1}	0.372×10^{-2}	0.199×10^{-3}
	$\dfrac{\lambda_1^{(1)}}{\lambda_0^{(1)}}$	0.246×10^2	0.169×10^3	0.265×10^4
$\sin 10°$	$\lambda_0^{(1)}$	0.317×10^{-1}	0.719×10^{-2}	0.798×10^{-3}
	$\dfrac{\lambda_1^{(1)}}{\lambda_0^{(1)}}$	0.178×10^2	0.837×10^3	0.863×10^4

TABLE 12.9-2   Eigenvalues of the 2-Attractor Model for $\rho = 4$ and $\gamma = 0$. For Comparison the Eigenvalues of the Renewal Process are also Indicated

| | $M = 2$ attractor model | | Renewal Process |
| | Relaxation Space | Transition Space | |
k	$\lambda_k^{(0)}$	$\lambda_k^{(1)}$	$\lambda_k^{(R)}$
0	0	1.99×10^{-4}	0.9925×10^{-4}
1	0.8335	0.7509	0.7509
2	1.5676	0.9838	0.7723
3	2.7780	1.269	0.8336
4	4.5156	2.085	0.9229

Renewal process	Absorbing boundary conditions
	$p(-2\pi, t) = p(2\pi, t) = 0$
Matrix eigenvalue	Periodic boundary conditions
approach	$p(-2\pi, t) = p(2\pi, t)$

The first eigenvalue of the renewal process equals $\lambda_0^{(R)} = 1/E(T_s) = r_1 + r_{-1}$ (see 11.1-133a), while the first eigenvalue of the bistable cyclic model equals $\lambda_0^{(1)} = 2a_1 = 2(r_1 + r_{-1})$ (12.3-7). The factor 2 in the two eigenvalues is confirmed by the numerical computation shown in the first row of Table 12.9-2. We observe that the eigenvalue $\lambda_1^{(R)}$ is also an eigenvalue ($\lambda_1^{(1)}$) of the $M = 2$ attractor model. Since the eigenfunction $p_1^{(R)}$ (if periodically extended) has a period of 4π the eigenvalue can only belong to the transition space. A similar plausibility argument holds for $\lambda_3^{(R)}$ which is associated with a 2π-periodic eigenfunction $p_3^{(R)}$. The eigenvalue $\lambda_1^{(0)} = \lambda_3^{(R)}$ belongs to the relaxation space.

Notice that only in the symmetric case $\gamma = 0$ do we find eigenvalues/eigenfunctions which are simultaneously solutions for both types of boundary conditions. The reason lies in the fact that (due to symmetry reasons) any $p_k(\phi)$ can be represented by sine functions alone

$$\{z_n(\phi)\} = \left\{ \sin\left[\frac{n}{2} (\phi + \pi) \right] \right\} \tag{12.9-1}$$

Indeed, for $\gamma = 0$ any image $L \sin[(n/2)(\phi + \pi)]$ can be represented as a linear combination of $\{\sin[(n/2)(\phi + \pi)]\}$. But since $\sin[(n/2)(\phi + \pi)]$ vanishes at $\phi = \pm\pi$ for any n this implies also that any $p_k(\phi)$ vanishes at $\phi = \pm\pi$. Hence, $p_k(\phi)$ is a solution of the boundary value problem with absorbing boundaries at $\phi = \pm\pi$. But since the absorbing boundaries are placed at $\phi = \pm\pi$ (and not at $\phi = \pm2\pi$ as required for the renewal process) we obtain

$$\lambda_0^{(R\pi)} = \frac{1}{E(T_s/2)} = 2a_1 \tag{12.9-2}$$

which explains the factor 2 of Table 12.9-2 in a different way.

For $\gamma \neq 0$ these correspondences are lost. Any image $L \sin[(n/2)(\phi + \pi)]$ contains cosine functions. This simply means that $\{z_n(\phi)\}$ must contain sine as well as cosine functions in order to comply with the basic equation (12.4-1). But since the $z_n(\phi) = \exp[j(n/2)\phi]$ do not vanish at $\phi = \pm\pi$, we can no longer claim that the eigenfunctions $p_k(\phi)$ vanish at these boundaries. Hence, the $p_k(\phi)$ are *not* eigenfunctions of the eigenvalue problem with absorbing boundaries.

Numerical Issues. The key parameter in the numerical analysis is the required order n_{max} of the truncated matrix $\mathbf{B}^{(1)}$. The order n_{max} was determined as that number for which the eigenvalue did not change significantly if the order was further increased. Figure 12.9-1 shows n_{max} as a function of the signal-to-noise ratio ρ. Also indicated is the value of $\lambda_0^{(1)}$ so that a correspondence between this value and n_{max} can be established. (This is useful as a first estimate of n_{max} for higher order systems.)

Example 2: First-Order PLL with Sinusoidal Phase Detector Characteristic ($M = 4$ attractors)
The same parameters are used as in the previous example. Table 12.9-3 gives the eigenvalues for $M = 4$ subspaces, where $\mathscr{S}^{(1)} \oplus \mathscr{S}^{(3)}$ denotes the direct sum of the subspaces $\mathscr{S}^{(1)}$ and $\mathscr{S}^{(3)}$. Notice that for $\gamma \neq 0$ the respective smallest eigenvalues $\lambda_0^{(1)}$ and $\lambda_0^{(3)}$ become complex. Furthermore,

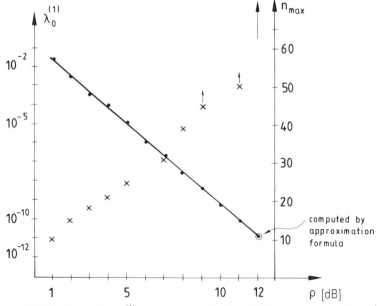

Figure 12.9-1. Eigenvalue $\lambda_0^{(1)}$ and order $[n_{max}] \times [n_{max}]$ of the truncated matrix $\mathbf{B}^{(1)}$.

TABLE 12.9-3 Smallest Two Eigenvalues $\lambda_0^{(i)}$ and $\lambda_1^{(i)}$ of the Subspaces $\mathscr{S}^{(i)}$, $i = 0, 1, 2, 3$ as a Function of ρ and Loop Detuning γ

Subspace	$\mathscr{S}^{(0)}$		$\mathscr{S}^{(2)}$		$\mathscr{S}^{(1\oplus3)}$	
ρ \ eigenvalues	$\lambda_0^{(0)}$	$\lambda_1^{(0)}$	$\lambda_0^{(2)}$	$\lambda_1^{(2)}$	$\lambda_0^{(1\oplus3)}$	$\lambda_1^{(1\oplus3)}$
			(a) $\gamma = 0$			
6 dB	0	0.834	0.199×10^{-3}	0.751	0.992×10^{-4}	0.772
4 dB	0	0.797	0.373×10^{-2}	0.629	0.185×10^{-2}	0.673
2 dB	0	0.890	0.235×10^{-1}	0.579	0.112×10^{-1}	0.664
			(b) $\gamma = 0.174$			
6 dB	0	0.805	0.780×10^{-3}	0.689	$0.3968 \times 10^{-3} \pm j0.3886 \times 10^{-3}$	$0.7171 \pm j0.4047 \times 10^{-1}$
4 dB	0	0.788	0.719×10^{-2}	0.602	$0.3469 \times 10^{-2} \pm j0.3155 \times 10^{-2}$	$0.6476 \pm j0.6845 \times 10^{-1}$
2 dB	0	0.888	0.317×10^{-1}	0.564	$0.1424 \times 10^{-1} \pm j0.1095 \times 10^{-1}$	$0.6544 \pm j0.8980 \times 10^{-1}$

they satisfy $\lambda_0^{(3)} = \lambda^{(1)*}$ i.e., in the space $\mathscr{S}^{(1)} \oplus \mathscr{S}^{(3)}$ the eigenvalues occur in conjugate pairs.

As we know, for $M = 4$ attractors we can obtain the positive- and negative-indexed cycle slip rates (r_1 and r_{-1}, respectively) individually. By virtue of (12.3-7) where $a_2 = 0$ (cycle slips only toward neighboring attractors) the rates r_1 and r_{-1} can be expressed in terms of the eigenvalue $\lambda_0^{(1)}$

$$r_1 = a_1 = \frac{1}{2} [\text{Re } \lambda_0^{(1)} - \text{Im } \lambda_0^{(1)}]$$

$$r_{-1} = a_3 = \frac{1}{2} [\text{Re } \lambda_0^{(1)} + \text{Im } \lambda_0^{(1)}]$$

These rates are plotted versus loop detuning γ in Figure 12.9-2.

Notice that as $|\gamma|$ increases, the magnitude of Im $\lambda_0^{(1)}$ also increases up to an asymptotic limit of $|\text{Re } \lambda_0^{(1)}|$. Consequently, small errors introduced by the numerical computation of $\lambda_0^{(1)}$ will lead to large errors for small transition rates. This explains the deviation of the computed rates from the theoretical results in Figure 12.9-2 for large $|\gamma|$.

Example 3: Second-Order Systems
Model. First-order PLL with random frequency modulation and additive noise disturbance

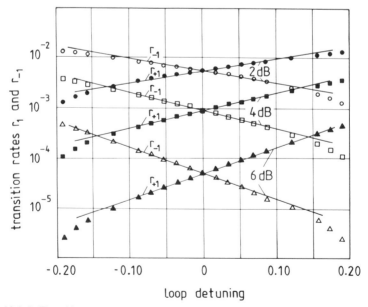

Figure 12.9-2. Transition rates $a_1 = r_1$ and $a_3 = r_{-1}$ as a function of loop detuning γ. First-order loop.

Stochastic differential equation

$$d\phi(\tau) = -[y(\tau) + \sin \phi(\tau)]\, d\tau - \sqrt{\frac{2}{\rho}}\, d\beta_1(\tau)$$

$$dy(\tau) = -\delta y(\tau)\, d\tau + \sigma_y \sqrt{2\delta}\, d\beta_2(\tau)$$

(12.9-3)

with

σ_y^2 variance of the random frequency modulation

δ bandwidth ratio of the random modulation process and the PLL,
   $\delta = B_y / B_L$

$E[d\beta_1^2(\tau)] = E[d\beta_2^2(\tau)] = d\tau$

$E[d\beta_1(\tau)\, d\beta_2(\tau)] = 0$

ρ signal-to-noise ratio in the loop in the absence of random frequency modulation

Fokker–Planck operator L

$$L = \frac{\partial}{\partial \phi}(y + \sin \phi) + \frac{\partial}{\partial y}(y\delta) + \frac{1}{\rho}\frac{\partial^2}{\partial \phi^2} + \delta\sigma_y^2 \frac{\partial^2}{\partial y^2}, \quad K_{12} = K_{21} = 0$$

(12.9-4)

Matrix Recursion. The following sets are used (see also examples in Section 12.7 for details)

$$\{z_n(\phi)\} = \left\{ \exp\left(j\,\frac{n}{M}\,\phi \right) \right\}$$

$$\{h_m(y, \alpha)\} = \left\{ \frac{1}{\left[(2\alpha)^m m! \sqrt{\frac{\pi}{\alpha}} \right]^{1/2}} \exp\left(-\alpha\,\frac{y^2}{2} \right) H_m(y, \alpha) \right\}$$

(12.9-5)

The image of $z_n(\phi)h_m(y, \alpha)$ for $|n| \le n_{max}$ and $0 \le m \le Q$ is found to be

$$Lz_n(\phi)h_m(y, \alpha) = -\frac{n - M}{2M}\, z_{n-M}(\phi)h_m(y, \alpha)$$

$$+ \frac{\delta}{2}\sqrt{m(m - 1)}(1 - \alpha\sigma_y^2)z_n(\phi)h_{m-2}(y, \alpha)$$

$$+ j\,\frac{n}{M}\sqrt{\frac{m}{2\alpha}}\, z_n(\phi)h_{m-1}(y, \alpha)$$

$$+ \left(\frac{\delta}{2} - \frac{1}{\rho}\left(\frac{n}{M} \right)^2 - \frac{\delta}{2}(2m + 1)\alpha\sigma_y^2 \right)z_n(\phi)h_m(y, \alpha)$$

$$+ j\,\frac{n}{M}\sqrt{\frac{m + 1}{2\alpha}}\, z_n(\phi)h_{m+1}(y, \alpha)$$

$$+ \frac{\delta}{2}\sqrt{(m + 1)(m + 2)}(-1 + \alpha\sigma_y^2)z_n(\phi)h_{m+2}(y, \alpha)$$

$$+ \frac{n + M}{2M}\, z_{n+M}(\phi)h_m(y, \alpha)$$

(12.9-6)

Notice that the image $Lz_n(\phi)h_Q(y, \alpha)$ contains terms with $h_{Q+1}(y, \alpha)$ and $h_{Q+2}(y, \alpha)$. These terms are truncated in the $(Q+1) \times (Q+1)$ matrices $\mathbf{A}_{n \pm M\,n}$ and $\mathbf{A}_{n\,n}$.

Numerical Results. We consider a bistable cyclic model $(M = 2)$. Figure 12.9-3 shows the transition rate $\lambda_0^{(1)} = 2a_1$ as a function of the modulation variance σ_y^2 for various values of ρ. The bandwidth of the loop is twice as large as the bandwidth of the frequency modulation, $\delta = 0.5$.

The most significant observation is the existence of a distinct threshold which can only be computed by means of the nonlinear theory. If the variance of the frequency modulation falls below a certain value, the transitions are entirely due to the additive noise since the PLL can track the frequency modulation. If σ_y^2 becomes larger than this threshold value the transition rate increases steeply. In this region the transitions are predominantly caused by the modulation.

Numerical Issues. The parameter α of the Hermite eigenfunction is chosen as $\alpha = 1/\sigma_y^2$. By this choice the eigenfunction $h_0(y, \alpha)$ agrees with the stationary probability density function of y in the linearized model and one can expect a minimum needed dimension of Q to achieve convergence of the eigenvalues.

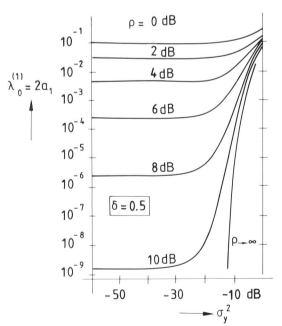

Figure 12.9-3. Transition rate as a function of the modulation variance σ_y^2. Signal-to-noise ratio ρ is parameter. The bandwidth ratio $\delta = B_y/B_L = 0.5$.

Tables 12.9-4 show the required n_{max} and Q for convergence of the eigenvalue $\lambda_0^{(1)}$ ($M = 2$). (Remember that the elements of the matrix $\mathbf{B}^{(1)}$ are submatrices of dimension $(Q + 1) \times (Q + 1)$. Thus, the matrix has order $[(Q + 1)n_{max}] \times [(Q + 1)n_{max}]$.)

The following observations can be made

- The maximum number n_{max} is increasing with decreasing $\lambda_0^{(1)}$. The value n_{max} is comparable with that of a first-order system for the same $\lambda_0^{(1)}$.
- The maximum Q is between 4 and 6 and independent of n_{max}. This is the case because σ_y^2 is independent of ρ.

The present example deals with a special case of a second-order system. When computing the eigenvalues of a second-order loop the state variables ϕ and y appear in both equations and σ_y^2 becomes a function of ρ. As a result, the convergence of the eigenvalues is critically dependent on a proper choice of the parameter α in the Hermite functions.

Acknowledgment: This example was worked out by L. Popken and S. Fechtel.

TABLE 12.9-4   Order $[n_{max} \times n_{max}]$ of the Truncated Matrix and Order $[Q + 1] \times [Q + 1]$ of the Submatrices

n_{max} \ Q	2	4	6	8
5	0.42517×10^{-2}	0.42594×10^{-2}	0.42594×10^{-2}	0.42594×10^{-2}
7	0.42807×10^{-2}	0.42885×10^{-2}	0.42885×10^{-2}	0.42885×10^{-2}
9	0.42807×10^{-2}	0.42886×10^{-2}	0.42886×10^{-2}	0.42886×10^{-2}
11	0.42807×10^{-2}	0.42886×10^{-2}	0.42886×10^{-2}	0.42886×10^{-2}
13	0.42807×10^{-2}	0.42886×10^{-2}	0.42886×10^{-2}	0.42886×10^{-2}
15	0.42807×10^{-2}	0.42886×10^{-2}	0.42886×10^{-2}	0.42886×10^{-2}

(a) $\sigma_y^2 = -20$ dB, $\rho = 4$ dB

n_{max} \ Q	2	4	6	8
5	-0.47207×10^{-2}	-0.48114×10^{-2}	-0.48087×10^{-2}	-0.48087×10^{-2}
7	-0.10122×10^{-3}	-0.10707×10^{-3}	-0.10675×10^{-3}	-0.10676×10^{-3}
9	0.35408×10^{-5}	0.38228×10^{-5}	0.38532×10^{-5}	0.38533×10^{-5}
11	0.45188×10^{-5}	0.49336×10^{-5}	0.49555×10^{-5}	0.49559×10^{-5}
13	0.45231×10^{-5}	0.49391×10^{-5}	0.49609×10^{-5}	0.49613×10^{-5}
15	0.45231×10^{-5}	0.49391×10^{-5}	0.49609×10^{-5}	0.49613×10^{-5}
17	0.45231×10^{-5}	0.49391×10^{-5}	0.49609×10^{-5}	0.49613×10^{-5}
19	0.45231×10^{-5}	0.49391×10^{-5}	0.49609×10^{-5}	0.49613×10^{-5}

(b) $\sigma_y^2 = -20$ dB, $\rho = 8$ dB

Main Points of the Chapter

- *M-attractor cyclic model* (*"coarse-grained"*)
  The model involves two kinds of coarse-graining: a *temporal* one which focuses on the slow transition phenomena; and a *spatial* one which, from the actual trajectories, only retains the information near which of the attractors they stay. In this model the transitions between the attractors is described by a markovian jump process (see also Section 10.5).
- *Transition rates.* The transition rates a_k in the M-attractor cyclic model are stationary. They are determined by the $M - 1$ smallest nonzero eigenvalues λ_k

$$a_k = -\frac{1}{M} \sum_{n=1}^{M-1} \kappa^{-kn} \lambda_n \qquad (12.3\text{-}7)$$

where $\kappa = \exp(j2\pi/M)$. The transition rates a_k are related to the average cycle slip rates r_k in the unrestricted phase error picture by

$$a_k = \sum_{\nu=-\infty}^{\infty} r_{k+\nu M}$$

The index k of r_k denotes the number of skipped cycles, and the sign of k denotes the cycle slip direction.
- *Eigenvalues.* The eigenvalues λ_n, $n = 1, \ldots, M - 1$ of the coarse-grained model are equal to the first $(M - 1)$ smallest (nonzero) eigenvalues of the Fokker–Planck operator (condition: moderate noise).

$$Lp_n(x) = -\lambda_n p_n(x) \qquad (12.1\text{-}2)$$

The eigenfunctions $p_n(x) := p_n(\phi, x_2, \ldots, x_N)$ are periodic in ϕ and fulfill natural boundary conditions for the remaining state variables.

The *periodicity* of $p_n(\phi, x_2, \ldots, x_N)$ is the *key property*. It is far simpler to solve an eigenvalue problem with periodicity constraints (or periodic boundary conditions) than it is to solve a boundary value problem with absorbing boundaries.

To find the eigenvalues λ_n, $n = 1, \ldots, M - 1$ we develop any $p_n(x)$ into a suitable set of functions $\{\Psi_n(x) := \Psi_n(\phi, x_2, \ldots, x_N)\}$. The function $\Psi_n(\phi, x_2, \ldots, x_N)$ must be periodic in ϕ and fulfill natural boundary conditions for the remaining variables. The function set $\{\Psi_n(x)\}$ must be such that the image $L\Psi_n(x)$ is expressible as a linear combination of $\{\Psi_n(x)\}$

$$L\Psi_n(x) = \sum_m \mathbf{A}_{m\,n} \Psi_m(x) \qquad (12.4\text{-}1)$$

The eigenvalues are then those of the matrix $\mathbf{A}$ with (submatrix) elements $\mathbf{A}_{m\,n}$ defined above.

Thus, the task of finding the eigenvalues has three steps:

1. Choice of a suitable function set $\{\Psi_n(x)\}$.
2. Mapping of $\Psi_n(x)$ by L.
3. Numerical evaluation of the eigenvalues of the truncated matrix $\mathbf{A}$.

REFERENCES

1. W. C. Lindsey, *Synchronization System in Communication and Control*, Prentice Hall, Englewood Cliffs, NJ, 1972.
2. H. Risken, *The Fokker–Planck Equation*, Springer-Verlag, Berlin, 1984.
3. H. D. Vollmer and H. Risken, Eigenvalues and Their Connection to Transition Rates for the Brownian Motion in an Inclined Cosine Potential, *Zeitschrift für Phys.* B, Condensed Matter, Vol. 52, pp. 259–266, 1983.
4. D. Ryter, On the Eigenfunctions of the Fokker–Planck Operator and of its Adjoint, *Physica*, Vol. 142A, pp. 103–121, 1987.
5. D. Ryter and H. Meyr, Multistable Systems with Moderate Noise: Coarse-Graining to a Markovian Jump Process and Evaluation of the Transition Rates, *Physica*, Vol. 141A, pp. 122–134, 1987.
6. B. J. Matkowsky and Z. Schuss, The Exit Problem for Randomly Perturbed Dynamical Systems, *SIAM Journal on Applied Mathematics*, Vol. 33, No. 2, pp. 365–382, September 1977.
7. D. Ryter, Coarse-Graining of Diffusion Processes with a Multistable Drift, *Zeitschrift für Phys.* B., Condensed Matter, Vol. 68, pp. 209–211, 1987.
8. B. J. Matkowsky, Z. Schuss and C. Knessl, Asymptotic Solution of the Kramers-Moyal Equation and First-Passage Times for Markov Jump Processes, *Physical Review* A, Vol. 29, pp. 3359–3369, June 1984.
9. C. Knessl, M. Mangel, B. J. Matkowsky, Z. Schuss and C. Tier, Solution of Kramers-Moyal Equations for Problems in Chemical Physics, *Journal of Chemical Physics*, Vol. 81, pp. 1285–1293, August 1984.
10. K. Nishiguchi and Y. Uchida, Transient Analysis of the Second-Order Phase-Locked Loop in the Presence of Noise, *IEEE Transactions on Information Theory*, Vol. 26, pp. 482–486, July 1980.
11. G. F. Miller, The Evaluation of Eigenvalues of a Differential Equation Arising in a Problem in Genetics, *Proceedings of the Cambridge Philosophical Society* (Mathematical and Physical Sciences), Vol. 58, pp. 588–593, 1962.
12. D. Ryter, Some Properties of the Backward and Forward Operators and of Their Eigenfunctions, Internal Report 713/24, 1988 (available on request).
13. H. A. Kramers, Brownian Motion in a Field of Force and the Diffusion Model of Chemical Reactions, *Physica*, Vol. 7, p. 284, 1940.
14. Z. Schuss, *Theory and Application of Stochastic Differential Equations*, John Wiley, New York, 1980.

15. M. I. Freidlin and A. D. Wentzell, *Random Perturbations of Dynamical Systems*, Springer-Verlag, New York, 1984.

16. B. Z. Bobrovsky and Z. Schuss, A Singular Perturbation Method for the Computation of the Mean First Passage Time in a Nonlinear Filter, *SIAM Journal on Applied Mathematics*, Vol. 42, p. 174, 1982.

17. D. Ryter and P. Jordan, A Way to Solve the Stationary Fokker–Planck Equation for Metastable Systems, *Physics Letters* A, Vol. 104, p. 193, 1984.

18. D. Ryter, Noise-Induced Transitions in a Double-Well Potential at Low Friction, *Journal of Statistical Physics*, Vol. 49, p. 751, 1987.

19. H. R. Jauslin, Nondifferentiable Potentials for Nonequilibrium Steady States, *Physica* Vol. 144A, p. 179, 1987.

APPENDIX 12.A.
A BRIEF ACCOUNT ON WEAK NOISE THEORIES

At very weak noise the eigenvalue approach requires large matrices, so that alternative methods [13–16] may become more appropriate. In practice, however, their validity is hard to check, and for higher order systems they typically involve numerical work. Therefore, we only give a short review.

The idea is to evaluate the mean exit time from a domain of attraction Ω of the drift (the conceptually different method of Kramers [13] was reduced to that problem by Ryter [4]). For a starting point x in Ω, the mean exit time $T(x)$ is determined by

$$L^+ T(x) = -1 , \quad T = 0 \text{ on the boundary} \tag{12A-1}$$

Multiplying by $p_0(x)$ and integrating over Ω gives

$$\frac{1}{2} \oint_{\partial\Omega} \sum_{i=1}^{N} dS_i \, p_0(x) \sum_{k=1}^{N} K_{ik} \frac{\partial T(x)}{\partial x_k} = -\int_\Omega dx \, p_0(x)$$

The details are given by Ryter [12]. Since in Ω the drift tends inwards and the noise is weak, trajectories typically fluctuate well inside Ω and thereby lose the memory of their starting point x, before reaching the boundary; therefore, $T(x)$ has a constant value $\hat{T}$ on Ω, except for a thin layer along the boundary $\partial\Omega$, where $T(x)$ rapidly decays to zero, due to direct exits. With $T(x) := \hat{T}\theta(x) \; (0 \le \theta(x) \le 1)$ it follows that

$$\hat{T} = \frac{\int_\Omega dx \, p_0(x)}{-\dfrac{1}{2} \oint_{\partial\Omega} \sum_{i=1}^{N} dS_i \, p_0(x) \sum_{i=1}^{N} K_{ik} \dfrac{\partial\theta(x)}{\partial x_k}} \tag{12A-2}$$

and

$$L^+\theta(x) = -1/\hat{T} \approx 0 , \quad \theta = 0 \text{ on } \partial\Omega, \; \theta = 1 \text{ well inside } \Omega \tag{12A-3}$$

Except for the first-order case (and for systems with "detailed balance" which are important in physics, but not in electronics) the crucial function $p_0(x)$ is to be approximated. The usual ansatz for low noise is

$$p_0(x) \sim z(x) \exp\left[-u(x)/\varepsilon\right] \tag{12A-4}$$

where ε measures the noise strength, so that $K_{ij}(x) = \varepsilon k_{ij}(x)$. Inserting this into $Lp_0 = 0$ and collecting the terms with the same order in ε yields

$$\sum_{i=1}^{N} \frac{\partial u}{\partial x_i}\left[K_i(x) + \frac{1}{2}\sum_{j=1}^{N} k_{ij}(x)\frac{\partial u(x)}{\partial x_j}\right] = 0 \tag{12A-5}$$

and a corresponding equation for $z(x)$, see [17]. Expanding (12A-5) around a stationary point of the drift gives

$$\mathbf{CB} + \mathbf{B}^T\mathbf{C} + \mathbf{CkC} = 0$$

where $\mathbf{C}$ is the matrix of the second derivatives of u, and $\mathbf{B}$ is the matrix of the first derivatives of the drift $\mathbf{K}$. This is equivalent to the linear equation

$$\mathbf{BC}^{-1} + \mathbf{C}^{-1}\mathbf{B}^T + \mathbf{k} = 0 \tag{12A-6}$$

for $\mathbf{C}^{-1}$, which in two dimensions is solved by

$$\mathbf{C}^{-1} = -(\mathbf{k} + \mathbf{B}^{-1}\mathbf{kB}^{-1T}\det\mathbf{B})/2\,\mathrm{tr}\,\mathbf{B}$$

The integrals in (12A-2) can now be worked out by approximating $p_0(x)$ as gaussians with parameters given by $\mathbf{C}$, for the numerator at the attractor of the drift, and for the denominator at the preferred exit point (typically a saddlepoint of the drift, i.e., a stationary point which is attractive within $\partial\Omega$, but repulsive transverse to it). The value of $\partial\theta/\partial x_k$ follows by a specific ansatz for (12A-3). See [18] where an improvement for low damping is also presented.

The difficult (but crucial!) problem is the computation of the p_0-values at the saddle and at the attractor, in view of their ratio, which in the leading order is given by the difference of the corresponding u-values. Since (12A-5) is first-order but nonlinear, the solution $u(x)$ can be constructed by characteristic curves (obtained by numerically integrating a system of $2N$ ordinary differential equations); it is even sufficient to focus on those curves that link the attractor with the saddlepoint. A practical problem arises from the fact that the characteristic curves are in a $2N$-dimensional space and that their projections into x-space may intersect for $N \geq 2$. As a consequence, the number of relevant curves (linking the attractor with the saddlepoint) is not a priori known, and only the one with the smallest difference in u really matters. The scanning of the continuously many curves emerging from the attractor would be extremely time consuming, particularly since the dependence on the initial direction may be as strong as in chaotic systems [19].

EPILOGUE: UNEXPLORED TOPICS

This volume of *Synchronization in Digital Communications* concentrates on phase-locked loops, amplitude control, and automatic frequency control. Yet not all the aspects in these fields have been discussed. Major unexplored topics are surveyed in this section.

DIGITAL SIGNAL PROCESSING

Two topics which have not been considered in this volume are digital phase-locked loops and digital automatic gain control. Discussion has been deferred to Volume 2 where digital signal processing is a major topic. This allows the comparison of performance and architectures derived from analog implementations with systematically derived algorithms aimed at digital signal processing implementation.

LONG LOOPS

A typical long loop is shown in Figure E-1. A major advantage of the long loop lies in the fact that the signal frequency in the IF path equals the crystal oscillator (XO) frequency when the loop is in lock. Therefore a narrowband IF filter may be used. Another advantage is that downconversion from very high frequencies may be carried out in several moderate steps by using additional mixers and frequency multipliers.

If the loop is in lock and the IF filter is wideband compared to the loop bandwidth, performance analysis is straightforward. But when these conditions are not met, in particular during acquisition, analysis can become very tedious [1].

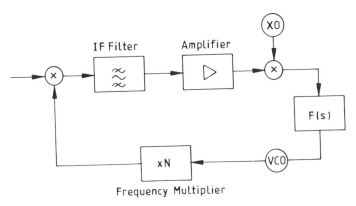

Figure E-1. Example of a long loop.

FREQUENCY SYNTHESIS

By including frequency dividers in the PLL as shown in principle in Figure E-2 any rational multiple of an input frequency may be generated. Frequency synthesis is an important application of phase-locked loops. There are several books devoted specifically to this topic, therefore a perfunctory discussion would not have been appropriate. The reader who is interested in frequency synthesis in more detail is referred to Chie and Lindsey [2] and Kroupa [3] for a comprehensive bibliography.

RANDOM ANGLE MODULATION

The influence of random angle modulation on loop tracking performance is examined in Volume 2. The effects are of particular interest when a large number of repeaters are chained (as in networks).

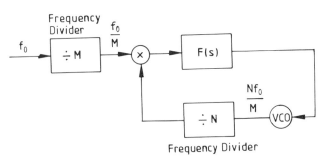

Figure E-2. Basic block diagram of a phase-locked loop including frequency division and multiplication.

REFERENCES

1. F. M. Gardner, *Phaselock Techniques*, Wiley, New York, 1979.
2. C. M. Chie and W. C. Lindsey, Phase-Locked Loops: Applications, Performance, Measures, and Summary of Analytical Results. In *Phase-Locked Loops*, W. C. Lindsey and C. M. Chie (eds.), IEEE Press, New York, 1986.
3. V. F. Kroupa, Noise Properties of PLL Systems. In *Phase-Locked Loops*, W. C. Lindsey and C. M. Chie (eds.), IEEE Press, New York, 1986.

INDEX